GLOBAL ENVIRONMENT: WATER, AIR, AND GEOCHEMICAL CYCLES

GLOBAL ENVIRONMENT: WATER, AIR, AND GEOCHEMICAL CYCLES

Elizabeth Kay Berner

Robert A. Berner

Yale University

Prentice Hall

Upper Saddle River, New Jersey 07458

Library of Congress Cataloging-in-Publication Data
Berner, Elizabeth Kay
Global environment : water, air, and geochemical cycles / Elizabeth Kay Berner, Robert A. Berner.
p. cm.
Includes index.
ISBN 0-13-301169-0
1. Atmospheric circulation. 2. Atmospheric chemistry. 3. Hydrologic cycle. 4. Water chemistry. 5. Geochemistry. I. Berner, Robert A. II. Title.
QC880.4.A8B47 1996
551.5—dc20 95-25174
CIP

Acquisitions editor: Robert A. McConnin
Editorial-production service: Electronic Publishing Services Inc.
Manufacturing buyer: Trudy Pisciotti
Cover design: Bruce Kenselaar
Cover photograph: Photo of Clark's Pond, Connecticut, by the authors.

Simon & Schuster/A Viacom Company
Upper Saddle River, New Jersey 07458

Printed in the United States of America

10 9 8 7 6 5 4 3 2 1

ISBN 0-13-301169-0

Prentice-Hall International [UK] Limited, *London*
Prentice-Hall of Australia, Pty. Limited, *Sydney*
Prentice-Hall Canada Inc., *Toronto*
Prentice Hall Hispanoamericana, S.A., *Mexico*
Prentice-Hall of India Private Limited, *New Delhi*
Prentice-Hall of Japan, Inc., *Tokyo*
Simon & Schuster Asia Pte. Ltd,, *Singapore*
Editora Prentice-Hall do Brasil, Ltda., *Rio de Janeiro*

To John, Susan, and James

CONTENTS

PREFACE

During the past decade there has been a veritable explosion of interest in global environmental problems. As a result, a number of books on the global environment and global change at the non-science background freshman level have been, or are about to be, published. At the other extreme, there are several books treating the chemistry of natural waters and the atmosphere at the graduate level in which detailed physical chemistry, hydrodynamics, etc., is discussed. We feel that these two levels of coverage leave an intermediate gap, which we hope to fulfill here.

This book is intended for those who have a fundamental understanding of elementary chemistry, but it requires no other background in science, whether in biology, geology, meteorology, oceanography, hydrology, soil science, or environmental science. Our approach is multidisciplinary and covers all of these fields, but we do it from an elementary standpoint. Mathematical complexity is held to an absolute minimum, with the only requirement being some previous training in chemistry at the freshman, or even the advanced high school, level. The book is appropriate as a primary or secondary text in junior or senior-level undergraduate courses, or beginning graduate courses, in environmental geochemistry, environmental geology, global change, biogeochemistry, water pollution, geochemical cycles, chemical oceanography,and geohydrology. Because we provide extensive data on natural fluxes of chemicals, the book is also of reference value to researchers of global environmental problems.

Much of this book is devoted to the natural behavior of the earth's surface. In other words, such subjects as solar radiation input, rainfall formation, chemical weathering, estuarine circulation, and the input and output of chemicals to the oceans are discussed. We attempt to quantify the rates by which the major constituents of rocks, water, air, and life are transferred from one reservoir to another, and to track down the sources of each constituent. We feel that a knowledge of geochemical cycles in the pre-human state is necessary before one can discuss how humans have perturbed these cycles. Also, our approach to human perturbation is global, or at least regional. Local problems such as waste dumps or sewage treatment are not covered.

The present book was originally intended to be a revised second edition of our book, *The Global Water Cycle: Geochemistry and Environment*. However, we decided that the book should be retitled, more emphasis be made on subjects other than the water cycle, and the book be published in paperback. This we have done here. A new chapter covering atmospheric environmental problems, such as the greenhouse effect and the ozone hole, has been added (Chapter 2). Discussion of additional atmospheric problems, specifically the atmospheric sulfur and nitrogen cycles and acid rain, are covered in Chapter 3. The remainder of the book follows the organization of *The Global Water Cycle*: Chapter 4, "Chemical Weathering and Water Chemistry"; Chapter 5, "Rivers"; Chapter 6, "Lakes"; Chapter 7, "Marginal Marine Environments: Estuaries"; Chapter 8, "The Oceans." Nevertheless, these chapters have been extensively updated and new sections added where appropriate.

This book has been used in courses taught over the past decade by E. K. Berner at Wesleyan University and the University of Connecticut, and by R. A. Berner at Yale University. We thank the various suggestions made to us by both students and teaching assistants during this period. Also, a review of a major portion of the manuscript by Fred T. Mackenzie and Raymond Siever has been helpful. Finally, the junior author acknowledges the musical inspiration of Howard Hanson, and the scientific inspiration of J. J. Ebelmen (1814–1852).

Elizabeth Kay Berner
Robert A. Berner
North Haven, Connecticut

GLOBAL ENVIRONMENT: WATER, AIR, AND GEOCHEMICAL CYCLES

CHAPTER 1

INTRODUCTION TO THE GLOBAL ENVIRONMENT

The Water Cycle and Atmospheric and Oceanic Circulation

INTRODUCTION

In this book we shall be concerned with the principal constituents of rocks, water, and life as they circulate through the land, the sea, and the air. In other words, our concern will be with the geochemistry of the earth's surface and how it operates naturally and how it has been perturbed by human activities. The approach is global in scope. Because of its special importance in geochemical cycles on the earth's surface, water will receive major attention. Accordingly much of the book is organized around the global water cycle, or *hydrologic cycle.* Water moves from the atmosphere to the land surface as rain and snow. From there it passes downward through the soil into the ground and eventually makes its way into rivers that flow to the ocean. From the ocean it is evaporated into the atmosphere. Some of this water is returned as rain to the oceans, while the remainder makes its way back to the continents, where it falls again as rain and snow. As it passes around and around in the hydrologic cycle, water undergoes chemical reactions with atmospheric gases, rocks, plants, and other substances, resulting in changes in its chemical composition as well as profound changes in the substances with which it reacts. It is these changes, along with nonaqueous reactions in the atmosphere, that act ultimately to maintain the overall chemical conditions at the earth's surface.

Throughout most of this book we shall concentrate on the principal constituents dissolved in and transported by natural waters. This automatically includes many major components of rocks, life, and the atmosphere: sodium, potassium, calcium, magnesium, silicon, carbon, nitrogen, sulfur, phosphorus, chlorine, and, of course, hydrogen and oxygen. It will be seen that global chemical cycles of these elements at the earth's surface are intimately interconnected with the hydrologic cycle. We shall also point out how the global cycles of some of these elements have been perturbed by humans, resulting in such things as acid rain and eutrophic lakes. Although they are of geochemical and environmental interest, we shall not be concerned with minor and trace elements (e.g., lead and mercury) or exotic synthetic chemicals (e.g., PCBs), since our goal is not that of an all-inclusive environmental coverage. [The interested reader is referred to books such as those by Laws (1993) for detailed discussion of environmental problems.] Rather, the approach is that of geochemists trying to understand how global chemical cycles operate and how they affect the major constituents of rocks, water, air, and life.

Water plays an important role in a variety of processes at the earth's surface. Much heat transfer on the earth occurs through oceanic currents such as the Gulf Stream and by the movement of water vapor in the atmosphere combined with condensation and evaporation. Water vapor also acts as a heat regulator in the atmosphere by absorbing outgoing earth radiation, thereby exerting a major influence on climate. Water is essential to life. The biological importance of water is manifested in photosynthesis, as a carrier of nutrients within vascular plants and as a principal component of the human body.

This book is not just about water. There are important global environmental phenomena that are not intimately connected to the hydrologic cycle, but rather are connected with other aspects of the chemistry of the atmosphere. This includes such things as the stratospheric ozone hole, global warming due to the atmospheric greenhouse effect, and global cooling due to the anthropogenic production of aerosols. Because of their environmental significance, these subjects are also covered in this book. However, because of the importance of water to geochemical cycles, the main purpose of the present chapter is to discuss the global water cycle.

Emphasis in this chapter is on the atmosphere and the oceans, and how their circulation takes place, because of their importance to the water cycle. This includes a discussion of the energy cycle of the earth and its role in meteorology and oceanography. This chapter, thus, helps to set the stage for discussions of such atmospheric and oceanic subjects as the atmospheric greenhouse effect (Chapter 2), rain water chemistry (Chapter 3), and the chemistry of the oceans (Chapter 8).

THE GLOBAL WATER CYCLE

Major Water Masses

Earth is the only planet in the solar system that has an abundance of liquid water on its surface; about 70% of the earth is covered by liquid water. Because of the particular combinations of temperature and pressure on the planet's surface, water can exist here in three states: as liquid water, as ice, and as water vapor. This is in sharp contrast to the surface of the planet Mars, for example, which is so cold and dry that water can exist there only as ice or as water vapor.

Water is by far the most abundant substance at the earth's surface. There are 1459×10^6 km^3 of it in its three phases: liquid water, ice, and water vapor. As shown in Table 1.1, most of the earth's water (96%) is stored as seawater in the oceans. The remaining 4% is either on the continents or in the atmosphere. The amount of water in the atmosphere, in the form of water vapor, is very small in comparison with the other reservoirs, only around 0.001% of the total. However, it plays a very important role in the water cycle, as we shall see.

Of the fresh water stored on the continents, around three quarters is in the form of ice in polar ice caps and glaciers. Most of the rest of the continental water is present either as subsurface groundwater or in lakes and rivers. It is this small part of the earth's total water (1%) that humans draw on for their water supplies. Here, we shall focus on water near the earth's surface and how it moves within and between the various reservoirs.

TABLE 1.1 Inventory of Water at the Earth's Surface

Reservoir	Volume 10^6 km^3 (10^{18} kg)	Percent of Total
Oceans	1400.	95.96
Mixed layer	50.	
Thermocline	460.	
Abyssal	890.	
Ice caps and glaciers	43.4	2.97
Groundwater	15.3	1.05
Lakes	0.125	0.009
Rivers	0.0017	0.0001
Soil Moisture	0.065	0.0045
Atmosphere total[a]	0.0155	0.001
Terrestrial	0.0045	
Oceanic	0.0110	
Biosphere	0.002	0.0001
Approximate total	1459.	

Sources: NRC 1986; Berner and Berner 1987.
[a]As liquid volume equivalent of water vapor.

Fluxes Between Reservoirs

Water does not remain in any one reservoir, but is continually moving from one place to another in the hydrologic cycle. This is illustrated in Figure 1.1. (For a more detailed discussion, see Penman 1970, Baumgartner and Reichel 1975, NRC 1986, Cahine 1992.) Water is evaporated from the oceans and the land into the atmosphere, where it remains for only a short time, on the average about 11 days, before falling back to the surface as snow or rain. Part of the water falling onto the continents runs off in rivers and, in some places, accumulates temporarily in lakes. Some also passes underground, only to emerge later in rivers and lakes. (Little groundwater flows

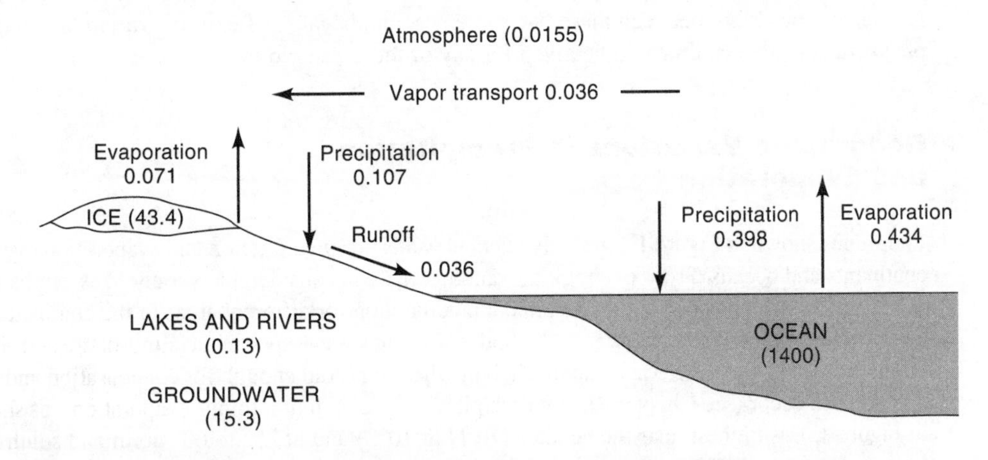

Figure 1.1 The hydrologic cycle. Numbers in parentheses represent inventories (in 10^6 km^3 = 10^{18} kg) for each reservoir. Fluxes are in 10^6 km^3/yr (10^{18} kg/yr). (Data from Table 1.1 and NRC 1986.)

directly to the ocean.) The remaining portion of the precipitation on the continents is returned directly to the atmosphere via evaporation. Over the oceans, evaporation exceeds precipitation, with the difference being made up by input via runoff from the continents. An idea of the sizes of these various fluxes of water (mass transported per unit time) is shown in Figure 1.1.

In order to conserve total water, evaporation must balance precipitation for the earth as a whole, since the total mass of water at the earth's surface is believed to be constant over time (see the later section on the origin of water). The average global precipitation rate, which is equal to the evaporation rate, is 0.505×10^6 km^3/yr. For any one portion of the earth, by contrast, evaporation and precipitation generally do not balance. On the land, or continental part of the earth, the precipitation rate (0.107×10^6 km^3/yr) exceeds the evaporation rate (0.071×10^6 km^3/yr), whereas over the oceans evaporation (0.434×10^6 km^3/yr) dominates over precipitation (0.398×10^6 km^3/yr). The difference in each case (0.036×10^6 km^3/yr) comprises water transported from the oceans to the continents as atmospheric water vapor, or that returned to the oceans as river runoff (see Figure 1.1). There is a minor amount of direct groundwater discharge to the oceans (Meybeck 1987).

The values given in Figure 1.1 have built-in errors. Precipitation is difficult to measure, rainfall measurements over the oceans are few, and oceanic evaporation is necessarily estimated from models. Also, because the evaporation rates over many land areas of the earth have not been measured, the value given in Figure 1.1 is based, by necessity, on the difference between measured worldwide precipitation on land and global river runoff. (Recently, attempts have been made to obtain evaporation rates by other means, including atmospheric water vapor balance and heat balance; see the section later on radiation and energy balance.)

Assumption of a constant volume of water in a given reservoir (water mass) enables the use of the concept of residence time. *Residence time* is defined as the volume of water in a reservoir divided by the rate of addition (or loss) of water to (from) it. It can be thought of as the average time a water molecule spends in a given reservoir. For the oceans, the volume of water present (1400×10^6 km^3; see Figure 1.1) divided by the rate of river runoff to the oceans (0.036×10^6 km^3/yr) gives a residence time of 39,000 yr. This long residence time, which can also be thought of as a filling or replacement time, reflects the very large volume of water in the oceans. By contrast, the residence time of water in the atmosphere, relative to evaporation from both the oceans and the continents, is only 11 days. Lakes, rivers, glaciers, and shallow groundwater have residence times lying between these two extremes, but, because of extreme variability, no simple average residence time can be given for any of these reservoirs.

Geographic Variations in Precipitation and Evaporation

The values shown in Figure 1.1 are only average values for precipitation and evaporation over the continents and oceans. From one region to another there is considerable variation, as can be seen in Figure 1.2, which shows the mean annual precipitation for different areas of the continents. In order to have rain or snow, there must be both sufficient water vapor in the atmosphere and rising air that can carry the water vapor up to a height where it is cold enough for condensation and precipitation to occur (see Chapter 3). Net precipitation (precipitation minus evaporation), as shown in Figure 1.3, is highest near the equator (10°N to 10°S) and at 35° to 60° north and south latitudes, where there is frequent storm activity with its accompanying air motion. Net precipitation is lowest in the subtropics (15° to 30° N and S), where the air is stable, and near the poles, which

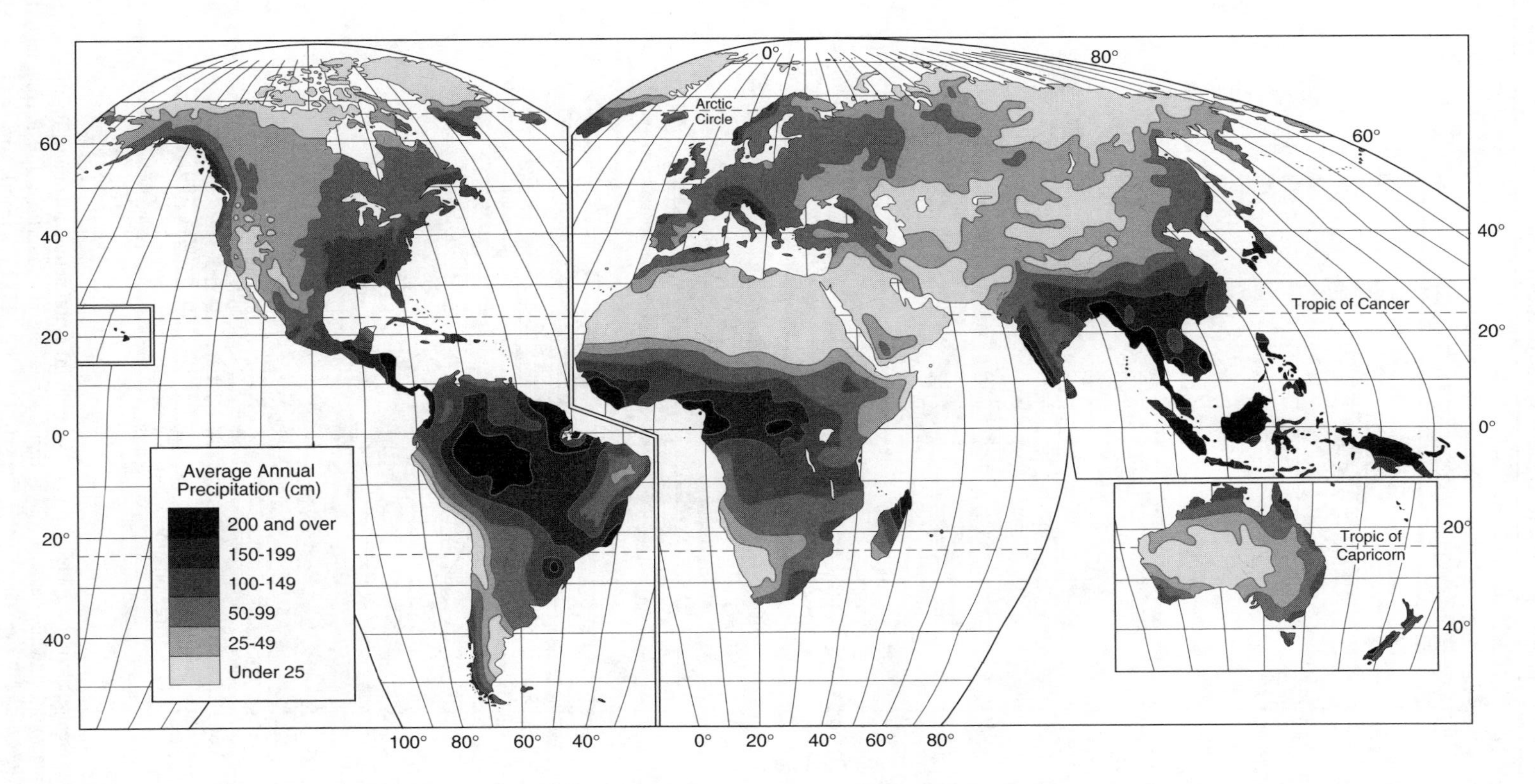

Figure 1.2 Global average annual precipitation. (From Tom L. Mc Knight, *Physical Geography: A Landscape Appreciation*, 5th ed. Copyright © 1996 Prentice Hall, Inc., Upper Saddle River, N.J.)

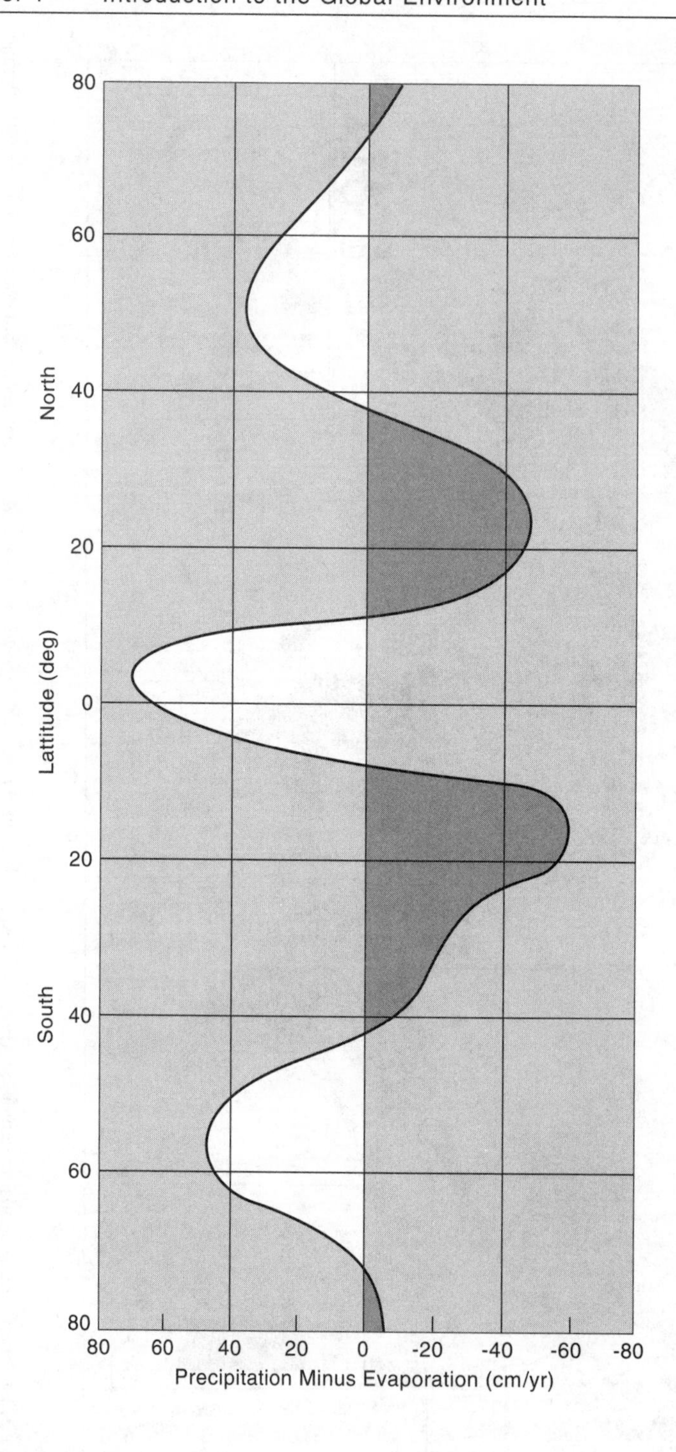

Figure 1.3 Net precipitation (precipitation minus evaporation) as a function of latitude. Positive values represent net precipitation, while negative values represent net evaporation. (From J. P. Peixoto and M. A. Kettani, "The Control of the Water Cycle." Copyright © April 1973 by Scientific American, Inc. All rights reserved.)

have both stable air and a very low moisture content due to low temperatures. (However, since there is also very low evaporation near the poles, precipitation can exceed evaporation in certain places, resulting in the formation of the ice caps of Greenland and Antarctica.) In continental areas of high rainfall, runoff is also high. Examples are the large rivers of equatorial regions (Amazon, Zaire, Orinoco) and those of middle latitudes such as the Mississippi and Hwanghe (Yellow).

Evaporation also varies considerably over the earth's surface. For net evaporation to occur, there must be a heat source (i.e., radiation from the sun), a low moisture content in the air, and the presence of water available for evaporation. In arid regions the evaporation rate is high but is limited by the availability of water. As shown in Figure 1.3, overall, evaporation exceeds precipitation at subtropical latitudes (15° to 30° N and S). Over the continents this leads to the formation of large deserts at these latitudes (for example, the Sahara Desert in Africa and the Great Desert of Australia). The greatest evaporation rates on the earth (more than 200 cm/yr), however, occur over the subtropical oceans such as over the Gulf Stream in winter, where warm water, which is carried northward, encounters cooler, drier air and evaporates. High evaporation rates can lead to locally higher ocean salinities due to the removal of almost pure water during evaporation, leaving dissolved sea salt behind. An outstanding example of this is the Mediterranean Sea, which has a salinity notably higher than that of the open ocean (e.g., see Chapter 7).

Origin of Water on Earth

As we noted earlier, it is believed that the total amount of water at the earth's surface has remained fairly constant over geologic time. This is based on our ideas about the history of the earth and the origin of water. The earth is thought to be around 4.6 billion years old, based on the ages of meteorites that presumably formed at the same time. Unfortunately, the oldest known rocks (metamorphic gneisses) are only about 4 billion years old (York 1993), with the oldest known sedimentary rocks being about 3.8 billion years old (Moorbath 1977). These sedimentary rocks were deposited under water, thus indicating that liquid water existed on the earth at that time. Of the period between 4.6 and 3.8 billion years ago, some 800 million years, we have no direct evidence for water because no sedimentary rocks of this age have been found; nor is it likely that they will be, since there is a high probability that they have been destroyed by erosion or metamorphism/magmatism since that time. Thus, although we know that water existed 3.8 billion years ago, we can only make an educated guess about its formation based on our ideas about the origin of the earth. (The following discussion is based on Kasting 1993, Holland 1984, Arrhenius, et al. 1974, Turekian 1976.)

The earth formed from a cloud of ionized gas around the sun. The gas condensed to form small globules or pieces, referred to as planetesimals, and once a number of these coalesced to form a protoplanet, more and more of them were drawn into it by gravitational attraction. The energy of motion of these pieces was changed into heat energy when they collided with the earth and, as a result, the earth heated up. Because rocks of the earth's crust (outer layers) do not contain appreciable amounts of water, scientists were originally at a loss to explain the origin of water. However, it has been discovered that one group of stony meteorites, the carbonaceous chondrites, do contain up to 20% H_2O. These meteorites have water locked up in the structure of minerals, particularly in clay minerals in the form of hydroxyl ion (OH^-). Since the carbonaceous chondrites are debris left over from when the solar system and the earth originally formed, they are believed to hold the clue as to how volatiles such as water were added to the earth.

Probably some of the bodies hitting the early earth had a composition like the carbonaceous chondrite meteorites (containing water). As a result of impact, and the internal generation of heat by the decay of radioactive elements within the earth, the accumulated chondritic material may have been heated to the point where water vapor was released from the clay minerals and other water-containing silicates (in the so-called *degassing process*), accompanied by other volatiles such as carbon dioxide, nitrogen, ammonia, and methane. Initially, the water vapor and volatiles may have formed a transient steam atmosphere during accretion (Kasting 1993). Once the main period of accretion ended about 4.5 billion years ago, the earth's surface would have cooled and the water vapor in the steam atmosphere could cool and condense, forming liquid water that could then accumulate on the earth's surface. The remaining atmosphere would then be mainly CO_2, CO, and N_2.

Even after the main accretionary phase ended, substantial numbers of meteorites continued to impact the earth, causing a period of heavy bombardment from 4.5 to 3.8 billion years ago (bya). If any of these bodies were carbonaceous chondrites, more water would have been released. Continual impacts would vaporize the surface layers of the ocean until 3.8 bya, when major bombardment ceased. By 3.5 bya, life had appeared.

The formation of water, according to the process we have described, took place mainly during the formation of the earth itself, and the ocean would have had essentially its present volume early in earth history. [Henderson-Sellers and Cogley (1982) reach a similar conclusion from models of the radiation balance of the early earth of the type shown in Figure 1.4.] Since that time, only very small additions and losses of water have occurred. Water is lost slowly from the earth by photodissociation to hydrogen and oxygen atoms in the upper atmosphere. This is brought about by

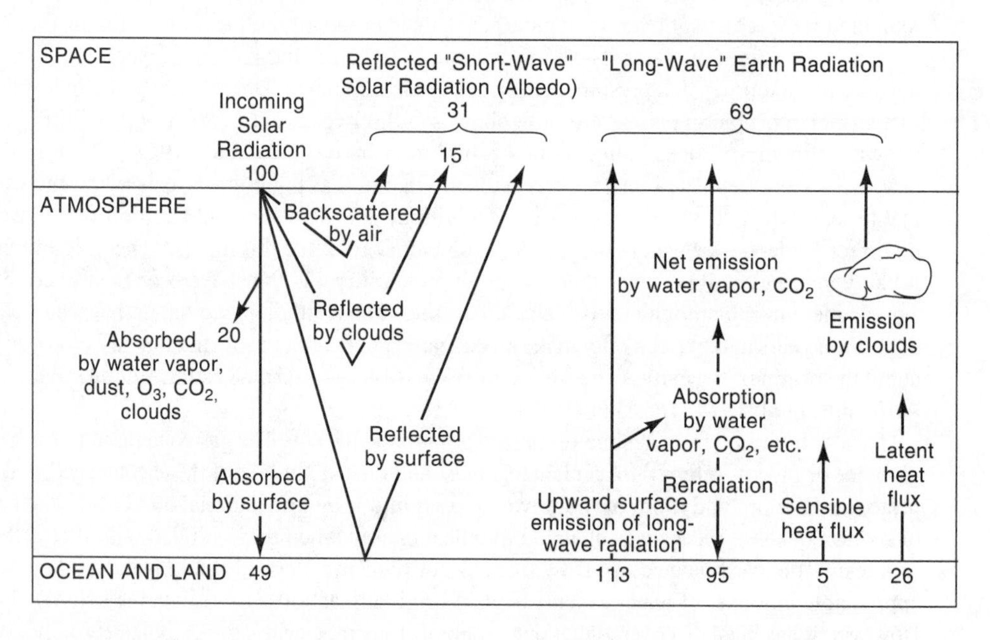

Figure 1.4 Mean annual radiation and heat balance of the atmosphere and earth. Units are assigned so that the incoming solar radiation (343 watts/m²) is set equal to 100 units; (i.e., one unit equals 3.43 watts/m²). "Short-wave solar radiation" has wavelength < 4 μm wavelength; "long-wave" earth radiation has wavelength >4 μm. Data from Ramanathan (1987), with cloud reflection of solar energy from Ramanathan et al. (1989a). The estimates of energy transfer between the top of the atmosphere and space are based on satellite measurements.

ultraviolet radiation, which is more intense at high altitudes, where it has not been absorbed by the earth's atmosphere. After the water breaks down, the hydrogen escapes to space, while oxygen remains behind. Almost all water vapor is prevented from escaping from the atmosphere via this process by the so-called *cold trap* at about 15 km above the ground. At this altitude the atmosphere becomes cold enough that practically all water vapor is condensed and thereby returned to lower altitudes, where it cannot undergo photodissociation. As a result, only about 4.8×10^{-4} km^3 of water (as water vapor) are destroyed each year by photodissociation (Walker 1977).

Very small amounts of water (juvenile water) are also added to the earth's surface by continued degassing from deep within the earth. The amount cannot be very great; Garrels and Mackenzie (1971) and Walker (1977) estimate that the extreme maximum rate of degassing is only 0.3 km^3/yr, which is negligible compared to other fluxes in the hydrologic cycle (see Figure 1.1). At most, degassing of water vapor can account for no more than 0.001% of the water transported each year from the continents to the oceans and back to the continents.

Relative constancy of total water at the earth's surface through geologic time does not imply constancy of the volume of water in any given reservoir. The distribution of water between land and sea has certainly varied with time. During glacial periods much more water was present on the continents, in the form of ice caps, than at present, whereas during warmer periods practically all of the continents were ice free. For example, during the last Pleistocene glacial maximum 18,000 years ago, sea level was lowered by around 130 m (Bloom 1971), accounting for a transfer of about 47×10^6 km^3 of water (equal to about 3.5% of the oceanic volume) from the oceans to the land. This is a very large amount of water and constitutes a doubling of the continental reservoir.

ENERGY FOR THE WATER CYCLE: THE ENERGY CYCLE

Introduction

Before considering the atmospheric and oceanic parts of the water cycle, in other words, the fluxes of water vapor in the atmosphere and liquid water in the oceans, we shall take a look at the energy cycle of the earth. This is what drives the water cycle, especially the movement of water vapor in the atmosphere. The atmosphere contains 0.0155×10^6 km^3 of water in the form of water vapor at any one time (Table 1.1), and this water remains in the atmosphere for only about 11 days on average. However, during this time it has travelled a mean distance of 1000 km (Peixoto and Kettani 1973), and this transport is controlled by the energy cycle of the earth. The energy cycle, in turn, is greatly influenced by the presence of water vapor in the atmosphere. Thus, the earth's energy and water cycles are intimately interconnected, and they exert strong influences on one another. (For more information on the earth's energy cycle, see Oort 1970, Sellers 1965, Ingersoll 1983, Ramanathan 1987.)

Radiation and Energy Balance

The primary energy sources for the earth's surface are summarized in Table 1.2. As can be seen, radiation from the sun is by far the most important source of energy (99.98%) and, consequently, it is the dominant influence on the circulation of the atmosphere and oceans. It is the only energy source that will be discussed here. The incoming solar radiation impinges on the

TABLE 1.2 Primary Energy Sources for the Earth

Source	Energy Flux (cal/cm²/min)	Percent of Total Energy Flux
Solar radiation	0.5[a]	99.98
Heat flow from interior of earth	0.9×10^{-4}	0.018
Tidal energy	0.9×10^{-5}	0.002

Sources: Hubbert 1971; Flohn 1977.

[a] 0.5 cal/cm²/min is approximately equal to 343 W/m², the solar flux used in Figure 1.4, after Ramanthan (1987); 1W = 0.2389 cal/sec.

top of the atmosphere, below which it is reflected, absorbed, and converted into other forms of energy. Apportionment of this energy into different forms is described in terms of the radiation or energy balance of the earth (Figure 1.4). (To simplify terminology, we express radiation in relative units, with 100 units set equal to the solar flux of 343 watts/m²).

As shown in Figure 1.4, incoming solar radiation at the top of the atmosphere is balanced by an equivalent outgoing energy flux from the earth. This must be true to avoid rapid heating up or cooling of the earth and has been verified by satellite measurements (Ramanathan 1987). This does not mean, however, that very slight imbalances over extended periods cannot bring about cooling or heating. An example of this is the large climate changes that occurred at northern mid-latitudes during the past 20,000 years, as evidenced by extensive glaciation and deglaciation that occurred during this time.

The earth's outgoing energy flux (Figure 1.4) is divided into two basic components. One component, consisting of 31 units, represents short-wave (< 4 μm) solar radiation that has simply been reflected back to space from air, clouds, and the land and oceans, without change in wavelength. This fraction of incoming solar radiation that is reflected, about 0.3 (30%), is known as the earth's *albedo.* It is a measure of how bright the earth would appear if viewed from outer space. Note that 15 units of this solar radiation are reflected by clouds (Ramanathan et al. 1989a), which translates into 50 watts/m² of solar radiation that does not reach the earth's surface. Thus, changes in cloud cover are very important to radiative heating of the surface. The remaining 69 units of outgoing earth energy represent solar radiation that has been absorbed and reradiated as long-wave (> 4 μm) infrared radiation by the earth's surface and various components of the atmosphere.

Because the sun is very hot (6000°C), most of the solar radiation is of shorter wavelength (less than 4 μm), with the peak radiation in the visible wavelengths (0.4–0.7 μm). Much of this visible radiation, or light, penetrates the atmosphere and ultimately reaches the ground. This is important because life is dependent on the absorption of light by photosynthetic organisms. By contrast, almost all of the ultraviolet solar radiation (< 0.4 μm) is absorbed in the upper atmosphere by ozone and O_2, which protects life from its harmful effects. (Because of the importance of ultraviolet absorption by ozone, there is growing concern about anthropogenic ozone depletion—see Chapter 2.) The sun's incoming long-wave, or infrared, radiation is also absorbed at certain wavelengths by atmospheric water vapor and CO_2 and by water droplets in clouds. Thus, the incoming solar radiation reaching the earth's surface has gaps at those wavelengths where atmospheric absorption occurs (see Figure 1.5). Overall, due to the absorption of 20 units by atmospheric ozone, water vapor, CO_2, and clouds, combined with reflection of 31 units, only 49 units or 49% of the incoming solar radiation actually reaches the earth's surface (Figure 1.4) to be absorbed there. The solar radiation reaching the ground is about 169 watts/m² (Ramanathan 1987).

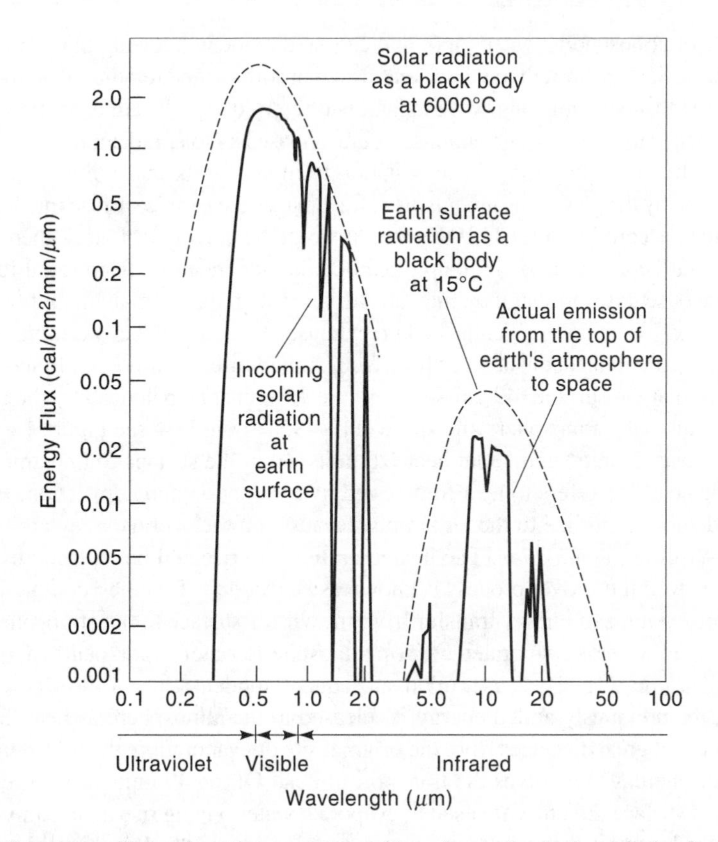

Figure 1.5 Incoming radiation from the sun at the earth's surface and outgoing radiation from the top of the earth's atmosphere as a function of wavelength. Also shown are the energy spectra expected for black bodies at the same temperature as the sun and the earth. (A black body emits the maximum radiation for its temperature.) (Adapted from W. D. Sellers, *Physical Climatology.* Copyright © 1965 by The University of Chicago Press. All rights reserved.)

The surface of the earth reradiates part of its absorbed solar energy back into space and the atmosphere, but since it is cooler than the sun (the average earth surface temperature is 15°C; Ramanathan 1987), the earth radiates at longer wavelengths (> 4 μm), with a maximum in the infrared at about 10 μm (see Figure 1.5). Atmospheric water vapor, carbon dioxide, and rarer gases such as methane are good absorbers of infrared energy in this range of wavelengths. Because of this, most of the infrared radiation originating from the earth surface is absorbed by these gases (mostly by water vapor), and very little of it escapes directly to space. As a result, the spectrum of outgoing radiation leaving the earth differs considerably from that expected for a black body at 15°C (see Figure 1.5).

Of the 113 units of long-wave (infrared) radiation emitted by the earth's surface, 95 units are both absorbed by atmospheric water vapor and other gases and reradiated back to the ground. There it is reabsorbed, keeping the earth warm. Clouds also contribute to warming of the earth by reducing long-wave emissions to space, because at their bases they absorb radiation emitted by the warmer earth surface and at their tops they emit to space at colder temperatures. However, clouds also reflect solar radiation as discussed above, and because this process outweighs their warming effect, the net effect of clouds is global cooling (Ramanathan et al. 1989b).

The role of atmospheric water vapor and carbon dioxide in allowing the incoming short-wave solar radiation to pass through to the ground, while absorbing and reradiating to the earth's surface most of the earth's outgoing long-wave radiation, is referred to as the atmospheric *greenhouse effect* by comparison to the glass in a greenhouse. A greenhouse lets solar radiation in, but keeps the greenhouse warm by preventing long-wave radiation from leaving because of absorption of the long-wave radiation by the glass. The greenhouse effect makes the earth's surface much warmer (around 30°C warmer, according to IPCC 1990) than it would be otherwise. Lately, there has been much concern about the atmospheric buildup of carbon dioxide, released from fossil fuel burning, and other greenhouse gases in that they may absorb more than the normal amount of the earth's outgoing radiation, resulting in an enhanced greenhouse effect and global warming (see Chapter 2).

In order to maintain a constant earth surface temperature, the amount of incoming solar radiation received at the surface (49 units; see Figure 1.4) must be balanced by the net loss of long-wave radiation to the atmosphere and space (113 – 95 = 18 units—see Figure 1.4) plus the fluxes of sensible heat (5 units) and latent heat (26 units) from the surface to the atmosphere.

Sensible heat flux refers to heat transferred by conduction and convection. Heat flows from the ground and sea surface to the air simply because, on the average, they are warmer than the air. This constitutes *conduction.* The heated air tends to rise and be replaced by cooler sinking air, and this overall turnover process is known as *convection.* Together, conduction and convection comprise 5 units of energy transfer from the earth's surface to the atmosphere (Figure 1.4).

When liquid water is evaporated to form atmospheric water vapor, heat (energy) is absorbed. This is called *latent heat* because, upon subsequent condensation of the water vapor into rain and snow, the previously added energy is released to the atmosphere as heat. Since condensation can occur at great distances from the original site of evaporation, the transport of water vapor in the atmosphere also involves the transport of heat. Of the 49 units of solar energy absorbed at the earth's surface, 26 units are used to evaporate water, giving rise to an equivalent latent heat flux from the land and oceans to the atmosphere (see Figure 1.4). It is this latent heat flux which drives the water cycle.

Knowledge of the latent heat flux to the atmosphere can be used to calculate the global rate of evaporation. Since it takes 588 cal to convert 1 cm^3 of liquid water to water vapor at the average earth surface temperature of 15°C, the rate of evaporation corresponding to the dissipation of 26 radiation units as latent heat over the earth's surface (510×10^6 km^2) is calculated from the total solar radiation received by the earth of 100 units (0.5 $cal/cm^2/min$—see Table 1.2) as follows (after Miller et al. 1983; see also Budyko and Kondratiev 1964):

$$\frac{(0.26) \times (0.5\ \text{cal/cm}^2/\text{min}) \times (510 \times 10^{16}\ \text{cm}^2)}{588\ \text{cal/cm}^3} = 1.13 \times 10^{15}\ \text{cm}^3/\text{min}$$

This is equivalent to 593,000 km^3 of water per year, which agrees fairly well with estimates of total annual evaporation from the earth (505,000 km^3 from Figure 1.1) based on setting the total evaporation equal to measured total precipitation in the global water balance.

Variations in Solar Radiation: The Atmospheric and Oceanic Heat Engine

The amount of solar radiation absorbed by the earth decreases with latitude from the equator to the poles. It is this variation in the earth's heating that drives the circulation of the ocean and atmosphere and, thus, much of the hydrologic cycle.

Latitudinal variations in the input of solar energy are due to two factors. First, the earth is a sphere, and the angle at which the sun's rays hit its surface varies from 90° (or vertical) near the equator to 0° (or horizontal) near the poles. This is shown in Figure 1.6. Less energy is received at the poles because the same amount of radiation is spread out over a much larger area at high latitudes (compare situation *C* with situations *A* and *B* in Figure 1.6) and because at high latitudes the sun's rays must travel through a much greater thickness of atmosphere, where more absorption and reflection occur.

The second factor affecting latitudinal variations in heating is the duration of daylight. Because the polar axis of the earth is tilted at an angle of 23.5° with respect to the ecliptic (the plane of the earth's orbit about the sun), we have a progression of seasons where the angle of the sun's rays striking any given point varies over the year. This is shown in Figure 1.7. In the Northern Hemisphere winter, no sunlight strikes the area around the North Pole during a full day's rotation of the earth because it is in the earth's shadow. Thus, little or no solar heating occurs in this area at this time. Conversely, at the South Pole there is continual daylight, but at a very low sun

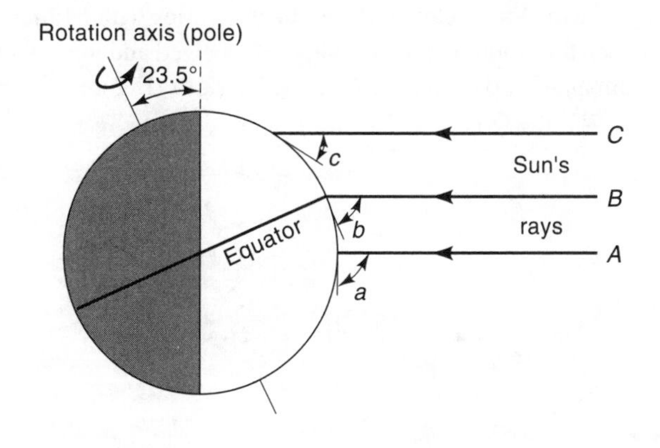

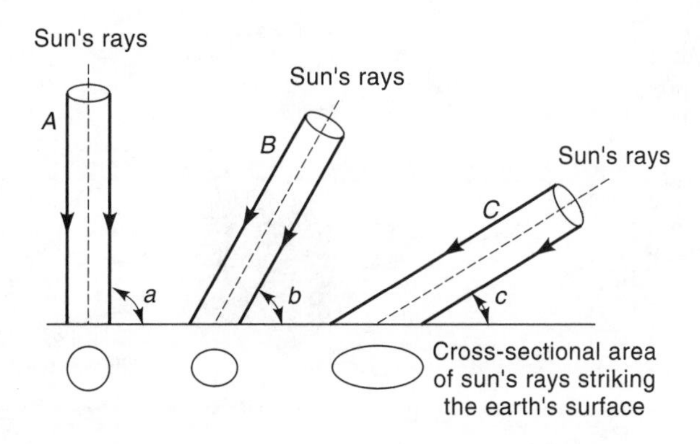

Figure 1.6 Schematic diagrams showing the variations of solar intensity (energy per unit area) with angle of incidence to the earth's surface. Lower angles (higher latitudes) result in the same energy spread out over a larger area and thus in a lower intensity of radiation. The scene depicted is for the Northern Hemisphere winter. (Adapted from Miller et al. 1983.)

angle, during the Northern Hemisphere winter. As the seasons shift, the South Polar region eventually becomes plunged into 24-hour darkness (during the Southern Hemisphere winter and Northern Hemisphere summer), just as the North Pole had been earlier. Low-latitude regions near the equator, by contrast, undergo little seasonal change in the duration of daylight, whereas intermediate latitudes are subjected to changes intermediate between those of the poles and the equator. Thus, because of seasonality, more annual radiation is received per unit area at lower as compared to higher latitudes.

The effects of variation in angle of the sun's rays with latitude and seasonal changes in the amounts of daylight result in strong variation with latitude in the total solar radiation received during a year, and this result applies equally to both the Northern and Southern Hemispheres. However, because the Northern Hemisphere is dominated by land and the Southern Hemisphere by water, one might expect hemispheric differences in received radiation due to differences in the albedo of land versus water. Nevertheless, according to satellite measurements, the annual mean albedo (relection of solar radiation) of both hemispheres is nearly the same. This is because of the dominant influence of clouds (over surface effects) in determining the mean albedo of each hemisphere (Ramanathan 1987).

The long-wave radiation leaving the earth also varies with latitude, but less strongly. The differences between the amount of solar radiation received and long-wave radiation emitted results in radiation imbalances over the surface of the earth (Figure 1.8). From 35° north and south

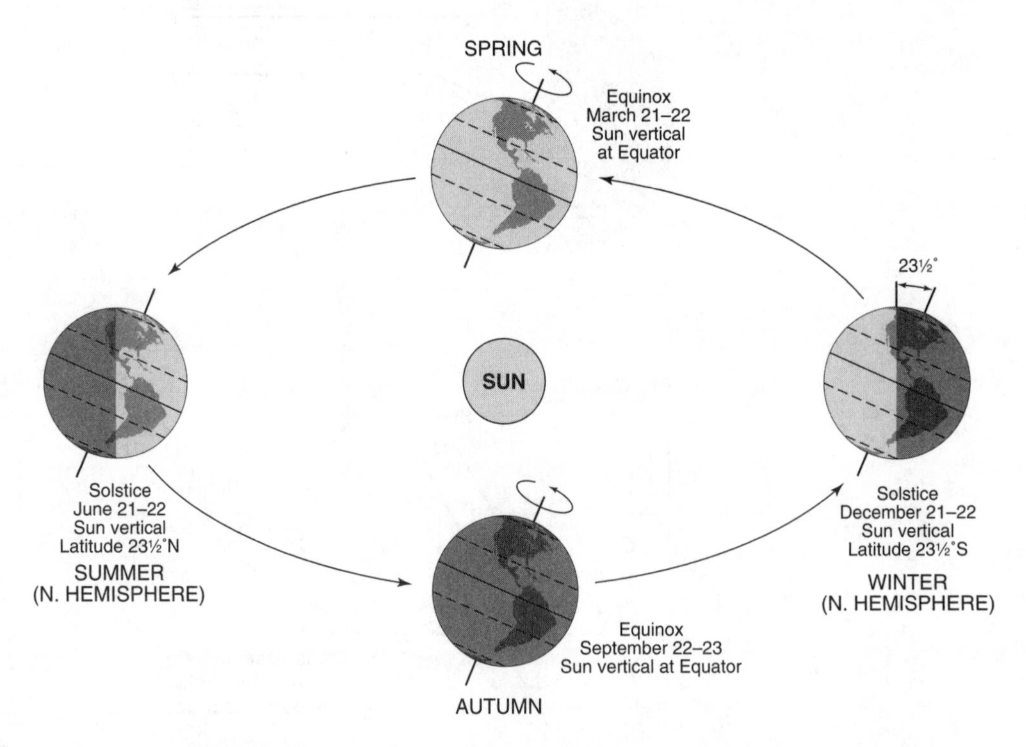

Figure 1.7 Revolution of the earth in its orbit around the sun, showing the changing seasons (also the length of the day). The seasons given are for the Northern Hemisphere; they are reversed in the Southern Hemisphere. (Frederick K. Lutgens/Edward J. Tarbuck, *The Atmosphere,* 5e, © 1992, p. 29. Adapted by permission of Prentice Hall, Englewood Cliffs, New Jersey.)

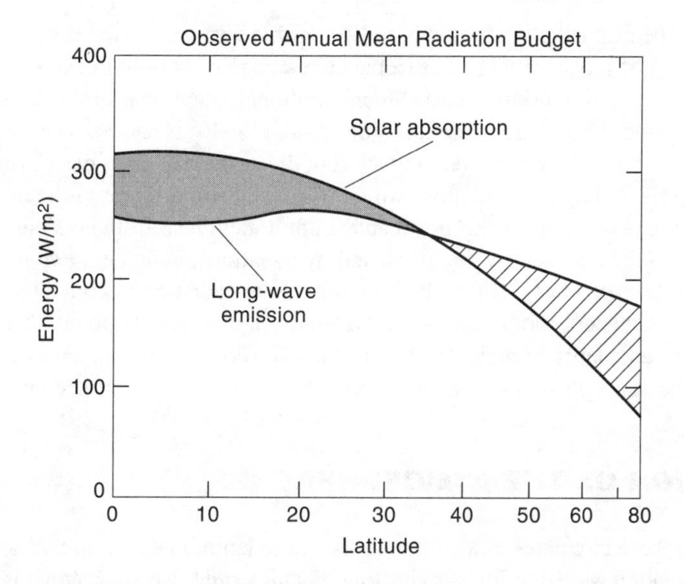

Figure 1.8 Annual zoned mean estimates for both hemispheres, which are nearly the same, of absorbed solar radiation and outgoing long-wave radiation emission obtained by satellites. Shaded regions denote net heating and dashed region denotes net cooling. (After V. Ramanathan, *Journal of Geophysical Research* 92, pp. 4076, 1987, copyright by the American Geophysical Union, based on data from Ellis and Vonder Haar 1976.)

latitudes to the poles (mid and polar latitudes), there is a net deficit of radiation (more leaves than enters); whereas from 35° to the Equator (tropical and subtropical latitudes), there is a net surplus (more solar radiation enters than the earth radiates back). To keep the poles from becoming colder and the tropics from becoming warmer, heat must be transported from lower to higher latitudes. This is accomplished by the circulation of the atmosphere and oceans. Thus the atmosphere and oceans act like a "heat engine" driven by latitudinal variations in solar radiation. By contrast, there is no requirement for transfer of heat across the Equator because both hemispheres have similar zoned energy balances (Ramanathan 1987).

Heat is transported from the Equator to the poles in three ways: (1) by ocean currents carrying warm water, (2) by atmospheric circulation (wind) carrying warm air, and (3) by atmospheric circulation carrying latent heat in the form of water vapor. The maximum meridional heat transport occurs around 30° latitude (Ramanathan 1987). In the Northern Hemisphere, the heat transport by the oceans and that by the atmosphere are roughly comparable in size, but the exact fluxes are not well known (Cahine 1992).

Warm, wind-driven ocean currents transport heat poleward from the zone between 20°N and 20°S; they are well known, examples being the Gulf Stream in the North Atlantic and the Kuroshio Current in the North Pacific. The poleward transport of warm water by the oceans tends to warm the overlying atmosphere at higher latitudes, especially during the winter. This provides a moderating influence on climate and helps to explain the relatively mild winters experienced, for example, by western Europe (Gulf Stream) and by Japan (Kuroshio Current).

The atmosphere carries heat poleward as warm air and latent heat. A major source of warm air is the tropical region between 10°N and 10°S, which is the zone of maximum surplus of solar radiation. The tropical air is heated both by sensible heat and by the release of latent heat upon condensation of moisture. (The hot tropics are a zone of both high evaporation and high rainfall, with the latter predominating). Much additional latent heat in the form of water vapor is injected into the atmosphere in the subtropic zones (15–30° N and S), where evaporation exceeds precipitation. From there, poleward transport of the latent heat takes place. The latent heat is subsequently released by condensation, which warms the atmosphere in the mid-latitude zones of intense storm activity at 30–50° north and south. Latent heat from condensation in clouds provides 30% of the heat energy that is carried by the atmospheric circulation (Cahine 1992).

A small part (about 0.7%) of the incoming solar radiation is converted into the energy of motion (kinetic energy) of ocean currents, winds, and waves. Although this is a small number, it represents an energy of major interest to the hydrologic cycle, that associated with the circulation of the atmosphere and oceans. This circulation will be discussed next.

CIRCULATION OF THE ATMOSPHERE

The atmosphere circulates as a consequence of the latitudinal heat imbalance discussed above. If the circulation were due solely to heating, hot air would rise at the equator and flow poleward at high levels. As the air was cooled in transit and piled up at the poles, it would tend to sink at the poles. To complete the cycle there would be a return flow near the earth's surface of cool air toward the Equator, where heating would produce two symmetric closed circuits, or cells, one in the Northern and one in the Southern Hemisphere. Such a circulation (incorporating the earth's rotation) was originally proposed in 1735 by George Hadley but, because of a number of factors, the actual circulation has turned out to be considerably more complicated.

The general circulation of the atmosphere, showing the mean annual winds, is depicted in Figure 1.9. Note that the winds do not simply blow along north–south, or meridional lines. This is because they are deflected by the rotation of the earth. The force that deflects moving objects—in this case moving air masses to the right in the Northern Hemisphere, and to the left in the Southern Hemisphere—is referred to as the *Coriolis force.* The general circulation differs from the simple Hadley circulation by being broken up into several latitudinal zones; however, there is still symmetry more or less between the Northern and Southern Hemispheres. For the sake of brevity, we shall discuss only the Northern Hemisphere, but what is said applies equally to the Southern Hemisphere (with a leftward- instead of rightward-directed Coriolis force). [For details of circulation not discussed here, the interested reader is referred to books on meteorology and climatology such as Lutgens and Tarbuck (1992) and Barry and Chorley (1987).]

Hot moist air rises at the Equator; as it rises, the moisture condenses and intense precipitation results, releasing latent heat. Here there are only very weak surface winds giving rise to the equatorial doldrums. After rising, the air flows northward at high levels, cools, and eventually sinks around 30°N. The descending air is very dry (having lost most of its moisture in the tropics) and, when it is warmed, its capacity to take up moisture is further increased. The resulting hot dry air causes intense evaporation at the earth's surface, and this gives rise to the subtropical belt of deserts centered between 15 and 30°N. After reaching the surface, the air flows southward, picking up moisture as it flows over the ocean, and being deflected to the right by the Coriolis force. This surface flow is known as the *northeast trade winds.* Upon reaching the Equator, the northeast trades converge with the southeast trades from the Southern Hemisphere, in the *Intertropical Convergence Zone* (ITC), and the air rises at the Equator to complete the low-latitude cycle known

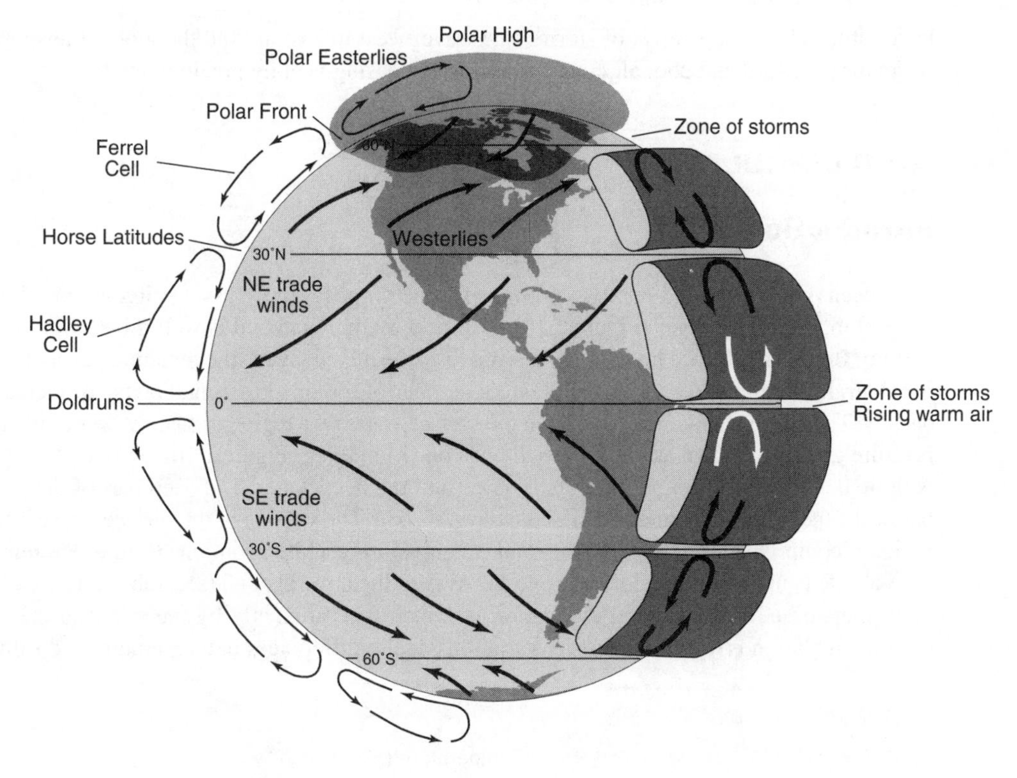

Figure 1.9 Schematic representation of the general circulation of the atmosphere. (Frederick K. Lutgens/Edward J. Tarbuck, *The Atmosphere,* 5th ed., Copyright © 1992, p. 170. Adapted by permission of Prentice Hall, Englewood Cliffs, New Jersey.)

as the *Hadley cell* (which behaves rather as Hadley expected the whole atmospheric circulation to behave).

At around 30°N, additional air descends and then flows north at the surface rather than south. This is the beginning of the *Ferrel cell.* The northward-flowing air is deflected to the right, forming the prevailing westerlies, which flow from southwest to northeast in the Northern Hemisphere. The westerlies continue until they encounter a cold mass of air moving south from the North Pole at about 50°N. This zone where the air masses meet is known as the *polar front,* and it is a region of unstable air, storm activity, and abundant precipitation. The polar jet stream (a very fast air stream) occurs at this boundary. The warmer air from the south rises over the polar air and then turns south at high altitude to complete the Ferrel cell. Meanwhile, the southward-flowing polar air (polar easterlies) becomes warmed by condensation at the polar front and by contact with the southern air. As a result, it too rises and then flows northward at high altitude to the pole, where it sinks, thus completing the polar cell.

In the mid-latitudes, the west-to-east flow (Northern Hemisphere westerlies) is subject to considerable turbulence because of the earth's rotation. At higher levels of the atmosphere, the flow forms waves (so-called planetary waves) that transport warm air from the surface to the top of the atmosphere (Goody and Walker 1972; Ingersoll 1983). These waves are expressed, in the lower atmosphere, in a series of storms that travel west to east around the globe, transporting warm air poleward and cool air equatorward and releasing heat by precipitation.

OCEANIC CIRCULATION

Introduction

The oceans can be divided into two portions for the purpose of discussing circulation. The top 50–300 m, or surface layer (Figure 1.10), is stirred by the wind and is well mixed from top to bottom. Below the surface layer (also referred to as the mixed layer), the remaining deeper water is colder, less well mixed, and divided into a number of roughly horizontal layers of increasing density. The deep water is separated from the surface water by a region of steeply decreasing temperature gradient, known as the *thermocline,* about 1 km thick (Figure 1.10), across which there is limited communication between the surface and the deep water. The deep part of the ocean, below the thermocline, is referred to as the *abyssal zone.* The volume of the surface (mixed) layer is small, comprising only 3.5% of the total ocean volume, while about one third of the remaining volume is in the thermocline and two thirds is in the abyssal zone (see Table 1.1).

In the surface ocean, lateral circulation is driven predominantly by the wind; in the deep ocean, circulation is driven by density variations due to differences in temperature and salinity,

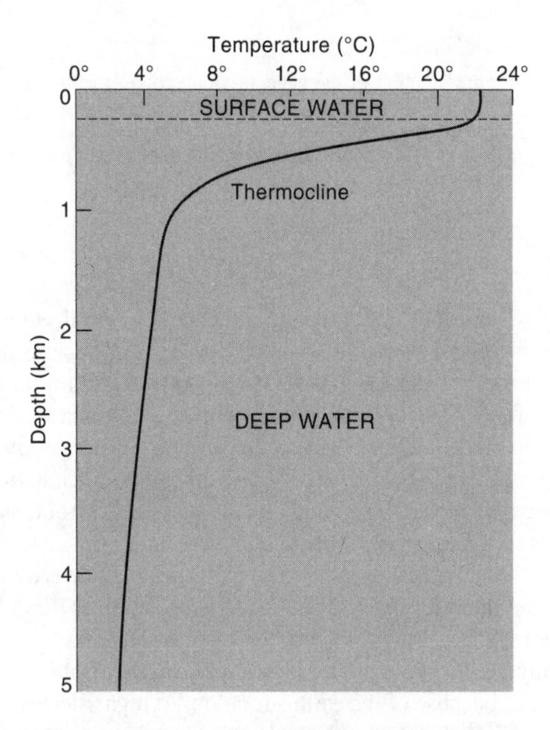

Figure 1.10 Generalized temperature-versus-depth profile for the oceans (at low to middle latitudes), showing vertical stratification into surface and deep water masses.

giving rise to the term *thermohaline circulation.* In this section the wind-driven and thermohaline circulations are discussed separately.

Wind-Driven (Shallow) Circulation

The circulation of the shallow ocean is driven by prevailing winds, which are in turn caused by uneven heating of the earth's surface. The circulation pattern can be summarized as a number of current rings, or *gyres,* that flow clockwise in the Northern Hemisphere and counterclockwise in the Southern Hemisphere due to the stresses imparted by the prevailing winds. These gyres extend poleward from about 10° north and south of the Equator to about 45° north and south. This is shown in Figure 1.11. Each gyre has a strong, narrow poleward current on the western side, with much weaker currents on the east. The most pronounced of these *westward intensification* currents are found in the Northern Hemisphere as the Gulf Stream in the Atlantic Ocean and the Kuroshio Current in the Pacific.

The circulation pattern shown in Figure 1.11 is not that expected if water were simply carried downwind. Transport of surface water, on the scale of the oceans, involves factors such as the earth's rotation and friction against continents, as well as wind stresses on the surface. The interaction of wind and water is complicated, and only a brief summary will be given here. [For details, the interested reader should consult references on physical oceanography, such as

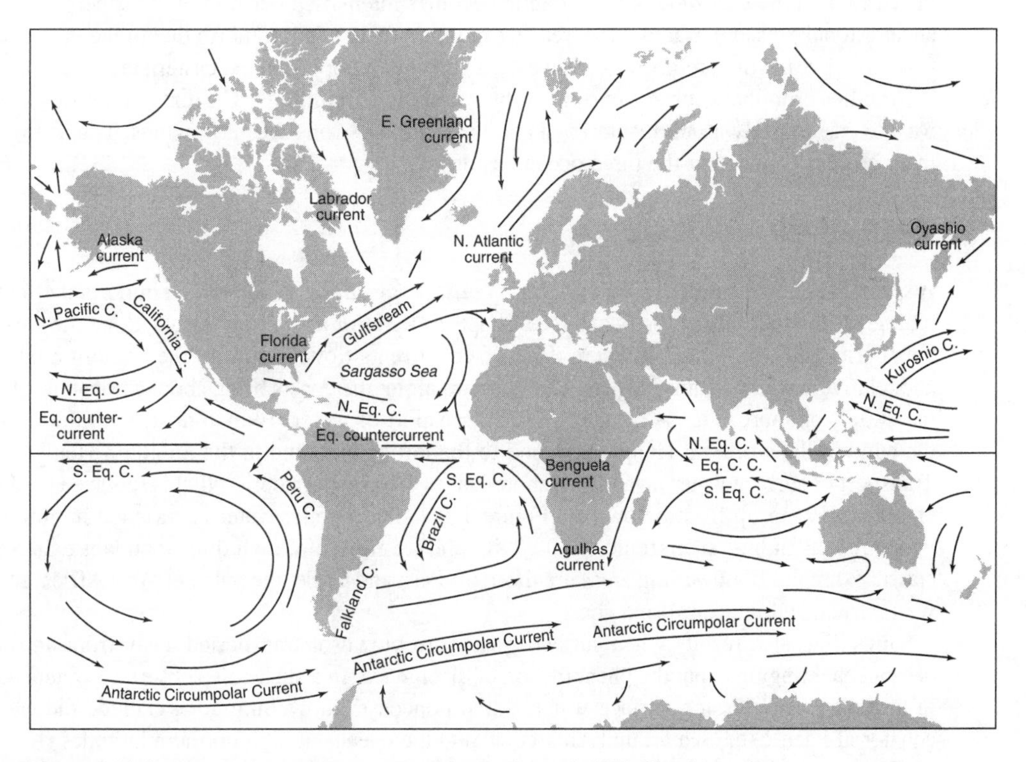

Figure 1.11 Surface currents of the oceans. (After Drake et al. 1978, Fig. 6.1, p. 88.)

Stommel (1965), Pickard and Emery (1982), and Pedlosky (1990), or general oceanography texts such as Thurman (1993) and Gross (1987).]

The origin of the prominent gyres can best be understood by reference to the North Atlantic. Here the prevailing winds are westerlies (from the west) at 40–50°N and trade winds (from the east) at 15–30°N. Because of the Coriolis force (see previous section), water in the top layer does not simply move downwind, but instead is moved to the right of the wind direction (Ekman flow). This brings about a convergence or piling up of water from both the north and south into the central portion of the North Atlantic in the Subtropical Convergence. (This area is also known as the Sargasso Sea.) The piled-up water then sinks and begins to return just below the surface, back to the north and south. As it does, it is turned to the right by the Coriolis force, which results in a strong east-flowing current on the north, and a strong west-flowing current on the south, just below the surface. These so-called *geostrophic currents* flow until they encounter the European continent at the northeast portion of the gyre and the North American continent along the southwest portion. Here they must turn. Since friction is strong along the continents and reduces the effect of the Coriolis force, the currents will flow from high pressure (where water is accumulating due to the current) to low pressure (where it is being removed), resulting in a north-to-south-flowing current on the European side and a south-to-north-flowing current (Gulf Stream) on the North American side. In this way a clockwise-flowing gyre results.

This type of explanation for the North Atlantic circulation can be applied to the other oceans except that in the Southern Hemisphere the Coriolis force is to the left of the direction of motion and, as a result, the gyres are counterclockwise (see Figure 1.11). In the Northern Hemisphere the current on the west side of the Atlantic becomes intensified because of the superposition of an additional pressure gradient (decreasing pressure from south to north due to the variation in Coriolis force), which reinforces the western current and opposes the eastern current.

Another prominent surface current is the Antarctic Circumpolar Current, which flows from west to east entirely around Antarctica and is driven by strong westerly winds. This is the primary current connecting the three ocean basins.

Coastal Upwelling

A special case of wind-driven circulation is *coastal upwelling,* which has an important effect on biological productivity in the ocean (see Chapter 8). Major upwelling occurs along the western boundaries of the continents, where the surface currents flowing toward the Equator are broad and relatively weak. Winds blowing equatorward along the coasts bring about a transport of surface water offshore. This is because the net transport of water (Ekman drift) is to the right of the wind in the Northern Hemisphere and to the left of the wind in the Southern Hemisphere. Transport of surface water away from shore leaves a near-shore deficit that is replaced by deeper water flowing up from below (see Figure 1.12). This deeper water is enriched in nutrients, which results in high planktonic productivity and teeming life, including abundant fish. Some classic examples of upwelling areas are those off Peru and Chile, the bulge of West Africa, southwest Africa, and the California coast.

Upwelling also results when surface waters are blown or transported away from an open-ocean area, bringing about the phenomenon of divergence. In a such case subsurface waters move in to replace the missing surface water. Some noncoastal upwelling areas include the eastern equatorial Pacific, the sea around Antarctica, and the oceans at high northern latitudes (Kennett 1982).

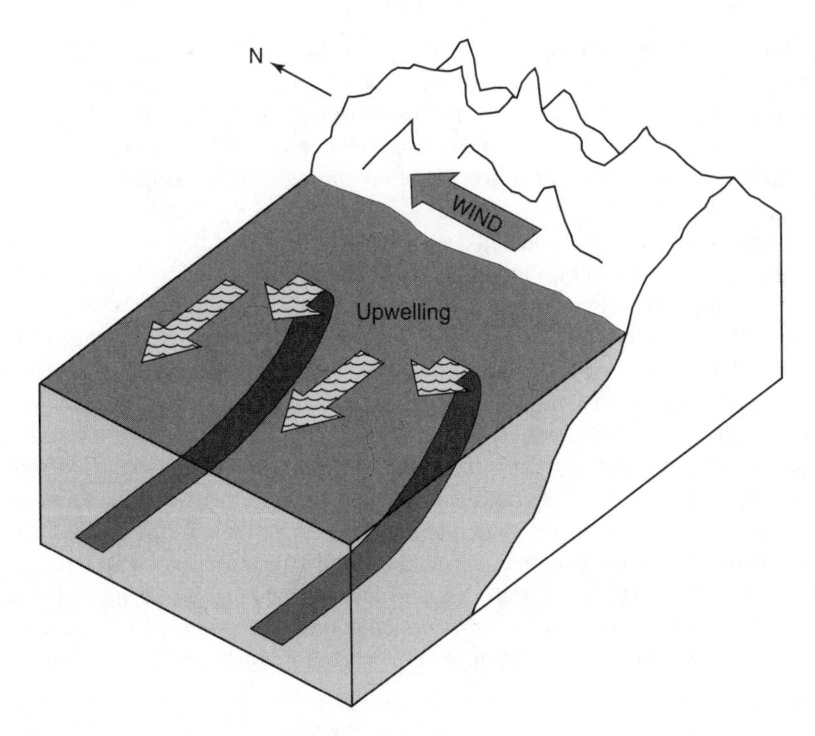

Figure 1.12 Upwelling, or the result of Ekman drift, in response to a north-blowing wind in the Southern Hemisphere. (After K. K. Turekian, *Oceans,* 2nd ed. Copyright © 1976, p. 35. Reprinted by permission of Prentice-Hall, Inc., Englewood Cliffs, N. J.)

Thermohaline (Deep) Circulation

Below the top few hundred meters, the oceans are not directly affected by the winds. Here circulation is brought about by density differences arising from differences in temperature and salinity. [The following discussion of deep ocean circulation is from Pickard and Emery (1982), Warren (1981), and Gordon (1986); see also Drake et al. (1978), Gross (1987), and Thurman (1993).] In seawater, density increases continuously with decrease of temperature and no density maximum, as is found in fresh water at 4°C (see Chapter 6), is encountered. Density also increases with increasing salt content, or salinity. Overall the deep ocean is vertically stratified into various water masses, with the densest water at the bottom and the lightest at the top, and the density stratification is due primarily to the decrease of temperature with depth. The stratification severely inhibits vertical motion; in other words, it is much easier to move water along surfaces of constant density (isopycnals) than it is to move it across them. Thus, the deep water circulation can be viewed as primarily horizontal.

The density stratification and density differences between water masses of the deep sea owe their origin to surface processes. Here changes in density are brought about by heating and cooling, evaporation, addition of fresh water, and freezing out of sea ice. Surface water migrating to high latitudes becomes more dense because of evaporation (which causes both cooling and increased salinity) and because of loss of sensible heat and sea-ice formation. (When sea ice forms, dissolved salts are excluded from the ice and the remaining water becomes more saline and, thus, more dense.) In certain locations in the far North and far South Atlantic, this surface density increase becomes so great that the cold, salty surface water occasionally becomes denser than the underlying water and sinks downward to replace it. In this way deep ocean water originates, and is replenished from above. Once it reaches great depths, the water tends to conserve its temperature and salinity as it flows laterally throughout the oceans, bringing about the deep water circulation.

An example of stratification and deep circulation for the Atlantic Ocean, is shown in Figure 1.13. Here each water mass is identified by its characteristic temperature and salinity. The deep water is dominated by two cold water masses, the North Atlantic Deep Water and the Antarctic Bottom Water. The North Atlantic Deep Water (NADW) originates in the Norwegian Sea off Greenland, from the cooling of Gulf Stream surface water mainly by evaporation. This water sinks and flows at depth southward, where it is joined by water sinking in the Labrador Sea off Canada. Ultimately this water crosses the Equator. In the Antarctic an even denser water, the Antarctic Bottom Water, is formed in the Weddell Sea by cooling and the freezing out of sea ice in the winter. This sinks to the bottom and flows north. In the Antarctic region farther north, at the Antarctic Convergence, the Antarctic Intermediate Water is formed, which also sinks but not to the bottom because it cannot displace the underlying denser North Atlantic Deep or Antarctic Bottom waters. This water thus occupies intermediate depths, as implied by its name.

Another intermediate-type water is formed in the Mediterranean Sea. Here intense evaporation causes the water to be sufficiently saline and dense that, upon passing out into the Atlantic

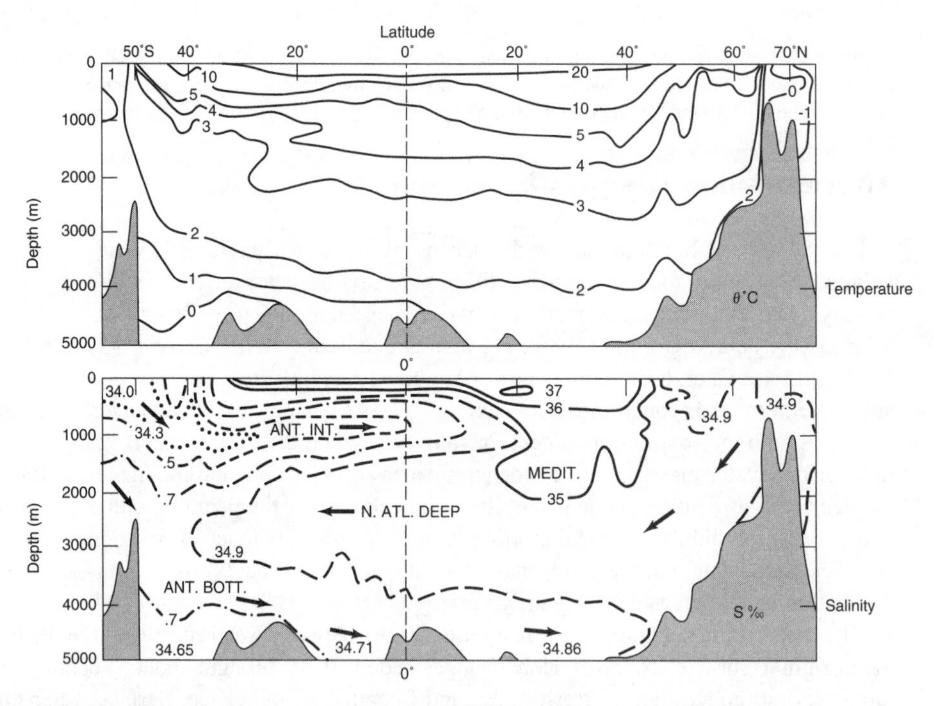

Figure 1.13 South–north vertical section of water properties of the Atlantic Ocean along the western trough as delineated by lines of constant temperature and salinity. N. Atl. Deep = North Atlantic Deep Water; Ant. Bott. = Antarctic Bottom Water; Ant. Int. = Antarctic Intermediate Water; Medit. = Mediterranean Water. [Adapted from Pickard and Emery (1982), based on data from Bainbridge (1976).]

through the Straits of Gibraltar, it sinks and fills intermediate depths. It does not sink to the bottom because, although it is more saline than North Atlantic Deep Water, it is also warmer, and the temperature difference counteracts the salinity effect to make it less dense than North Atlantic Deep Water (NADW). For the same reason, surface water at lower latitudes remains at the surface, even though it is more saline (see Figure 1.13, bottom); that is, the density-lowering effect of higher temperature overpowers the density-raising effect of higher salinity. Ultimately the Mediterranean water moves west and south across the Atlantic and joins the NADW in the western boundary current (Gordon 1986).

The lateral deep water circulation over the entire ocean, which was originally modelled by Stommel (1958), is shown in Figure 1.14 (after Gordon 1986). (Deep water circulation is not well documented and maps, such as that shown in Figure 1.14, are based largely on theoretical models.) North Atlantic Deep Water flows away southward from its source as an intense bottom current on the western side of the North Atlantic. This meets a strong northward-flowing current of Antarctic Bottom Water in the South Atlantic, and as a result they merge and flow east through the Antarctic into the deep Indian Ocean and ultimately into the deep Pacific Ocean (Gordon 1986; Warren 1981). Thus, the deep water originates only in two areas, and both are in the Atlantic Ocean.

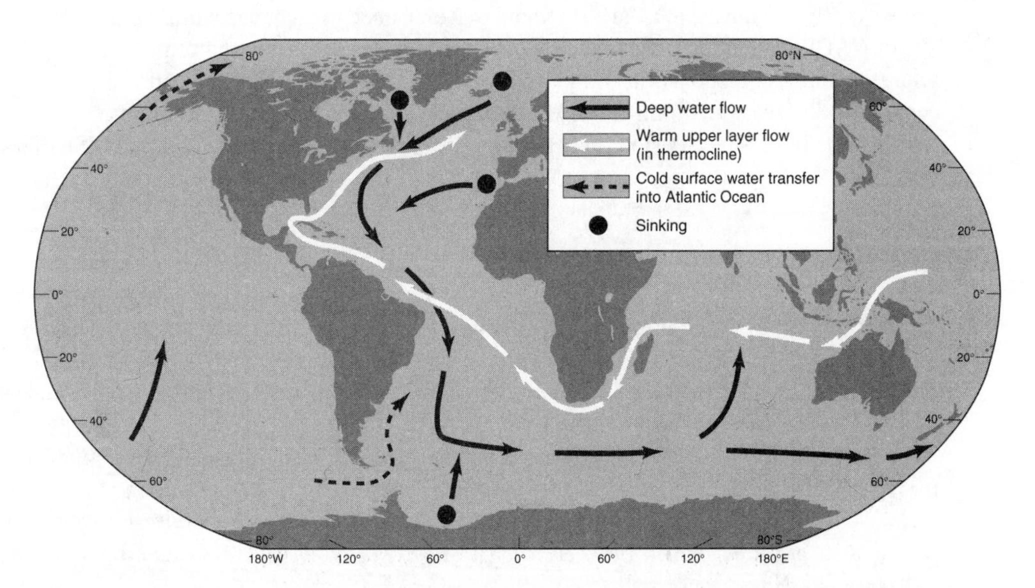

Figure 1.14 Global cycle of thermohaline circulation. Deep water circulation (solid arrows) originates in the North Atlantic by sinking of North Atlantic Deep Water (NADW). The NADW starts with water from the Norwegian Sea, which is joined by water that sinks in Labrador Bay. This water then flows south at depth as intensified currents along the western side of the ocean basin to the South Atlantic, where it is joined by Antarctic Bottom Water that sinks in the Weddell Sea. The deep water then flows into the deep Indian Ocean and Pacific Ocean. There is upwelling to the surface in all oceans. In addition, a warm-water return circulation in the thermocline layer (shown by dashed arrows) has been proposed (see Gordon 1986) as a way to return water to the North Atlantic. There is also return flow of cold surface water to the Atlantic through the Drake Passage (south of South America) and also from the Pacific through the Bering Sea and Arctic Sea into the North Atlantic. (Modified from Gordon 1986.)

During the thermohaline circulation, deep waters remain out of contact with the atmosphere for long periods of time, which can result in appreciable changes in their chemical composition (see Chapter 8). The residence time of deep water, or the average time it spends out of contact with the atmosphere, is about 200–500 years for the Atlantic and 1000–2000 years for the Pacific (see, e.g., Turekian 1976).

As the deep water traverses the ocean floor, there is very slow, diffuse upwelling (at the rate of about 1 m/yr) of deep water into surface layers. This slow upwelling is supplemented by the more intense but localized coastal and open-ocean upwelling discussed in the previous section. In addition, three direct routes are proposed to return water to the Atlantic Ocean: One is the flow of cold surface water through the Bering Strait and Arctic Sea into the North Atlantic, and another is the flow of cold surface water through the Drake Passage (south of South America) into the South Atlantic. The third is a proposed return flow of warm water toward the North Atlantic (white arrows in Figure 1.14) in the ocean's thermocline layer to feed North Atlantic Deep Water formation (Gordon 1986).

There is considerable interest in the thermohaline circulation of the North Atlantic and the rate of formation of the NADW because of its climatic implications. Possible changes in NADW formation have been cited as a cause of rapid climate change over the last 18,000 years (e.g., Street-Perrott and Perrott 1990), and increased global temperatures from greenhouse warming (see Chapter 2) may affect the rate of formation of NADW in the future (e.g., Gates et al. 1992; Broecker 1987). Manabe et al. (1991) found that enhanced greenhouse warming may reduce the rate of NADW formation, thus reducing the surface water flow from the south and delaying the future greenhouse response in the northern North Atlantic. NADW formation is sensitive to changes in surface water temperature and salinity (the latter due to the addition of fresh water from ice melting; Street-Perrott and Perrott 1990) and changes in ocean currents (Shaffer and Bendtsen 1994).

REFERENCES

Arrhenius, G., B. R. De, and H. Alfven. 1974. Origin of the ocean. In *The Sea,* vol. 5, ed. E. Goldberg, pp. 839–861. New York: Wiley-Interscience.

Bainbridge, A. E. 1976. *GEOSECS Atlantic Expedition, v. 2:* (198). *Sections and Profiles.* Washington, D.C.: National Science Foundation.

Barry, R. G., and R. J. Chorley. 1987. *Atmosphere, Weather, and Climate,* 5th ed. London: Methuen.

Baumgartner, A., and E. Reichel. 1975. *The World Water Balance, Mean Annual Global, Continental and Maritime Precipitation, Evaporation and Runoff,* trans. by R. Lee. New York: Elsevier.

Berner, E. K., and R. A. Berner. 1987. *The Global Water Cycle: Geochemistry and Environment.* Englewood Cliffs, N.J.: Prentice-Hall.

Bloom, A. L. 1971. Glacial-eustatic and isostatic controls of sea level since the last glaciation. In *The Late Cenozoic Ice Ages,* ed. K. K. Turekian, pp. 355–379. New Haven, Conn.: Yale University Press.

Broecker, W. S. 1987. Unpleasant surprises in the greenhouse, *Nature* 328: 123–126.

Budyko, M. I., and K. Y. Kondratiev. 1964. The heat balance of the earth. In *Research in Geophysics,* vol. 2, pp. 529–554. Cambridge, Mass.: MIT Press.

Cahine, M. T. 1992. The hydrological cycle and its influence on climate, *Nature* 359: 373–380.

Drake, C. L., J. Imbrie, J. A. Knauss, and K. K. Turekian. 1978. *Oceanography.* New York: Holt, Rinehart & Winston.

Ellis, J. S., and T. H. Vonder Haar. 1976. Zonal average earth radiation budget measurements from satellites for climate studies, Atmos. Sci. Rep. 240, Fort Collins, Colo.: Dept. Atmospheric Science, Colorado State University.

Flohn, H. 1977. Man-induced changes in the heat budget and possible effects on climate. In *Global Chemical Cycles and Their Alterations by Man,* ed. W. Stumm, pp. 207–224. Berlin: Dahlem Konferenzen.

Garrels, R. M., and F. T. Mackenzie. 1971. *Evolution of Sedimentary Rocks.* New York: W. W. Norton.

Gates, W. L., J. F. B. Mitchell, G. L. Boer, U. Cubasch, and V. P. Meleshko. 1992. Climate modelling, climate prediction and model validation. In *Climate Change 1992. The Supplementary Report to the IPCC Scientific Assessment,* ed. J. T. Houghton, B. A. Callendar, and S. K. Varney, pp. 97–134. Cambridge, U.K.: Cambridge University Press.

Goody, R. M., and J. C. G. Walker. 1972. *Atmospheres.* Englewood Cliffs, N.J.: Prentice-Hall.

Gordon, A. L. 1986. Interocean exchange of thermocline water, *J. Geophys. Res.* 91(C4): 5037–5046.

Gross, M. G. 1987. *Oceanography: A View of the Earth,* 4th ed. Englewood Cliffs, N.J.: Prentice Hall.

Henderson-Sellers, A., and J. G. Cogley. 1982. The Earth's early hydrosphere, *Nature* 298: 832–835.

Holland, H. D. 1984. *The Chemical Evolution of the Atmosphere and Oceans.* Princeton, N.J.: Princeton University Press.

Hubbert, M. K. 1971. The energy resources of the earth, *Sci. Am.* 224(3): 60–70.

Ingersoll, A. P. 1983. The atmosphere, *Sci. Am.* 249(3): 162–175.

IPCC (Intergovernmental Panel on Climate Change). 1990. *Climate Change. The IPCC Assessment,* ed J. T. Houghton, G. J. Jenkins, and J. J. Ephraums. Cambridge, U.K.: Cambridge University Press.

Kasting, J. F. 1993. Earth's early atmosphere, *Science* 259: 920–926.

Kennett, J. P. 1982. *Marine Geology.* Englewood Cliffs, N.J.: Prentice-Hall.

Laws, E. A. 1993. *Aquatic Pollution,* 2nd ed. New York: John Wiley.

Lutgens, F. K., and E. J. Tarbuck. 1992. *The Atmosphere: An Introduction to Meteorology,* 5th ed. Englewood Cliffs, N.J.: Prentice-Hall.

Manabe, S., R. J. Stouffer, M. J. Spelman, and K. Bryan. 1991. Transient responses of a coupled ocean-atmosphere model to gradual changes of atmospheric CO_2. Part I: Annual mean response, *J. Clim.* 44: 785–818.

Meybeck, M. 1987. Global chemical weathering of surficial rocks estimated from river dissolved loads, *Am. J. Sci.* 287: 401–428.

Miller, A., J. C. Thompson, R. E. Peterson, and D. R. Haragan. 1983. *Elements of Meteorology,* 4th ed. Columbus, Ohio: Charles. E. Merrill.

Moorbath, Stephen. 1977. The oldest rocks and the growth of continents, *Sci. Am.* 236(3): 92–104.

National Research Council (NRC). 1986. *Global Change in the Geosphere-Biosphere.* Washington, D.C.: National Academy Press.

Oort, A. H. 1970. The energy cycle of the earth, *Sci. Am.* 223(3): 54–63.

Pedlosky, J. 1990. The dynamics of the oceanic subtropical gyres, *Science* 248: 316–322.

Peixoto, J. P., and M. Kettani. 1973. The control of the water cycle, *Sci. Am.* 228(4): 46–61.

Penman, H. L. 1970. The water cycle, *Sci. Am.* 223(3): 98–108.

Pickard, G. L., and W. J. Emery. 1982. *Descriptive Physical Oceanography,* 4th ed. New York: Pergamon Press.

Ramanathan, V. 1987. The role of earth radiation budget studies in climate and general circulation research, *J. Geophys. Res.* 92: 4075–4095.

Ramanathan, V., B. R. Barkstron, and E. F. Harrison. 1989a. Climate and the earth's radiation budget, *Physics Today,* May 1989: 22–32.

Ramanathan, V., R. D. Cess, E. F. Harrison, P. Minnis, B. R. Barkstrom, E. Ahmad, and D. Hartmann. 1989b. Cloud-radiative forcing and climate: Results from the Earth Radiation Budget Experiment, *Science* 243: 57–63.

Sellers, W. D. 1965. *Physical climatology.* Chicago: University of Chicago Press.

Shaffer, G., and J. Bendtsen. 1994. Role of the Bering Strait in controlling North Atlantic ocean circulation and climate, *Nature* 367: 354–357.

Stommel, H. 1958. Circulation of the abyss, *Sci. Am.* 199(1): 85–90.

Stommel, H. 1965. *The Gulf Stream,* 2nd ed. Berkeley, Calif.: University of California Press.

Street-Perrott, F. A., and R. A. Perrott. 1990. Abrupt climate fluctuations in the tropics: The influence of the Atlantic circulation, *Nature* 343: 607–612.

Thurman, H. V. 1993. *Essentials of Oceanography,* 4th ed. New York: Macmillan.

Turekian, K. K. 1976. *Oceans,* 2nd ed. Englewood Cliffs, N.J.: Prentice-Hall.

Walker, J. C. G. 1977. *Evolution of the Atmosphere.* New York: Macmillan.

Warren, B. A. 1981. Deep circulation of the world ocean. In *Evolution of Physical Oceanography: Scientific Surveys in Honor of Henry Stommel,* ed. B. A. Warren and C. Wunsch, pp. 6–41. Cambridge, Mass.: MIT Press.

York, D. 1993. The earliest history of the earth, *Sci. Am.* 268(1): 90–96.

CHAPTER 2

AIR CHEMISTRY

The Greenhouse Effect and the Ozone Hole

In order to better understand the chemistry of rainwater and to discuss some major global environmental problems, it is necessary to delve into several aspects of air chemistry. Air consists of a mixture of gases and suspended particles, and the composition of this mixture has been perturbed in recent decades by human activities. This leads us to a discussion of such subjects as the atmospheric greenhouse effect, the ozone hole, and cooling by the scattering of sunlight by particle layers. (Air chemistry as it leads to the formation of acid rain is discussed in Chapter 3.) In keeping with the general approach of this book, emphasis is on global or large-scale regional problems.

ATMOSPHERIC GASES

Air consists mainly of three gases, nitrogen, oxygen, and argon, which make up over 99.9% of the total volume. These major gases, along with a number of minor inert gases (helium, neon, krypton), because of their long residence times, occur in constant ratios to one another throughout the atmosphere—these ratios stay constant over the human time scale. This is not true for the other gases, which are affected by changes in rates of input by natural processes, but more importantly, by human processes that have brought about major perturbations in their concentrations, both locally and globally. A major example is the rise of CO_2 over the past century, due to the burning of fossil fuels, a subject that will be dealt with in detail later in this chapter. A listing of gas concentrations in air is shown in Table 2.1.

Because of their long residence times, the two major atmospheric gases, N_2 and O_2, cannot be affected by anthropogenic activities. For instance, to remove all O_2 from the atmosphere would take over 80,000 years of fossil fuel burning at the present rate. On a larger scale, the natural processes of global photosynthesis (O_2 production) and respiration (O_2 consumption) are almost perfectly balanced (to within 0.4%). If human activities caused photosynthesis to cease and respiration continued at its present rate (a drastic and incredibly unrealistic scenario), it would

TABLE 2.1 Concentration of Gases in Air at Sea Level

Gas	Volume Percent in Air
Nitrogen (N_2)	78.084
Oxygen (O_2)	20.948
Argon (Ar)	0.934
Carbon dioxide (CO_2)	0.036
Neon (Ne)	0.0018
Helium (He)	0.0005
Methane (CH_4)	≃0.0002
Sulfur dioxide (SO_2)	0–0.0001
Krypton (Kr)	0.0001
Hydrogen (H_2)	≃0.00005
Nitrous oxide (N_2O)	≃0.00003
Carbon monoxide (CO)	≃0.00001
Nitrogen dioxide (NO_2)	0–0.000002
Ammonia (NH_3)	≃0.000001
Ozone (O_3)	0–0.000001

Sources: Modified from Turekian (1972) and Walker (1977), with CO_2 updated to 1994.

still take over 8000 years to consume all the atmospheric O_2. Similarly, if the production of N_2 by global denitrification (see Chapter 3) ceased tomorrow and consumption of N_2 by nitrogen fixation (see also Chapter 3) continued at its present rate, it would take more than 9 million years to strip all the N_2 from the atmosphere.

Gases of lesser abundance, but which are impacted anthropogenically, are carbon dioxide (CO_2), methane (CH_4), nitrous oxide (N_2O), ammonia (NH_3), ozone (O_3), carbon monoxide (CO), sulfur dioxide (SO_2), and nitrogen oxides (a combination of NO_2 + NO, represented as NO_x). All of these gases have increased in atmospheric concentration as the result of human activities. Carbon dioxide, methane, and N_2O are the most long lived (residence times on the scale of several years), so their concentrations at any one time are relatively uniform over the globe. Carbon dioxide is the major contributor to the enhanced atmospheric greenhouse effect, and because of this, major emphasis is given CO_2 in this chapter. Because of their influence on the composition of rain, especially acid rain, SO_2, NO_2, and NH_3 are discussed in Chapter 3 in terms of the atmospheric cycles of sulfur and nitrogen.

An important atmospheric gas, water vapor, is omitted from Table 2.1. Its concentration is highly variable, both spatially and temporally, and varies from a low of less than 0.01% to as much as 3%. Water vapor is omitted only because it is a topic of special consideration that is discussed extensively in Chapter 1 as it affects the global energy cycle and in Chapter 3 as it affects rain formation.

THE ATMOSPHERIC GREENHOUSE EFFECT

Anthropogenic Carbon Dioxide—Present, Past, and Future

Although carbon dioxide is the fourth most abundant gas in the atmosphere, it constitutes only some 0.036% (by volume) (see Table 2.1). Atmospheric CO_2 is important for two reasons:

(1) It strongly absorbs infrared (long-wave) radiation given off by the earth and reradiates energy back to the earth, thus helping to maintain the earth's surface temperature (the so-called greenhouse effect—see Chapter 1); and (2) it is a source of carbon, which is the dominant element in life and in the biogeochemical cycles of the earth.

Beginning in 1958, the atmospheric concentration of CO_2 has been measured at the Mauna Loa Observatory in Hawaii (Figure 2.1). There is an obvious annual oscillation in atmospheric CO_2 concentration of around 6 parts per million (ppm), which is a result of terrestrial biological cycling, with uptake of CO_2 by plants during the spring and summer due to excess photosynthesis and its release during the fall and winter due to excess respiration. Superimposed on the annual oscillation, one can see from Figure 2.1 that the yearly average value of CO_2 has clearly increased from 315 ppm in 1958 to 357 ppm in 1993, and the CO_2 concentration has been rising faster in the last 15 years (at an average rate of 0.4% per year) than it did earlier.

The rise in atmospheric CO_2 has been attributed mainly to the burning of fossil fuels (coal and oil), which release CO_2 to the atmosphere. A much smaller amount of CO_2 comes from the production of cement. Estimates of contributions from fossil fuel combustion plus cement production over the past two centuries is shown in Figure 2.2. The rate of fossil fuel CO_2 release has risen on the average 2.5% per year over the last 100 years, and it is estimated that it will rise over the next 100 years to about three times its present rate (6.0 ± 0.5 Gt C/yr in 1990; IPCC 1992). However, estimates of the increase in emissions rate are dependent on assumptions about population growth and economic growth and vary from 4.6 Gt C/yr to 35.8 Gt C/yr in 2100 for reasonable scenarios. Thus, although this source of CO_2 is presently increasing, its future growth rate is only poorly known.

Deforestation by humans, resulting in the burning or oxidation of organic carbon stored in plant material, is another anthropogenic source of atmospheric CO_2, the present magnitude of which is not well agreed upon (see discussion below) but which is expected to decline in the future. Burning comes about mainly from the clearing of land in the tropics for agriculture.

From the data of Boden et al. (1992), one can calculate that the amount of fossil fuel CO_2 released from 1959 through 1990 was 135 Gt C. During the same period, the increase in atmospheric CO_2-C was about 81 Gt C. Thus, only about 60% of the fossil fuel carbon added has remained in the atmosphere. Where has the rest of the CO_2 gone? The answer to this question lies in the carbon cycle.

Two major carbon reservoirs could take up the missing CO_2; both are much larger than the atmosphere and exchange rapidly with it on a time scale of years to hundreds of years. These reservoirs are the oceans and the terrestrial biosphere plus soils (see Figure 2.3). [Carbonate rocks and buried organic matter represent much larger carbon reservoirs that are important over geologic time, but they are not important on a human time scale; see, e.g., Berner (1994).] The oceans represent the largest of the rapidly exchanging reservoirs, and most of the missing atmospheric fossil fuel CO_2-C has probably been stored in them. Carbon is present in the oceans primarily as inorganic carbon in the form of dissolved bicarbonate ion (HCO_3^-), and carbonate ion (CO_3^{2-}). When atmospheric CO_2 is added to ocean surface water, the following reaction occurs:

$$CO_2 + CO_3^{2-} + H_2O \rightarrow 2HCO_3^-$$

Thus, atmospheric CO_2 is converted to HCO_3^- ion and stored in the oceans.

The surface oceans (top 75 m) are well mixed with the atmosphere. Below this lies the thermocline (1000 m thick), where exchange with atmospheric CO_2 is on a time scale of several

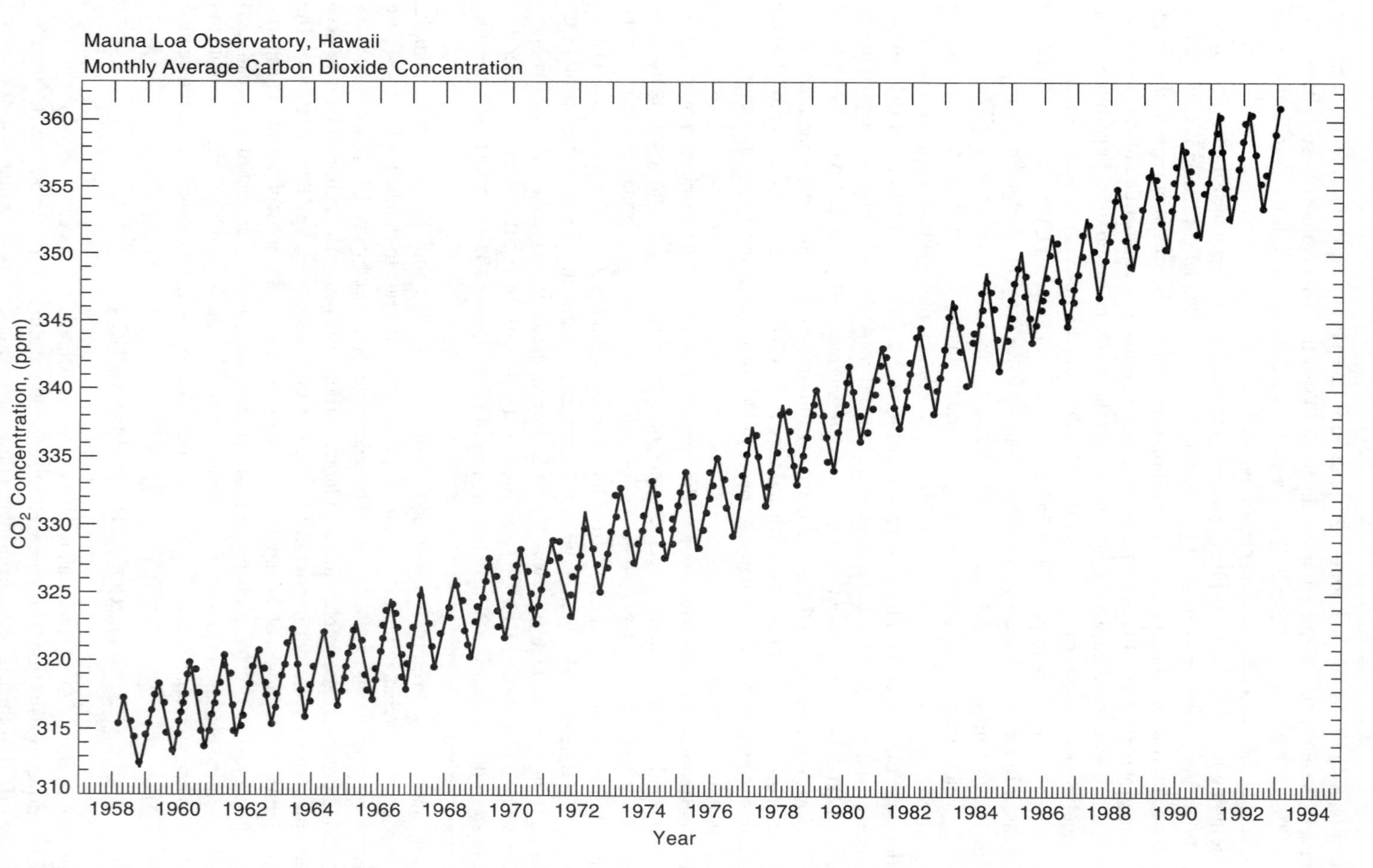

Figure 2.1. Mean monthly concentration of atmospheric CO_2 at Mauna Loa, Hawaii, 1958–late 1993 (C. D. Keeling, personal communication, 1994). The yearly oscillation is explained mainly by the annual cycle of photosynthesis and respiration of plants in the Northern Hemisphere. (Note: 1 ppm CO_2 = 2.12 Gt C, where 1 Gt C = 10^9 tons C = 10^{15} g C.)

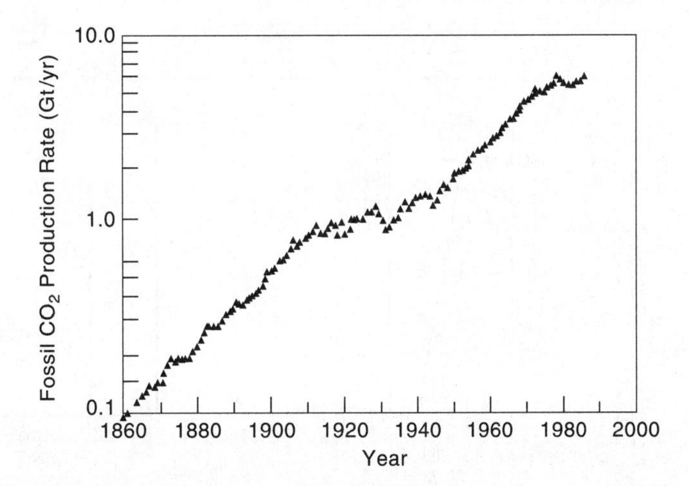

Figure 2.2. Global annual emissions of CO_2 from fossil fuel combustion and cement production in gigatons of carbon per year (plotted on a log scale). Gt = 10^9 tons. [After Watson et al. (1990), Greenhouse gases and aerosols. In *Climate change: The IPCC scientific assessment,* ed. J. T. Houghton et al., p. 10, copyright World Meteorological Organization, from Rotty and Marland (1986) and Marland (1989).]

decades, and below the thermocline lies the deep sea, which is isolated from the atmosphere and mixes very slowly with it, on a time scale of several hundred to a thousand years [Broecker et al. (1979); for further details on ocean structure, see the oceanic section of Chapter 1]. Thus, on a time scale of decades, only the surface ocean, and certain other regions extending down to 1000 m, take up CO_2. Because of the solubility of CO_2 and the large amount of carbon in the oceans, the increase in atmospheric CO_2 over the past century of about 25% should have been accompanied by a rise of about 2% in the CO_2 stored in the oceans (Sarmiento 1993a). Natural oceanic variability is of the same order of magnitude as 2%, and accurate measurements of total CO_2 concentration have been made only over recent decades. Thus, it has been impossible to measure the increase in oceanic carbon directly.

In the absence of direct measurements, the amount of excess fossil fuel CO_2 taken up by the oceans has been estimated using various tracers of gas exchange, such as radiocarbon measurements combined with ocean circulation modeling (Siegenthaler and Sarmiento 1993) or by the use of changes in the $^{13}C/^{12}C$ ratio of seawater (Quay et al. 1992). Both approaches agree on an oceanic uptake flux of about 2 Gt C per year. This value is also in general agreement with measurements of total (anthropogenic plus natural) CO_2 exchange between the atmosphere and oceans (Tans et al. 1990), once correction is made for the natural carbon cycle (Sarmiento and Sundquist 1992).

Another means by which excess carbon dioxide can be stored in the oceans is known as the *biological pump* (Sarmiento 1993a). Carbon dioxide is taken up in surface waters by photosynthesis to produce living planktonic organic matter. When the plankton die, their remains fall into deep water, where they decay back to carbon dioxide. This carbon dioxide dissolves in the deep water, which is not in rapid exchange with the atmosphere. Thus, the biological pump results in a net transfer of CO_2 from the atmosphere to deeper layers of the ocean. Some of this transfer may take place by intense biological production in coastal waters, with the organic

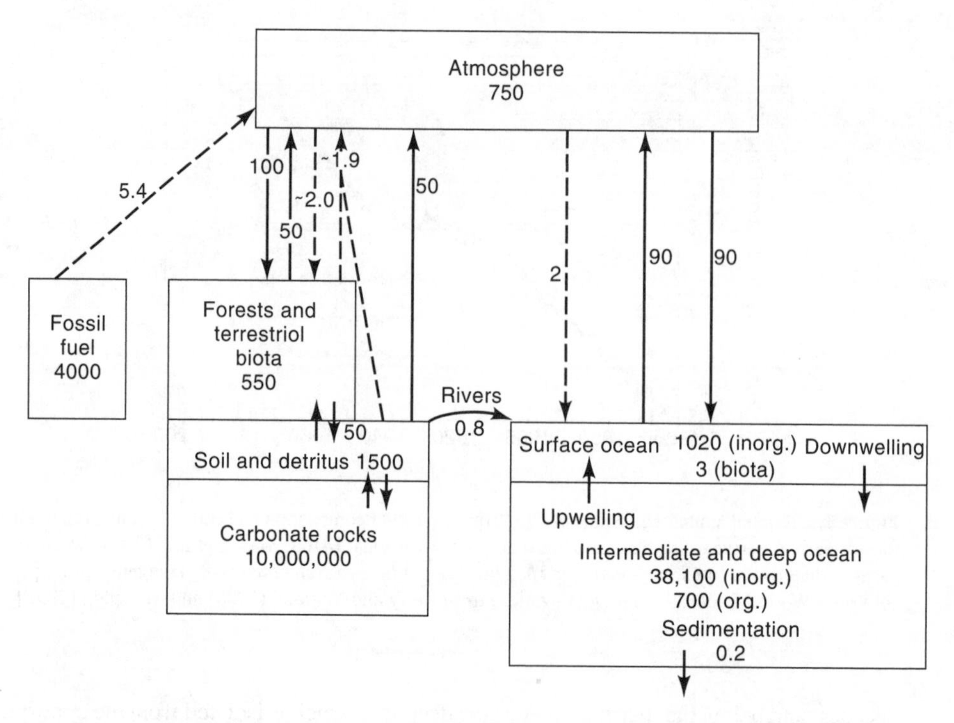

Figure 2.3. The carbon cycle. Reservoirs in 10^{15} g C = 10^9 t C = Gt C. Fluxes in Gt C/yr (dashed fluxes due to human activity; solid fluxes, natural). (Modified from Siegenthaler and Sarmiento 1993; NRC 1983; Bolin et al. 1979.)

carbon transported off the continental shelves into deeper waters (Wollast 1991). Since the production of organic matter in the oceans is generally limited by the availability of nutrients other than CO_2 (see Chapter 8), an increase in atmospheric CO_2 does not necessarily mean an increase in the rate of downward biological transfer. To increase this rate and to try to accommodate excess CO_2 produced by fossil fuel burning, it has been suggested that extra nutrients, specifically iron, be added to the Antarctic Ocean as a corrective (Martin et al. 1990). However, this proposal has proven to be controversial.

Some atmospheric CO_2 removed by marine organisms during photosynthesis is eventually deposited in bottom sediments. However, most of this CO_2 is released fairly rapidly by organic matter decay, and only a very small amount is eventually buried and thereby permanently removed from the ocean–atmosphere system. Calculations indicate that sediment burial probably is not a major mechanism for the removal of anthropogenic CO_2 (Berner 1982).

The other large carbon reservoir and possible missing CO_2 sink is forests and the terrestrial biosphere (550 Gt C) plus soils and detritus (1500 Gt C). There is rapid annual exchange between the terrestrial biosphere and soils and the atmosphere of approximately 100 Gt C/yr, and this causes the annual oscillation in atmospheric CO_2 concentration mentioned previously. The amplitude of the annual oscillation of atmospheric CO_2 seems to be increasing (see Figure 2.1), which leads some to suspect that the size of the terrestrial biosphere may be growing in response to increasing atmospheric CO_2. A growing biosphere would constitute increased CO_2 storage.

Experiments [see Bazzaz (1990) for a review] have shown that many plants grow faster when exposed to higher levels of CO_2; this is often referred to as the *CO_2 fertilization effect.*

Woodwell et al. (1983) and others, however, believe that the terrestrial biosphere is not a sink for CO_2 but a source. They have calculated that the rate of net deforestation (deforestation minus forest regrowth) leads to an addition of CO_2 to the atmosphere of 1.8–4.7 Gt C/yr. Further, Schlesinger (1991) notes that deforestation, besides loss of trees, is accompanied by the loss of soil carbon during the conversion of forests soils to plowed fields. However, more recent estimates of atmospheric CO_2 gain from changes in land use (mainly deforestation) by the Intergovernmental Panel on Climate Change (IPCC) are much lower, namely, 0.6–2.5 Gt C/yr (Siegenthaler and Sarmiento 1993). This CO_2 gain is approximately balanced by CO_2 loss to an unidentified sink, possibly the terrestrial biosphere (see later discussion).

The annual balance for anthropogenic CO_2 can be expressed as follows:

$$I = F - A - Oc + B$$

where I = any imbalance in the calculation; F = fossil fuel addition per year; A = annual increase in atmospheric CO_2; Oc = oceanic storage per year; and B = net addition by deforestation minus reforestation. If all sources and sinks were adequately accounted for, the value of I would be zero. Recent estimates of average values for 1980–1989 (Siegenthaler and Sarmiento 1993) (in Gt C/yr) are

$$\begin{aligned} A &= 3.2 \pm 0.2 \\ F &= 5.4 \pm 0.5 \\ Oc &= 2.0 \pm 0.6 \\ B &= 1.6 \pm 1.0 \end{aligned}$$

From these values and the above equation, $I = 1.8 \pm 1.3$ Gt/yr (error obtained by quadratic addition—see Siegenthaler and Sarmiento 1993). This means that there is a definite imbalance in the calculation due to the presence of a missing sink.

The most likely missing sink is greater storage of carbon in forest regrowth than that considered in the above calculation or storage in soils. Siegenthaler and Sarmiento based their value of B on the difference between deforestation and forest regrowth on abandoned land given by the IPCC (1990, 1992). However, there have been suggestions of a Northern Hemisphere terrestrial biosphere carbon sink. If there is in fact an increased accumulation of carbon in terrestrial vegetation and soils, there are several possible causes (IPCC 1990, 1992; Siegenthaler and Sarmiento 1993). Two possibilities are (1) the CO_2 fertilization effect—enhanced vegetative growth due to more atmospheric CO_2 (as suggested by the increased seasonal amplitude of the CO_2 curve); and (2) nitrogen fertilization from enhanced N deposition due to air pollution (Kauppi et al. 1992). Quay et al (1992) conclude that there has been little net release by the biosphere between 1970 and 1990 based on oceanic uptake of CO_2 from fossil fuel and deforestation and that deforestation must have been nearly equalled by biosphere growth. (In Figure 2.2 it is assumed that the difference between deforestation and reforestation is only about 0.1 Gt C/yr.)

At any rate, the problem of the exact amount of CO_2 uptake or loss from the biosphere is an unsettled issue, but the net effect is probably considerably smaller than the values for the other major fluxes.

Apparently there have been small natural fluctuations in the concentration of atmospheric CO_2 over the past 30 years. An input eqivalent to about 1.0 ppm has been observed during the changes in oceanic and atmospheric circulation that accompany El Nino–Southern Oscillation events (Sarmiento 1993b). In addition, a maximum removal equivalent to 1.5 ppm occurred during the 2 years following the volcanic eruption of Mt. Pinatubo in the Phillippines in June 1991 (Sarmiento 1993b). The exact mechanism of these changes in atmospheric CO_2 is not known but either involves changes in the CO_2 uptake by the oceans or net biosphere release of CO_2 or both.

In order to make projections of future rises in atmospheric CO_2, and in order to estimate the size of the biosphere and fossil fuel contributions of CO_2 over the past two centuries, it is necessary to know what the concentration of CO_2 in the atmosphere was before human intervention. Measurements of CO_2 concentrations of air bubbles trapped in buried Antarctic ice have shown that, during the millenium prior to the industrial revolution, the concentration of CO_2 was relatively constant at 280 ± 10 ppm (Siegenthaler and Sarmiento 1993). This shows that natural variations in CO_2, at least over the past 1000 years, did not approach the human-created changes of the past century. The air bubble data also show that there was a continual rise from about 270 ppm to 295 ppm from 1800 to 1900 (Siegenthaler and Sarmiento 1993). If these values are correct, this implies that during the nineteenth century about 25 ppm of CO_2 was added and much of it, especially prior to 1850, was probably due to deforestation.

Over much longer periods, natural CO_2 variations have been much larger. For example, during the past 150,000 years, from the study of air bubbles in deeply buried Antarctic ice (Raynaud et al. 1993), the CO_2 concentration was found to range between 180 and 280 ppm (Figure 2.4). Very rapid atmospheric CO_2 increase and climatic change occurred at the end of the last ice age, between around 18,000 years ago when CO_2 was about 200 ppm and 11,000 years ago when CO_2 had risen to 280 ppm. However, the rate of rise of CO_2 was only about 10% of the recent anthropogenic rise in CO_2 (Sundquist 1993). Also, during the distant geologic past, hundreds of millions of years ago, CO_2 could have been more than 10 times higher than at present (Berner 1994). [For a general discussion of CO_2 in the geologic past, see Appenzeller (1993a) and Culotta (1993).]

Other Greenhouse Gases: Methane and Nitrous Oxide

Measurements during the past two decades have shown that the concentrations of certain trace gases, specifically methane, nitrous oxide, and chlorofluorocarbons (CFCs), in the atmosphere have been increasing along with CO_2 (see Figure 2.5), and these increases are also likely due to human activities. (CFCs, which are discussed later in this chapter with regard to the ozone hole, are entirely artificial in origin.) These gases are excellent absorbers of infrared radiation, in fact much stronger absorbers per unit of concentration than CO_2, and thus serve to enhance the atmospheric greenhouse effect. Because of their high efficiency in absorbing long-wave radiation, the combined effect of methane and the other trace gases in raising the earth's surface temperature is about two thirds the effect of CO_2 (Dickinson and Cicerone 1986; Shine et al. 1990).

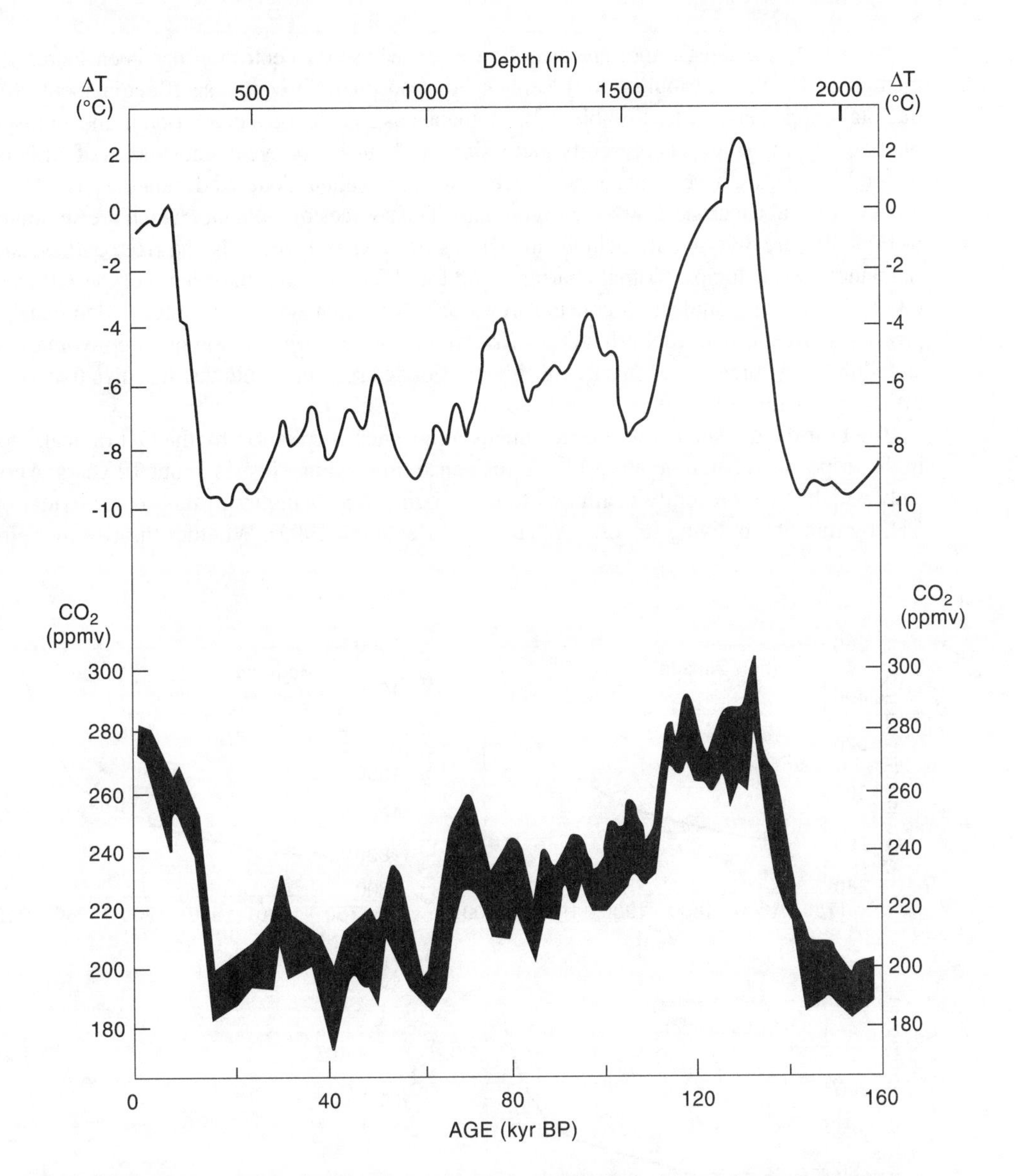

Figure 2.4. Variations in CO_2 concentration (bottom curve) and estimated Antarctic temperature changes (top curve) during the past 160,000 years, as determined from air bubbles in an ice core from Vostok, Antartica, by Barnola et al. (1987). Temperature changes were estimated from measurements of deuterium concentrations in the ice (Lorius et al. 1990). (After R. T. Watson et al. 1990, Greenhouse gases and aerosols. In *Climate change: The IPCC scientific assessment* ed. J. T. Houghton et al., p. 11, copyright World Meteorological Organization.)

This is shown in Table 2.2. [Ramaswamy et al. (1992) suggest that, because ozone is a minor greenhouse gas, its loss in the lower stratosphere between 1979 and 1990, which was caused largely by CFCs, has resulted in net negative radiative forcing or cooling, which counterbalances to a considerable extent the role of CFCs in causing global warming in the same period.]

After CO_2, the most important greenhouse gas whose concentration has been increasing is methane, CH_4. Its contribution to the enhanced atmospheric greenhouse effect is about 20% of the total trapped energy flux (Table 2.2). Methane has a present concentration in the atmosphere of about 2 ppm, its level is presently increasing by about 1% per year, and studies of air bubbles trapped in polar ice have shown that concentrations have about doubled during the past 150 years (Dickinson and Cicerone 1986; see Figure 2.5). The sources of methane, which are summarized in Table 2.3, are diverse and include emanations from natural wetlands and rice paddies, animal flatulance, waste dumps, biomass burning, and fossil fuel production (several of these fluxes are poorly known and subject to constant revision). Note that no single process dominates. All processes, except fossil fuel production, involve the anoxic fermentation of carbohydrates, such as cellulose, by microorganisms living either in flooded soils or within the digestive tracts of animals (e.g., cattle, termites).

The best-documented sink for methane is atmospheric oxidation by the OH radical, mostly in the troposphere, and the atmospheric residence time against loss is about 10 years. Another sink, which is very recently beginning to be recognized as being important, is the oxidation of CH_4 by microbiota living in soils (Whalen and Reeburgh 1992). Whether the rise in methane

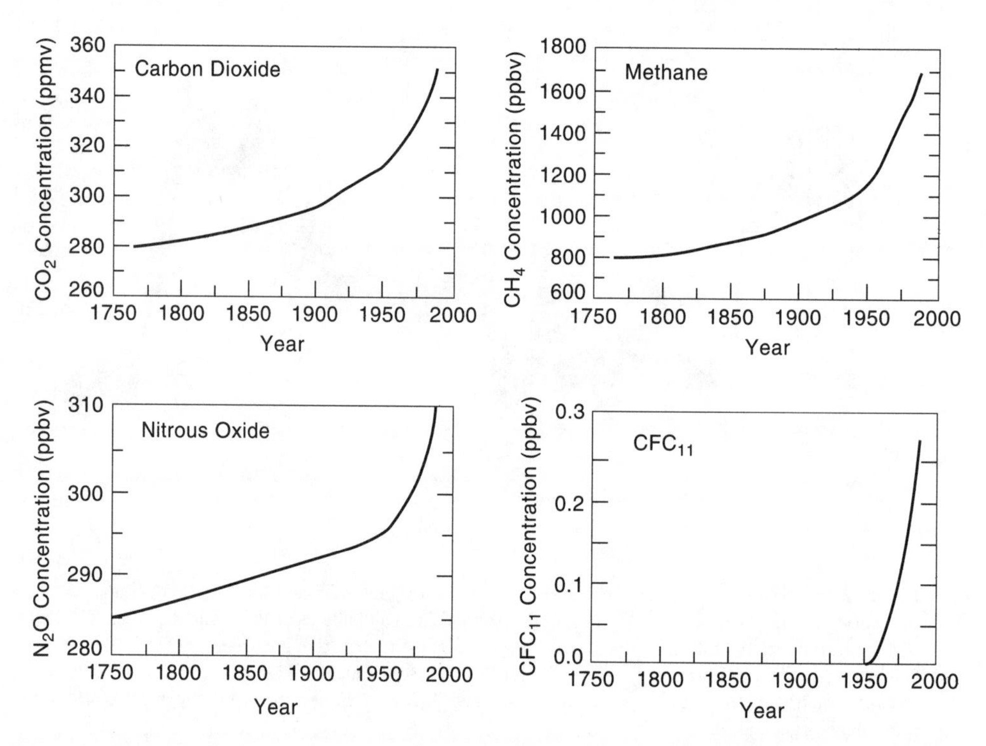

Figure 2.5. Increases in concentrations of greenhouse gases since 1750. Concentrations of CO_2 and methane, which were relatively constant up until the 1700s, have increased steeply since then due to human activities. Nitrous oxide concentrations have risen since about 1750, with the steepest increases after 1950. CFCs, which are entirely anthropogenic in origin, appeared initially in the 1930s and have increased steeply since 1950. (*Source:* IPCC 1990, *Climate change: The IPCC scientific assessment,* ed. J. T. Houghton et al., Executive summary p. xvi, Fig. 3, copyright World Meteorolgical Organization.)

TABLE 2.2 Extra Trapping of Infrared Radiation ΔQ (Positive Radiative Forcing) by Excesses of Trace Atmospheric Gases above Their Preindustrial Concentration

Gas	Preindustrial Concentration 1765 (ppm)	Concentration 1990 (ppm)	Concentration Change per year[a] %	ΔQ (W/m^2)	Percent	Lifetime (yr.)[a]
CO_2	279	354	0.5	1.5	61	50–200
CH_4	0.8	1.72	0.9	0.42	17	10
Strat. H_2O[b]	—	—	—	0.14	6	
N_2O	0.285	0.310	0.25	0.1	4	150
CFC–11	0	0.00028	4.0	0.062	2.5	65
CFC–12	0	0.000484	4.0	0.14	6.	130
Other CFCs	0			0.085	3.5	
Total				2.45	100	
Strat. O_3 loss[c]				–0.08	–3.3	

[a]IPCC 1990.

[b]Indirect effect of changes in methane.

[c]After Ramaswamy et al. (1992); net negative forcing from ozone destruction of CFCs, 1979–1990.

Source: After Shine et al. 1990 (IPCC 1990).

TABLE 2.3 Sources and Sinks for Atmospheric Methane

Source or Sink	$TgCH_4$–C per Year
Sources	
Natural	
Wetlands	86
Termites	15
Oceans	7.5
Lakes	4
Methane hydrates	4
Anthropogenic	
Animal flatulence (enteric fermentation), domestic and wild	60
Rice paddies	45
Biomass burning	30
Waste dumps (landfills)	23
Coal mining, natural gas and petroleum industry	75
Animal wastes	19
Domestic sewage treatment	19
Total	388
Sinks	
Atmospheric removal (troposphere + stratosphere)	353
Removal by soils	23
Atmospheric increase	24
Total	400

Sources: Data from Watson et al. 1990 (IPCC 1990); Watson et al. 1992 (IPCC 1992).

concentration is due to a lowering of the rate of consumption, such as reduced oxidation by OH (Schlesinger 1991), or to increased fluxes of CH_4 to the atmosphere, is not well established.

The effects on atmospheric CH_4 due to processes occurring in high-latitude tundra may be of special importance. Recent estimates by Whalen and Reeburgh (1992) of the contribution from microbial fermentation in tundra are much higher (42 ± 26 Tg C/yr) than that given in Table 2.3. However, the authors emphasize that tundra is also a major locus for the removal of methane via soil oxidation. The possibility then exists that tundra may be a net sink. In addition, there is probably a large mass of methane presently trapped as crystals of methane hydrate at several hundred meters depth in high-latitude permafrost (Kvenvolden 1988; MacDonald 1990). (Methane hydrate is a highly concentrated form of CH_4 but is stable only at high pressures and low temperatures.) If global warming continues, the methane hydrate should eventually be liberated but, because of its depth and the time it takes for a thermal signal to be conducted downward, the release should be delayed for several centuries. Methane has also varied considerably over the geologic past (Raynaud et al. 1993).

Nitrous oxide (N_2O) is not only an important greenhouse gas, it is also involved in the destruction of stratospheric ozone (see later, with regard to the ozone hole). N_2O is produced from various biological sources in the soil and in water (see Table 2.4). A major source is denitrification in soils, dominantly in tropical regions but also in temperate forests. Combustion, biomass burning, and industrial processes are fairly minor anthropogenic sources. The production of nitrous oxide from fertilizer use is difficult to quantify. Because there are great uncertainties in the fluxes, and quite possibly missing sources, it is difficult to account for the rise of N_2O in the atmosphere. Perhaps agricultural development has increased biological emissions from the soil (IPCC 1990, 1992). N_2O is removed from the stratosphere by photochemical breakdown (photolysis).

TABLE 2.4 Sources and Sinks for Nitrous Oxide

Source or Sink	Tg N_2O–N per Year
Sources	
Natural	
Oceans	1.4–2.6
Tropical soils	
In wet forests	2.2–3.7
In dry savannah	0.5–2.0
Temperate soils	
In forests	0.05–2.0
In grasslands	?
Anthropogenic	
Fertilizer use in cultivated soils	0.03–3.0
Biomass burning	0.2–1.0
Combustion	0.3–0.9
Industrial acid production	0.5–0.9
Sinks	
Removal by soils	?
Photolysis in the stratosphere	7–13
Atmospheric increase	3–4.5

Sources: Data from Watson et al. 1990 (IPCC 1990); Watson et al. 1992 (IPCC 1992).

This process oxidizes N_2O to NO, which then reacts with ozone to form NO_2, thereby destroying the ozone.

Another greenhouse gas of lesser importance is ozone. Changes in ozone concentration, in both the lower stratosphere and the upper troposphere (see next section) can change the radiative forcing of the troposphere–surface system (Wang et al. 1993; Ramaswamy et al. 1992). Increases in ozone concentrations in the upper troposphere cause positive radiative forcing (warming) of the troposphere–surface system, particularly at mid-latitudes. On the other hand, decreases in stratospheric ozone can result either in warming or cooling. The combined effect of the lower stratospheric and upper tropospheric ozone changes appears to be net warming at mid-latitudes, with the effect at high latitudes being less certain (Wang et al. 1993).

Climatic and Hydrologic Effects of CO_2 and Trace Gas Increase: Global Warming

Increases in atmospheric CO_2, CH_4, etc., should lead to the enhancement of the atmospheric greenhouse effect (Chapter 1) and thus global warming. The climatic effects of greenhouse gases are often modelled for a doubling of atmospheric CO_2, from 330 ppm to 660 ppm. Estimates of the time when doubling of atmospheric CO_2 is attained range from years 2050 to 2090 (IPCC 1992) and are dependent on projections of future fossil fuel use.

The climatic response to increases in greenhouse gases is calculated by means of highly complex general circulation models (GCMs). For a doubling of atmospheric CO_2, the equilibrium warming of the earth predicted by ocean–atmosphere GCM models in 1990 was an increase of approximately 2.5°C in the global mean surface temperature, with a range of 1.5 to 4.5°C depending on the model (IPCC 1990). An update of these predictions by the IPCC in 1992 (Gates et al. 1992) gives a similar range (1.5–4.5°C) with an average warming around 4°C (mean of all models of 3.8°C). The models differ mainly in their predictions of the response of climate, including temperature, to changes in radiative forcing. The term *radiative forcing* refers to the absorption of long-wave radiation (in watts per square meter) by the various greenhouse gases, and there is better agreement on how radiative forcing responds to changes in greenhouse gas concentrations than how global mean temperature responds to radiative forcing (IPCC 1990). When Cess et al. (1993) compared 15 atmospheric general circulation models, they found a range in CO_2 radiative forcing of 3.3–4.8 W/m^2, as compared to a temperature range of 1.5–4.8°C, for the response to a doubling of CO_2.

There has been an approximate global warming of about 0.5°C over the past century (IPCC 1990; Jones and Wigley 1990), and it is tempting to combine this observation with changes in greenhouse gas concentrations over the same period to test the various climate model predictions. Unfortunately, there are too many complications to allow this to be done with any confidence. Natural time variations in temperature, due to such things as changes in solar radiation, injections of volcanic aerosols, changes in ocean circulation, and the recent input of aerosols to the atmosphere from human activities, preclude an adequate test.

Warming of the earth's surface due to changes in greenhouse gases is due to radiative forcing combined with a number of climatic and chemical factors known as feedbacks. Important feedbacks include the following. (1) Increases of atmospheric water vapor, which absorbs earth radiation even more strongly than CO_2, should accompany global warming and thus amplify CO_2-induced global warming (by an average factor of 1.6—IPCC 1990). (2) Increased surface temperatures cause melting and decreased areal extent of surface ice and snow at high latitudes, which reduces the earth's albedo—the amount of solar radiation reflected from the earth (see Chapter 1). Ice and snow are much stronger reflectors of solar radiation than vegetation or bare ground, so increased absorption of solar radiation accompanying a lower albedo further increases surface temperature. (3) The warming of soils, especially at high latitudes (Schlesinger 1991; Kvenvolden 1988), should cause greater microbiological activity in the soils, with the release of extra CO_2 by decay. All three of these feedbacks are positive; in other words, they enhance the temperature rise to increases in greenhouse gases.

Other feedbacks may be important and need to be taken into consideration. The effects of clouds is complex. The net effect of high-level clouds is to reduce the terrestrial radiation leaving the top of the atmosphere (warming effect), while low-level clouds reflect more entering solar radiation (cooling effect). At present clouds produce a net cooling effect (IPCC 1990). The cloud feedback from global warming—i.e., changes in cloud coverage, altitude, and water content (and therefore reflectivity)—is a source of uncertainty in climate models (Cess et al. 1989) and can cause either a positive or a negative effect.

The oceans tend to slow down a rise in earth surface temperature by 20–30 years because of their ability to absorb heat (NRC 1983). Thus, even if the earth's energy balance has already been changed, changes in the earth's surface temperature may be delayed because there has been insufficient time for the oceans to equilibrate thermally with the atmosphere.

Although the overall rise in global mean earth surface temperature predicted for an atmospheric CO_2 doubling is 1.5 to 4.5°C, the change varies both with latitude and with season (IPCC 1990). The predicted temperature rise is greatest at high latitudes in late autumn and winter, ranging up to more than 12°C in the Arctic and Antarctic. The predicted warming in North America and Europe in winter is 4–6°C, with much higher warming in the northeast of North America. These variations are due to changes in sea ice and snow cover. The warming at mid-latitude continental interiors in summer is higher than average. At lower latitudes the predicted warming is 2–3°C and there will be little seasonal differences (IPCC 1990).

A major effect of a rise in surface temperature will be to speed up the hydrologic cycle—with increases in both the global mean evaporation and precipitation rates. However, the model-predicted hydrologic changes are variable in different zones on the earth's surface. In middle latitudes of the Northern Hemisphere, in a zone from 35°N to 50°N, there will be a tendency to greater summer dryness on the order of 20% less soil moisture for a doubling of CO_2 (IPCC 1990). Areas such as the north central and western United States and central Asia would be particularly affected. The cause of the reduction in soil moisture in summer is primarily stronger evaporation coupled with an earlier occurrence of the summer period of low rainfall and an earlier end of the snowmelt season. In areas of the United States where agriculture is dependent upon rain, the positive effects of increased CO_2 and temperature (i.e., increased photosynthesis, improved plant hydration, and longer growing season) are less important than the negative effects of a lack of water (Waggoner 1983). In arid western areas of the United States that are dependent upon irrigation, decreases in runoff could cause severe water shortages (Revelle and Waggoner 1983).

Greenhouse radiative forcing varies areally and seasonally. Kiehl and Briegleb (1993) state that the largest radiative forcing (but not necessarily temperature increase) occurs in warm dry regions because in these areas, due to a paucity of water vapor, the effect of spectral overlap of water vapor with other greenhouse gases is small, and because the difference between surface and atmospheric temperature is greatest. Tropics with high cirrus clouds show the least radiative forcing, due to the shielding effects of these clouds.

Global warming should cause a gradual rise in sea level through two processes: (1) slow partial melting of ice and snow in the Greenland and (possibly) Antarctic ice caps, along with mountain glaciers, which transfers water to the ocean; and (2) expansion of ocean water due to heating. Wigley and Raper (1992) predict a gradual rise in sea level over the next century of 0.5 m from a combination of these effects.

Revelle (1983) suggested that global warming might cause the West Antarctic Ice Sheet to disintegrate, abruptly adding large masses of floating ice to the oceans, which would cause a much larger additional rise of 5–6 m over the next several hundred years. However, current evidence suggests that future warming would more likely lead to increased Antarctic ice accumulation and a corresponding reduction in global sea-level rise (IPCC 1990). Enhanced Antarctic ice accumulation would result from an increase in moisture transfer to the continent from warming of the surrounding seas.

A reduction in the areal extent and thickness of sea ice in the Arctic and circum-Antarctic oceans is another hydrologic change expected from rising temperatures. (This change would not effect sea level, however.) There may also be changes is surface atmospheric circulation and weakening in the meridional Atlantic ocean circulation.

Thus, to summarize, a rise in the concentration of atmospheric CO_2, CH_4, etc., from anthropogenic processes alters the earth's energy balance through the greenhouse effect and should result in global warming. Since the hydrologic cycle is so intimately connected with the earth's energy cycle, increases in the earth's surface temperature should cause modification in the hydrologic cycle. Although the average rate of precipitation is increased globally, in some areas a drier climate should result. In addition, the amount of ice stored on the earth should decrease, resulting in a rise in ocean level. The main deleterious consequences of global warming to humans will be changes in the hydrologic cycle and a rise in sea level. Increased aridity could affect global food production, and a sea-level rise would inundate some coastal lowlands.

OZONE AND THE OZONE HOLE

Although ozone, O_3, is a very minor gas in the atmosphere, it plays a major role in its effects on life on earth, with both favorable and unfavorable aspects. On the good side, ozone in the stratosphere absorbs harmful ultraviolet (UV) radiation from the sun and serves as a shield against the life-threatening action of UV. (For example, excessive UV radiation interferes with photosynthetic processes and promotes the incidence of skin cancers in humans.) On the other hand, ozone in the troposphere is a pollutant produced as a consequence of the burning of fossil fuels. Because of this dichotomy, the discussion of ozone naturally can be divided into two parts: stratospheric ozone and tropospheric ozone.

Stratospheric Ozone: The Ozone Hole

Most ozone is located in the stratosphere. Here the ozone is produced mainly as a result of the photodecomposition of O_2. Because of the abundance of O_2, a large fraction of incoming solar energy is used to break the bonds of the O_2 molecule to produce atomic oxygen (Cicerone 1987). One oxygen atom then reacts with an O_2 molecule to produce O_3:

$$O_2 + h\nu \rightarrow O + O$$
$$O + O_2 + M \rightarrow O_3 + M$$

where $h\nu$ represents ultraviolet radiation and M is any inert third-body molecule, such as N_2 or O_2, that enables the reaction to take place by carrying away excess momentum from the collision of O with O_2.

Ozone is destroyed by a large number of reactions. The simplest destruction reaction, one that occurs naturally, is the photodissociation of ozone, which is the reaction that actually absorbs harmful UV radiation. The approprate reaction scheme is

$$O_3 + h\nu \rightarrow O + O_2$$
$$O + O_3 \rightarrow 2O_2$$

so that the overall net reaction is

$$2O_3 + h\nu \rightarrow 3O_2$$

The standing level of ozone represents a balance between production and destruction, and this level, and consequently, the ability of O_3 to absorb UV radiation, is lowered if the rate of destruction is increased. Increased destruction is brought about by a large number of processes in addition to that given above. All of the additional reactions are catalytic. In other words, certain molecules interact with ozone, accelerating its destruction, but are regenerated (via a series of reactions) so that there is no net consumption of these catalyst molecules. Principal catalysts are gases that have long lives in the stratosphere. This includes N_2O, CH_4, chlorofluorocarbons or CFCs (Cicerone 1987) and bromine compounds (methyl bromide and halogens) (Cicerone 1994; Mano and Andrae 1994). Both N_2O and CH_4 are produced naturally, but their emission to the atmosphere has increased with human activity. The CFCs are entirely artifical in origin, being manufactured for use as components of refrigerators, aerosol sprays, foam insulation, and so on. Bromine compounds are both anthropogenic and natural.

The CFCs are of special interest in that they have been implicated in the destruction of ozone over Antarctica, leading to the well-known "ozone hole" (Rowland 1989; Solomon 1990; Anderson et al. 1991; Stolarski et al. 1992). Normally there is a loss of ozone in the stratosphere over Antarctica that reaches a maximum during early October, when, after the long dark austral winter, sunlight is able to bring about photodissociation of O_3. Later (by early December), the ozone recovers as Antarctic air mixes northward. The concentration of O_3 at the minimum in October has been found to have decreased dramatically since the 1970s. This is shown in Figure 2.6. In other words, the "hole" has been found to have deepened and widened to the point that it touches the tip of South America. This decrease has been attributed mainly to the action of CFCs in supplying chlorine atoms to the upper atmosphere.

Atomic chlorine, formed by the photodecomposition of CFCs, readily combines with ozone to produce chlorine monoxide, ClO, and then, via a series of reactions, the Cl is regenerated, proving it to be a catalyst. A simplified scheme is

$$2Cl + 2O_3 \rightarrow 2ClO + 2O_2$$
$$ClO + ClO \rightarrow Cl_2O_2$$
$$Cl_2O_2 \rightarrow 2Cl + O_2$$

which results in the same net overall reaction as that given above:

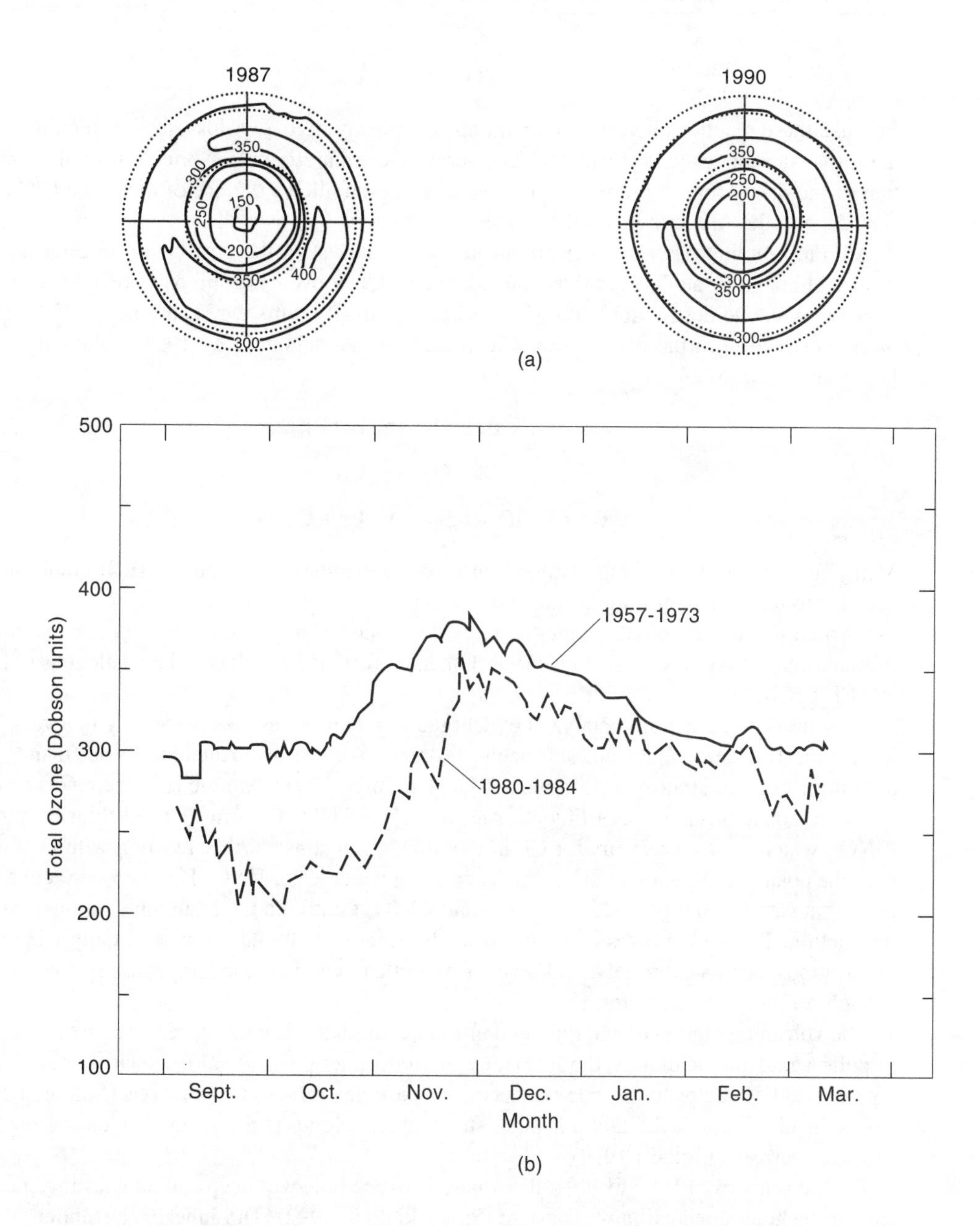

Figure 2.6. Development of the Antarctic ozone hole. (a) Maps of the Antarctic ozone hole in October 1987 and 1990 from polar orthographic projections of TOMS (Total Ozone Measuring Spectrometer) data from the *Nimbus 7* satellite. The South Pole is at the center; the Equator, 30°S, and 60°S latitude circles are shown by dotted circular lines; and Greenwich is to the top. Contours are in Dobson units which are a measure of the total integrated ozone concentration in a vertical column of air measured from the ground up. (Adapted from Stolarski et al. 1992, Fig. 1, p. 343.) (b) The seasonal variation of ozone (in Dobson units) at Halley Bay, Antarctica, for two different time periods. Note the rapid drop during the 1980s in ozone in September, with low values persisting through October into November, and the deepening of the October minimum. (Reprinted with permission from *Nature* 347, p. 348, S. Solomon, Progress toward a quantitative understanding of Antarctic ozone depletion, copyright 1990 Macmillan Magazines Limited.)

$$2O_3 \rightarrow 3O_2$$

Because these reactions involve ClO buildup accompanying the O_3 loss, the concentrations of these two substances tend to anticorrelate across the Antarctic polar vortex during the formation of the ozone hole (Figure 2.7). This anticorrelation allows the use of increased ClO concentration in the Arctic to be used as a measure of ozone loss as well.

Atomic bromine has also been implicated as a catalyst in the destruction of atmospheric ozone. Although Br atoms are far less abundant than Cl atoms in the stratosphere, they are more effective in ozone destruction, and when coupled with Cl atoms they cause about 25% of the ozone destruction in the ozone hole (Cicerone 1994; Anderson et al. 1991; Solomon 1990). Important reactions are

$$Br + O_3 \rightarrow BrO + O_2$$

$$Cl + O_3 \rightarrow ClO + O_2$$

$$BrO + ClO + \text{light} \rightarrow Br + Cl + O_2$$

Methyl bromide (CH_3Br) is the largest source of stratospheric Br atoms. CH_3Br comes about equally from biomass burning (Mano and Andrae 1994), from agricultural pesticide (fumigant) use, which will be phased out in the future by the United States (Cicerone 1994), and naturally from marine plankton (Khalil et al. 1993). Other sources of Br include mixed halogens (CF_3Br and CF_2BrCl).

As it turns out, reactions catalyzed by Cl (and Br) themselves take place only to a very limited extent. The reason that wholesale ozone destruction occurs over Antarctica is due to the presence there of polar stratospheric clouds (Toon and Turco 1991). Atomic chlorine reacts readily with common atmospheric constituents, such as CH_4 and NO_2, to form HCl and chlorine nitrate, $ClNO_3$, which act as reservoirs for Cl, immobilizing it against further reaction with O_3. However, the polar stratospheric clouds, which consist of water ice or $HNO_3 \cdot H_2O$, provide active surfaces that catalyze the breakdown of HCl and $ClNO_3$ to atomic Cl, enabling continued ozone destruction. These clouds explain why ozone holes are not found elsewhere. Only where the upper atmosphere is extremely cold, as over Anarctica in winter, can temperatures become low enough for these clouds to form.

The volcanic eruption of Mt. Pinatubo in 1991 caused an ozone drop because, in the absence of polar stratospheric clouds, the formation of stratospheric liquid sulfuric acid aerosols from volcanic activity helps to provide the necessary particles for surface reactions (Solomon et al. 1993). In addition, it is thought that polar stratospheric clouds (PSCs) can form on stratospheric sulfate aerosols (Tolbert 1994).

For the years from 1987 to 1991, the Antarctic ozone hole was deep and long lasting, and the area of the hole remained fairly constant (Stolarski et al. 1992). The June 1991 eruption of Mt. Pinatubo, which put sulfuric acid aerosols into the lower stratosphere, which were subsequently trapped in the south polar vertex, enhanced the development of the ozone hole in October 1992, when anomalously low ozone readings of 105 Dobson units (see Figure 2.6) were obtained (Lathrop et al. 1993). In October 1993 about 70% of the ozone disappeared in the Antarctic ozone hole, a record amount, continuing the trend from 1992. The minimum ozone reading in 1993 was 85 Dobson units (Newman 1994; Lathrop et al. 1993), the lowest ever measured to that time.

There is also evidence that limited stratospheric ozone depletion (of the order of 5%) and the buildup of ClO are ocurring in the Arctic in the northern hemisphere (Proffitt et al. 1990; Waters

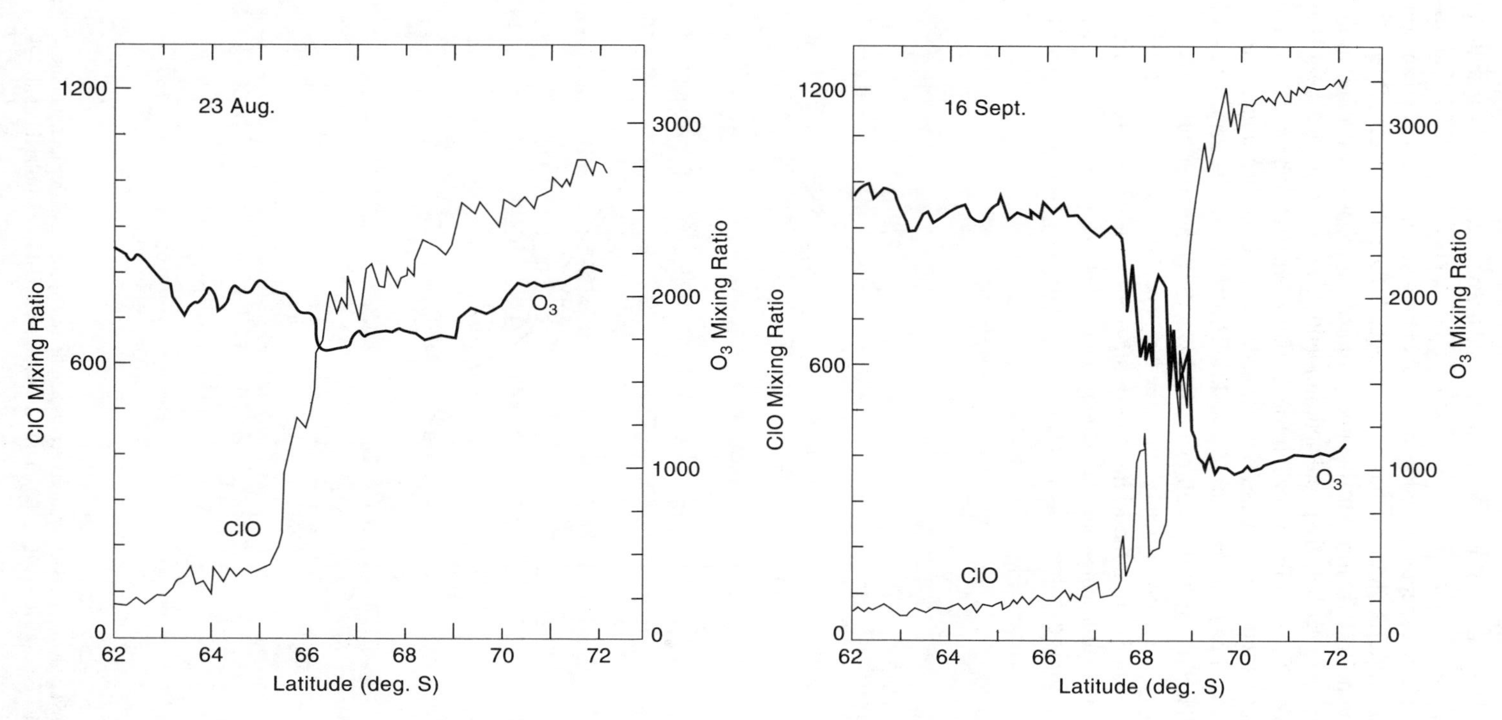

Figure 2.7. Evolution of the anticorrelation between ClO and O_3 across the Antarctic polar vortex from August 23, 1987, to September 16, 1987. This relationship occurs because the destruction of ozone by reaction with Cl produces ClO as a product. (Adapted from Anderson et al. 1991, Fig. 5, p. 43.)

et al. 1993). The Arctic polar vortex is smaller and the occurrence of polar stratospheric clouds is episodic because the stratosphere is not as cold as that over Antarctica. Thus, the maximum ozone depletion would be expected to be less than that in the Antarctic (Schoeberl and Hartmann 1991).

Ozone loss on a smaller scale has occurred outside the polar regions. During the period from 1970 to March 1991 (before Mt. Pinatubo erupted), there was a decline in the total column amount of ozone over heavily populated mid-latitude areas of the Northern Hemisphere in all seasons. Most of the decline in ozone was in the lower stratosphere at 20–25 km (see Figure 2.8). The mid-latitude decline was 2% per decade from ground-based measurements and 3% per decade from satellite measurements (Stolarski et al. 1992), with greater decreases (6–8%) at high latitudes (Gleason et al. 1993). Following the 1991 Mt. Pinatubo eruption, the 1992 global average ozone was 2–3% lower than in any previous year (Gleason et al. 1993). Ozone was about 13% below normal during the winter and spring of 1992–1993 over the United States, and 9% below normal the following summer (Komhyr et al. 1994). However, in the winter of 1993–1994, stratospheric ozone over the United States recovered from the record lows of the previous winter,

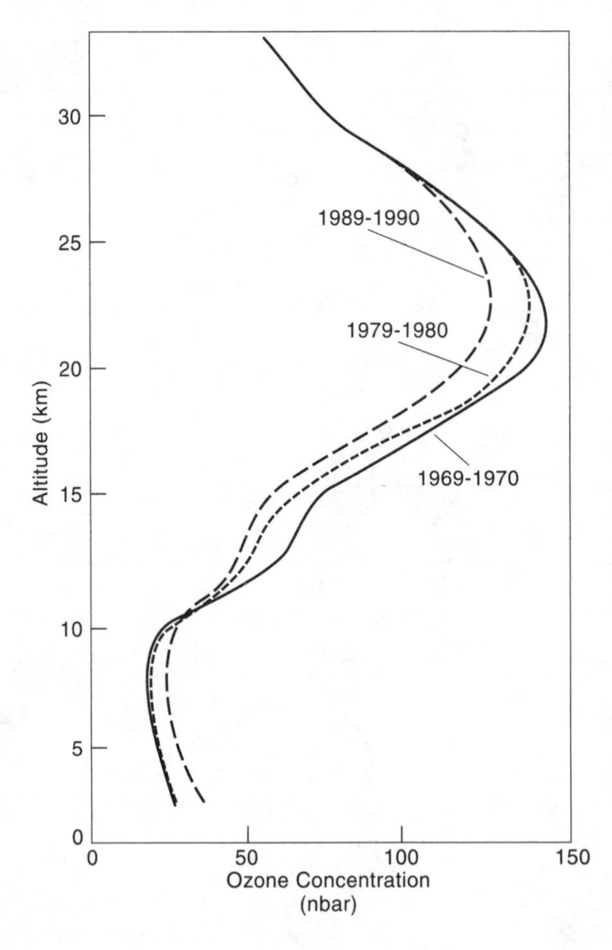

Figure 2.8. Average ozone concentrations versus height for Payerne, Switzerland, latitude 47°N, for three 2-year periods, 1969–1970, 1979–1980, and 1989–1990 (Adapted from Stolarski et al. 1992.)

which had been due primarily to the Mt. Pinatubo eruption (Hoffmann et al. 1994). In the period 1989 to 1993 in Toronto, Canada, the ozone drop was accompanied by a measurable increase in UV-B radiation in the wavelengths affected by ozone (Kerr and McElroy 1993; Appenzeller 1993b).

Loss of atmospheric ozone should lead to increases in the frequency of skin cancers caused by extra UV radiation not trapped by the missing ozone. Since ozone loss has been correlated with ClO buildup from the injection of CFCs into the atmosphere, there are efforts underway to ban the global production of CFCs. Current projections are that most production of CFCs will cease before the year 2000. However, because of continued use of these compounds and their long life in the stratosphere, cessation of production will not be matched by a drop in stratospheric CFC concentration for several decades. Also, there are other anthropogenic sources of stratospheric Cl atoms, such as methyl chloride (CH_3Cl) from biomass burning (Mano and Andrae 1994), which are not subject to simple controls.

Tropospheric Ozone: Air Pollution

Ozone in the troposphere, in contrast to that in the stratosphere, is a pollutant with adverse effects on plants and animals. It is a principal component of photochemical air pollution (smog) arising from the burning of fossil fuels. It forms from a combination of volatile organic carbon compounds (VOCs), nitrogen oxides (NO_x), and carbon monoxide (CO), which are produced during combustion by automobiles and industry and by biomass burning. As a result of photochemical reactions promoted by the hydroxl radical (OH), the VOCs are converted to reactive organic radicals, which then react with atmospheric O_2 to form organic peroxy radicals. It is these peroxy radicals that ultimately bring about O_3 formation.

The nitrogen oxides serve as intermediaries or catalysts in the conversion of organic peroxy radicals to ozone. Simplified reactions (NRC 1991) are

$$RO_2\cdot + NO \rightarrow NO_2 + RO\cdot$$

$$NO_2 + h\nu \rightarrow NO + O$$

$$O + O_2 + M \rightarrow O_3 + M$$

so that the net overall reaction is

$$RO_2\cdot + O_2 \rightarrow O_3 + RO\cdot$$

Here $h\nu$ is ultraviolet radiation, $RO_2\cdot$ represents peroxy radicals, and M is an inert third molecule that enhances the reaction of O with O_2 by carrying away excess momentum.

Atmospheric VOCs arise from natural processes such as emanations from conifers, but they are heavily supplanted in urban environments by hydrocarbon gases from incomplete coal and oil burning, and emissions from the production and use of organic chemicals (e.g., solvents). Nitrogen oxides form naturally, but they are supplanted by high-temperature reactions between N_2 and O_2 gases inside automobile engines and in furnaces (see Chapter 3). Thus, by producing excessive levels of VOCs and NO_x in the urban environment, excess levels of O_3 result.

Biomass burning is another source of human atmospheric ozone pollution, primarily in less developed countries (Crutzen and Andrae 1990; Ciceronė 1994). Gases released by biomass burning, such as carbon monoxide (CO) and nitrogen oxides (NO_x) produce O_3 downwind from regions of burning via photochemical reactions similar to those that occur in polluted urban smog.

Tropospheric ozone (particularly in the upper troposphere) increased in Northern Hemisphere mid-latitudes by 10% per decade from 1970 to 1991 (Stolarski et al. 1992) and has been increasing since 1800. Models predict increases of 0.3–1.0% per year over the next 50 years (Thompson 1992). However, the amount of ozone in the troposphere is still only about 10% of total column ozone (Wang et al. 1993) (see Figure 2.8), and thus the trend of *total ozone* is dominated by stratospheric ozone decreases. The cause of the tropospheric ozone increases probably is due to greater anthropogenic production of ozone precursor trace gases such as NO_x, CO, CH_4, and other hydrocarbons (Thompson 1992).

High concentrations of ozone, along with high concentrations of similar but lesser oxidants resulting from urban air pollution (NO_2, NO_3, H_2O_2, etc.) pose a serious problem to human health because of their adverse effects on the respiratory system. Because of this, maximum levels have been set by the U.S. government for urban air (NRC 1991). When it is not possible to prohibit such high levels, smog alerts are issued. In addition to affecting humans, ozone, at high levels for short periods, has an adverse effect on vegetation. Since air masses move fast enough that some ozone is retained before it can be destroyed by photodissociation, there have been problems with crops, trees, and so on, affected by urban ozone in rural environments near large cities.

AEROSOLS

In addition to major and trace gases, the atmosphere contains *aerosols,* small particles of solid or liquid ranging in size from clusters of a few molecules to about 20 μm in radius. (Particles larger than 20 μm do not remain in the atmosphere very long because they are heavy enough to settle out rapidly.) Some aerosols, such as those consisting of sulfuric acid, reflect sunlight, and therefore serve to cool the earth. Further, because atmospheric water vapor always condenses on a particle or nucleus (see Chapter 3), the chemical composition of rain is dependent to a large extent on the presence and composition of aerosols. For these and other reasons, we shall now consider aerosols in the atmosphere and their origin and composition.

Types of Aerosols

There are two main types of particles or aerosols in the atmosphere: primary particles emitted directly into the atmosphere (such as wind-blown dust, sea salt, plant fragments), and secondary particles formed from gaseous emissions that subsequently condense in the atmosphere. Gas-to-particle conversion results in the formation of fine particles (<1 μm), whereas directly emitted particles are dominantly coarse (>1 μm) (Prospero et al. 1983).

In chemical composition, aerosols may consist of one or more fractions (Rahn 1976): (1) water-soluble ions (such as sulfate, nitrate, ammonium, and several sea-salt-derived ions); (2) a mostly insoluble inorganic part (silicates, oxides, etc.); and (3) a carbonaceous part (soluble and insoluble organic matter). In form, aerosols range from dry dust particles to sea-salt particles, which are sometimes drops of salty water (at high relative humidity). Most continental aerosols are a mixture of soluble and insoluble components (mixed particles), whereas most marine aerosols are soluble, consisting of sea salt and sulfate from oceanically produced reduced sulfur gases (Junge 1963; Fitzgerald 1991).

Condensation of water vapor occurs preferentially on the larger particles (radius 0.1–20 μm), and most of the smaller particles are never used in condensation. The type of particles used for condensation is apparently controlled not only by the sizes available, but also by preferential use of certain chemical compositions, such as hygroscopic or soluble particles (see Chapter 3). Precipitation tends to be quite efficient in removing particles from the atmosphere; that is, most particles in the atmosphere end up in rain, although not necessarily in the same relative concentrations as in the atmosphere, because of their different solubilities and sizes (SMIC 1971; Junge 1972). The residence time (the average time that particles remain in the atmosphere) of fine particles such as those formed from gaseous emissions is much longer (several days) than that of coarse sea salt and soil dust particles (a day or less).

The main sources of particles or aerosols in the atmosphere are given in Table 2.5. Although the particle fluxes are not well known and represent approximate estimates, it is still apparent that the main particle sources are natural: soil dust and sea salt. Overall, roughly, about two thirds of the particles are natural and one third are anthropogenic (considering only long-distance transport of sea salt and soil dust).

Sea-Salt Particles

Sea salt, produced by the oceans, makes a large contribution to atmospheric particles. The bursting of small air bubbles in the foam of breaking waves or "white caps" forms sea-salt particles, as shown in Figure 2.9 (Junge 1963). A bubble breaks when it reaches the ocean surface and the water rushes in to make a rapidly upward-moving jet that projects into the air. This jet breaks to form about 10 droplets, which are ejected about 15 cm above the ocean surface and then are carried upward by air currents. The droplets of seawater that are thrown into the air in this manner evaporate to produce sea-salt particles of a radius ranging from 2 to 20 μm (Blanchard and Woodcock 1957). Numerous smaller particles, which range from 0.1 to 1 μm in radius, are also produced from the bubble film itself.

Over the oceans, sea-salt particles make up a major part of the particles in the coarse size range (0.5–20 μm). By contrast, the smaller particles (< 0.3 μm), known as the background aerosol, are predominantly from marine sulfur gases (DMS) converted to sulfate particles, and sulfate particles from pollutive SO_2 from the continents. The coarse sea-salt particles make up only 5–10% of the total number of particles (CCN) over the oceans but 90–95% of the total mass (Fitzgerald 1991).

The average residence time in the atmosphere for sea salt is 3 days (Junge 1972); thus, it can be transported considerable distances to and over the continents. Since some sea-salt ions are common constituents of continental rains, it is important to ascertain how much sea salt is carried to the continents. Eriksson (1960) calculated that the yearly deposition of sea-salt Cl on the continents is 100 Tg/yr (Tg = 10^{12} g), which represents 10% of his estimate of total marine sea-salt Cl production (1000 Tg Cl/yr). Since sea salt is 55% Cl by weight, about 1800 Tg of total sea salt would be generated over the oceans and 180 Tg carried to and deposited on land. Eriksson's calculation of sea-salt Cl deposition on the continents is problematic, however, in that both higher and lower estimates have been made by others. (See Chapter 5 for a discussion of estimates of atmospherically derived chloride in rivers.) Erickson and Duce (1988) estimate global (land plus sea) wet deposition of sea salt to be 1500–4500 Tg/yr, with maximum deposition at high latitudes.

Several authors (among them Duce and Hoffman 1976; Junge 1972; Maclntyre 1974; Buat-Menard 1983; and Keene et al. 1986) have discussed the question of possible chemical fractionation of sea salt at the sea–air surface before emission to the atmosphere. This can be confused

TABLE 2.5 Fluxes of Particles Smaller than 20μm Radius Emitted into or Formed in the Atmosphere (in Tg/yr; Tg = 10^{12} g)

	Natural	Anthropogenic	Data Source
Soil and rock dust:			
Flux to oceans	910		Duce et al. 1991
Total production	3000–4000	?	Graham and Duce 1979
Sea salt:			
Total wet deposition	1500–4500		Ericson and Duce 1988
Land deposition	180		Ericksson 1960
Biogenic particulates	100–500		Nriagu 1989
Biomass burning (soot)	6–11	36–154	Crutzen and Andrae 1990
Volcanic particles	15–90	—	Jaenicke 1993
Particles from anthropogenic direct emissions: fuel, incinerators, and industry	—	15–90	Prospero et al. 1983; World Resources for 1988–1989
Particles formed from gaseous emissions:			
Sulfate from biogenic DMS, etc.	51	—	Table 3.9
Sulfate from volcanic SO_2	18–27	—	Table 3.9
Sulfate from fossil fuel[a]	—	105	Table 3.9; Charlson and Wigley 1994
Nitrate from NO_x	62	128	Table 3.1
Ammonium from NH_3	28	37	Table 3.16
Biogenic hydrocarbons	20—250	—	Nriagu 1989
Anthropogenic hydrocarbons[b]	—	100	Watson et al. 1990
Total	1390–2109	421–614	

[a] Half of the SO_2 from fossil fuel is removed in rain and dry deposition and the rest becomes sulfate aerosol (Charlson and Wigley 1994).

[b] From fossil fuel and biomass burning.

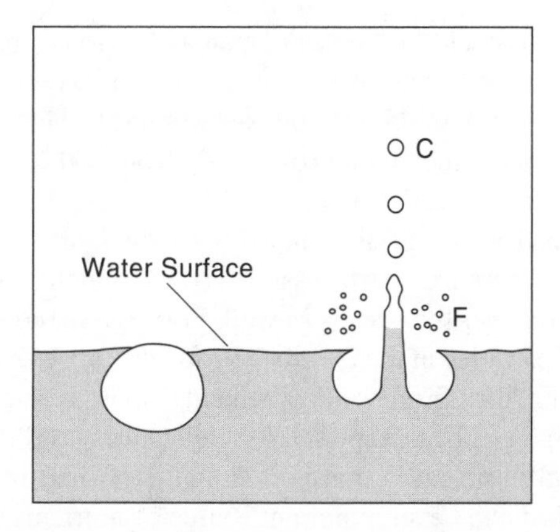

Figure 2.9. The formation of sea-salt particles from the bursting of bubbles. Large droplets originating from the jet are designated as C, whereas smaller particles produced from the bubble are designated as F. (After Junge 1963.)

with the presence of particles from nonmarine sources or with changes in the chemical composition of sea-salt particles by interaction with gases in the atmosphere (such as HCl or SO_2 gas). The reason for expecting that chemical fractionation might occur is that certain substances are more concentrated in the top 10–20 cm of the ocean than they are at greater depth. However, recent work has shown that these substances are not major ions, but are primarily surface-active organic compounds and other chemical substances associated with them (particulate and dissolved organic nitrogen and phosphate and trace metals such as iron, lead, and manganese). Bubbles scavenge surface-active material from lower layers and transport it to the sea surface. When these bubbles break, ejecting sea salt into the air, they essentially skim off a very thin and enriched microlayer at the ocean surface. In this way the substances enriched at the ocean surface also become enriched, along with organic matter, in sea-salt particles (MacIntyre 1974).

The enrichment in sea-salt aerosols, relative to seawater, of the alkali and alkaline earth metals, whose ions are important in rainfall (Na, K, Mg, and Ca), appears to be slight or nonexistent (MacIntyre 1974; Duce and Hoffman 1976; Buat-Menard 1983; Keene et al. 1986). Chloride and sulfate also probably do not undergo appreciable fractionation during the formation of sea salt. Changes observed in the ratio of Cl/Na relative to seawater are probably due to loss of HCl upon the acidification of sea-salt aerosols by reaction with NO_2 (see further discussion in Chapter 3 in the section about rain).

Soil and Mineral Dust: Continental Aerosols

Another major source of atmospheric particles is soil and rock debris. Natural weathering breaks down rocks to produce soil and mineral dust, which is then transported by the wind. Typically, the wind-blown dust consists of yellow-brown aggregates of quartz (SiO_2), mica, and clay minerals (hydrous cation alumino-silicates; for a further description of clay minerals see Chapter 4), the latter formed in part by weathering. The yellow-brown color of wind-blown dust is from iron oxides, which are strongly enriched in the dust.

Since Al, Fe, and Si come almost exclusively from soil dust, their presence in aerosols is a strong indication of a soil dust origin. Al is most commonly used as an aerosol tracer for soil dust, and the ratio of another soil element, such as Ca, to Al in an aerosol is compared to the known ratio in soils, rocks, or the earth's crust to determine enrichment of the element. In general, the element ratios in wind-blown soils are similar to bulk crustal rock (Rahn 1976). The major soluble ions in rain which come commonly from soil dust include Ca, K, and Na, while SO_4 and Cl are sometimes locally important. The sources of soil dust in the United States are wind erosion, unpaved roads, and tilled fields (Gillette et al. 1992).

Although wind-blown mineral particles are derived from the continents, they are found in considerable concentration over the oceans as well. For example, over the North Atlantic there is a large contribution of particles in the 0.1- to 20-μm size range from Saharan dust storms (Junge 1972). Over the Pacific, dust from Asian continental sources, such as the Gobi Desert, and agricultural areas in China, can be transported long distances (10,000 km). Duce et al. (1991) estimated the flux of soil dust that is transported and deposited in the oceans as 910 Tg/yr. The soluble part of this dust can be an important source of nutrients such as Fe and P to the remote oceans. The total dust production over the continents is considerably larger, 3000–4000 Tg/yr (Graham and Duce 1979). Also, some part of this is anthropogenic, resulting from roads, construction, farming, and other human activities.

Gaseous Emissions

Fine secondary aerosols formed from gases emitted to the atmosphere include sulfuric acid formed from volcanic sulfur gases, biogenic DMS, and anthropogenic SO_2; $(NH_4)_2SO_4$ and NH_4HSO_4 formed by the reaction of NH_3 with sulfuric acid aerosols; nitric acid formed from emissions of nitrogenous gases; and organic matter formed from the oxidation of biogenic gases. Nitrate is also found on coarse aerosols over the ocean, suggesting oxidation of gaseous NO_2 on sea-salt droplets (Fitzgerald 1991). These secondary aerosols are important both to the thermal budget of the earth and to the atmospheric cycles of sulfur and nitrogen. Here the role of aerosols in reflecting solar radiation is emphasized. Their role in the formation of acid rain and in the atmospheric cycles of sulfur and nitrogen is discussed in Chapter 3.

Sulfate Aerosols and Climate

Sulfate aerosols in the atmosphere have several potential effects on climate (Charlson and Wigley 1994; Kiehl and Briegleb 1993; Hobbs 1993; Charlson et al. 1992) First, they have a *direct effect* by scattering or reflecting incoming solar radiation in the atmosphere, and cooling the earth. This effect occurs in clear sky. Second, sulfate aerosols may have on *indirect effect* on clouds. Because sulfate aerosols are water soluble (hygroscopic), converting to sulfuric acid in the atmosphere at low relative humidities, and because they are in the size range of 0.1–1 μm, they make good cloud condensation nuclei (CCN). An increase in the number of CCN increases the number of cloud droplets, raising the albedo (solar reflectance or brightness) of clouds, and cooling the earth. Also, for a given amount of condensible water vapor, more cloud droplets mean smaller droplets, leading to longer-lived clouds relative to removal of the droplets as rain. This should amplify cooling.

Large areas of the earth are covered by thin stratiform clouds, which exhibit variable albedo due to differences in the number of cloud droplets. This is also true of clouds in areas where the number of CCN are low, such as over the remote oceans. In either case, cloud albedo is sensitive to the abundance of CCN and can be affected by the input of sulfate aerosols. The magnitude of the indirect cooling effect of sulfate aerosols by increasing cloud albedo is not well known (Charlson and Wigley 1994) because there is no accepted relationship between the sulfate mass concentration and the number of CCN and cloud droplets. Langner et al. (1992) and Ghan et al. (1989) believe that the indirect cloud effect has been overestimated because most of the SO_2 that is oxidized to sulfate in cloud droplets becomes associated with preexisting particles and does not form new particles.

It has been suggested that the cooling from anthropogenic sulfuric aerosols, which is similar in magnitude to the warming expected from greenhouse gases, has delayed greenhouse warming (Charlson et al. 1992; Charlson and Wigley 1994). The estimated size of these effects is listed in Table 2.6. Charlson et al. (1991) assume that half of the SO_2 from fossil fuel burning is removed by precipitation and dry deposition and the rest becomes sulfate aerosol available for global cooling. Langner et al. (1992), by contrast, estimate that only 6% of anthropogenic SO_2 forms new sulfate particles. They agree that half of anthropogenic SO_2 is removed in precipitation, but they believe that only the small part of the sulfate formed by OH oxidation becomes new particles.

There is evidence for local net cooling where the cooling effect from sulfate aerosols is greater than the heating effect from greenhouse gases. This net cooling occurs in the eastern United States, south central Europe, and eastern China (Kiehl and Briegleb 1993). Carbon-rich aerosols are also produced by combustion in urban areas (Ghan et al. 1989), but their warming effect by absorption of solar radiation is thought to be less than the cooling effect of sulfate aerosol.

If the greenhouse effect due to CO_2, CH_4, and so on, were the sole mechanism for global warming, then the Northern Hemisphere should warm faster than the Southern Hemisphere because the Northern Hemisphere contains a much greater industrialized population and the Southern Hemisphere has most of the oceans with more thermal inertia. The reverse seems to be true. The delay in warming in the Northern Hemisphere may be due, at least to some extent, to anthropogenic sulfate aerosols, which are concentrated in the industrial areas. Another argument for

TABLE 2.6 Changes in Solar Radiation Balance Due to Various Means of Forcing

	W/m^2
Loss due to direct anthropogenic sulfate aerosol forcing	−0.3[a]; (−0.3 to −0.5)[b]
Loss due to direct natural sulfate aerosol forcing	−0.26[a]
Loss due to increase in cloud albedo from sulfate aerosol	? (Negligible)[c]
Loss due to biomass burning aerosols	? (See text)
Increase from anthropogenic CO_2 from 280 ppm to 354 ppm (1990)	1.5[d]
Increase from other greenhouse gases (methane, nitrous oxide, etc.)	0.9[d]

[a] After Kiehl and Briegleb (1993).

[b] After Charlson and Wigley (1994); loss of 0.2–0.3% of solar radiation reaching the ground (assumed to be 169 W/m^2); after Ramanathan (1987).

[c] From Langner et al. (1992).

[d] From Table 2.2.

sulfate aerosol cooling is that the model predictions for temperature change resulting from increased emissions of greeenhouse gases over the last 100 years are considerably larger than the observed temperature change of only 0.5°C (Charlson and Wigley 1994). However, there are a number of problems in the models, which could have caused this discrepancy. One factor that is important in considering the long-term effects of sulfate aerosols versus greenhouse gases is that aerosols have an atmospheric lifetime of only about a week as compared to the much longer lifetime of greeenhouse gases (hundreds of years in some cases).

There has been a suggestion that naturally formed sulfate aerosols over the ocean might also regulate climate (Charlson et al. 1987). Marine phytoplankton release dimethylsulfide (DMS), which oxidizes in the air to from SO_2, part of which is converted to sulfate aerosol. As we have already mentioned, sulfate aerosol can increase the number of CCN and therefore presumably the number of cloud droplets, increasing the cloud albedo and cooling the earth. Remote marine stratus clouds, which cover 25% of the oceans, are a particularly sensitive source of this increase in albedo. The size of the effect of direct radiative forcing from natural sulfate aerosol from DMS is estimated by Kiehl and Briegleb (1993) to be similar in magnitude to forcing from anthropogenic sulfate aerosol (see Table 2.6). However, Langner et al. (1992) estimate that only 14% of the natural DMS emissions become new particles, not 50% as assumed by Charlson et al. (1991).

If one assumes [as Charlson et al. (1987) did] that DMS emissions increase with increasing global temperature, then the sequence of processes could result in a global thermostat with higher temperatures resulting in more DMS and greater clouds and cooling. However, the global thermostat feedback idea is not supported by the observations of Antarctic ice cores, which show an *increase* of 20–46% in non-sea-salt sulfate (presumably derived from DMS) during glacial conditions and a drop during interglacial periods (Legrand et al. 1988).

Other Aerosols: From Vegetation, Biomass Burning, Volcanism, and Fuel Combustion

Vegetation can emit volatile organic matter (mainly hydrocarbons) directly to the atmosphere, which are converted to fine organic aerosols by gas-to-particle conversion (Went 1960; Rasmussen and Khalil 1988; Zimmerman et al. 1988). Sources include direct emanations, particularly from tropical forests and conifers in general, and the decomposition of dead organic matter. Plants in the Amazon rain forest and in West Africa (even in the absence of burning) also directly emit coarse and fine aerosol particles, which consist predominantly of organic carbon and are enriched in potassium (Lawson and Winchester 1979; Crozat 1979; Araxto et al. 1988; Talbot et al. 1988). In addition, biogenically produced NH_3 gas is emitted in sufficient amounts in tropical forests to neutralize H_2SO_4 from biogenic DMS and HNO_3 from NO_x (Talbot et al. 1988).

Biomass burning due to forest fires set by humans (the dominant cause), or occurring naturally, is another source of particles. Large-scale forest burning, which is used seasonally to clear the land in the Amazon Basin and West Africa, is a source of fine particulate organic carbon and soot carbon that is enriched in NH_4, NO_3, K, and SO_4 (Crozat et al. 1978; Lawson and Winchester 1978; Andrae et al. 1988). In fact, the ratio of K to black soot carbon is used as a tracer for biomass-burning aerosol (Andrae et al. 1988). Crutzen and Andrae (1990) estimate that 36–154 Tg of smoke (soot) particles are produced annually by biomass burning, predominantly in the tropics.

Biomass burning has the potential to change the radiation budget, particularly in the tropics. Smoke particles are similar in size to sulfate particles and also partly hygroscopic, so they reflect solar radiation and act as cloud condensation nuclei, increasing the reflectivity of clouds (Crutzen and Andrae 1990; Penner et al. 1992). The estimate given by Penner et al. of the cooling effect of smoke (–2 watts/m^2) is believed by others to be too large (Kiehl and Briegleb 1993). However, carbon aerosols in smoke also absorb solar radiation, warming the earth (by about 0.2 watts/m^2), which could at least partly counterbalance their cooling effect.

Volcanic debris accounts for a small part of total global particle formation, but the production of volcanic material tends to be episodic and therefore dramatic right after major eruptions. For example, the eruption of Krakatoa in the East Indies in 1883 ejected an estimated 25,000 Tg of material into the atmosphere, or around 300 times the estimated normal yearly production of volcanic material (Goldberg 1971).

The airborne particles produced by volcanoes consist of finely divided ash consisting of silicate minerals and sulfuric acid aerosols, the latter originating from the oxidation of SO_2 from the volcanic plume. The sulfuric acid aerosols can reach the stratosphere during unusually violent eruptions, and because of a much longer residence time for particles in the stratosphere due to a lack of rainout, the sulfuric aerosols can bring about appreciable global cooling for a few years until they fall out. This was the case for Krakatoa mentioned above, El Chichon (1982), and more recently for the eruption of Mt. Pinatubo (1991). Hansen et al. (1992) calculate a mean global surface temperature change of –0.5°C as a result of the input of aerosols from Mt. Pinatubo. An additional effect of sulfuric acid particles in the stratosphere is to participate in reactions that destroy ozone (Solomon et al. 1993; Tolbert 1994; see section on ozone above). Volcanic aerosols may also absorb solar radiation, warming parts of the stratosphere and changing stratospheric wind patterns (Kerr 1993).

Direct injection of primary aerosols also results from fuel combustion (e.g., coal fly ash and fuel oil soot). The aerosols often contain trace elements, such as lead and vanadium, which can be used as tracers of their anthropogenic origin. Certain combinations of trace elements in fine primary aerosols, whose proportions vary from area to area, are used as signatures of different regional pollution sources and can be correlated with secondary sulfate aerosols over long transport distances (>200 km) (Rahn and Lowenthal 1984).

Dry Deposition of Aerosols

In addition to aerosol removal from the atmosphere in precipitation by condensation processes (*rainout*) and by impaction with falling raindrops (*washout*), there is also *dry deposition* from the atmosphere. The processes of dry aerosol removal include (1) *sedimentation,* which involves gravity settling of larger (>20 μm), and thus heavier, particles; and (2) *dry impaction* of aerosol particles on trees and foliage. If dry deposition is ignored, estimates of the amount of a substance delivered on land from precipitation may seriously underestimate the total amount being delivered.

The amount of aerosols removed by sedimentation or gravitational settling is sometimes measured separately and referred to as *dry fallout.* Certain types of precipitation collectors that are continuously open to the atmosphere will include contributions by dry fallout in their precipitation collections, and this combination of dry fallout and rainfall is referred to as *bulk precipitation* (Whitehead and Feth 1964). Because only large particles are heavy enough to settle out of the air rapidly, dry fallout is greatly subject to local influences. In areas away from the ocean or industry, the elements most represented in dry fallout are soil elements: K, Na, Ca, and Mg from wind-blown soil dust.

Dry impaction of aerosols on plants and trees is another way in which aerosols are removed from the atmosphere besides by precipitation. Foliage, particularly evergreens, scavenges aerosols from the air as the air passes by. Material also settles out on trees, and this material, combined with biological exudates, is washed off by subsequent precipitation. The combination of rainfall plus soluble exudates and captured aerosols that wash off in passing through trees is known as *throughfall* (see Chapter 4).

In the marine atmosphere, dry deposition is largely from the recycling of coarse sea-salt aerosols. For large particles over the oceans (mainly sea salt and soil dust), wet and dry deposition are about equal. However, wet deposition dominates for small particles, which represent most of the atmospheric net particle flux from the continents to the oceans (Buat-Menard 1983).

REFERENCES

Anderson, J. G., D. W. Toohey, and W. H. Brune. 1991. Free radicals within the Antarctic vortex: The role of CFCs in the Antarctic ozone loss, *Science* 251: 39–46.

Andrae, M. O., E. V. Brownell, M. Garstang, G. L. Gregory, R. C. Harriss, G. F. Hill, D. J. Jacob, M. C. Pereira, G. W. Sachse, A. W. Setzer, P. L. Silva Dias, R. W. Talbot, A. L. Torres, and S. C. Wofsy. 1988. Biomass-burning emissions and associated haze layers over Amazonia, *J. Geophys. Res.* 93: 1509–1527.

Appenzeller, T. 1993a. Searching for clues to ancient carbon dioxide, *Science* 259: 906–909.

Appenzeller, T. 1993b. Finding a hole in the ozone argument, *Science* 262: 990–991.

Araxto, P., H. Storms, F. Bruynseels, and R. Van Grieken. 1988. Composition and sources of aerosols from the Amazon Basin, *J. Geophys. Res.* 93: 1605–1615.

Barnola, J. M., D. Raynaud, Y. S. Korotkevitch, and C. Lorius. 1987. Vostok ice core: A 160,000 year record of atmospheric CO_2, *Nature* 329: 408–414.

Bazzaz, F. A. 1990. Response of natural ecosystems to the rising CO_2 levels. *Ann. Rev. Ecol. Syst.* 21: 167–196.

Berner, R. A. 1982. Burial of organic carbon and pyrite sulfur in the modern ocean and its geochemical and environmental significance, *Am. J. Sci.* 282: 451–473.

Berner, R. A., 1994. GEOCARB II: A revised model of atmospheric CO_2 over Phanerozoic time, *Am. J. Sci* 294: 56–91.

Blanchard, D. C., and A. H. Woodcock. 1957. Bubble formation and modification in the sea and its meteorological significance, *Tellus* 9: 145–158.

Boden, T. A., R. J. Sepanski, and F. W. Stoss (eds.). 1992. Trends '91: A compendium of data on global change. U.S. Dept. of Energy Carbon Dioxide Information Analysis Center Publ. ORNL/CDIAC-49.

Bolin, B., E. T. Degens, P. Duvigneaud, and S. Kempe. 1979. The global biogeochemical carbon cycle. In *The Global Carbon Cycle,* SCOPE Report No. 13, ed. B. Bolin, pp. 1–56. New York: John Wiley.

Broecker, W. S., T. Takahashi, H. J. Simpson, and T. H. Peng. 1979. Fate of fossil fuel carbon dioxide and the global carbon budget, *Science* 206: 409–418.

Buat-Menard, P. 1983. Particle geochemistry in the atmosphere and ocean. In *Air-Sea Exchange of Gases and Particles,* ed. P. N. Liss and W. G. N. Slinn, pp. 455–532. Boston: D. Reidel.

Cess, R. D., et al. (19 authors). 1989. Interpretation of cloud-climate feedback as produced by 14 atmospheric general circulation models, *Science* 243: 513–516.

Cess, R. D., et al. (29 authors). 1993. Uncertainties in carbon dioxide radiative forcing in atmospheric general circulation models, *Science* 262: 1252–1255.

Charlson, R. J., J. E. Lovelock, M. O. Andrae, and S. G. Warren. 1987. Oceanic phytoplankton, atmospheric sulphur, cloud albedo and climate, *Nature* 326: 655–661.

Charlson, R. J., J. Langner, H. Rodhe, C. B. Leovy, and S. G. Warren. 1991. Perturbation of the northern hemisphere radiative balance by backscattering from anthropogenic sulfate aerosols, *Tellus* 43AB: 152–163.

Charlson, R. J., S. E. Schwartz, J. M. Hales, R. D. Cess, J. A. Coakley, Jr., J. E. Hansen, and D. J. Hoffmann. 1992. Climate forcing by anthropogenic aerosols, *Science* 255: 423–430.

Charlson, R. J., and T. M. L. Wigley. 1994. Sulfate aerosol and climatic change, *Sci. Am.* 270(2): 48–57.

Cicerone, R. J. 1987. Changes in stratospheric ozone, *Science* 237: 35–42.

Cicerone, R. J. 1994. Fires, atmospheric chemistry and the ozone layer, *Science* 263: 1243–1244.

Crozat, G. 1979. Sur l'emission d'un aerosol riche en potassium par la foret tropical, *Tellus* 31: 52–57.

Crozat, G., J. L. Domerque, J. Baudet, and V. Bongi. 1978. Influence des feux de brousse sur la composition chimique des aerosols atmospheriques en Afrique de l'ouest, *Atmos. Environ.* 12(9): 1917.

Crutzen, P. J., and M. O. Andrae. 1990. Biomass burning in the tropics: Impact on atmospheric chemistry and biogeochemical cycles, *Science* 250: 1669–1678.

Culotta, E. 1993. Is the geological past a key to the (near) future? *Science* 259: 906–908.

Dickinson, R. E. and Cicerone, R. J. 1986. Future global warming from atmospheric trace gases, *Nature* 319: 109–115.

Duce, R. A., and E. J. Hoffman. 1976. Chemical fractionation at the air/sea interfaces. In *Annual Review of Earth and Planetary Sciences,* pp. 187–228. Palo Alto, Calif.: Annual Revues, Inc.

Duce, R., P. S. Liss, J. T. Merrill, E. L. Atlans, P. Buat-Menard, B. B. Hicks, J. M. Miller, J. M. Prospero, R. Atimoto, T. M. Church, W. Ellis, J. N. Galloway, L. Hansen, T. D. Jickells, A. H. Knap, K. H. Reinhardt, B. Schneider, A. Soudine, J. J. Tokos, S. Tsunogai, R. Wollast, and M. Zhou. 1991. The atmospheric input of trace species to the world ocean, *Global Biochem. Cycles* 5: 193–259.

Erickson, D. J. III, and R. A. Duce. 1988. On the global flux of atmospheric sea salt, *J. Geophys. Res.* 93: 14,079–14,088.

Eriksson, E. 1960. Yearly circulation of chloride and sulfur in nature, meteorological, geochemical and pedological implications, Part 2, *Tellus* 12: 63–109.

Fitzgerald, J. W. 1991. Marine aerosols: A review, *Atmos. Environ.* 25A: 533–545.

Gates, W. L., J. F. B. Mitchell, G. L. Boer, U. Cubasch, and V. P. Meleshko. 1992. Climate modelling, climate prediction and model validation. In *Climate Change 1992. The Supplementary Report to the IPCC Scientific Assessment,* ed. J. T. Houghton, B. A. Callendar, and S. K. Varney, pp. 97–134. Cambridge, U.K.: Cambridge University Press.

Ghan, S. J., J. E. Penner, and K. E. Taylor. 1889. Sulphate aerosols and climate, *Nature* 340: 438.

Gilette, D. A., G. J. Stensland, A. L. Williams, W. Banard, D. Gatz, P. C. Sinclair, and T. C. Johnson. 1992. Emissions of alkaline elements calcium, magnesium, potassium, and sodium from open sources in the contiguous United States, *Global Biogeochem. Cycles* 6: 437–457.

Gleason, J. F., P. K. Bhartia, J. R. Herman, R. McPeters, P. Newman, R. S. Stolarski, L. Flynn, G. Labow, D. Larko, C. Seftor, C. Wellemeyer, W. D. Komhyr, A. J. Miller, and W. Planet. 1993. Record low global ozone in 1992, *Science* 260: 523–526.

Goldberg, E. 1971. Atmospheric dust, the sedimentary cycle and man. In *Comments on Earth Science: Geophysics,* vol. 1: 117–132.

Graham, W. F., and R. F. Duce. 1979. Atmospheric pathways of the phosphorus cycle, *Geochim. Cosmochim. Acta* 43: 1195–1208.

Hansen, J., A. Lacis, R. Ruedy, and M. Sato. 1992. Potential climate impact of Mount Pinatubo eruption, *Geophys. Res. Lett.* 19: 215–218.

Hobbs, P. V. 1993. Aerosol-cloud interactions. In *Aerosol-Cloud-Climate Interactions,* ed. P. V. Hobbs, pp. 33–73. San Diego, Calif.: Academic Press.

Hoffman, D. J., S. J. Oltmans, J. M. Harris, J. A. Lathrop, G. L. Koenig, W. D. Komhyr, R. D. Evans, D. M. Quiney, T. Deschler, and B. J. Johnson. 1994. Recovery of stratospheric ozone over the United States in the winter of 1993–1994, *Geophys. Res. Lett.* 21: 1779–1782.

IPCC (Intergovernmental Panel on Climate Change). 1990. *Climate Change. The IPCC Assessment,* ed. J. T. Houghton, G. J. Jenkins, and J. J. Ephraums. Cambridge, U.K.: Cambridge University Press.

IPCC (Intergovernmental Panel on Climate Change). 1992. *Climate Change 1992. The Supplemetary Report to the IPCC Scientific Assessment,* J. T. Houghton, B. A. Callendar, and S. K. Varney, eds. Cambridge, U.K.: Cambridge University Press.

Jaenicke, R. 1993. Tropospheric aerosols. In *Aerosol-Cloud-Climate Interactions,* ed. P. V. Hobbs, pp. 1–31. San Diego, Calif.: Academic Press.

Jones, P. D., and T. M. L. Wigley. 1990. Global warming trends, *Sci. Am.* 263: 84–91.

Junge, C. 1963. *Air Chemistry and Radioactivity.* New York: Academic Press.

Junge, C. 1972. Our knowledge of the physico-chemistry of aerosols in the undisturbed marine environment, *J. Geophys. Res.* 77: 5183–5200.

Kauppi, P. E., K. Mielikanen, and K. Kuusela. 1992. Biomass and carbon budget of European forests, 1971 to 1990, *Science* 256: 70–74.

Keene, W. C., A. A. A. Pszenny, J. N. Galloway, and M. E. Hawley. 1986. Sea-salt corrections and interpretation of constituent ratios in marine precipitation, *J. Geophys. Res.* 91: 6647–6658.

Kerr, J. B., and C. T. McElroy. 1993. Evidence for large upward trends of ultraviolet-B radiation linked to ozone depletion, *Science* 262: 1032–1034.

Kerr, R. A. 1993. Ozone takes a nose dive after the eruption of Mt. Pinatubo, *Science* 260: 490–491.

Khalil, M. A. K., R. A. Rasmussen, and R. Gunawardena. 1993. Atmospheric methyl bromide: Trends and global mass balance, *J. Geophys. Res.* 98: 2887–2896.

Kiehl, J. T., and B. P. Briegleb. 1993. The relative role of sulfate aerosols and greenhouse gases in climate forcing, *Science* 260: 311–314.

Komhyr, W. D., R. D. Grass, R. D. Evans, R. K. Leonard, D. M. Quincy, D. J. Hofmann, and G. L. Koenig. 1994. Unprecedented 1993 ozone decrease over the United States from Dobson spectrophotometer observations, *Geophys. Res. Lett.* 21: 201–204.

Kvenvolden, K. 1988. Methane hydrates and global climate. *Global Biogeochem. Cycles* 2: 221–229.

Langner, J., and H. Rodhe, P. J. Crutzen, and P. Zimmermann. 1992. Anthropogenic influence on the distribution of tropospheric sulphate aerosol, *Nature* 359: 712—716.

Lathrop, J. A., S. J. Oltman, and D. J. Hofmann. 1993. Record low ozone at the South Pole in 1992 and preliminary results on the 1993 ozone hole (abstract), *EOS* 74: 166.

Lawson, D. R., and J. W. Winchester. 1978. Sulfur and trace element relationships in aerosols from the South American continent, *Geophys. Res. Lett* 5: 195–198.

Lawson, D. R., and J. W. Winchester. 1979. Sulfur, potassium, and phosphorus associations in aerosols from South American tropical rain forests, *J. Geophys. Res.* 84 (C7): 3723–3727.

Legrand, M. R., R. J. Delmas, and R. J. Charlson. 1988. Climate forcing implications from Vostok ice-core sulphate data, *Nature* 334: 418—420.

Lorius, C., J. Jouzel, D. Raynaud, J. Hansen, and H. L. Trent. 1990. The ice core record: climate sensitivity and future greenhouse warming, *Nature* 347:139-145.

MacDonald, G. J. 1990. Role of methane in past and future climates, *Climatic Change* 16: 247—281.

MacIntyre, F. M. 1974. The top millimeter of the ocean, *Sci. Am.* 230: 62–77.

Mano, S., and M. O. Andrae. 1994. Emission of methyl bromide from biomass burning, *Science* 263: 1255—1257.

Marland, G. 1989. Fossil fuel CO2 emissions: Three countries account for 50% in 1988. *CDIAC* (Carbon Dioxide Information Analysis Center) *Commun.,* Winter 1989, p. 1–4.

Martin, J. H., S. R. Fitzwater, and R. M. Gordon. 1990. Iron deficiency limits phytoplankton growth in Antarctic waters, *Global Biogeochem. Cycles* 4: 5–12.

National Research Council (NRC). 1991. *Rethinking the Ozone Problem in Urban and Regional Air Pollution.* Washington, D.C.: National Academy Press.

National Research Council Board on Atmospheric Sciences and Climate (NRC). 1983. *Changing Climate.* Report of the Carbon Dioxide Assessment Committee. Washington, D.C.: National Academy Press.

Newman, P. A. 1994. Antarctic total ozone in 1958, *Science* 264: 543–546.

Nriagu, J. O. 1989. A global assessment of natural sources of atmospheric trace metals, *Nature* 338: 47–49.

Penner, J. E., R. E. Dickinson, and C. A. O'Neill. 1992. Effects of aerosol from biomass burning on the global radiation budget, *Science* 256: 1432—1434.

Proffitt, M. H., J. J. Margitan, K. K. Kelly, M. Loewenstein, J. R. Podolske, and K. R. Chan. 1990. Ozone loss in the Arctic polar vortex inferred from high altitude aircraft measurements, *Nature* 347: 31–36.

Prospero, J. M., R. J. Charlson, V. Mohnen, R. Jaenicke, A. C. Delany, J. Moyers, W. Zoller, and K. Rahn. 1983. The atmospheric aerosol system: An overview, *Rev. Geophys. Space Phys.* 21: 1607–1929.

Quay, P. D., B. Tilbrook, and C. S. Wong. 1992. Oceanic uptake of fossil fuel CO_2: Carbon-13 evidence, *Science* 256: 74–79.

Rahn, K. A. 1976. The Chemical Composition of the Atmospheric Aerosol, Tech. Rep., Graduate School of Oceanography, University of Rhode Island, Kingston.

Rahn, K. A., and D. H. Lowenthal. 1984. Elemental tracers of distant regional pollutive aerosols, *Science* 223: 132–139.

Ramanathan, V. 1987. The role of earth radiation budget studies in climate and general circulation research, *J. Geophys. Res.* 92: 4075–4095.

Ramaswamy, V., and M. D. Schwarzkopf, and K. P. Shine. 1992. Radiative forcing of climate from halocarbon-induced global stratospheric ozone loss, *Nature* 355: 810–812.

Rasmussen, R. A., and Khalil, M. A. K. 1988. Isoprene over the Amazon basin, *J. Geophys. Res.* 93: 1417–1421.

Raynaud, D., J. Jouzel, J. M. Barnola, J. Chappelaz, R. J. Delmas, and C. Lorius. 1993. The ice record of greenhouse gases, *Science* 259: 926–934.

Revelle, R. 1983. Probable future changes in sea level resulting from increased atmospheric CO_2. In *Changing Climate.* Report of the Carbon Dioxide Assessment Committee, NRC Board on Atmospheric Sciences and Climate, pp. 433–448. Washington, D.C.: National Academy Press.

Revelle, R. R., and P. E. Waggoner. 1983. Effects of a carbon dioxide induced climatic change on water supplies in the western United States. In *Changing Climate,* Report of the Carbon Dioxide Assessment Committee, NRC Board on Atmospheric Sciences and Climate, pp. 419–432. Washington, D.C.: National Academy Press.

Rotty, R. H. and G. Marland. 1986. Production of CO_2 from Fossil Fuel Burning by Fuel Type. 1860–1982. Report NDP-006 Carbon Dioxide Information Center. Oak Ridge, Tenn.: Oak Ridge Natl. Lab.

Rowland, F. S. 1989. Chlorofluorocarbons and the depletion of stratospheric ozone, *Am. Sci.* 77: 36–45.

Sarmiento, J. L. 1993a. Ocean carbon cycle, *Chem. Eng. News,* 71: 30–43.

Sarmiento, J. L. 1993b. Atmospheric CO_2 stalled, *Nature* 365: 697–698.

Sarmiento, J. L., and E. T. Sundquist. 1992. Revised budget for the oceanic uptake of anthropogenic carbon dioxide, *Nature* 356: 589–593.

Schlesinger, W. H. 1991. *Biogeochemistry: An Analysis of Global Change.* New York: Academic Press.

Schoeberl, M. R., and D. L. Hartmann. 1991. The dynamics of the stratospheric polar vortex and its relation to springtime ozone depletions, *Science* 251: 46–52.

Shine, K. P., R. G. Derwent, D. J. Wuebbles, and J. J. Morcette. 1990. Radiative forcing of climate. In *Climate Change: The IPCC Scientific Assesment,* ed. J. T. Hougton, G. J. Jenkins, and J. J. Ephraums, pp. 41–68. Cambridge, U.K.: Cambridge University Press.

Siegenthaler, U., and J. L. Sarmiento. 1993. Atmospheric carbon dioxide and the ocean, *Nature,* 365: 119–125.

SMIC 1971. *Study of Man's Impact on Climate: Inadvertent Climate Modification.* Cambridge, Mass.: MIT Press.

Solomon, S. 1990. Progress towards a quantitative understanding of Antarctic ozone depletion, *Nature* 347: 347–354.

Solomon, S., R. W. Sanders, R. R. Garcia and J. G. Keys. 1993. Increased chlorine dioxide over Antarctica caused by volcanic aerosols from Mount Pinastubo, *Nature* 363: 245–248.

Stolarski, R., R. Bojkov, L. Bishop, C. Zerefos, J. Staehelin, and J. Zawodny. 1992. Measured trends in stratospheric ozone, *Science* 256: 342–349.

Sundquist, E. T. 1993. The global carbon dioxide budget, *Science* 259: 934–941.

Talbot, R. W., M. O. Andrae, T. W. Andrae, and R. C. Harriss. 1988. Regional aerosol chemistry of the Amazon Basin during the dry season, *J. Geophys. Res.* 93: 1499–1508.

Tans, P. P, I. Y. Fung, and T. Takahashi. 1990. Observational constraints on the global atmospheric CO_2 budget. *Science* 247: 1431–1438.

Thompson, A. M. 1992. The oxidizing capacity of the earth's atmosphere: Probable past and future changes, *Science* 256: 1157–1165.

Tolbert, M. A. 1994. Sulfate aerosols and polar cloud formation, *Science* 264: 527–528.

Toon, O. B., and R. P. Turco, 1991. Polar stratospheric clouds and ozone dpletion. *Sci. Am.* 264: 68–74.

Turekian, K. K. 1972. *Chemistry of the Earth.* New York: Holt, Rinehart & Winston.

Waggoner, P. E. 1983. Agriculture and a climate changed by more carbon dioxide. In *Changing Climate.* Report of the Carbon Dioxide Assessment Committee NRC Board on Atmospheric Sciences and Climate, pp. 383–418. Washington, D.C.: National Academy Press.

Walker, J. C. G. 1977. *Evolution of the Atmosphere.* New York: Macmillan.

Wang, W.-C., Y.-C. Zhuang, and R. D. Bojkov. 1993. Climate implications of observed changes in ozone vertical distributions at middle and high latitudes of the Northern Hemisphere, *Geophys. Res. Lett.* 20: 1567–1570.

Waters, J. W., L. Froidevaux, W. G. Read, G. L. Manney, L. S. Elson, D. A. Flower, R. F. Jarnot, and R. S. Harwood. 1993. Stratospheric ClO and ozone from the microwave limb sounder on the upper atmosphere research satellite, *Nature* 362: 597–602.

Watson, R. T., L. G. Mena Filho, E. Sanhueza, A. Janetos. 1992. Greenhouse gases: sources and sinks. In *Climate Change 1992. The Supplementary Report to the IPCC Scientific Assessment,* ed. J. T. Houghton, B. A. Callendar, and S. K. Varney, pp. 23–46. Cambridge, U.K.: Cambridge University Press.

Watson, R. T., H. Rodhe, H. Oeschger, and U. Siegthaler. 1990. Greenhouse gases and aerosols. In *Climatic Change: The IPCC Scientific Assessment,* ed. J. T. Houghton, G. J. Jenkins, and J. J. Ephraums, pp.1–40. Cambridge, U.K.: Cambridge University Press.

Went, F. W. 1960. Organic matter in the atmosphere, *Proc. Natl. Acad. of Sci. USA* 46: 212–221.

Whalen, S. C., and W. S. Reeburgh. 1992. Interannual variations in tundra methane emission: A 4 year time series at fixed sites, *Global Biogeochem. Cycles* 6: 139–140.

Whitehead, H. C., and J. H. Feth. 1964. Chemical composition of rain, dry fallout and bulk precipitation, Menlo Park, Calif., 1957–1959. *J. Geophys. Res.* 69: 3319–3333.

Wigley, T. M. L. and S. C. B. Raper. 1992. Implications for climate and sea level of revised IPCC enissions scenarios, *Nature* 357: 293–300.

Wollast, R. 1991. The coastal organic carbon cycle: fluxes, sources, and sinks. In *Ocean Margin Processes in Global Change,* ed. R. F. C. Mantoura, J.-M. Martin, and R. Wollast, pp. 365–381. New York: John Wiley.

Woodwell, G. M., J. E. Hobbie, R. A. Houghton, J. M. Melillo, B. Moore, B. J. Robertson, and G. R. Shaver. 1983. Global deforestation: Contribution to atmospheric carbon dioxide, *Science* 222: 1081–1086.

World Resources 1988–1989, 1988. World Resources Institute and International Institute for Environment and Development in Collaboration with the United Nations Environment Programme. New York and Oxford: Basic Books.

Zimmerman, P. R., J. P. Greenberg, and C. E. Westberg. 1988. Measurements of atmospheric hydrocarbons and biogenic emission fluxes in the Amazon boundary layer, *J. Geophys. Res.* 93: 1407–1416.

CHAPTER 3

RAINWATER AND ATMOSPHERIC CHEMISTRY

INTRODUCTION

Most of the water in the atmosphere (more than 95%) is present in the form of water vapor. How does this water become liquid rain or snow, and what chemical changes have occurred in it in the meantime? Even though we often think of rainwater as being very pure, it is, in fact, no longer just H_2O. It has become a dilute solution of a number of substances picked up during its trip through the atmosphere. We shall discuss these "impurities" in detail in this chapter: how they get into the atmosphere and how they become dissolved in rainwater.

In addition to its intrinsic importance, the chemistry of rainwater is also interesting from other points of view. Rainfall provides a major input of several elements to the earth's surface, and the importance of the rain input can be determined only if its composition is well known. Likewise, in attempting to determine the effects of rock weathering or biological processes on the concentration of a given element in a lake, river, or groundwater, one must first correct for the concentration of this element in rainwater arriving at the ground.

From a practical viewpoint, rainwater composition is of interest to those concerned with air pollution and the role of humans in altering the chemistry of the atmosphere. An outstanding example of this is the formation, in recent years, of acid rain downwind from industrial areas. Because of intense public interest in acid rain and air pollution in general, much attention will be given in the present chapter to these subjects.

FORMATION OF RAIN (AND SNOW)

How is atmospheric water vapor transformed into the precipitation (rain and snow) that arrives at the earth's surface? In this section we shall address this problem, starting with a discussion of water vapor itself.

Water Vapor in the Atmosphere

The amount of water vapor that is present in any given volume of air varies from place to place. One way of expressing the quantity of water vapor in air is the *absolute humidity* or density of water vapor as grams of water vapor in a unit volume of air (g/m^3). (A similar measure, which can be derived from the absolute humidity, is the *mixing ratio,* or grams of water vapor per kilogram of dry air.) The other commonly used expression for the water vapor content of the atmosphere is the *water vapor pressure,* that part of the total atmospheric pressure that is due to water vapor. Since water vapor is a minor constituent of air, its pressure is much lower than that of the atmosphere (about 2% on the average of the atmospheric pressure at the earth's surface).

A volume of air can hold just so much water vapor before the air becomes saturated and the water vapor condenses as a liquid or sublimates as ice. The saturation vapor pressure, or maximum amount of water that air can hold before condensation or sublimation, is a function of temperature, as shown in Figure 3.1. As can be seen, if air that is saturated with water vapor at a certain temperature is cooled, it becomes supersaturated and water condenses (or sublimates). A measure

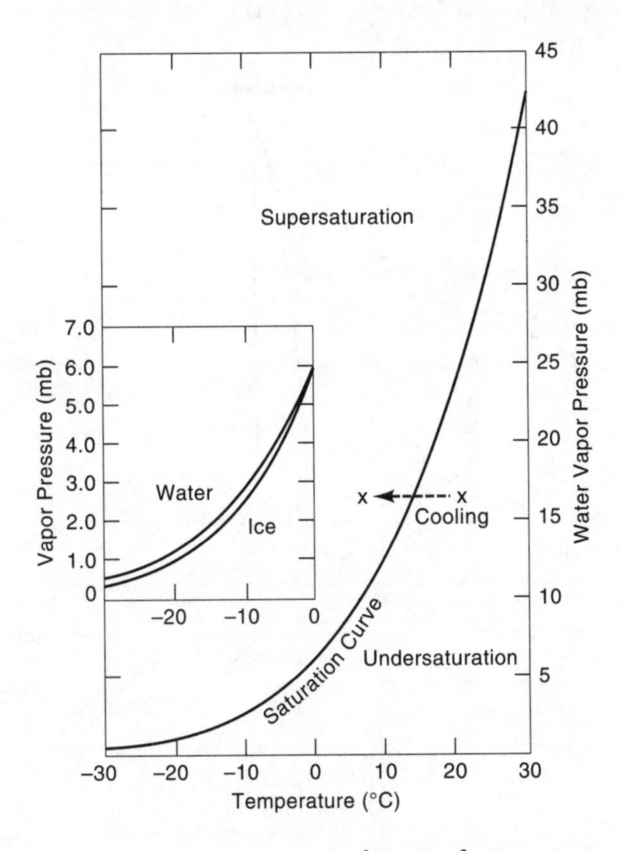

Figure 3.1. Saturation vapor pressure in millibars (1 mb = 10^3 dynes/cm^2) of pure water as a function of temperature (°C). The dashed arrow represents the change from undersaturation to supersaturation upon cooling of air with a constant water vapor pressure. Inset: Saturation vapor pressure (mb) over water ice at temperatures below 0°C. (After H.R. Byers, *Elements of Cloud Physics.* ©1965 by the University of Chicago Press. All rights reserved.)

of how close a given air mass is to saturation is given by the familiar term *relative humidity,* which is often mentioned in weather reports. Relative humidity is the ratio between the actual water vapor pressure and the saturation vapor pressure for the same temperature. It is usually expressed in terms of percent. (For further discussion, see Miller et al. 1983, or Neiburger et al. 1973.)

Because temperature decreases with height in the atmosphere (at an average rate of –6.5°C/km), water vapor content, due to condensation and sublimation, also decreases. This is shown in Figure 3.2. From the ground to an altitude of about 10 km, water vapor content and temperature continually decrease. At this altitude, one encounters the *tropopause,* the boundary between the lower atmosphere or *troposphere* and the upper atmosphere or *stratosphere.* (The height of the tropopause varies from about 5 km at the poles to 15 km at the Equator.) Air in the stratosphere contains a very small and nearly constant amount of water vapor (see Figure 3.2), because any air travelling to this height has already lost most of its moisture by condensation and sublimation in the troposphere. In this way the temperature minimum at the top of the troposphere serves as a cold trap that prevents loss of water from the earth to space (Newell 1971; see also Chapter 1).

The total amount of water vapor in the atmosphere over the whole globe is 13×10^{15} kg (which still represents only 0.001% of the total water on the earth; see Table 1.1). In a column

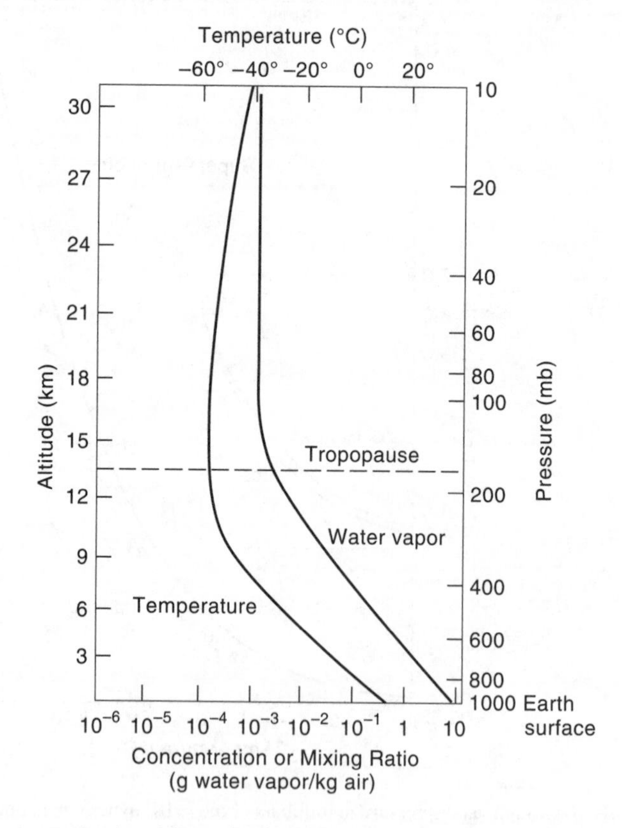

Figure 3.2. Decrease in concentration (or mixing ratio) of water vapor in the atmosphere with height (after Newell 1971) and decrease in temperature (°C) with height (data from Miller et al. 1983). The approximate height of the tropopause (which varies with latitude) is shown by the dashed line. Pressure is in millibars (1 mb = 10^3dynes/cm^2).

of air overlying a square meter of the earth's surface from the ground to about 10 km (roughly the tropopause), this would amount to an average of 25 kg of water per square meter if the water were evenly spread over the land surface or 2.5 cm of water if the water vapor fell as rain or snow. However, the water vapor content of air changes with latitude, as a result of temperature changes, from about 5 kg/m^2 at the North Pole through 20 kg/m^2 at 45°N, to 45 kg/m^2 over the Equator. These values also change with the seasons. Over the continental United States the average water content varies from 9 kg/m^2 in the winter to about 27 kg/m^2 in the summer, with the change again being due to temperature variation (Miller 1977).

As was pointed out in Chapter 1, the average *residence time* of water vapor in the atmosphere is only about 11 days. This is the length of time the average water vapor molecule spends in the atmosphere between its evaporation from the earth's surface and its precipitation as rain or snow. Such rapid turnover is important for the removal of atmospheric pollutants that can be "washed out" by rain. It is not so rapid, however, that appreciable transport does not take place. Because of the speed of atmospheric winds, during a period of 11 days a given mass of water vapor can be transported over great distances. For instance, the mean water vapor travel distance in temperate latitudes is around 1000 km (Peixoto and Kettani 1973).

Because of rapid transport, the amount of water vapor passing over a given landmass can be very large. For example, the average inflow of water vapor across the Gulf of Mexico coast of the United States, during the summer of 1949, was almost 10 times higher than the flow of the Mississippi River during the same period (Benton and Estoque 1954). Much of the water vapor passing over a landmass, however, is not lost via precipitation. Benton and Estoque (1954) report that, on a mean annual basis, only 63% of the water vapor passing eastward from the Pacific Ocean across the entire continental United States and Canada is delivered to the land as precipitation. This can be compared to values for smaller regions of 40% for European Russia, 20% for the Mississippi Basin, and only 11% for Arizona (Sellers 1965). (Low removal percentages, as shown here for Arizona, pose a local problem for arid regions, and efforts have been made to convert more water vapor to precipitation by means of such methods as cloud seeding. This is discussed later, in the section on cloud formation.) Because of the large-scale transport of water vapor, most continental precipitation does not come from locally evaporated water.

Condensation

The transformation of water vapor in the atmosphere into rain involves two processes. First, the water vapor gas must *condense* to form water droplets (or *sublimate* to form ice crystals). This is cloud or fog formation. However, condensation does not necessarily lead to precipitation or the fall of water droplets or snowflakes. In order for cloud droplets to fall to the ground as rain, they must become large and heavy enough to reach the ground without evaporation. The average lifetime of a cloud is around 1 hour, while on the average water spends 11 hours in the atmosphere in the form of droplets before being removed as rain. Thus, cloud water evaporates and condenses several times before actually forming rain (Pruppacher 1973).

In order for condensation to occur, air must become supersaturated with water vapor, but this is not enough. Nuclei are needed to begin the condensation process. The nuclei can be any one of a number of small bodies suspended in air as aerosols, chief of which are soil dust particles, combustion products, and sea salt. (For a detailed discussion of serosols, consult Chapter 2.) In air that is completely free of condensation nuclei, it would be possible to have relative humidities of as high as 800% without condensation taking place, whereas in actuality relative humidities (for condensation) never exceed 102% (Miller et al. 1983). This is because condensation nuclei promote the formation of water droplets, and such nuclei are always present.

Different suspended nuclei bring about condensation of water with differing effectiveness. First, larger particles serve as better nuclei because they have "flatter" surfaces, which favor condensation (Neiburger et al. 1973; Pruppacher 1973). *Hygroscopic* particles, which are substances that readily absorb water and are very soluble in it, serve as the best nuclei. They are so efficient that they can initiate condensation at less then 100% relative humidity (Neiberger et al. 1973; Pruppacher 1973). Examples are NaCl from sea spray, H_2SO_4 and HNO_3 from the burning of fossil fuels, and H_2SO_4 from oceanic DMS. Condensation may start at less than 100% because the hygroscopic particles are highly soluble in water and at high concentrations they lower the saturation vapor pressure of water (but because of dilution of the salts, continued condensation still requires relative humidities of more than 100%). Thus, hygroscopic nuclei require much lower degrees of supersaturation than their nonhygroscopic equivalents, and it is for this reason that they are more efficient in forming water droplets.

Over the oceans, condensation nuclei are dominated by hygroscopic particles derived from sea spray and from the oxidation of reduced-sulfur gases. On land, the nuclei are both hygroscopic and nonhygroscopic, with the latter being derived from soils and combustion products. The greater concentration of all kinds of particles over land, especially larger particles, results in competition for water and the formation of more but smaller water droplets in clouds. Over the oceans, the few but highly efficient hygroscopic particles produce relatively large droplets. Since larger cloud droplets favor rain formation (see below), rainout over the oceans (and coastal regions) is relatively easier than over land (For a more detailed discussion, see Hobbs 1993). This aids in the efficient removal of marine aerosols and helps explain the high sea-salt content of marine and coastal rain (Junge 1963).

Sublimation

Instead of forming water droplets, water vapor is transformed to ice crystals via the process of sublimation whenever the temperature is sufficiently cold, as often occurs at the tops of clouds. However, because ice crystals nucleate with much greater difficulty than water droplets, temperatures considerably lower than 0°C are necessary for sublimation to occur. For ice to form, a nucleus is needed that promotes crystal growth by having interatomic spacings similar to those found in ice. Only a few substances are suitable for this purpose, principally clay minerals from soils. Hygroscopic particles and combustion products are much less important in sublimation than they are in condensation. The degree of supercooling necessary to bring about crystallization is variable from cloud to cloud, the highest temperature at which ice crystallizes being about –10°C. Between –10 and –20°C, clouds form that consist of both ice crystals and water droplets, whereas below –20°C only ice clouds are present (Neiburger et al. 1973). An example of the latter are the high, wispy ice clouds known as cirrus clouds.

Rain (and Snow) Formation

The water droplets in most clouds average about 5–10 μm in diameter, with the largest being about 20 μm in diameter. Because of constant updrafts, these sizes are too small for the droplets to fall to the ground. In order to have rain, there must be a process whereby the droplets can become big enough, on the average about 1000 μm in diameter, to fall as raindrops (Neiburger et al. 1973). Further condensation on existing droplets in clouds is not an efficient mechanism. Instead, the most commonly cited processes of rain formation are *collision-coalescence* and *ice crystal growth followed by melting.* For snow formation the major process is ice crystal growth.

In collison-coalescence, droplets somewhat larger than the average begin to fall and, as they collide with smaller droplets, they grow in size by incorporating the small droplets. As a result of growth, they fall faster. After a large number of collisions (about 1 million), a rain-sized drop may be produced. If the drop becomes big enough, it can split into two drops, and these too can grow by further coalescence. Continued splitting and growth by collision plus coalescence produces a chain reaction resulting ultimately in the formation of rain. This process is especially effective over the oceans where, on the average, larger cloud droplets are found, which can begin to fall and initiate the chain reaction. It is also operative wherever air masses are too warm (greater than –10°C) to allow ice crystal formation (Mason 1971), and it serves as a process for enlarging raindrops originally formed by ice crystal growth plus melting.

Ice crystal growth (or the Bergeron process) occurs in the upper or colder parts of clouds, which contain both water droplets and ice crystals at temperatures of about –10 to –20°C (Neiburger et al. 1973). At such temperatures the saturation vapor pressure of water is greater than that of ice. In other words, air containing a given amount of water vapor can be supersaturated over ice (so that sublimation will occur) even though it is saturated or even undersaturated with respect to water (see Figure 3.1). As a result, ice crystals grow at the expense of coexisting water droplets. Eventually, the ice crystals may become big enough that they fall to the ground. If they melt on the way down, rain is produced; if they do not, snow results. If the ice grows very rapidly, it may fall as hail. [For further details on rain, snow, and hail formation, see Miller et al. (1983), Neiburger et al. (1973), and Pruppacher (1973).] Because of the smaller droplets found in clouds over the continents, much rain over land forms via the process of ice crystal growth followed by melting and enlargement by collision-coalescence.

In areas of deficient rainfall, humans have attempted to produce rain by artificial methods. Such cloud seeding proceeds along the lines of the two major processes of rain formation discussed above. Sometimes artificial ice nuclei, such as silver iodide or solid carbon dioxide (dry ice), are introduced at the cold tops of clouds in an attempt to induce the formation of ice crystals of a size large enough to fall. In other situations, hygroscopic nuclei (for example, $CaCl_2$) or large water droplets are added to clouds to initiate rain via the collision-coalescence process.

Air Motion in Cloud Formation

In considering the processes within the cloud that convert water vapor in the atmosphere to precipitation falling on the ground, we should not neglect the fact that larger-scale air motions are important. Basically, we need to cool moist air masses, since the amount of water vapor the air can hold will decrease with cooling (see Figure 3.1). Cooling can occur either directly or indirectly. Direct cooling comes about by the movement of warmer air over a colder land or water surface, whereas indirect or *adiabatic cooling* occurs when an air mass is lifted vertically. Since air pressure decreases with height, an ascending air mass expands upon uplift, and this causes a drop in temperature. An example of adiabatic cooling is when an advancing cold front forces warm air upward in front of it.

Once the moist air has been cooled by one of these mechanisms, its relative humidity increases and, when supersaturation occurs, condensation can proceed around particles present in the air. Air motions are also involved in the formation of rain in clouds; they affect where and how fast the rain falls, and how much moist air is being brought in and converted to rain. Thus, although we are concerned primarily with the processes of condensation and rain formation within the cloud, because of their influence on the chemical composition of rain, we should not neglect the fact that cooling of moist air by air motion is essential to these processes. Certain areas of the earth receive more rain than others because of these effects (see Chapter 1).

CHEMICAL COMPOSITION OF RAINWATER: GENERAL CHARACTERISTICS

The major dissolved element composition of a large number of rainfalls is presented in Table 3.1. As can be seen, rainwater can be characterized as being dilute (with average total dissolved salt contents of a few milligrams per liter) and weakly acidic (pH 4–6). Dilution is brought about by the way rain forms. Evaporation into the atmosphere involves extensive separation of water molecules from dissolved salts in surface waters. The resulting water vapor ultimately condenses to form rain, and the overall process can be viewed as purification by natural distillation. However, rainwater is not totally pure. Solid particles and gases in the atmosphere are dissolved by rainwater, which results in a wide range in chemical composition, as well as in variations of pH. This section briefly summarizes, in tabular form, compositional variation, and ensuing sections cover in detail the origin of each major element in rain. In general, two characteristics of rain data will be considered: (1) concentrations of the various ions, and (2) relative amounts of ions (i.e., ion ratios).

The dissolved chemical components of rainwater can be divided into two groups: (1) those derived primarily from particles in the air (Na^+, K^+, Ca^{2+}, Mg^{2+}, and Cl^-), and (2) those derived mainly from atmospheric gases (SO_4^{2-}, NH_4^+, and NO_3^-). The particles and gases, in turn, have a variety of sources, and the element associations in rain that result from these sources are given in Table 3.2. In addition, Tables 3.3 and 3.4 list the sources and typical ranges in concentration for each of the major dissolved components of rain.

The composition of condensation that ultimately falls to the ground as rainwater is determined by the composition of nucleating aerosols and soluble trace gases that react with water during both the condensation process and the fall to the ground. In the former case the process is called *rainout,* referring to reactions occurring within the clouds, while in the latter case the process is called *washout,* referring to reactions occurring below the clouds. Elements in rain that result from rainout will show little change or a slight rise in concentration with time. By contrast, elements contributed to rain by washout exhibit a sharp drop in concentration with time because the air becomes essentially cleansed (Junge 1963). Brief showers are washout dominated. A drop in the concentrations of ions in rain with time due to washout always occurs for terrestrially dominated species (Ca^{2+}, K^+, NO_3^-), which are concentrated in the lower atmosphere near the ground. Species derived from marine aerosols (Cl^-, Na^+, Mg^{2+}) show washout near the coast, where they are concentrated in the lower atmosphere, but inland, as marine aerosols become dispersed through the atmosphere, washout does not occur for these species (Stallard 1980).

Cl^-, Na^+, Mg^{2+}, Ca^{2+}, AND K^+ IN RAIN

The primary sources of dissolved Cl^-, Na^+, Mg^{2+}, Ca^{2+}, and K^+ in rain are marine (sea-salt aerosols), terrestrial (soil dust, biological emissions), and anthropogenic (industrial, biomass

TABLE 3.1 Composition of Precipitation—World (in mg/l)

Area	Na^+	K^+	Mg^{++}	Ca^{++}	Cl^-	SO_4^{--}	NO_3^-	NH_4^+	pH	Reference
Coastal Europe										
S.W. Sweden coast 1967–1969	1.96	0.27	0.36	0.84	3.48	4.9	2.0	0.91	4.65	Granat 1972
S. Norway coast (polluted) 1972	11.0	0.59	1.58	0.90	20.38	7.87	3.35	0.43	4.15	Likens et al. 1979
W. Ireland coast (unpolluted) 1967	21.3	0.94	2.59	1.52	36.42	6.29	0.06	0.02	5.8	Likens et al. 1979
World average coastal (<100 km inland)	3.45	0.17	0.45	0.29	6.0	1.45	—	—	—	Maybeck 1983
Inland Eurasia										
W. Sweden 1956 (unpolluted)	0.16	0.12	0.10	0.70	0.36	1.39	0	0	5.4	Likens et al. 1979
N. Sweden 1967–1969	0.30	0.20	0.12	0.64	0.39	2.0	0.31	0.12	—	Granat 1972
S. & c. Sweden 1973–75	0.35	0.12	0.17	0.52	0.64	3.31	1.92	0.56	4.3	Granat 1978
S. Norway 1974–1975 (polluted)	0.21	0.12	0.15	0.16	0.39	2.5	1.61	0.39	4.32	Likens et al. 1979
Belgium 1967–1969	0.97	0.23	0.36	1.32	1.95	6.0	2.23	0.48	4.42	Granat 1972
France 1967–1969	0.92	0.16	0.39	0.68	2.13	2.8	1.9	0.29	4.8	Granat 1972
Switzerland 1977	0.18	0.27	0.11	0.82	0.82	4.0	3.1	0.003	4.47	Zobrist & Stumm 1980
N. Europe 1955–1956 (average)	2.05	0.35	0.39	1.42	3.47	2.19	0.27	0.41	5.47	Carroll 1962
USSR (average pptn from cloud fronts)	0.4	0.2	0.3	0.4	0.8	2.7	0.2	0.5	—	Petrenchuk 1980
USSR–European (pptn-57cm)	2.4	0.7	0.5	2.0	1.8	5.7	0.8	0.6	5.9	Zverev & Rubeikin 1973
USSR–Asian (pptn-45 cm)	1.55	0.7	0.2	2.1	1.5	4.35	0.7	0.8	6.0	Zverev & Rubeikin 1973
USSR–European										
North	1.6	0.5	0.4	0.7	2.5	4.4	0.6	0.7	5.4	Petrenchuk & Selezneva 1970
Northwest	1.2	0.7	1.4	1.2	1.4	7.4	0.7	0.9	5.2	Petrenchuk & Selezneva 1970
U. Lena R., Russia	1.0	—	0.55	1.0	1.4	1.7	—	—	—	Gordeev & Siderov 1993
L. Lena R., Russia	0.9	0.3	1.0	1.2	3.2	2.0	—	—	—	Gordeev & Siderov 1993
Miscellaneous Land										
S. E. Australia (average)	2.46	0.37	0.50	1.20	4.43	Trace	—	—	—	Hutton and Leslie 1958
Jabiru, Australia	0.09	0.035	0.014	0.012	0.27	0.25	0.2	0.031 (0.43 organic)	4.89	Post & Bridgman 1991
Barrington, Australia (60 km from coast)	0.44	0.86	0.073	0.22	0.85	0.42	0.45	0.031 (0.48 organic)	5.8	Post & Bridgman 1991

TABLE 3.1 Composition of Precipitation—World (in mg/l), *continued*

Area	Na^+	K^+	Mg^{++}	Ca^{++}	Cl^-	SO_4^{--}	NO_3^-	NH_4^+	pH	Reference
Katherine, N. Central Australia, 1980–1984	0.10	0.04	0.02	0.03	0.27	0.19	0.25	0.05	4.74	Likens et al. 1987
Bankipur, India (monsoon 100 cm rain)	0.47	0.23	0.23	1.4	0.92	0.63	—	—	—	Handa, 1971
Tavapur, India (70 km from Bombay)	2.4	0.16	0.32	1.4	4.4	1.3	—	0.13	6.15	Sequeira, 1976
Japan (average)	1.1	0.26	0.36	0.97	1.2	4.5	—	—	—	Sugawara 1967
Beijing, N. China	3.24	1.57	—	3.68	5.59	13.11	3.11	2.54	6.8	Zhao & Sun 1986
Tianjiin, N. China	4.03	2.31	—	5.74	6.5	15.25	1.81	2.26	6.26	Zhao & Sun 1986
Chonqing, S. China	0.39	0.58	—	2.01	0.54	13.58	1.33	1.47	4.14	Zhao & Sun 1986
Guiyang, S. China	0.23	0.37	—	2.98	0.32	16.56	0.59	1.15	4.02	Zhao & Sun 1986
Kampala, Uganda (near L. Victoria)	1.7	1.7	—	0.05	0.9	1.8	1.7	0.63	7.9	Visser 1961
Ivory Coast, Africa	0.3	0.26	0.05	0.26	1.033	0.84	1.26	—	4.2	Lacaux et al. 1987
Greenland (ice and snow)	0.007	—	—	0.007	0.021	0.12	—	0.006	—	Busenberg & Langway 1979
Iceland MYRI 1982–83	0.74	0.19	0.08	0.2	0.91	1.03	—	—	5.5	Gislason & Eugster 1987
Marine										
Pacific Ocean (34°46'N 117°115'W)	24.	1.0	—	4.3	43	8.0	—	—	—	Gambell & Fisher 1966
Hawaii (near ocean)	5.46	0.37	0.92	0.47	9.63	1.92 (0.57)[a]	0.2	0.1	4.8	Eriksson 1957
N. Atlantic Ocean (120 m off NC)	2.8		0.2	0.2	5.1	1.2 (0.61)[a]	0.2	0.1	—	Gambell & Fisher 1964
N. W. Atlantic:										
Westward source	2.41	0.2	0.24	0.19	4.58	1.20 (0.87)[a]	0.42	0.07	4.66	Galloway et al. 1983
Eastward source	3.62	0.2	0.38	0.19	6.46	1.22 (0.29)[a]	0.26	0.045	5.07	Galloway et al. 1983
Bermuda 1955–1981	7.23	0.36	—	2.91	12.41	2.12	0.10	0.10	—	Junge & Werby 1958
Bermuda 1980–1981	3.38	0.17	0.41	0.19	6.2	1.74 (0.88)[a]	0.34	0.04	4.8	Galloway et al. 1982
S. Atlantic (250 km off Brazil)	5.34	0.18	0.73	0.17	10.26	2.87 (1.53)[a]	—	—	—	Stallard & Edmond 1981

TABLE 3.1 Composition of Precipitation—World (in mg/l), *continued*

Area	Na^+	K^+	Mg^{++}	Ca^{++}	Cl^-	SO_4^{--}	NO_3^-	NH_4^+	pH	Reference
S. Atlantic (85 km off Brazil)	2.99	0.17	0.39	0.15	5.01	2.30	—	—	—	Stallard & Edmond 1981
Amsterdam Is., Indian Ocean 1980-87	6.18	0.14	0.72	0.24	11.28	(1.63)[a] 1.79	0.10	0.04	5.08	Moody et al. 1991
South America										
Amazon R. Basin (mean)	0.285	0.039	0.029	0.044	0.49	0.49	0.13	—	5.03	Stallard & Edmond 1981
Over Amazon River (670 km inland)	0.50	0.020	0.036	0.028	0.87	0.64	0.19	0.002	4.71	Stallard & Edmond 1981
Over Amazon River (1700 km inland)	0.23	0.039	0.024	0.056	0.30	0.55	0.25	0.007	5.32	Stallard & Edmond 1981
Over Amazon R. (1930 km inland)	0.21	0.035	0.034	0.060	0.41	0.70	0.18	0.00	4.97	Stallard & Edmond 1981
Over Amazon R. (2050 km inland)	0.12	0.094	0.012	0.056	0.24	0.56	—	—	5.04	Stallard & Edmond 1981
Over Amazon R. (2230 km inland)	0.23	0.012	0.012	0.008	0.39	0.28	0.056	—	5.31	Stallard & Edmond 1981
Peru (3000 km from Atlantic)	0.039	0.039	0.020	0.184	0.12	0.18	—	—	5.67	Stallard & Edmond 1981
Venezuela (near coast)	2.2	0.6	0.7	1.14	2.6	2.2	0.2	0.3	—	Lewis 1981
Venezuela (San Carlos rain forest; 400 cm rain)	0.04	0.03	0.01	0.01	0.09	0.14	0.16	0.04	4.81	Galloway et al. 1983
Torres del Paine, Chile	0.43	0.055	0.052	0.024	0.78	0.211	0.031	0.013	5.31	Likens et al. 1987
U.S. Coastal (and Canada)										
Bodie, Is., N.C. 1955–1956	7.16	0.1	1.3	1.02	15.8	3.41	0.59	—	5.4	Gambell & Fisher 1966
Cape Hatteras, N.C. 1955–1956	4.49	0.24	—	0.44	6.9	1.22	0.04	0.01	—	Junge & Werby 1958
Cape Hatteras, N.C. 1962–1963	4.36	0.1	0.59	0.41	8.2	1.97	0.23	—	5.4	Gambell & Fisher 1966
N.J. Pine Barrens 1970–1972	1.39	0.32	0.23	1.10	2.82	5.09	0.39	—	—	Means et al. 1981
Stevensville, Nfld., 1955–1956	5.16	0.32	—	0.78	8.85	2.16	0.29	0.05	—	Junge & Werby 1958

TABLE 3.1 Composition of Precipitation—World (in mg/l), *continued*

Area	Na^+	K^+	Mg^{++}	Ca^{++}	Cl^-	SO_4^{--}	NO_3^-	NH_4^+	pH	Reference
Menlo Park, Calif. 1957–1959	2.0	0.25	0.37	0.79	3.43	1.39	0.16	—	6.0	Whitehead & Feth 1964
Tatoosh Is., Wash. (remote)	14.30	0.59	—	0.73	22.58	3.40	0.38	0.02	—	Junge & Werby 1958
Coastal Washington	1.81	0.12	0.22	0.08	3.49	0.73	0.14	0.04	5.1	Vong et al. 1988
Brownsville, Tex., 1955–1956	22.3	1.0	—	6.5	22.0	10.68	0.13	0.01	—	Junge & Werby 1958
San Diego, Calif., 1955–1956	2.17	0.21	—	0.67	3.31	3.35	1.5	0.05	—	Junge & Werby 1958
U.S. coastal (average)	3.68	0.24	—	0.58	4.83	2.45	—	—	—	Whitehead & Feth 1964; (data Junge & Werby 1958)
U.S. Inland (and Canada)										
N.E. U.S. average 1978–1979:										
All	0.36	—	—	—	0.40	2.81	1.58	0.31	4.2	Pack 1980
Noncoastal	0.32	—	—	—	0.29	2.70				
N.E. U.S. 1958–1968 (average)	0.27	0.16	0.11	0.60	0.45	4.3	0.34	0.22	4.4	Pearson & Fisher 1971
Hubbard Brook, N.H., 1975–1987 (average)	0.08	0.04	0.02	0.07	0.22	2.08	1.46	0.15	4.24	Butler & Likens1991
Ithaca, N.Y., 1977–1987	0.05	0.04	0.02	0.11	0.21	2.78	1.88	0.30	4.18	Butler & Likens1991
Whiteface Mt., N.Y. 1977–1987	0.04	0.05	0.02	0.09	0.17	2.07	1.31	0.24	4.34	Butler & Likens1991
Penn. State, PA., 1977–1987	0.07	0.05	0.02	0.14	0.24	3.14	1.93	0.32	4.15	Butler & Likens1991
Charlottesville, Va., 1977–1987	0.12	0.05	0.02	0.07	0.32	2.44	1.49	0.50	4.27	Butler & Likens1991
N.C. & Va. average 1962–1963	0.54	0.11	0.14	0.65	0.57	2.18	0.62	0.1	4.9	Gambell & Fisher 1966
Gatlinburg, Tenn., 1973	0.05	0.07	0.03	0.20	0.15	3.19	1.24	0.19	4.19	Cogbill & Likens 1974
Tallahassee, Fla., 1978–1979:										
N. air	—	0.16	—	0.37	—	1.69	—	—	4.4	Tanaka et al. 1980
S. air	—	0.10	—	0.38	—	0.65	—	—	5.3	
Average	—	0.12	—	0.38	—	1.09	—	—	—	
Central Illinois, 1978–1987	0.07	0.05	0.03	0.26	0.24	3.03	1.66	0.39	4.27	Butler & Likens 1991
Oxford, Ohio, 1975–1987	0.07	0.06	0.03	0.17	0.22	2.95	1.54	0.34	4.24	Butler & Likens 1991
Huron, S.D., 1980–1981	0.07	0.04	0.07	0.35	0.16	1.27	1.48	0.76	5.75	NADP (Stensland & Semonin 1984)

TABLE 3.1 Composition of Precipitation—World (in mg/l), *continued*

Area	Na^+	K^+	Mg^{++}	Ca^{++}	Cl^-	SO_4^{--}	NO_3^-	NH_4^+	pH	Reference
Glasgow Mont., 1955–1956	0.40	0.26	—	1.72	0.17	2.62	0.71	0.24	—	Junge & Werby 1958
Grand Junction, Colo., 1955–1956	0.26	0.17	—	3.41	0.28	4.76	0.98	0.26	—	Junge & Werby 1958
Columbia, Mo., 1955–1956	0.33	0.31	—	2.82	0.15	3.6	0.6	0.17	—	Junge & Werby 1958
Tewaukon, N.D., 1978–1979	0.27	0.23	0.27	1.05	0.20	1.74	1.59	0.86	5.27	Munger 1982
Ithasca, W. Minn., 1978–1979	0.20	0.17	0.23	0.69	0.15	1.53	1.24	0.60	5.0	Munger 1982
Hovland, E. Minn., 1978–1979	0.14	0.13	0.13	0.40	0.10	1.89	1.18	0.67	4.67	Munger 1982
Amarillo, Tex., 1955–1956	0.22	0.23	—	2.7	0.14	1.86	0.68	0.05	—	Junge & Werby 1958
Tom Green Co., Tex., 1972–1973	0.86	0.15	0.05	0.14	0.61	3.17	1.5	1.5	5.98	Miller 1974
Bishop, Calif., 1972–1973	0.84	0.42	0.08	0.67	0.64	2.26	1.03	0.47	6.1	Miller 1974
Ely, Nev., 1955–1956	0.69	0.22	—	3.28	0.3	2.84	1.44	0.35	—	Junge & Werby 1958
Albuquerque, N.M., 1955–1956	0.24	0.18	—	4.74	0.09	2.39	0.86	0.09	—	Junge & Werby 1958
Tesuque Mtn., N.M., 1975–1976	0.07	0.12	0.08	0.70	0.33	3.29	1.12	—	5.0	Graustein 1981
Santa Fe., N.M., 1975–1976	0.06	0.08	0.15	3.62	0.33	2.95	0.99	—	6.7	Graustein 1981
U.S. inland (average)	0.40	0.20	0.10	1.4	0.41	3.0	1.20	0.30	—	(From above)
U.S. (average), 1955–1956	0.90	0.23	(0.15)	1.0	1.13	2.02	0.70			Garrels & Mackenzie, 1971; (Lodge et al. 1968)
Poker Flat, Alas. (N. of Fairbanks)	0.02	0.02	0.002	0.002	0.09	0.35	0.12	0.02	5.0	Galloway et al. 1982
Experimental Lakes Area, Ont.	0.19	0.13	0.11	0.45	0.35	4.32	0.11	0.38	5.0	Schindler et al. 1976
Haney, B.C., 1972–1973	0.3	0.1	0.1	0.2	0.6	1.3	0.7	0.1	4.5	Feller & Kimmins 1979

[a] Numbers in parentheses refer to excess sulfate—see text.

TABLE 3.2 Primary Associations in Rain

Origin	Associations
Marine inputs	$Cl - Na - Mg - SO_4$
Soil inputs	Al – Fe – Si – Ca – (K, Mg, Na)
Biological inputs	$NO_3 - NH_4 - SO_4 - K$
Biomass burning	$NO_3 - NH_4 - P - K - SO_4 -$ (Ca, Na, Mg)
Industrial pollution	$SO_4 - NO_3 - Cl$
Fertilizers	$K - PO_4 - NH_4 — NO_3$

Sources: Modified after Stallard 1980; Lewis 1981.

TABLE 3.3 Sources of Individual Ions in Rainwater

	Origin		
Ion	Marine Input	Terrestrial Inputs	Pollutive Inputs
Na^+	Sea salt	Soil dust	Biomass burning
Mg^{++}	Sea salt	Soil dust	Biomass burning
K^+	Sea salt	Biogenic aerosols Soil dust	Biomass Burning Fertilizer
Ca^{++}	Sea salt	Soil dust	Cement manufacture Fuel burning Biomass burning
H^+	Gas reaction	Gas reaction	Fuel burning
Cl^-	Sea salt	—	Industrial HCl
SO_4^{--}	Sea salt DMS from biological decay	DMS, H_2S etc., from biological decay Volcanoes Soil dust	Fossil fuel burning Biomass burning
NO_3^-	N_2 plus lightning	NO_2 from biological decay N_2 plus lightning	Auto emissions Fossil fuels Biomass burning Fertilizer
NH_4^+	NH_3 from biological activity	NH_3 from bacterial decay	NH_3 fertilizers Human, animal waste decomposition (Combustion)
PO_4^{3-}	Biogenic aerosols adsorbed on seasalt	Soil dust	Biomass burning Fertilizer
HCO_3^-	CO_2 in air	CO_2 in air Soil dust	—
SiO_2, Al, Fe	—	Soil dust	Land clearing

Sources: Junge 1963; Mason 1971; Miller 1971; Granat et al. 1976; Stallard and Edmond, 1981.

TABLE 3.4 Typical Concentrations of Major Ions in Continental and Marine Rainfall (in mg/l)

Ion	Continental Rain	Marine and Coastal Rain
Na^+	0.2–1	1–5
Mg^{++}	0.05–0.5	0.4–1.5
K^+	0.1–0.3[a]	0.2–0.6
Ca^{++}	0.1–3.0[a]	0.2–1.5
NH_4^+	0.1–0.5[b]	0.01–0.05
H^+	pH = 4–6	pH = 5–6
Cl^-	0.2–2	1–10
SO_4^{--}	1–3[2,b]	1–3
NO_3^-	0.4–1.3[b]	0.1–0.5

[a] In remote continental areas; K^+ = 0.02–0.07; Ca^{++} = 0.02–0.20; SO_4 = 0.2–0.8

[b] In polluted areas; NH_4^+ = 1–2; SO_4^{--} = 3–8; NO_3^- = 1–3.

Sources: See Table 3.1.

burning) (see Table 3.3). The relative importance of marine sources varies with distance from the coast and levels off at a fairly constant low level inland. A "hierarchy of ions" can be established (after Means et al. 1981 and Stallard and Edmond 1981) based on the relative importance of marine sea-salt sources and continental (terrestrial or pollutive) sources:

$$Cl^- = Na^+ > Mg^{2+} > K^+ > Ca^{2+} > SO_4^{2-} > NO_3^- = NH_4^+$$

mostly marine mostly continental

Of the principal cations, Na^+ is the dominant cation in areas of marine-influenced rain, while Ca^{2+} is the dominant cation in inland rain. In terms of both cations and anions, for an area near the ocean, the rain is basically a NaCl solution but it rapidly changes to a $CaSO_4$ solution inland. In addition, there is a much greater content of total dissolved salts in marine-influenced rain than is usual for inland rains, except in arid areas.

Ratios of major ions can be used to compare rainfall compositions with that of sea salt. Sea-salt weight ratios with respect to the concentration of Na^+, taken from the data for seawater composition given in Chapter 8, are shown in Table 3.5. Using known seawater ratios, the contribution of sea-salt ions to precipitation (assuming no fractionation on aerosol formation from seawater) can be determined by assuming that all Cl^- or Na^+ (reference species) and the proportionate amount of other ions are derived from sea salt. The amount of the concentration of an ion that is greater than sea-salt proportions is referred to as *excess ion* or *non-sea-salt ion* (e.g., excess SO_4^{2-}, etc).

Unfortunately, only a few measurements of marine rain have been made over the oceans, but rain falling on oceanic islands such as Bermuda, Hawaii, and Amsterdam Island (Indian Ocean) has a composition much like marine rain. The choice of Na^+ over Cl^- as the reference species is preferred for marine rain because of Cl gain from pollution and HCl loss from acidified marine sea-salt aerosols (Keene et al. 1986). As shown in Table 3.5, the ratio of Cl/Na is close to sea-salt proportions for marine and island rains. The ratios of Mg^{2+}, Ca^{2+}, and K^+ to Na^+ are very similar to sea-salt proportions for South Atlantic rain (Stallard and Edmond 1981) and

TABLE 3.5 Major Element Composition, in Terms of Weight Ratios to Na^+, for Seawater and Marine and Island Rains

Ion	Seawater	South Atlantic Rain	N. Atlantic Rain (1981) West Source	N. Atlantic Rain (1981) East Source	Pacific Rain	Bermuda Rain	Hawaii Rain	Amsterdam Is. Rain
Na^+	1.00	1.00	1.00	1.00	1.00	1.00	1.00	1.00
Cl^-	1.797	1.92	1.90	1.78	1.79	1.83	1.76	1.83
SO_4^{2-}	0.252	0.266	0.622	0.337	0.333	0.515	0.352	0.290
Mg^{++}	0.12	0.137	0.100	0.105	—	—	0.168	0.117
Ca^{++}	0.038	0.032	0.052	0.052	0.179	0.056	0.086	0.039
K^+	0.037	0.034	0.083	0.055	0.042	0.05	0.068	0.023

Sources: See Chapter 8 for seawater data and Table 3.1 for rainwater data.

Amsterdam Island rain (Galloway et al. 1982). There are varying small amounts of contamination by continental cations (Ca^{2+}, Mg^{2+}, and K^+) in the other island and marine rains. The ratio of SO_4/Na in most of these rains is greater than the sea-salt ratio, ranging up to more than twice the sea-salt ratio; this is presumably due either to transported sulfate pollution or to marine sulfate sources other than sea salt (see the discussion in the section on sulfate in rain).

Marine coastal rainwater generally has close to sea-salt ratios to Na^+ for Cl^- and Mg^{2+}, but Ca^{2+} and K^+ ratios may be greater than in sea salt due to varying continental inputs (see Table 3.1). The definition of "coastal" rain varies considerably, from less than 40 km from the coast in the northeastern United States (Pearson and Fisher 1971) to less than 100 km from the coast in Europe (Meybeck 1983).

Over the oceans, the Cl content of rain is around 10–15 mg/l, but over land the concentration drops rapidly within 10–20 km of the coast (Junge 1963). In the United States, Cl drops off to a fairly constant concentration of 0.15–0.20 mg/l by 600 km inland. This is shown in Figure 3.3. A similar pattern is shown by Munger and Eisenreich (1983). The concentration of chloride in precipitation over the Amazon Basin also decreases sharply with increasing distance from the Atlantic Coast. The average value levels off at 1200 km inland to 0.35 mg/l (Stallard and Edmond 1981), which is somewhat higher than the U.S. value, and declines gradually thereafter. A larger and deeper-penetrating marine contribution might be expected here because the prevailing winds in the Amazon Basin are from off the Atlantic Ocean, while they are predominantly from the continental interior in the eastern United States.

The decrease in Cl^- content inland is usually interpreted as rapid deposition of sea salt in precipitation near the coast. However, Junge (1963) has offered an additional explanation involving air mixing. Sea salt over the oceans is concentrated in the lower 500 m of air, but when the maritime air moves over the land, there is intense vertical mixing of the air up to 7000 m. Thus, even if the amount of sea salt in the air remains constant, the concentration of sea salt in the air is considerably reduced because the sea salt is mixed through a much greater volume of air.

There is often a change in the Cl/Na ratio from the seawater value on going inland. This could be due to (1) fractionation-differential removal of sea-salt components; (2) addition of terrestrial ions to the rain (particularly soil dust sodium, which would lower the ratio); or (3) anthropogenic Cl^- (which would increase the ratio).

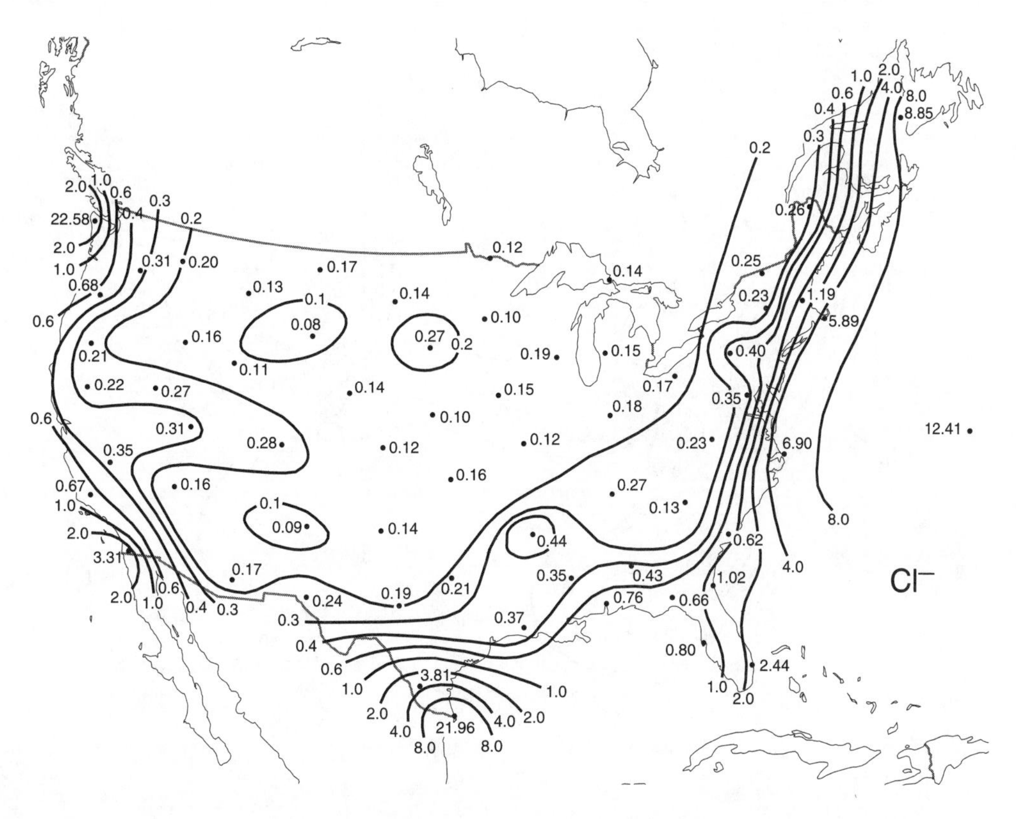

Figure 3.3. Average Cl^- concentration (mg/l) in rain over the United States, July 1955–June 1956. [After C. E. Junge and R. T. Werby, "The Concentration of Chloride, Sodium, Potassium, Calcium and Sulfate in Rainwater over the United States," *Journal of Meteorology* 15 (October 1958): 418, © 1958 by the American Meteorological Society.]

In the Amazon Basin, where the prevailing winds are off the Atlantic Ocean and where pollution is minimal and soil dust is low, Stallard and Edmond (1981) found that the Cl/Na ratio (and the Cl/Mg ratio) of inland rain remains close to sea-salt proportions even 2000 km inland. This confirms that there are not major changes due to fractionation of the marine component going inland, and that explanation 1 above is generally unimportant.

The marine influence on U.S. inland rain is considerably less. The prevailing winds are from the west and, although they bring marine air onshore along the west coast, much of the marine aerosols are not carried inland because of rainout and washout over the Pacific coastal mountains. On the eastern coast of the United States, the prevailing winds are from over the continent, particularly in the summer, although in the winter storms move up the coast from the Gulf of Mexico, bringing marine air onshore. The United States is also considerably dustier than the Amazon Basin. All these factors cause concentration ratios of U.S. rain to vary considerably from that of sea salt. Junge and Werby (1958) measured the Cl/Na ratio of U.S. precipitation and found the rain ratios to be close to the seawater ratio (1.8) at the coast but to decrease rapidly inland, levelling off at a value of 0.5–0.8 by 800 km inland (Figure 3.4). From one half to two thirds of the Na in interior U.S. rain is believed to be from soil dust.

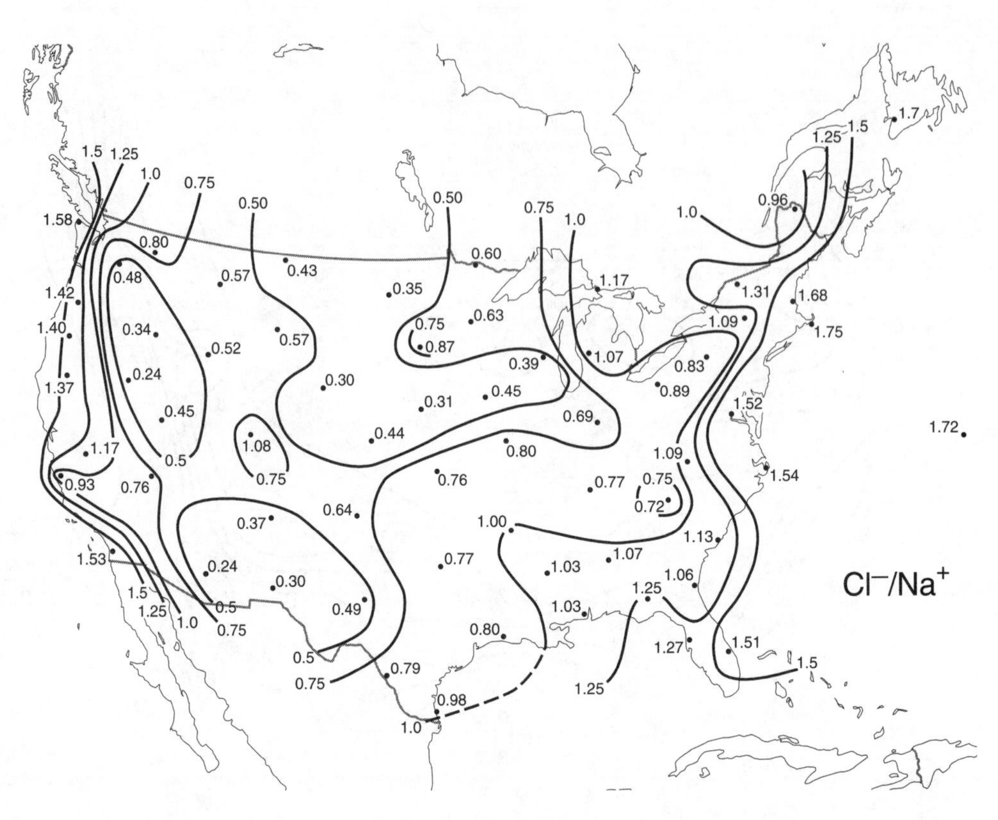

Figure 3.4. Average weight ratio of Cl^-/Na^+ in rain over the continental United States, July 1955–June 1956. Sea salt $Cl^-/Na^+ = 1.8$. [After C. E. Junge and R. T. Werby, "The Concentration of Chloride, Sodium, Potassium, Calcium and Sulfate in Rainwater over thc United States," *Journal of Meteorology* 15 (October 1958): 419, © 1958 by the American Meteorological Society.]

Maps of total Na^+, and the "excess" not contributed by sea salt, are shown in Figure 3.5. Note particularly the area of excess Na^+ (using Cl^- as the reference seawater ion) in the Great Basin, an arid, dusty area of igneous rocks relatively enriched in Na^+ (over most sedimentary rocks), and one with many saline (Na-rich) dry lake beds (Gambell 1962), presumably good sources of Na-rich dust. [This high-Na^+ area also appears on Munger and Eisenreich's (1983) map of Na^+ in rain.] Further evidence for a soil-dust origin for Na comes from a decrease of Na in rain going from agriculturally developed (and presumably dusty) North Dakota to heavily forested Minnesota (Munger 1982), and in the high excess Na shown in Figure 3.5b for south Texas. Gambell and Fisher (1966) found that in North Carolina and Virginia, rain excess Na^+, excess Ca^{2+},

Figure 3.5. Dissolved Na^+ (in mg/l) in continental United States rainfall, June 1955–June 1956. (a) Average Na^+ concentration. (b) Average excess Na^+ concentration, calculated as the difference between measured values shown in (a) and the values expected for seawater Cl^-/Na^+ ratios and the Cl^- concentrations shown in Figure 3.3. [After C. E. Junge and R. T. Werby, "The Concentration of Chloride, Sodium, Potassium, Calcium and Sulfate in Rainwater over the United States," *Journal of Meteorology* 15 (October 1958): 418, 420, © 1958 by the American Meteorological Society.]

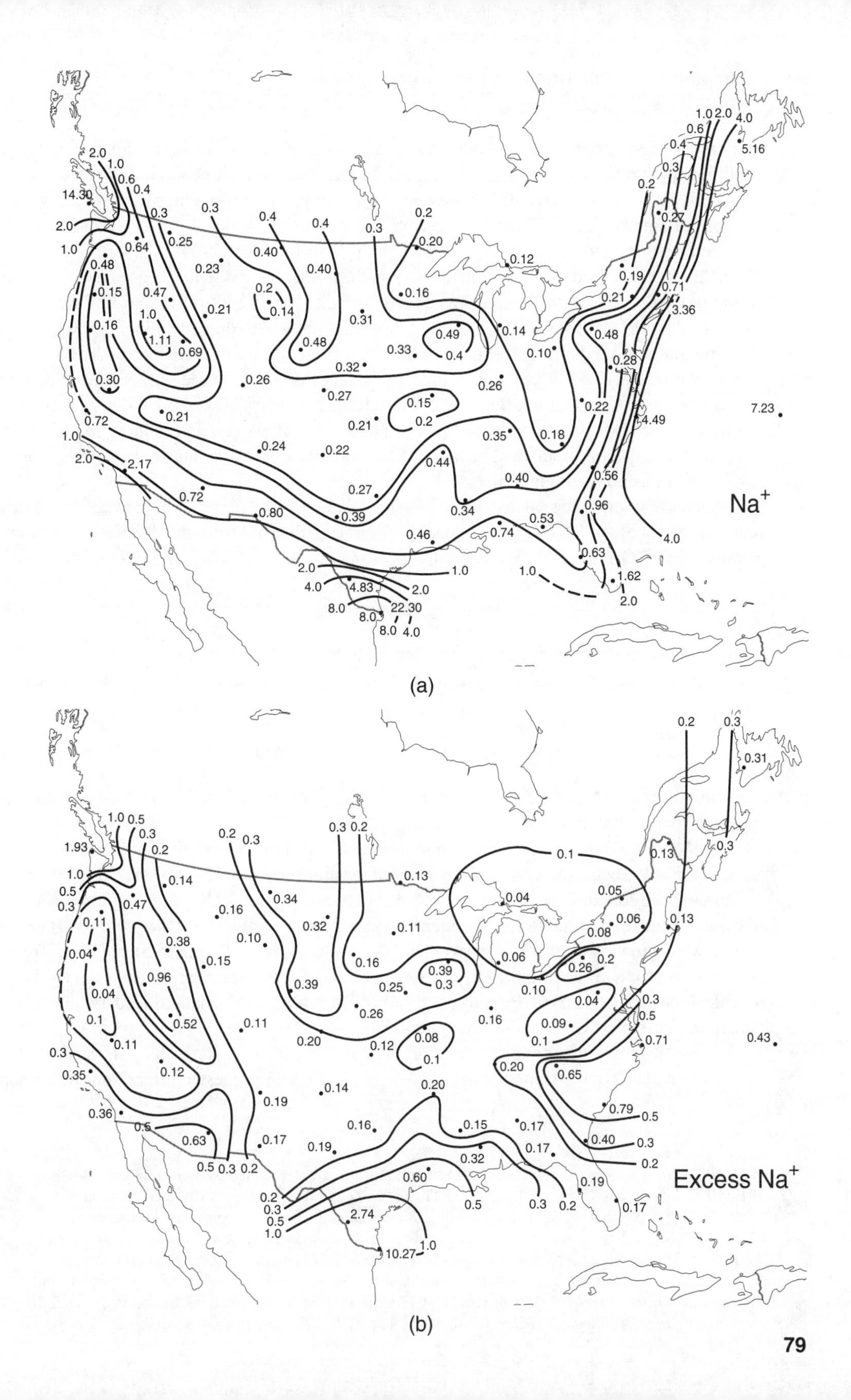

(a)

(b)

and excess Mg^{2+} (excess being above sea-salt ratios based on Cl^-) all had similar monthly patterns and areal distributions, again suggesting a common soil-dust origin for all three ions.

All Cl in rain is not derived from sea salt; some may come from chlorine-containing gases. Sources of anthropogenic Cl, as HCl and other gases, in the atmosphere include automobiles, coal combustion, and burning of polyvinyl chloride in incinerators (Paciga and Jervis 1976). There has been an increase in pollutive Cl from these sources in the northeastern United States in the last 20 years (Cogbill and Likens 1974), which is reflected in Cl/Na ratios greater than the sea-salt ratio. Values of Cl/ Na ratio in European rain, where the prevailing winds are off the Atlantic, range from 2.3 to 1.3, showing considerably more marine influence than U.S. interior values, with the higher Cl/Na ratios (which are greater than those in sea salt) occurring in industrial areas due to Cl pollution. Because of pollutive Cl, workers often tend to determine sea-salt contributions to precipitation on the basis of Na, not Cl. However, determining sea salt on the basis of Na is also problematical because of the variable soil dust contributions of Na (particularly inland), as pointed out earlier.

The contribution of sea salt to Ca^{2+} in continental rain is very small, since the Ca/Cl ratio in sea salt is only 4% of the Na/Cl ratio. Instead, Ca^{2+} comes primarily from the dissolution of calcium carbonate ($CaCO_3$) in soil dust. $CaCO_3$ dissolves in rain to form HCO_3^- and Ca^{2+} by the reaction

$$H^+ + CaCO_3 \rightarrow Ca^{2+} + HCO_3^-$$

In this way neutralization of rain acidity comes about. In the case of strong acids such as H_2SO_4 (when SO_2 is present in the air), this neutralization results in the production of CO_2 and acid anions such as SO_4^{2-}:

$$H_2SO_4 + CaCO_3 \rightarrow CO_2 + H_2O + SO_4^{2-} + Ca^{2+}$$

Ca^{2+} in rain can also come from $CaSO_4$ (gypsum) soil dust and rarely from $CaCl_2$, but neither of these sources will neutralize acidity (Butler et al. 1984).

In general, Ca^{2+} is the dominant cation in inland U.S. rain. The Ca^{2+} concentrations in U.S. rain in 1955–1956 are shown in Figure 3.6a (Junge and Werby 1958). The maximum Ca^{2+} concentrations are found in (1) arid zones of the southwest, where $CaCO_3$ is a major constituent of the topsoil due to its formation from intense evaporation of soil water at the ground surface, and where wind-blown soil dust is common (soil dust is transported on the average about 300–650 km downwind); and (2) in the windy western Canadian-U.S. prairie areas from the cultivation of calcareous soils. Large Ca^{2+} (and other ion) concentrations in arid areas also are the result of small and infrequent amounts of precipitation.

The mid-1950s were a time of severe drought and dust storms, particularly in the central and lower Great Plains of the United States. As a result, it has been suggested (Stensland and Semonin

Figure 3.6. (a) Dissolved Ca^{2+} concentrations (in mg/l) in rainwater over the continental United States, June 1955–June 1956 [After C. E. Junge and R. T. Werby, "The Concentration of Chloride, Sodium, Potassium, Calcium and Sulfate in Rainwater over the United States," *Journal of Meteorology* 15 (October 1958): 421, © 1958 by the American Meteorological Society.] (b) Weighted-average Ca^{2+} concentrations (in mg/l) in 1978–1981 contoured for NADP wet deposition data. The other points with numerical data are calcium concentrations determined in earlier years by PHS/NCAR (1960–1966) (bold numbers) and WMO/NOAA/EPA (1972–1973) (italic numbers). [Source: G. J. Stensland and R. G. Semonin (1984), *Bulletin American Meteorological Society* 65: 6, p. 642, copyright 1984 American Meteorological Society.]

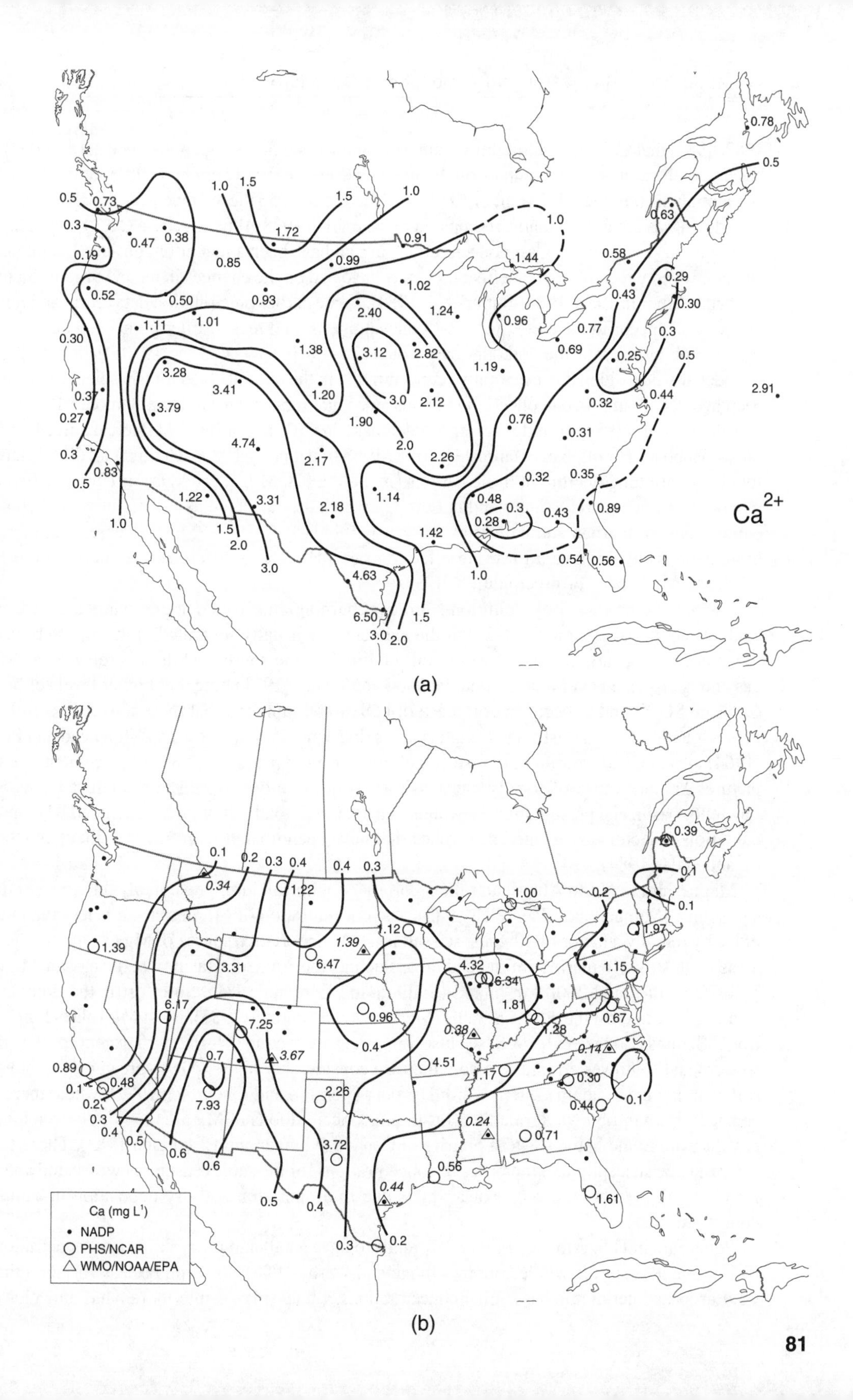
Ca^{2+}
0.78
0.5
1.0
1.5
1.5
1.0
0.5
0.73
0.3
0.47
0.38
1.72
0.91
1.0
0.63
0.19
0.85
0.99
1.44
0.58
0.52
0.50
0.93
1.02
0.29
0.43
0.30
2.40
1.24
1.01
0.96
0.77
1.11
0.30
1.38
3.12
2.82
0.69
0.3
3.28
0.25
0.5
3.41
1.20
1.19
0.37
3.0
2.12
0.32
0.44
2.91
3.79
0.27
1.90
0.76
0.31
4.74
2.0
0.3
2.17
2.26
0.5
0.83
0.32
0.35
1.22
3.31
1.14
0.48
0.3
0.43
0.89
2.18
0.28
1.0
1.5
1.42
2.0
3.0
0.54
0.56
1.0
4.63
6.50
1.5
3.0
2.0
(a)

0.39
0.1
0.2
0.3
0.4
0.4
0.3
0.34
1.22
1.00
0.2
0.1
0.1
1.12
1.97
1.39
1.39
6.47
3.31
4.32
1.15
6.34
6.17
0.96
1.81
7.25
0.67
1.28
0.38
0.4
3.67
0.14
0.89
0.7
4.51
0.1
0.48
1.17
0.30
0.2
7.93
2.26
0.3
0.44
0.1
0.4
0.24
0.5
0.6
0.6
0.71
3.72
0.56
1.61
0.44
0.5
0.4
0.3
0.2
Ca (mg L^{-1})
NADP
PHS/NCAR
WMO/NOAA/EPA
(b)

1982, 1984) that Ca^{2+} concentrations in rain were about three times higher than normal at this time because of the drought. For comparison, Figure 3.6b (Stensland and Semonin 1984) shows the Ca^{2+} concentrations (contoured) in rain in 1978–1981. In 1955–1956 the average Ca^{2+} concentration in inland rain was around 1.0 mg/l as compared to 0.3 mg/l in 1978–1981. Because $CaCO_3$ neutralizes acid rain, these changes in Ca^{2+} concentrations in rain have been a matter of considerable interest and controversy regarding what effect they may have had on the changes in the pH of rain (Stensland and Semonin 1982, 1984; Butler et al. 1984). Some earlier precipitation data (early and mid-1970s) also appear to have elevated Ca^{2+} concentrations, and these earlier points can be seen in Figure 3.6b along with the contours based on 1978–1981. However, the 1970s data may have included dry deposition and evaporative concentration in the rainfall collectors, problems that have been avoided by improved collection methods since that time (Stensland and Semonin 1984).

Likens et al. (1984) found a sevenfold decrease in Ca^{2+} and a fivefold decrease in Mg^{2+} in rain at Hubbard Brook, New Hampshire, from 1963 to the early 1980s. Hedin et al. (1994) have found a continuing decline in basic cations (including Ca, Mg, K, and Na) in Hubbard Brook rain (and also in other U.S. areas and in Europe), which they attribute to less pollutive particulate emmissions from urban and industrial sources, particularly Ca^{2+} and Mg^{2+}. (Decreases in these basic cations should have a lowering effect on rain pH.) These differences do not seem to be related to wind speed or precipitation effects.

Ca^{2+} can be produced by pollution from coal burning and from cement manufacture. Coal contains about 0.4% Ca and 3% S, but the elements are usually separated in the air after burning, since Ca appears in ash and would fall in rain near the source, while S is released as SO_2 gas and is dispersed over a large area. Pearson and Fisher (1971) note that higher levels of both Ca^{2+} and SO_4^{2-} tend to occur in northeastern U.S. urban rain, and that these ions also correlate in precipitation from industralized English areas. In Menlo Park, California, Whitehead and Feth (1964) attribute Ca^{2+} in rain to industrial pollution from gypsum processing and cement manufacture. Another cause of locally elevated Ca^{2+} concentrations in rain in the Great Lakes-St. Lawrence basin, and presumably elsewhere, is from $CaCl_2$ road salt, which is included when precipitation collectors are located near roads that have been treated with $CaCl_2$ to melt ice (Barrie and Hales 1984).

Magnesium in North American rain is correlated with calcium, presumably due to a similar soil source (Munger and Eisenreich 1983). The concentration of Mg in sea salt is less than that of Na by roughly a factor of 10; thus, sea salt is not as important a source of Mg except in coastal areas or in strongly marine-influenced areas such as the Amazon Basin, where sea-salt Mg/Cl ratios are still found 2000 km inland (Stallard and Edmond 1981). Soil dust is the dominant source of Mg in most interior U.S. rains (Munger and Eisenreich 1983). Most Mg values for interior U.S. rains (Lodge et al. 1968; see also Table 3.1) are around 0.1–0.2 mg/l except for the arid western and southwestern areas, where they are considerably higher due to greater wind-blown soil dust. For example, in eastern North Dakota (Thornton and Eisenreich 1982), where there is considerable wind-blown agricultural dust, rain concentrations of Mg are 0.33 mg/l; even higher values are found in the western prairies of Canada (Munger and Eisenreich 1983). These values would be susceptible to increases in concentrations for the same reasons as was stated above for Ca^{2+} values, that is, drought or artifacts due to evaporation and dry deposition in sample collectors.

Potassium in U.S. rain (see Figure 3.7) tends to have small and fairly uniform concentrations (0.1–0.2 mg/l) over the whole country (Junge and Werby 1958), showing far less variation than Ca. European interior rain has similar concentrations, while Amazon interior rain has much lower

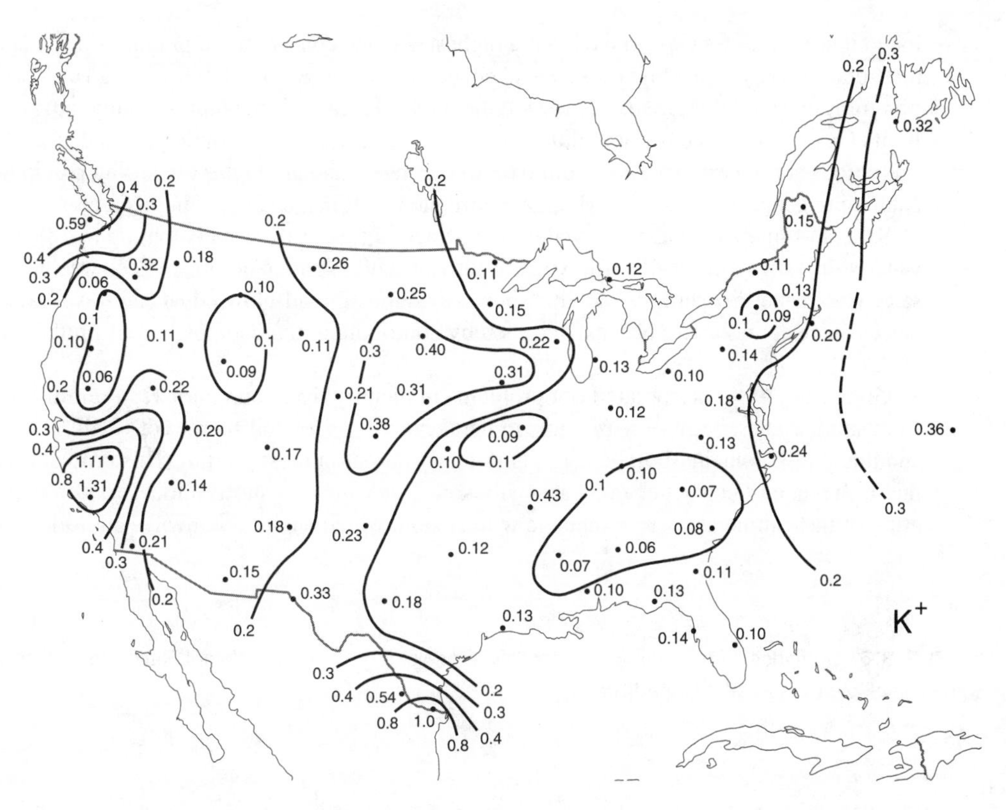

Figure 3.7. Dissolved K concentrations (in mg/l) in rainwater over the continental United States, June 1955–June 1956. [After C. E. Junge and R. T. Werby, "The Concentration of Chloride, Sodium, Potassium, Calcium and Sulfate in Rainwater over the United States," *Journal of Meteorology* 15 (October 1958): 421, © 1958 by the American Meteorological Society.]

K concentrations (0.03–0.07 mg/l), about one fourth of U.S. and European concentrations (see Table 3.1). Based on the chloride content, sea salt contributes only about 10% of K in U.S. interior rain, whereas in Amazon rain, with its much lower K concentrations, it may account for as much as 25% of the K.

There are several possible nonmarine origins for K in continental rain, including (1) dissolution of soil dust; (2) K-containing fertilizers, which contribute K to soil dust; (3) pollen and seeds; (4) biogenic aerosols; and (5) forest burning, particularly in tropical areas. The relative importance of these sources varies strongly from area to area.

A dissolved soil-dust origin for potassium in U.S. interior rains has been suggested by Junge (1963), who points out that the average Na/K ratio of such rains (1.7) is more like the average soil ratio (0.5) than like sea salt (27). Similarly, the Mg/K ratio of interior rain is more like soil than sea salt (see Table 3.6). Gillette et al. (1992) show that for a large part of the western United States, K^+ makes up 20–30% of the soil composition.

Munger (1982) measured K^+, Ca^{2+}, and P in precipitation for three sites on a 600-km transect going from agricultural-prairie (eastern North Dakota) to mixed forest (eastern Minnesota). He found that K^+ follows Ca^{2+} in having the highest average concentration in rain over the agricultural-prairie area and the highest seasonal concentration there in the fall (0.86 mg/l). Concentrations of K and Ca decrease sharply toward the forested area. This trend presumably indicates that K and Ca are dominated by agricultural soil dust from wind erosion. In the forested area, K concentrations are lower, vary less (from 0.08 to 0.2 mg/l), and are highest in spring precipitation, paralleling concentrations of P, which are attributed to pollen and seeds in this area.

Saharan soil dust contributes a significant fraction (5%–10% usually, but up to 38%) of the water-soluble K in tropical North Atlantic aerosols, although the dominant K source is still sea salt (Savoie and Prospero 1980). This shows, again, that wind-blown dust can travel great distances, even across entire oceans, and thereby contribute to the composition of rainfall (Junge 1972).

Graustein (1981) has pointed out problems in identifying a source for K in rain if it comes from soil particles, because reasonable mineral candidates for soil sources of K (illite, feldspar, and mica) also contain SiO_2, but SiO_2 is virtually absent in most rainfall. This criticism can be met if the time of contact of rainwater with soil-dust sources is short. Initial reaction in weathering of these minerals is K^+ exchange with H^+ ion, which would not involve addition of silica

TABLE 3.6 Composition of Various Sources of Aerosols, Expressed as Weight Ratios to K^+ Compared with Rainfall over the Interior United States

Material	Na/K	Mg/K	Ca/K	S/K	Al/K	Reference
Rain:						
Interior U.S.	1.66	0.81	9.26	7.4	—	(Table 3.1)
Eastern interior U.S.	1.0	0.83	4.2	9.6	—	(Table 3.1)
Western interior U.S.	2.1	0.79	12.6	5.7	—	(Table 3.1)
Soil:						
U.S. average	0.52	0.4	1.04	—	2.5	Shacklette et al. 1971
Eastern U.S.	0.35	0.31	0.43	—	4.5	Shacklette et al. 1971
Western U.S.	0.6	0.46	1.06	—	3.2	Shacklette et al. 1971
U.S. from igneous rocks (less–more mature	0.5–0.3	1.0–0.3	2.0–0.7	0.05	—	Bohn et al. 1979
World average soil	0.46	0.36	1.0	0.05	5.1	Bowen 1966
Crustal rock	1.1	0.81	1.4	0.01	3.14	Mason 1966
Average platform sedimentary rock	0.34	1.1	4.7	0.26	3.0	Holland 1978
Seawater	27.0	3.2	1.0	2.25	—	(Table 8.1)
Angiosperms	0.09	0.23	1.29	0.24	0.04	Bowen 1966
Trees (U.S.)	0.01	0.25	2.0	—	0.03	Connor & Shacklette 1975
Crops (U.S.)	0.001	0.12	0.14	—	0.003	Connor & Shacklette 1975
Soil ash of burned tropical vegetation (cane and grass)	0.42	0.25	0.8	0.2	—	Lewis 1981
Urban aerosol:						
Toronto	0.75	1.6	6.1	—	2.4	Paciga & Jervis 1976
Chicago	0.5	1.13	2.5	—	1.25	Gatz 1975

to solution. In addition, K may be enriched over usual soil concentrations and appear in a more soluble form in agricultural soil dust due to the addition of K fertilizers.

Biogenic aerosols are another possible K source. As noted in Chapter 4, plants exude waste products on their leaf surfaces, which can then escape as particles into the atmosphere. This would produce a soluble salt (K_2SO_4) that could contribute K^+ and SO_4^{2-} to rain. Lawson and Winchester (1979) believe that these plant exudates explain the presence of K (associated with P and S) in coarse aerosols of the Amazon Basin of South America, which is a humid, heavily forested area where dust production is minimal. Stallard and Edmond (1981) also believe that K in interior Amazon Basin rain is derived primarily from plant exudates and point out that K aerosol concentrations observed in the atmosphere would yield rain concentrations close to those observed. The concentrations of soluble K found by Crozat (1979) in biogenic aerosols of the Ivory Coast would give about 0.04 mg/l K if converted to rain. According to Crozat, the formation of biogenic aerosols should be favored by high heat and humidity, tropical conditions that do not generally exist in the United States. However, Graustein (1981) has evidence from his work in New Mexico to suggest that K may be given off locally by trees to the atmosphere and thus appear in rain.

In tropical areas such as the Amazon Basin, Venezuela, and the Ivory Coast, forest burning is used during the dry season to clear the land, and this can contribute to K in rain. Lewis (1981) found that fine atmospheric particulates produced by forest burning were flushed out in considerable quantities at the onset of the rainy season in coastal Venezuela, resulting in high rain concentrations of K^+, Ca^{2+}, SO_4^{2-}, Na^+, and Mg^{2+} followed by a rapid decline as the season progressed. In contrast to the tropics, there is little effect on rainfall composition by forest burning in North America because of its much lower frequency there.

GASES AND RAIN

Because atmospheric water droplets and raindrops are small, they have large specific surface areas available for the exchange of gases between the droplets and the atmosphere. As a result, atmospheric gases become dissolved in rainwater, both during the condensation process (rainout) and while the rain is on its way to the ground (washout). Washout occurs if the concentration of the gas below a cloud is greater than it is within the cloud or if the earlier reaction within the cloud, is not complete.

The amount by which a gas dissolves in rainwater depends on (1) its partial pressure or concentration in the atmosphere, and (2) its solubility in water. (Solubility, in turn, is a function of temperature.) Partial pressure is pressure exerted by each gaseous component in a mixture, so that together the sum of partial pressures equals total pressure. The ratio of partial pressure to total pressure is a measure of concentration and is equivalent to the volume fraction of a given gas in air. Volume fractions of atmospheric gases were listed in Chapter 2, in Table 2.1. From these data one can calculate that for $T = 25°C$ and a total pressure of 1 atm, the concentrations in aqueous solution of some major gases for equilibrium with air are: nitrogen (N_2) 13.5 mg/l; oxygen (O_2) 6.7 mg/l; carbon dioxide (CO_2) 0.5 mg/l.

Some gases not only dissolve in water, but also react with H_2O and with other gases, before and after dissolving, to form new species. This includes carbon dioxide (CO_2), ammonia (NH_3), sulfur dioxide (SO_2), nitrogen dioxide (NO_2) and hydrogen chloride (HCl). In the case of CO_2,

SO_2, NO_2, and HCl, acids are produced: H_2CO_3 (carbonic acid), H_2SO_4 (sulfuric acid), HNO_3 (nitric acid), and HCl (hydrochloric acid); whereas for NH_3, a base is formed: NH_4OH (ammonium hydroxide). (The details of these reactions are discussed later in this chapter.) Sulfuric, nitric, and hydrochloric are strong acids that completely dissociate to H^+, SO_4^{2-}, NO_3^-, and Cl^-, whereas carbonic acid and ammonium hydroxide are weak and only partially dissociate:

$$H_2CO_3 \rightarrow H^+ + HCO_3^-$$

$$NH_4OH \rightarrow NH_4^+ + OH^-$$

Either way, the ions SO_4^{2-}, NO_3^-, Cl^-, HCO_3^-, NH_4^+, and H^+ result, and these are major components of rainwater.

SULFATE IN RAIN: THE ATMOSPHERIC SULFUR CYCLE

As can be seen in Table 3.1, sulfate is the most abundant ion in almost all but purely marine rains. It is also the principal indicator of worldwide atmospheric pollution and the major culprit in the formation of acid rain. These observations demand that there be detailed discussion of both the natural and anthropogenic processes affecting sulfate concentrations in rain, as well as discussion of general atmospheric sulfur pollution. This is done in the present section. Our ultimate goal will be to derive quantitative estimates of the major fluxes in the atmospheric cycle of sulfur.

The two largest sources of sulfate in rain are (1) sea-salt aerosols and (2) sulfur dioxide from fossil fuel combustion. Other important sources include (3) biogenic reduced sulfur gases, such as H_2S and $(CH_3)_2S$, (4) biomass burning, and (5) volcanic emissions of sulfur dioxide (see also Table 3.9). Sea-salt aerosols are injected into the atmosphere in coarse particulate form (>1 μm), while the other inputs are gases that become converted to fine aerosols and eventually to sulfate in rain.

Sea-Salt Sulfate

Sulfate is a major seawater constituent. Thus, the amount of rain sulfate of sea-salt origin generally is determined from the known weight ratio of SO_4^{2-}/Cl^- in seawater (which equals 0.14) and the measured concentration of Cl^- in rain. In polluted areas, because there is excess rain Cl^- of non-sea-salt (industrial) origin, and because some Cl^- may be lost as HCl gas, the ratio of SO_4^{2-}/Na in seawater and the measured concentration of Na in rain is often used instead. The use of the sea-salt ratio of SO_4/Cl (or SO_4/Na) assumes that there is no fractionation between seawater and sea-salt aerosol. This assumption has been challenged for sulfate by Garland (1981), who claims limited enrichments of approximately 10% in SO_4/Cl and SO_4/Na ratios during the formation of sea-salt aerosol from seawater, but his results can be explained without invoking fractionation (see Duce et al. 1982) and may be due to reaction of sea salt with atmospheric SO_2 of either anthropogenic or biogenic origin in clouds (Andrae et al. 1986; Savoie et al. 1989). The amount of rain sulfate remaining, after sea-salt sulfate has been subtracted from total sulfate, is referred to as *excess sulfate* or non-sea-salt (NSS) sulfate.

The amount of sea-salt sulfate cycled through the atmosphere is poorly known. Its production has been estimated by Eriksson (1960) as approximately 44 Tg S/yr based on a total sea-salt production rate of 1000 Tg/yr. (Recall that 1 Tg = 10^6 metric tons = 10^{12} g). Most of this is simply redeposited on the ocean, but a small part also falls on land. Eriksson estimated that about 100 Tg Cl^- from sea salt falls on land annually, based on rain Cl^- deposition rates. Using the ratio in sea salt, this would mean that about 4 Tg of sea-salt-derived S is deposited annually on land, with the remaining 40 Tg S/yr deposited on the oceans. Based on his estimates of Cl deposition on land and the sea-salt production rate, Eriksson concluded that only about 10% of seasalt aerosol falls on land. His estimate of land deposition of sea-salt S, which is based on the amount of sea-salt Cl^- that falls on land, is *independent* of his estimate of total sea-salt production over the oceans. Unfortunately, other workers have adopted the value of land sea-salt deposition being 10% of sea-salt production, and because their estimates of sea-salt production vary widely [e.g., 1300 Tg/yr (Petrenchuk 1980); 5200 Tg/yr (Warneck 1988)], their resulting values for land deposition also vary widely. However, their is no reason that the 10% value should be applied to these other estimates. In this chapter, for lack of as complete a study by others, we adopt the Eriksson values given above.

Most sulfur deposited on land does not arise from sea-salt aerosol. Specifically, the percentage of total land sulfur deposition due to sea spray is less than 5%. Over the oceans sea spray is a major source of sulfur deposition (~50%), but even here as much as half of the sulfur comes from other sources.

Anthropogenic SO_2

Much of the sulfate, as well as excess sulfate, in rain in heavily populated areas arises from the oxidation of sulfur dioxide gas, which is given off to the atmosphere during the combustion of fossil fuels (coal and oil). The SO_2 forms during combustion from the oxidation of sulfur contaminants present as pyrite (FeS_2) in coal and organic sulfur compounds in both coal and oil. Fossil fuel combustion (primarily coal) for electricity generation accounted in 1980 for 67% of anthropogenic sulfur oxide emissions in the United States, heating for 6%, and industries such as refining and smelting accounted for 14% (see Table 3.7). In the western United States, however, metal smelters, primarily in Arizona and New Mexico, accounted for 70% of the SO_2 emissions and electric power plants 20% (Oppenheimer et al. 1985). Automobiles accounted for only 2% of the sulfur oxides produced in 1975 in the United States (Nader 1980), in contrast to their much greater contribution to pollutive nitrogen oxides (40%; see Table 3.13). In 1985, electricity generation continued at a high percentage accounting for about 70% of the sulfur oxides emitted in the United States (*World Resources 1988–89*), similar to the United Kingdom and Germany, but much greater than that in France (31%) and Canada (17%), which use more alternative sources (such as nuclear fission and water power) for energy generation.

The choice of fossil fuel type for energy greatly affects sulfur emissions. Most coal has a higher sulfur content (mean 2% S by weight) than either oil (mean 0.3%–0.8% S) or natural gas (0.05% S) (Möller 1984). This is due, to a large extent, to pyrite in coal, which does not occur in oil or gas. Figure 3.8 shows U.S. SO_2 (and NO_x) emissions over the period 1940–1988. Note

TABLE 3.7 U.S. Emissions of Sulfur in 1980 as Sulfur Oxides

Source	Tg S/yr	Percent of total
Electricity generation	8.75	67
Heat (commercial & residential	0.75	6
Industry (smelting[a] & refining)	1.9	14
Other[b]	1.75	13
Total	13.15	100

[a]Smelting produced 0.5 Tg S in 1980 (Husar 1986).
[b]Includes auto emissions, which were 0.3 Tg S in 1975 (Nader 1980).
Source: data from Gschwandtner et al. 1986.

that U.S. SO_2 emissions peaked in 1970 and have been dropping ever since to below 1940 levels, attaining 10.4 Tg SO_2-S per year in 1988 (*World Resources 1992–93*). The decrease in U.S. SO_2 emissions is due mainly to the lower sulfur content of coal that is burned and emission reductions brought about by environmental regulations (Gschwandtner et al. 1986).

The total world fossil fuel emission of SO_2-S for 1986 amounted to 67 Tg S/yr (Hameed and Dignon 1992). This makes fossil fuel-derived sulfur a dominant input to the world sulfur cycle. The world trend in fossil fuel SO_2-S emissions from 1940 to 1986 is shown in Figure 3.9. While North American emissions have declined since 1970, steady increases have occurred in the former USSR and Asia—particularly in China, where coal consumption has increased. The major emitters of SO_2 in 1986 were the USSR (20% of global emissions), the United States and China (15% each), and India and Poland (about 3% each) (Hameed and Dignon 1992). On a global scale, fossil fuel SO_2 emissions are not evenly distributed, but are concentrated in the Northern

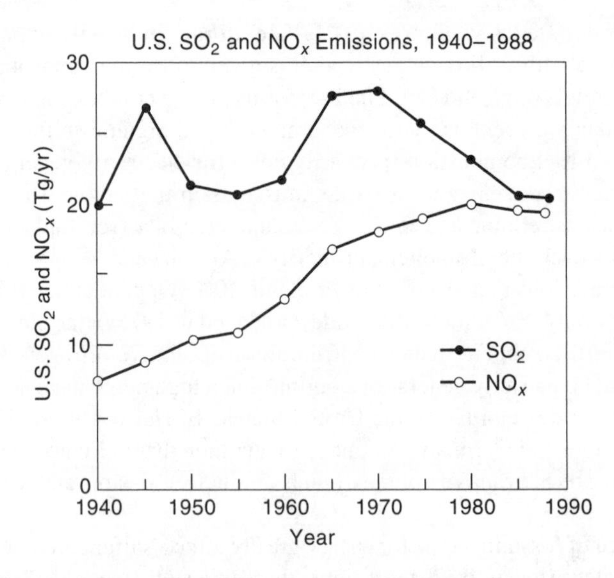

Figure 3.8. Rise in U.S. emissions of SO_2 and NO_x from 1940 to 1988. Top curve: U.S. SO_2; bottom curve: U.S. NO_x. To compare SO_2 data here with data given elsewhere in the present chapter, divide by 2 to obtain Tg S/yr. (*Sources:* Gschwandtner et al. 1986; *World Resources 1992–93,* p. 351.)

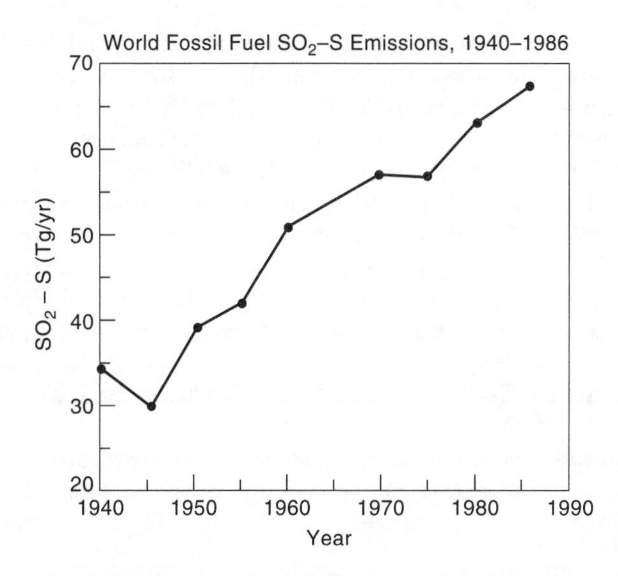

Figure 3.9. World fossil fuel emissions from 1940 to 1986 in Tg SO_2-S/yr. [Data from Möller (1984) for prior to 1970; data for 1970 and thereafter from Hameed and Dignon (1992).]

Hemisphere (see Figure 3.12). In northwestern Europe and the eastern United States, fuel combustion is the dominant sulfur source. For example, Galloway and Whelpdale (1980) calculated that anthropogenic sulfur emissions around 1980 made up 90% of the total atmospheric sulfur input for eastern North America.

Sulfur dioxide gas has a residence time in the atmosphere of 2–7 days before it is removed by dry deposition or is converted to sulfate (e.g., Tanaka and Turekian 1991; Lelieveld 1993). In addition, particulate sulfate often stays in the atmosphere for an additional 5–12 days (Tanaka and Turekian 1991). During this time atmospheric sulfur may be transported large distances before it is removed in precipitation (Charlson et al. 1992), and consequently, the effects of anthropogenic sulfur dioxide production tend to be felt over a large area, often as far as 1000 km from the source of pollution. This is particularly true in Scandinavia, where a large part of the pollutive sulfate in rain emanates from central Europe and the eastern British Isles. Likewise, the northeastern United States receives sulfur pollution from the industrialized portions of the Middle West, and the western U.S. Rocky Mountain states experience pollutive sulfate in rain that is transported more than 1000 miles to the northeast from Arizona and New Mexico metal smelters (Oppenheimer et al. 1985).

To avoid local sulfur pollution, both as SO_2 in air and sulfate in rain, U.S. power plants have built taller smoke stacks. However, the taller stacks, while reducing local SO_2 concentrations, increase SO_2 and sulfate transport over longer distances, thereby sending pollution downwind to more distant locales.

Sulfur oxide pollution by humans has been going on since the Industrial Revolution, and the greater use of fossil fuels in the last 20–30 years has in many areas greatly increased the amount of SO_2 in the air and sulfate (and acidity) in rain. An example of this is shown in Figure 3.10. Here we compare excess sulfur deposition in 1955–1956 (Figure 3.10a) for U.S. rain with total sulfur deposition in 1987 (Figure 3.10b). (Although the figures are not strictly comparable, the main difference between total sulfate-sulfur deposition and excess sulfate-sulfur deposition is at the coast; inland, the differences are negligible.) The bull's-eye pattern in the northeastern United States is similar in both years, and the 1960 central area of high sulfur deposition (>10 kg S/ha/yr = 1.0 g S/m²/yr) is also similar in 1987, indicating a levelling off of sulfate concentrations (see also NRC 1983). However, the area covered in 1960 by high sulfur deposition, defined here as greater than 7 kg S/ha/yr in Figure 3.10a, has expanded considerably into the southeastern United States in 1987 as defined by the area enclosed by 0.67 g S/m²/yr in Figure 3.10b (which is equivalent to 7 kg S/ha/yr). The changes in sulfate-sulfur deposition reflect regional changes in the SO_2 emissions: All regions showed increases in SO_2 in the 1960s, but in the 1970s the northeast declined, the southeast rapidly increased, and the midwest had a moderate increase (NRC 1986).

Conversion of Sulfur Dioxide to Sulfate in Rain

Two main processes, whose relative importance varies, are involved in the conversion (oxidation) of SO_2 gas to sulfate (SO_4^{2-}) in rain or dry deposition. First, SO_2 gas reacts with neutral

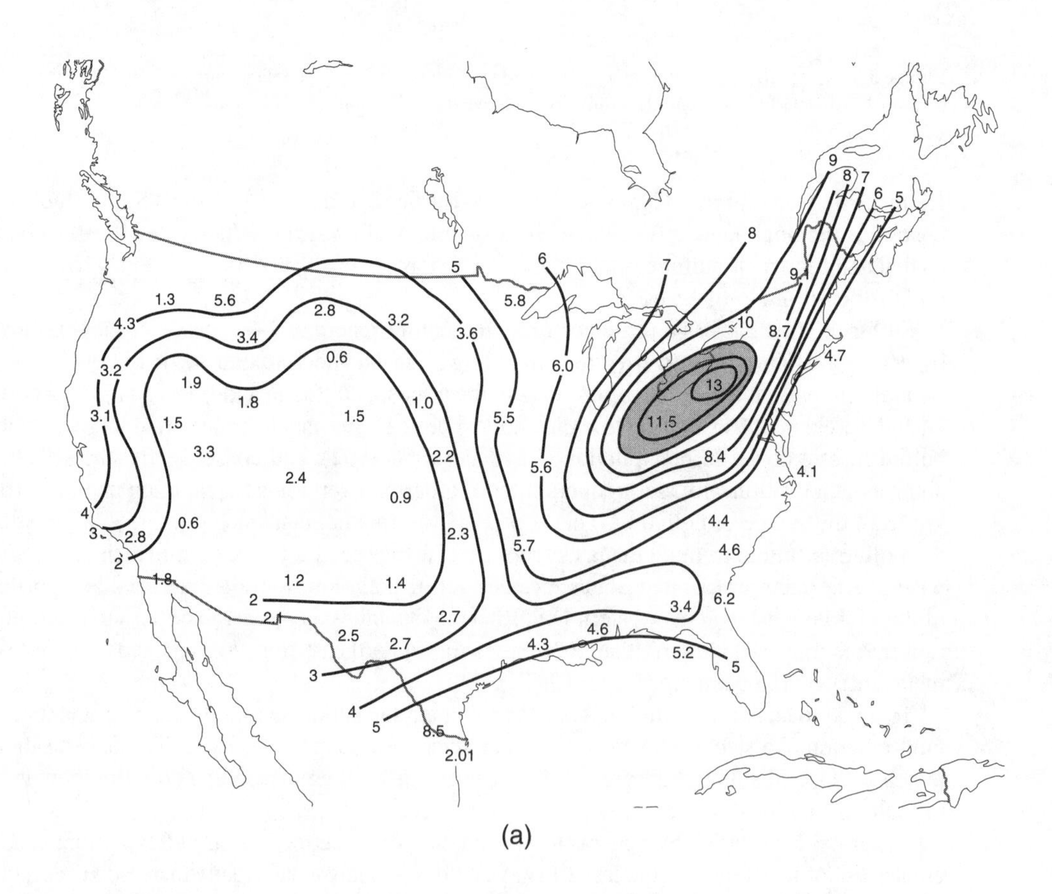

Figure 3.10. (a) Excess sulfur deposition in precipitation over the United States in 1955–1956 in kg S/ha/yr (g × 10^{-1} S/m²/yr)—i.e., 10 kg S/ha/yr = 1 g S/m²/yr. [From Eriksson (1960), based on data from Junge and Werby (1958).]

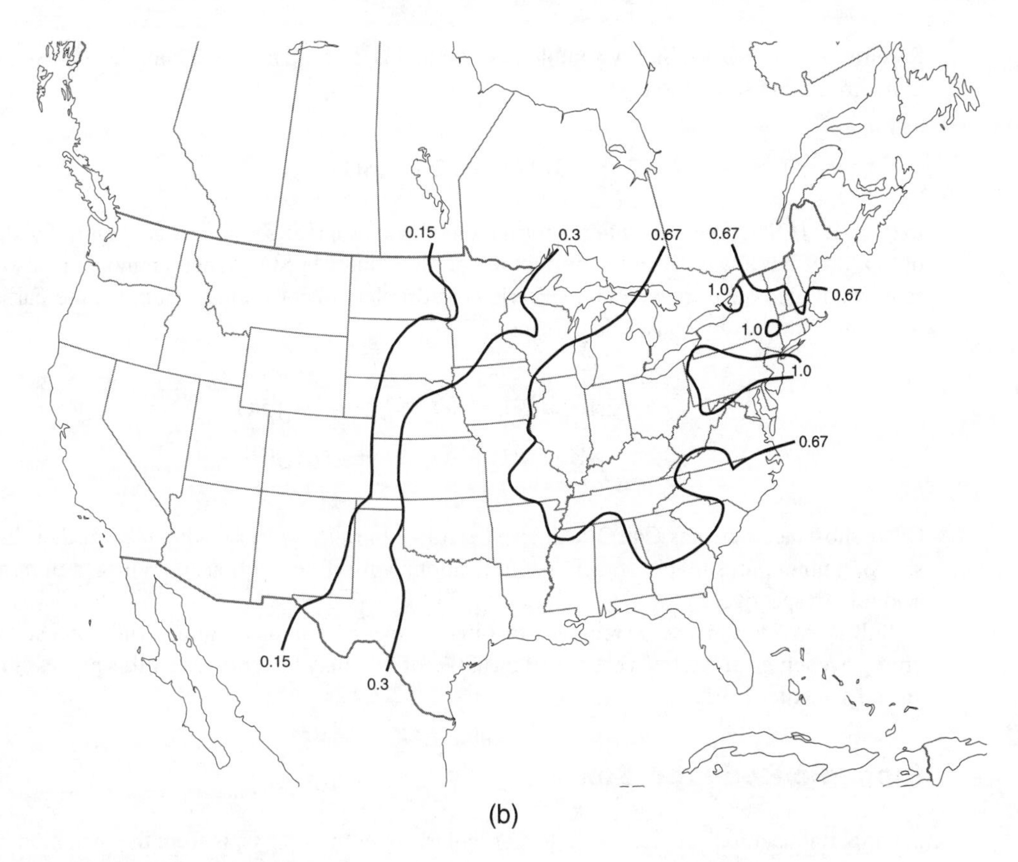

(b)

Figure 3.10. (b) Wet deposition of sulfur (g S/m²/yr) in 1987. (Adapted from Charlson et al. 1992.)

OH radicals in gaseous form in the atmosphere (via intermediate stages, represented here by . . .) to form sulfuric acid aerosol (NRC 1983):

$$SO_2 + OH \rightarrow \ldots \rightarrow H_2SO_4 \quad \text{(in gas phase)}$$

Reactions of SO_2 with CH_3O_2 (Rodhe and Isaksen 1980) or with HO_2 (Galbally et al. 1982) apparently also occur but are less important. In the second process, SO_2 dissolves in cloud droplets (or liquid aerosol particles) and is rapidly converted to sulfuric acid by reaction with H_2O_2 (Rodhe et al. 1981):

$$SO_2 + H_2O_2 \rightarrow \ldots \rightarrow H_2SO_4 \quad \text{(in liquid cloud droplets)}$$

Sulfur dioxide in cloud droplets also may react with O_3 to form H_2SO_4 (Charlson et al. 1992). The mean oxidation time of SO_2 has been estimated as 7 days by Tanaka and Turekian (1991) in the summer at mid-latitudes. Lelieveld (1993) estimates an average residence time of SO_2 of 2 days via aqueous-phase conversion (85% of total oxidation) and a minimum of 8 days for gas-phase conversion.

Sulfuric acid (H_2SO_4) is highly soluble in water and is a strong acid that completely dissociates to hydrogen and sulfate ions:

$$H_2SO_4 \rightarrow 2H^+ + SO_4^{2-}$$

In this way H_2SO_4 contributes to the formation of acid rain (see section on acid rain). Oxidation of SO_2 does not always result simply in the production of H_2SO_4. When ammonia is also present in polluted air, it reacts with sulfuric acid droplets to form ammonium sulfate particles, which raise the pH by consuming H^+ ions:

$$2NH_3 + 2H^+ + SO_4^{2-} \rightarrow (NH_4)_2SO_4$$

$$NH_3 + 2H^+ + SO_4^{2-} \rightarrow (NH_4)HSO_4$$

Other substances such as $CaCO_3$ can also react to neutralize sulfuric acid. (For further discussion of ammonium sulfate and acid neutralization, consult the sections on nitrogen in rain and acid rain, respectively.)

Sulfur dioxide can also be removed by direct scavenging in precipitation (in-cloud and raindrop scavenging). It has been estimated that 30% of SO_2 may be removed by this process (Tanaka and Turekian 1991).

Biogenic Reduced Sulfur

An important natural source of sulfate, particularly over the oceans, is from the oxidation in the atmosphere of biologically produced reduced-sulfur gases: dimethyl sulfide, $(CH_3)_2S$, usually abbreviated as DMS; hydrogen sulfide, H_2S; and other organic sulfur compounds (COS and CS_2). Oxidation probably proceeds via a large number of intermediate steps to the formation of sulfur dioxide gas (Rodhe and Isaksen 1980; Galbally et al. 1982), for example,

$$H_2S + OH \rightarrow \ldots \rightarrow SO_2 \text{ (4 days)}$$

$$(CH_3)_2S + OH \rightarrow \ldots \rightarrow SO_2 \text{ (3 days)}$$

and the SO_2 is subsequently oxidized to H_2SO_4 as discussed above.

The amount of natural biogenic reduced sulfur (DMS, H_2S, and COS) added to the atmosphere over land is difficult to measure, extremely variable especially with temperature, and thus difficult to quantify. Recently, Bates et al. (1992) have made a global inventory of terrestial sulfur emissions, and they estimate a total release of 0.35 Tg S/yr. This is an order of magnitude less than many previous estimates. The sources of terrestrial biogenic sulfur include vegetation, soils and cropland, and wetlands. Vegetation, deciduous trees, and to a lesser extent conifers contribute two thirds of the terrestrial biogenic S emissions. Sulfur is an essential nutrient for plants and is contained within plant biomolecules such as proteins. Reduced-S emissions by trees are highly temperature dependent and have a seasonal cycle at higher latitudes. Soil emissions, which contribute another 30% of the total, are apparently not dependent on soil microorganisms but rather on S desorption from soil surfaces. Soil emissions are also temperature dependent. Wetlands, including marshy areas and tidelands, contribute less than 4% of the total terrestrial biogenic emissions. Coastal emissions are highly variable with the tidal cycle. Bates et al. (1992) modelled these biogenic sulfur emissions on the basis of temperature and latitudinal distribution of land types and found that the tropics (20°N to 20°S) generate 60% of all coastal emissions but show little seasonal variation. Summertime (due to the temperature dependence) accounts for 60% of the annual coastal flux and is particularly important in the Northern Hemisphere.

In areas such as the United States, the natural background biogenic sulfur production is swamped by anthropogenic SO_2 production as a contributor to sulfur emissions; biogenic sulfur is estimated to contribute only 0.13% of U.S. sulfur emissions and is at a maximum in summer in certain areas such as Vermont, where it reaches 7% of total emissons (Guenther et al. 1989). Biogenic sulfur, by contrast, is relatively more important in more remote areas unaffected by pollution, particularly in the tropics.

No direct measurements have been made of reduced-sulfur release from the open ocean; it is calculated from the measured concentration of atmospheric sulfur gases dissolved in seawater and in the atmosphere overlying the ocean. Marine phytoplankton (floating organisms, primarily algae), produce DSMP (dimethylsulfonium propionate), which regulates osmosis and may be broken down by enzymes to produce DMS. DMS is released to the marine atmosphere, where it is oxidized to SO_2 and sulfate aerosol. (This aerosol may have important climatic significance because of its ability to serve as cloud condensation nuclei—see Chapter 2). The resulting concentrations of DMS gas and sulfate aerosol are higher in areas with a greater flux of solar radiation at the ocean surface and in zones of high surface biological productivity such as coastal upwelling areas. Production of DMS is highly dependent on plankton species, with some species being efficient emitters and others producing very little. Open-ocean DMS fluxes away from areas of high productivity are highest in summer months at higher latitudes (up to 50°) and lowest in winter.

Bates et al. (1987) and Bates et al. (1992) estimate a marine DMS emission to the atmosphere of about 16 ± 10.5 Tg S/yr with an additional 1.5 Tg S from H_2S emission (10% of DMS flux; Saltzman and Cooper 1988), for a total marine biogenic production of 17 Tg S/yr. Other recent estimates range from 6 to 15 Tg S/yr from marine DMS. These estimates are considerably lower than earlier estimates such as 39 Tg S/yr (Andrae and Raemdonck 1983).

Other Sulfur Sources: Biomass Burning, Volcanism, and Soil Dust

Biomass or forest burning occurs extensively in tropical areas. Wood is burned as fuel, burning is used to clear the land, and wildfires also consume a small amount of biomass. Biomass burning is predominantly anthropogenic (96%; Logan et al. 1981), with 80% of the emissions occurring in the tropical dry season (winter) (Bates et al. 1992). An estimated 1–4 Tg SO_4-S/yr is produced by the burning of sulfur-containing biomass (Crutzen and Andrae 1990). Bates et al. (1992) similarly estimate 1–3 Tg S/yr. Although a much smaller anthropogenic source overall than fossil fuel combustion, biomass burning is an important S source in remote continental areas such as the Amazon Basin and West Africa, where fossil fuel combustion is small, as are natural S input sources. In fact, savanna fires in West Africa are an important sulfate aerosol source for the tropical Atlantic Ocean (Savoie et al. 1989).

Sulfur in the atmosphere can also be produced by volcanic activity. Volcanic emissions are mainly SO_2 (with less than 1% sulfate or H_2S). Stoiber et al. (1987) estimate that 9.3 Tg SO_2-S/yr are produced by volcanoes (see Table 3.9). Of this, 4.9 Tg S are from nonerupting degassing volcanoes, and 4.5 Tg S are from erupting, explosive volcanoes. Based on satellite measurements, Bluth et al. (1993) estimate emissions from explosive volcanoes as 2 Tg S/yr, giving total volcanic emissions of 7 Tg S/yr. Berresheim and Jaeschke (1983) similarly estimate 7.6 Tg SO_2-S/yr from volcanoes. Although most volcanic sulfur goes into the troposphere, highly explosive volcanoes (7%) have eruption clouds that reach the stratophere. Large volcanic eruptions increase the number of sulfate particles in the so-called Junge layer at the bottom of the stratosphere for several years and cause cooling of the earth of around 0.5°C. Volcanoes are generally related to tectonic plate boundaries, and two thirds are in the Northern Hemisphere (Bates et al. 1992).

Windblown dust can contribute sulfate to precipitation as $CaSO_4$ (gypsum). The top layers of soil in arid areas may contain precipitated $CaSO_4$ from the evaporation of soil water and, because of the dryness, the sulfate-containing dust is easily picked up by the wind. Most of the soil particles are probably large enough to have a short residence time in the atmosphere. However, on a regional or local scale in arid regions, dust may be an important source of sulfate in rain. Lodge et al. (1968), as well as Junge (1963) and Munger (1982), suggest that the high concentration of sulfate in rain in the arid areas of the western United States may be due to $CaSO_4$ dust. Russian sulfate measurements of rain in semiarid and arid areas are very high, presumably also due to dust.

An alternative explanation of $CaSO_4$-derived sulfate in rain is from the reaction of wind blown $CaCO_3$ with H_2SO_4 in droplets. Savoie et al. (1989) found that non-sea-salt sulfate associated with Saharan dust transported to the Barbados on the trade winds was anthropogenic in origin and that H_2SO_4 had reacted with $CaCO_3$ dust (in the presence of hygroscopic sea-salt aerosols) to give $CaSO_4 \cdot 2H_2O$ (gypsum). The original Saharan dust did not contain gypsum or other forms of $CaSO_4$. It has also been suggested that $CaSO_4$ in remote marine atmospheres may originate from the reaction of $CaCO_3$ (coccoliths) with atmospheric SO_2 (Andrae et al. 1986). Because of the unimportance of rain in arid areas and because many $CaSO_4$-containing rains may be derived from the reaction of H_2SO_4 with $CaCO_3$, we assume that $CaSO_4$ dust globally is not an important source of sulfate in rain.

Sulfur Deposition on Land

In order to construct an atmospheric sulfur budget, it is necessary to know how much sulfate is delivered to the earth's surface in rain. Sulfate concentrations in rain vary both with time and with location (see Table 3.1). (There are also analytical problems in accurately measuring sulfate at the low concentrations often found in rain.) The usual range in excess sulfate over land is from less than 1 to 10 mg SO_4/l, with higher values being due to pollution. The unpolluted remote-land background concentration of excess sulfate in rain has been estimated as 0.5–0.6 mg/l SO_4 (Kramer 1978; Granat et al. 1976), and about 0.5 mg/l SO_4 has been measured in Amazon Basin rain (Stallard and Edmond 1981). These values are considerably less than most North American or European values, which reflect varying degrees of pollution.

Estimates of the total amount of excess sulfate deposited on the continents in rain range from 43 to 84 Tg S/yr (Robinson and Robbins 1975; Kellogg et al. 1972; Friend 1973; Granat et al. 1976; Ryaboshapko 1983; Warneck 1988). Differences are due to different estimates of the average excess sulfate concentration in rain. Warneck's estimate of 55–60 Tg S/yr is based on a summation within latitude bands and includes some aerosol sulfate that is trapped with precipitation (up to 20% aerosol sulfate in polluted areas). We have independently calculated excess sulfate deposition on land to be 53 Tg S/yr based on the average concentrations of excess sulfate in rain and average amount of precipitation on different types of land area (see Table 3.8). Adding 4 Tg S/yr as due to cyclic sea salt to the excess sulfate flux gives 57 Tg S/yr for total land deposition in precipitation.

Besides sulfur deposition in precipitation ("wet" deposition), there is also "dry" deposition of sulfur. Sulfur dioxide gas may be absorbed or dissolved directly on land by standing water, vegetation, soil, and other surfaces. Also, sulfate particles may settle out of the air or be trapped by vegetation (dry fallout). The size of these fluxes in the sulfur cycle are even less well known than those involving precipitation because they are harder to measure. Dry deposition of SO_2 gas over land is dependent on the characteristics of the surface, small-scale meteorological effects, and the atmospheric concentration of SO_2 (NRC 1983). Dry deposition is greater near the emission sources of the gas and decreases more rapidly with distance from the source than does wet deposition. Annual dry deposition of SO_2 was measured directly in New Haven, Connecticut, using radioisotopes as 25% of total deposition (Tanaka and Turekian 1995). If this is taken as representative of polluted areas, dry deposition in these areas equals 8 Tg S/yr. [Another recent estimate of dry deposition in polluted areas is 37% of total sulfur deposition at forested Hubbard Brook, New Hampshire (Likens et al. 1990).] Because dry deposition of SO_2 decreases away from gas sources, we assume that there is little dry depositon in remote areas.

Marine Sulfur Deposition

Estimates of excess sulfate deposition in rain over the oceans also vary a great deal because the concentration of excess sulfate in ocean precipitation is even less well known than that over land.

TABLE 3.8 Rate of Excess Sulfate Deposition on Land by Precipitation (Predominantly Rainfall)[a]

Continent	Mean annual Precipitation[b] (cm)	Nondesert Area[c] (10^{12} m^2)	Excess SO_4^{--} (mg/1)	Deposition Rate (Tg SO_4– S/yr)
Africa	69	21.8	1.0	5.0
Asia:	73			
Clean		24.8	1.5	9.1
Polluted		9.0	4.0	8.8
Europe: Polluted	73	10.0	4.0	9.5
North America:				
Clean	67	18.0	1.5	6.0
Polluted[d]	100	5.2	3.0	5.2
South America	165	17.1	0.9	8.5
Antarctica	17	12.2	0.25	0.2
Australia	44	4.9	0.8	0.5
Total world				52.8

[a] Excess sulfate defined as total sulfate minus sea-salt sulfate. Tg = 10^{12}g.

[b] From Lvovitch 1973.

[c] From Meybeck 1979.

[d] North America polluted area = eastern United States and Canada (Galloway and Whelpdale 1980).

(Values range in Table 3.1 from 0.07 to 1.5 mg/l.) Junge (1963), using 0.5-mg/l excess SO_4 in rain, calculated excess sulfur deposition in rain over the ocean as 57 Tg S/yr, while Granat et al.'s (1976) estimate of 23 Tg S/yr corresponds to 0.2 mg/l. By comparison, Galloway (1985) has estimated an excess sulfur deposition over the oceans of 40 Tg S/yr (0.11 g S/m^2/yr = 0.33 g SO_4/m^2/yr).

Calculating the amount of excess sulfate deposition in rain over the oceans based on excess sulfate in marine rain does not seem very practical for several reasons: (1) A small change in excess rain sulfate concentrations results in a large change in sulfate deposition because of the large area involved; (2) there are very few observations of marine rain sulfate (see Table 3.1), and they do not give very reliable excess sulfate estimates. As a result, we shall try here to make a rough estimate of marine rain excess sulfate deposition based on balancing the sulfur cycle. (See the section on the atmospheric sulfur cycle.)

Concerning dry deposition of sulfur over the oceans, seawater, with a pH of 8, is sufficiently basic to be a good absorber of SO_2. Several authors (Friend 1973; Varhelyi and Gravenhorst 1981, quoted in Church et al. 1982) consider dry deposition of SO_2 to constitute 20% of oceanic sulfur deposition (with the remainder in wet deposition). Mészáros (1982) calculated dry deposition of SO_2-S as 8 Tg S/yr based on SO_2 concentrations and deposition velocities over the oceans. Liss and Galloway (1993) give examples of anthropogenic sulfur deposition over the oceans as 30% dry and 70% wet, and sulfur deposition in remote ocean areas as 20% dry and 80% wet. This overall gives 9 Tg of dry deposition (25%) and 26 Tg of wet deposition (75%) when set to balance net excess (non-sea-salt) sulfur input to the marine atmosphere (35 Tg S). This is the value used in our budget (Figure 3.11 and Table 3.9). Dry deposition of SO_2 is several times more important than dry deposition of sulfate. Sea-salt sulfate aerosols are coarse and do undergo dry deposition, but this flux is included in the return cyclic seasalt flux (wet plus dry) of 40 Tg S/yr to the oceans that was discussed above.

Flux of Sulfur Between Continental and Marine Air Masses

In constructing an overall atmospheric sulfur cycle we have not considered the transport of sulfur between continental and marine air masses. The transfer of sulfur from the continental air mass to the oceanic air mass is calculated here to be about 22 Tg S/yr. This number is not well known. It has been estimated that one third of U.S. and Canadian SO_2-S production is exported away from the continents (NRC 1983). Granat et al. (1976) estimated that, globally, one fourth of anthropogenic emissions of SO_2-S are transported from land to sea; this would be 16 Tg S. In addition, if it is assumed that two thirds of volcanic S (or 6 Tg S/yr) is transported over the oceans since volcanoes are commonly located at plate margins (Stoiber et al. 1987), total transport of sulfur to the oceans then would equal 22 Tg S/yr.

We have also assumed a reverse transport to land of marine sulfur from biogenic sources of 4 Tg S/yr, which is equal to, and in addition to, the sea-salt sulfur flux. This is based on measured ratios of SO_4/Cl in unpolluted marine rain and aerosols that are approximately twice the sea-salt ratio (see Table 3.5; Bonsang et al. 1980; Stallard and Edmond 1981; Galloway et al. 1982).

Atmospheric Sulfur Cycle

After all these transport considerations, in order to balance inputs to and outputs from the atmosphere, we need a depositional flux over the oceans of non-sea-salt (excess) sulfur of 18 (22 – 4)

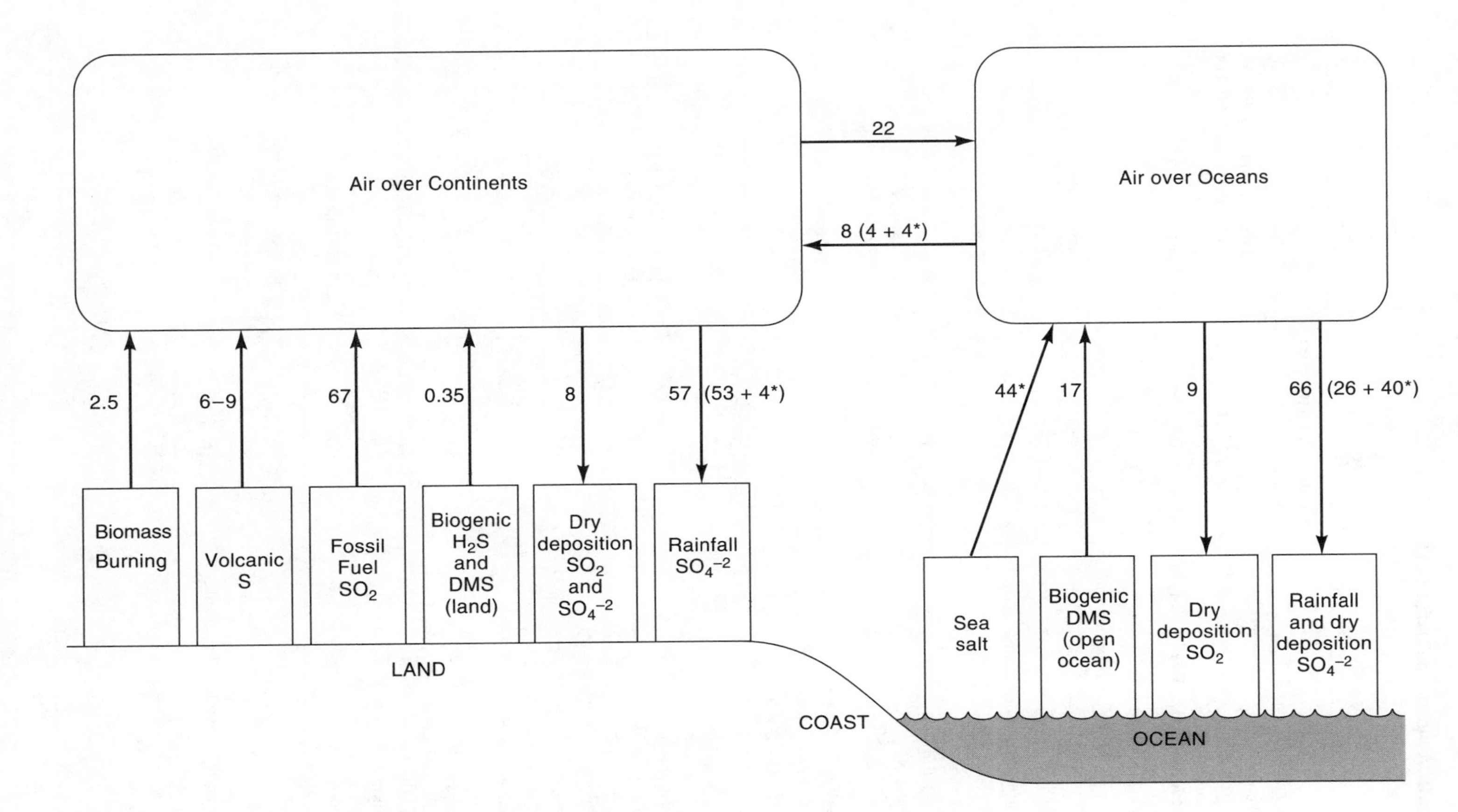

Figure 3.11. The atmospheric sulfur cycle. Values are fluxes in Tg S/yr. Values denoted by an asterisk refer to sea salt.

TABLE 3.9 Atmospheric Sulfur Cycle[a]

	Flux (Tg S/yr)	Source
	Continental Atmosphere	
A. *Input from land*		
Fossil fuel combustion	67	Hameed & Dignon 1992 (for 1986)
Biomass burning	2.5 ± 1.5	Crutzen & Andrae 1990; Bates et al. 1992
Volcanic emissions	6–9	Bluth et al. 1993; Stoiber et al. 1987
Biogenic reduced S (H_2S, DMS) from land	0.35	Bates et al. 1992
Total input from land	76–79	
B. *Input via transport from marine atmosphere*		
Sea salt	4	Eriksson 1960
Marine biogenic S	4	(See text)
Total transport from ocean atmosphere	8	
C. *Deposition over land*		
Rainfall		
As excess sulfur	53	(See text)
As sea salt	4	Eriksson 1960
Dry deposition as SO_2	8	(See text)
Total deposition over land	65	
D. *Transport to marine atmosphere*	22	(See text)
Balance: A + B – C – D = 0		
	Ocean Atmosphere	
A. *Input from ocean*		
Marine biogenic sulfur emissions (mainly DMS)	17	Bates et al. 1992
Sea spray (particulate)	44	Eriksson 1960
Total input from oceans	61	
B. *Input via transport from continental atmosphere*	22	(See text)
C. *Deposition over ocean*		
Rainfall		
As excess sulfate-sulfur	26	(See text)
As sea salt (includes dry deposition of sea salt)	40	Eriksson 1960
Dry deposition as SO_2	9	(See text)
Total deposition over oceans	75	(To balance)
D. *Transport to marine atmosphere*		
Sea salt	4	Eriksson 1960
Marine biogenic S	4	(See text)
Total transport to continental atmosphere	8	
Balance: A + B – C – D = 0		

[a] See also Figure 3.11

Tg S to balance the S input to the marine atmosphere from the continents plus 17 Tg S/yr to balance the input from marine biogenic production. The total of 35 Tg S/yr S/yr is apportioned 9 Tg SO_2-S to dry deposition of SO_2 and 26 Tg S to wet deposition (see previous discussion). Marine rainfall deposition of 26 Tg of excess sulfur per year would amount to 0.22 mg/l of excess sulfate in rain (based on an ocean area of 3.6×10^{14} m^2 and 100 cm of rain per year). This falls within the range of other previously mentioned estimates (0.2–0.5 mg/l SO_4^{2-}). Redeposition of sulfate on the ocean as sea salt in rain (and dry deposition) is about 40 Tg S/yr (Eriksson 1960). Thus, total oceanic sulfur deposition in rain and dry deposition is about 75 Tg S/yr (see Table 3.9).

Our global atmospheric sulfur budget is summarized in Figure 3.11 and Table 3.9. In summary, for the world as a whole (land plus sea), the inputs to the atmosphere for non-sea-salt emissions are 26 Tg S/yr (27%) for natural emissions (primarily marine biogenic reduced sulfur and volcanic sulfur), and 69.5 Tg S/yr (73%) for pollutive sulfur (including biomass burning, which is mainly anthropogenic). This totals to 95.5 Tg S/yr for excess sulfur input to the atmosphere. If sea salt is included (44 Tg S/yr) in the sulfur totals, the result is 31% from seasalt, 19% from non-sea-salt natural emissions, and 50% from anthropogenic emissions.

Ignoring sea salt, our estimate for natural production of 27% is lower than the 35–65% of non-sea-salt natural emissions estimated in some earlier atmospheric sulfur budgets (Table 3.10). This is due primarily to better information about natural fluxes that is now available. For example, in our present budget the biogenic reduced sulfur flux from the oceans is lowered to about half that from sea spray. Our new totals are very similar to the estimates used by Langner and Rodhe (1991) and by Bates et al. (1992).

The pollutive contribution to the total atmospheric sulfur input is large, about 73% (88% over the land alone) and illustrates, as is the case for nitrogen, the importance of human activities as they affect the composition of major reservoirs and fluxes at the earth's surface, in this case the entire atmosphere. For sulfur, atmospheric pollution is a global as well as a local problem. As we shall see, pollution also plays a major role as it affects sulfur in continental waters and in

TABLE 3.10 Annual Global Emissions of Gaseous Sulfur to the Atmosphere by Natural and Anthropogenic Processes (in Tg S/yr)

Natural Emissions[a]	Anthropogenic Emissions	Natural as a Percent of Total	Reference
92	50	65	Kellogg et al. 1972
108	65	62	Friend 1973
98	64	60	Robinson & Robbins 1975
35	65	35	Granat et al. 1976
50–100	69	42–59	Möller 1984
100	100	50	Andrae & Galbally 1984
25.5	72.5	26	Langner & Rodhe 1991
25	79	24	Bates et al. 1992
26	69.5[b]	27	Table 3.9

[a]Includes biological decay and volcanism, but does not include sea salt.
[b]Includes anthropogenic biomass burning of 2.5 Tg S/yr.

the oceans. However, the distribution of anthropogenic sulfur globally is very uneven (see Figure 3.12). In the Northern Hemisphere, anthropogenic emisssions are about 90% of total emissions, while in the Southern Hemisphere they reach a maximum of 70% and more often are 40% or less. This reflects partly the lower anthropogenic production rate in the Southern Hemisphere and partly the greater ocean area (with its marine biogenic source) relative to land area there.

NITROGEN IN RAIN: THE ATMOSPHERIC NITROGEN CYCLE

Nitrogen is both an essential nutrient for plants and animals (and thus is strongly involved in biogeochemical cycles) and a major pollutant. It occurs in gaseous form in the atmosphere, where the most abundant nitrogen gases are elemental nitrogen (N_2), nitrous oxide (N_2O), nitrogen dioxide (NO_2), nitric oxide (NO), and ammonia (NH_3). Combinations of the gases NO and NO_2 are referred to collectively as NO_x. Because nitrogen occurs in so many different forms in the atmosphere, and has so many sources, we shall be concerned in this section not only with rainwater composition, but also with the rates by which nitrogen is transported to and from the atmosphere

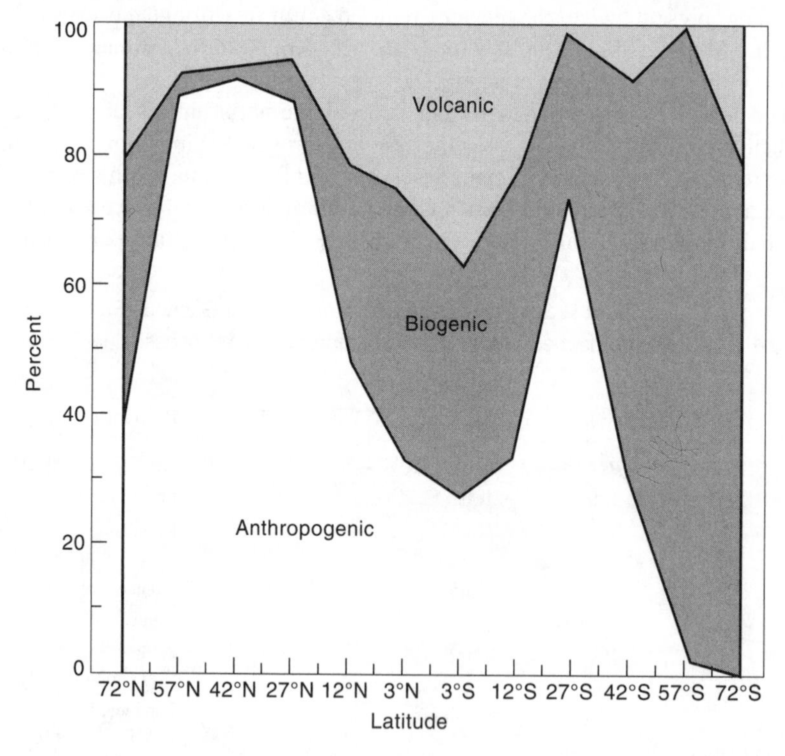

Figure 3.12. Sulfur emissions by latitude and type. The percentage of sulfur emissions from volcanic, biogenic and anthropogenic (fossil fuel combustion + biomass burning) sources in each latitude zone. Latitude zones are 15° wide starting with 80°N–65°N, represented on the graph by its approximate midpoint 72°N, and 65°N–50°N, represented by its midpoint 57°N. The exception is the two zones around the Equator, which are 5°N–0°N and 0°S–5°S. (Data from a model by Bates et al. 1992, Table 6, p. 329.)

and converted from one form to another. In other words, we shall be concerned with the *atmospheric nitrogen cycle,* particularly the conversion of nitrogen to and from gases that react with rainwater, namely, NO, NO_2, and NH_3.

N_2, Nitrogen Fixation, and Total Nitrogen Fluxes

Elemental nitrogen (N_2) is by far the most abundant atmospheric nitrogen gas, making up nearly 80% of the atmosphere by volume, but nitrogen in the form of N_2 is very unreactive because of the strong bonds between the nitrogen atoms. The conversion of N_2 into chemically reactive and biologically available compounds by the combination of nitrogen with hydrogen, carbon, and/or oxygen is called *nitrogen fixation.* This is a very important process because nitrogen is an essential nutrient for life.

The major nitrogen-fixation processes are listed and evaluated quantitatively in Table 3.11. In biological fixation, N_2 is combined with H, C, and O to form proteins and other essential organic compunds by marine organisms, particularly blue-green algae (cyanobacteria), and on land by plants such as legumes and lichens. This is the dominant natural nitrogen-fixation process. However, some 30% of nitrogen fixation by plants results from human cultivation of legumes, such as peas and beans, which contain root microorganisms that are excellent nitrogen fixers. Minor natural fixation of atmospheric nitrogen (to form NO_x and NO_3^- in rain) also occurs in the heat generated by lightning bolts. Humans fix nitrogen in internal combustion engines and power plants by heating N_2 and O_2 to high temperatures and also by the production of nitrogen fertilizers. A final source of N fixation is forest fires (biomass burning), which can occur naturally but are predominantly set by humans.

Since humans cause more than 50% of total nitrogen fixation, one might expect to see changes in the atmospheric nitrogen cycle. However, only around one quarter of anthropogenic fixation is due to combustion that involves direct release to the atmosphere of fixed nitrogen. The rest of anthropogenic nitrogen fixation involves the biogeochemical cycles of plants and animals, and the effect of this fixation on atmospheric nitrogen is more complex and harder to quantify.

TABLE 3.11 Summary of Major Nitrogen Fluxes Involving N_2 (in Tg N/yr; Tg = 10^{12}g)

Process	Flux	Anthropogenic Flux	Reference
Nitrogen fixation			
Land (biological)	139	44	Burns and Hardy 1975
Ocean (biological)	25		Codispoti & Christensen 1985
Atmosphere (lightning)	3		Borucki & Chameides 1984
Fertilizers and industrial	85	85	FAO 1989
Fossil fuel combustion			
NO_3–N (1986)	24	24	Hameed & Dignon 1992
NH_4–N	0.03	0.03	See Table 3.16
Biomass burning (NH_3–N + NO_x–N)	5	5	Crutzen & Andrae 1990
Total	281	158	
Denitrification			
Land	>150		Schlesinger & Hartley 1992
Ocean	141		Christensen et al. 1987

Atmospheric N_2, once fixed, is released back to the atmosphere via the process of *denitrification,* or bacterial nitrate reduction, which is necessary to maintain its atmospheric concentration. Estimated denitrification fluxes are also shown in Table 3.11. These values, as pointed out in Chapter 5, are not well established, and better values are needed before one can say whether denitrification does or does not balance fixation worldwide.

The forms of nitrogen in rain that are derived from fixed nitrogen gases include nitrate (NO_3^-) from NO and NO_2, and ammonium (NH_4^+,) from NH_3. Together NO_3^- and NH_4^+ are delivered to the surface of the land in an annual rain flux of about 40 Tg N/yr, with an additional 21 Tg N/yr from dry deposition (see Table 3.12). If we compare the rain and dry-deposition N flux to land with other terrestrial N sources (Table 3.11), we see that rain and dry deposition (61 Tg N/yr) are a smaller N source than fertilizers and industry (85 Tg N/yr), and a much smaller source than biological N fixation on land (139 Tg N/yr).

Nitrate in Rain: Natural Sources

Nitrate (NO_3^-) in rain results both from natural processes and from human activities. Most nitrate in rain comes from nitrogen oxide gases (NO_x), directly from nitrogen dioxide (NO_2) and indirectly from nitric oxide (NO). Nitrate is an important rain component. It is a major contributor to acid rain, and it is a limiting nutrient in the oceans so that increases in its concentration can stimulate ocean productivity (Fanning 1989).

Nitric oxide is formed by the reaction of nitrogen and oxygen in the air at high temperatures, in excess of 2000°C. This process occurs in lightning and during combustion and can be summarized as

$$N_2 + O_2 \rightarrow 2NO \qquad \text{(nitric oxide)}$$

Nitric oxide is quite reactive and combines easily in the atmosphere with ozone (O_3), or with peroxides (HO_2 or organic peroxides) to form nitrogen dioxide (NO_2):

$$NO + O_3 \rightarrow NO_2 + O_2$$

TABLE 3.12 Fluxes of Nitrogen Forms in Rain and Dry Deposition (in Tg N/yr)

	Flux		
Process	NO_3–N	NH_4–N	Total N
To land:			
Rain	17	23	40
Dry deposition	13	8	21
Total land			61
To the oceans:			
Rain	9	14	23
Dry deposition	4	2	6
Total oceans			29

Sources: See text for land; for ocean, Duce et al. 1991.

or

$$NO + HO_2 \rightarrow NO_2 + OH$$

The NO_2 in turn reacts with OH in the air (catalytically) to form nitric acid (HNO_3) (Logan 1983):

$$NO_2 + OH \rightarrow HNO_3$$

Once formed, nitric acid is removed in rain (or surface deposition). Nitric acid is a strong acid and completely dissociates in rain water to form NO_3^- and H^+:

$$HNO_3 \rightarrow H^+ + NO_3^-$$

In this way nitric acid can lower the pH of rain. The conversion of NO_x to HNO_3 in the air is fairly rapid, being about 1 day according to Warneck (1988), and the HNO_3 is removed from the air in rain or surface deposition after about 5 days.

The sources of NO_x gas (and NO_3^- in rain) (see also Table 3.14) include four processes that are predominantly natural: (1) lightning; (2) photochemical oxidation in the stratosphere of N_2O gas to NO and NO_2; (3) chemical oxidation in the atmosphere of ammonia to NO_x; and (4) soil production of NO by microbial processes. (The other two major sources, which are largely anthropogenic and are discussed below, are fossil fuel combustion and biomass burning.)

Atmospheric lightning can heat up the air enough to cause nitrogen and oxygen to combine to form nitric oxide (NO) and ultimately NO_3^- in rain by the reactions given above. Borucki and Chameides (1984) estimate that the global production rate of NO-N by lightning is about 3 Tg N/yr, which is the value used in our summary budget (Table 3.14). Borucki and Chameides state a range of 1–8 Tg N/yr. Other estimates are similar: 3–8 Tg N/yr (Chameides et al. 1987; Logan 1983; Dawson 1980; Hill et al. 1980). Penner et al. (1991), in modelling natural sources of NO_x, found that lightning is an insignificant source at ground level in continental areas because most NO_x formation by lightning occurs at higher atmospheric levels.

Before considering the other three natural sources of NO_x listed above, it is helpful to review briefly the major features of the biological nitrogen cycle (see Figure 3.13) as it affects atmospheric nitrogen chemistry. (For further details on the nitrogen cycle, see Chapters 5 and 8.) Organic nitrogen formed by N_2 fixation and/or photosynthetic uptake of NO_3^-, NO_2^-, or NH_4^+ upon the organism's death undergoes bacterial decomposition in soils and sediments resulting in the liberation of ammonia to solution. The ammonia may escape to the atmosphere or may remain in solution as NH_4^+, where it can be oxidized to dissolved NO_2^- and NO_3^- (nitrification). The produced nitrate may, in turn, be taken up by plants or undergo denitrification (nitrate reduction) by soil bacteria to N_2 and N_2O gases, which also escape to the atmosphere. This is the usual path in the nitrogen cycle. However, sometimes the soil conversion of NH_4^+ to NO_3^- (nitrification) by microorganisms is incomplete and part of the NH_4^+ is instead converted to NO (and N_2O) gas. In sum, as a result of all these soil processes, the gases N_2, NH_3, NO, and N_2O are all released to the atmosphere.

Nitrous oxide gas (N_2O) released to the atmosphere by denitrification, or during ammonia oxidation, is relatively inert chemically. However, it is oxidized to NO in the stratosphere by exposure to the sun's radiation. NO reacts with ozone (O_3) to form NO_2, which can then find its way back into the troposphere and be converted to nitrate in rain. However, the proportion of

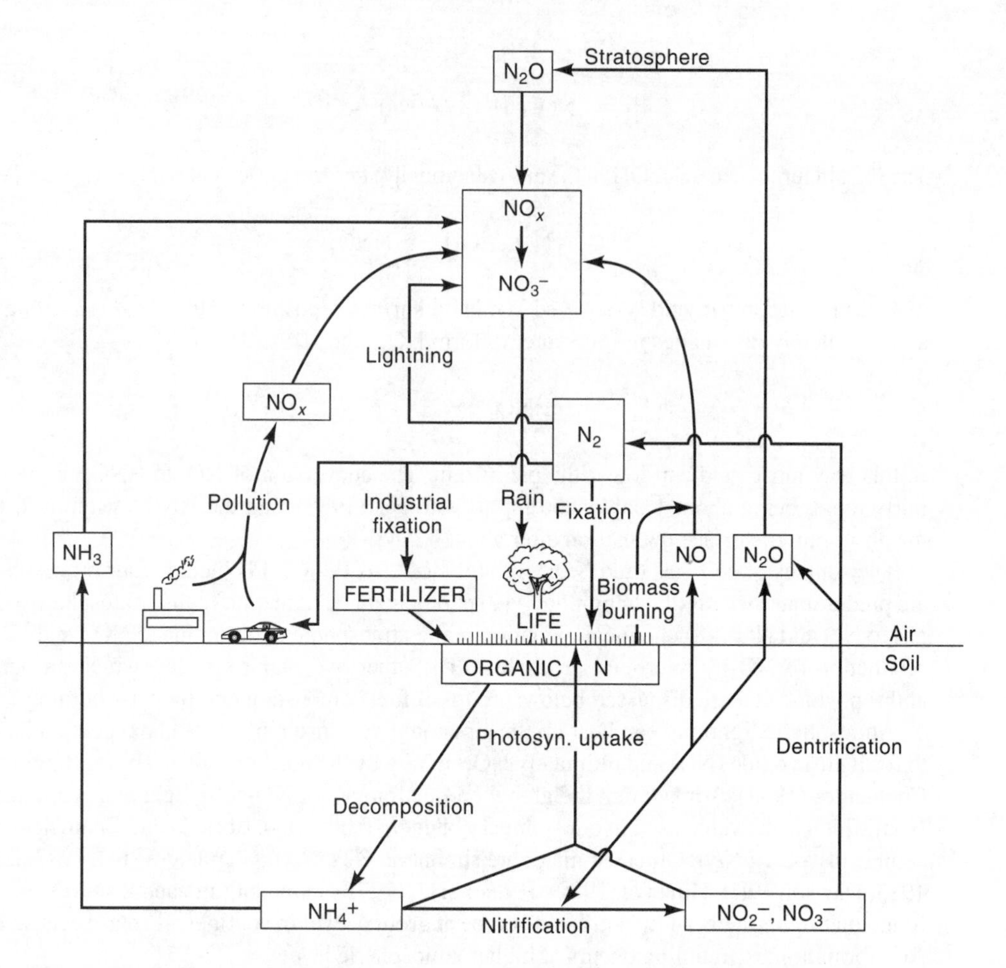

Figure 3.13. The atmosphere-soil nitrogen cycle, emphasizing sources of NO_x and NO_3^- in rain.

NO_x originating from the breakdown of N_2O is only about 1% (0.5 Tg N/yr; Logan 1983), and thus N_2O is not a major contributor to rain nitrate. The importance of these reactions is not NO_2 or nitrate formation, but the destruction of stratospheric ozone during the process of converting the NO (produced from N_2O) to NO_2. Stratospheric ozone blocks harmful solar ultraviolet radiation from reaching the earth's surface and its destruction results in a greater flux of ultraviolet radiation to the surface (see Chapter 2).

As mentioned above, NH_3 also escapes to the atmosphere during the soil N cycle. The oxidation of atmospheric NH_3 has been suggested as an NO_x source. Warneck (1988) estimates that 1 Tg NH_3-N is oxidized in the atmosphere.

Another natural NO_x source is direct soil loss of NO to the atmosphere, where it is oxidized to NO_2. One mechanism for the formation and release of NO gas is as a by-product (along with N_2O) of the bacterial oxidation (nitrification) of NH_4^+ in soils (see Figure 3.13), which has been observed in the laboratory (Lipschultz et al. 1981; Levine et al. 1984). Measurements have been made of the flux of NO-N from a number of temperate ecosystems (Galbally and Roy 1978; Johansson and Granat 1984; Slemr and Seiler 1984; Anderson and Levine 1987; Williams et al. 1987; Kaplan et al. 1988). Soil losses of NO are higher at higher temperatures (Williams et al. 1987), in wet soil, or as a result of prior burning (Levine et al. 1988). Losses are reduced by a plant cover (Slemr and Seiler 1984). Measured fluxes from tropical rain forests (Kaplan et al. 1988) and savannas (Johansson et al. 1988) are larger than those from temperate areas.

We have used a total global flux of 10 Tg NO-N/yr from biogenic soil emissions. This amounts to some 24% of total NO_x-N production and is a major natural NO_x source. Most estimates of the biogenic soil flux are close to this value (Galbally and Roy 1978; Johansson and Granat 1984; Slemr and Seiler 1984; Logan 1983; Levine et al. 1984). In addition, this flux was used in a model by Penner et al. (1991) of NO_x emissions from natural and fossil fuel sources, and gave reasonable results. Some part of soil production is probably indirectly anthropogenic due to increased fertilizer use; Anderson and Levine (1987) estimate 0.8% of fertilizer N (79 Tg N) is lost as NO; this translates to a flux of 0.6 Tg N/yr.

Release of NO_x to the atmosphere may be important as a local source of nitrate in rain. Junge and Werby (1958) found a maximum of rain nitrate in warm weather in unpolluted north-central U.S. areas, which they felt pointed to a soil source there (see Figure 3.14a).

Overall, the natural processes discussed above, predominantly soil production of NO_x and lightning, account for 15 Tg N/yr or about 35% of total NO_x-N production . However, as already noted, since part of lightning production is at higher tropospheric levels, soil production dominates at the surface level (Penner et al. 1991). From N deposition in remote areas, Lyons et al. (1990) estimate that natural sources of NO_3-N are about 19 Tg N/yr, a value reasonably close to that adopted here.

Nitrate in Rain: Anthropogenic Sources

Natural sources can account for only 35% of the input of NO_x to the atmosphere and, thus, to NO_3^- in rain. The other 65% comes from anthropogenic sources: fossil fuel combustion and biomass burning. Humans undoubtedly provide the largest source of nitrate in rain (56%) by the production of NO_x from fossil fuel combustion, mainly in automobile engines and power plants. The NO_x gases are formed by the high temperature reaction between atmospheric N_2 and O_2 (see above). Table 3.13 gives the relative importance of various combustion sources (and industrial sources) in the 1979 world production of nitrogen oxides. In 1985, vehicle transportation in the United States contributed 45% of fossil fuel NO_x (*World Resources 1988–89*), similar to the 40% contribution of transportation in the United States and Canada in 1979. Power and heating combustion sources generate most of the rest. (This can be compared to SO_2 from combustion, which is produced predominantly by power plants, with only 2% from vehicles.)

The major sources of fossil fuel combustion NO_x are in urban areas, and thus there tends to be a much greater concentration of NO_x in urban air than in rural areas (Logan 1983). This is reflected in U.S. rain NO_3^-, which, for example, exhibited an area of high concentrations over the heavily populated region of the Great Lakes both in 1955–1956 and in 1980 (Figures 3.14a and 3.14b). In addition, Smith et al. (1987) found highest U.S. levels of atmospheric NO_3^- deposition (0.5–0.6 g NO_3/m²/yr) in the Ohio River basin, the mid-Atlantic region, and the Great

TABLE 3.13 Production of Nitrogen Oxides from Fossil Fuel Combustion and Industry in 1979 (in Tg NO_x – N/yr)[a]

Source	U.S. and Canada	Europe[b]	Rest of World	Total
Power and heating combustion:				
Coal	1.6	2.4	2.4	6.4
Oil	0.8	1.3	1.1	3.2
Gas	1.1	0.9	0.3	2.3
Subtotal power + heating	3.5	4.6	3.8	11.9
Transportation[c]	2.5	2.8	2.5	8.0
Industrial sources[d]	0.2			1.2
Total	6.2			21.1

Sources: After J. A. Logan, "Nitrogen oxides in the troposphere: global and regional budgets," *Journal of Geophysical Research,* 88(C15), 10792, ©1983 by the American Geophysical Union.

[a] NO_x is the sum of NO_2 and NO. Tg = 10^{12}g.

[b] Europe includes 80% of USSR fuel.

[c] Total includes 0.2 Tg from air traffic.

[d] Petroleum refining and manufacture of nitric acid and cement in the United States.

Lakes. This area correlates with high sulfate concentrations, supporting an anthropogenic source for both nitrate and sulfate.

High nitrate concentration due to excessive inputs of anthropogenic NO_x are spread out over large areas (see Figure 3.13b) because modern power plants have been located in more remote, less populated areas and have taller smokestacks. In this way nitrogen oxide pollutants, like SO_2, are dispersed over wider and less urban areas. In addition, appreciable dispersal is possible because the residence time of NO_x in the atmosphere before removal as HNO_3 is about 6 days (Warneck 1988).

The U.S. emissions of NO_x tripled from 1940 until 1973 (see Figure. 3.8), due to more highway vehicles and (since 1960) the use of gas-fired power plants (Gschwandtner et al. 1986). There was a decline in U.S. NO_x emissions in 1973 because of the oil embargo followed by the 1974–1975 recession. NO_x emissions then rose slowly until 1979, when they peaked, and declined slightly in the 1980s. This is probably due to the institution of emission control devices on new vehicles.

It appears that there have been considerable increases overall in the nitrate concentration of rain over the United States in the past 30 years, reflecting increased NO_x emissions. For example,

Figure 3.14. Dissolved nitrate (mg/l NO_3^-) in rain over the continental United States and its change with time. (a) Average concentrations, July–September 1955. (After C. E. Junge, "The Distribution of Ammonium and Nitrate in Rain Water over the United States," *Transactions of the American Geophysical Union* 39: 244, © 1958 by the American Geophysical Union.) (b) Average concentrations in 1980 for North America. Open circles indicate stations for which less than 20 weeks' data were available. (Data from National Atmospheric Deposition Program.) [After J. A. Logan, "Nitrogen Oxides in the Troposphere: Global and Regional Budgets," *Journal of Geophysical Research* 88 (C15): 10795, © 1983 by the American Geophysical Union.]

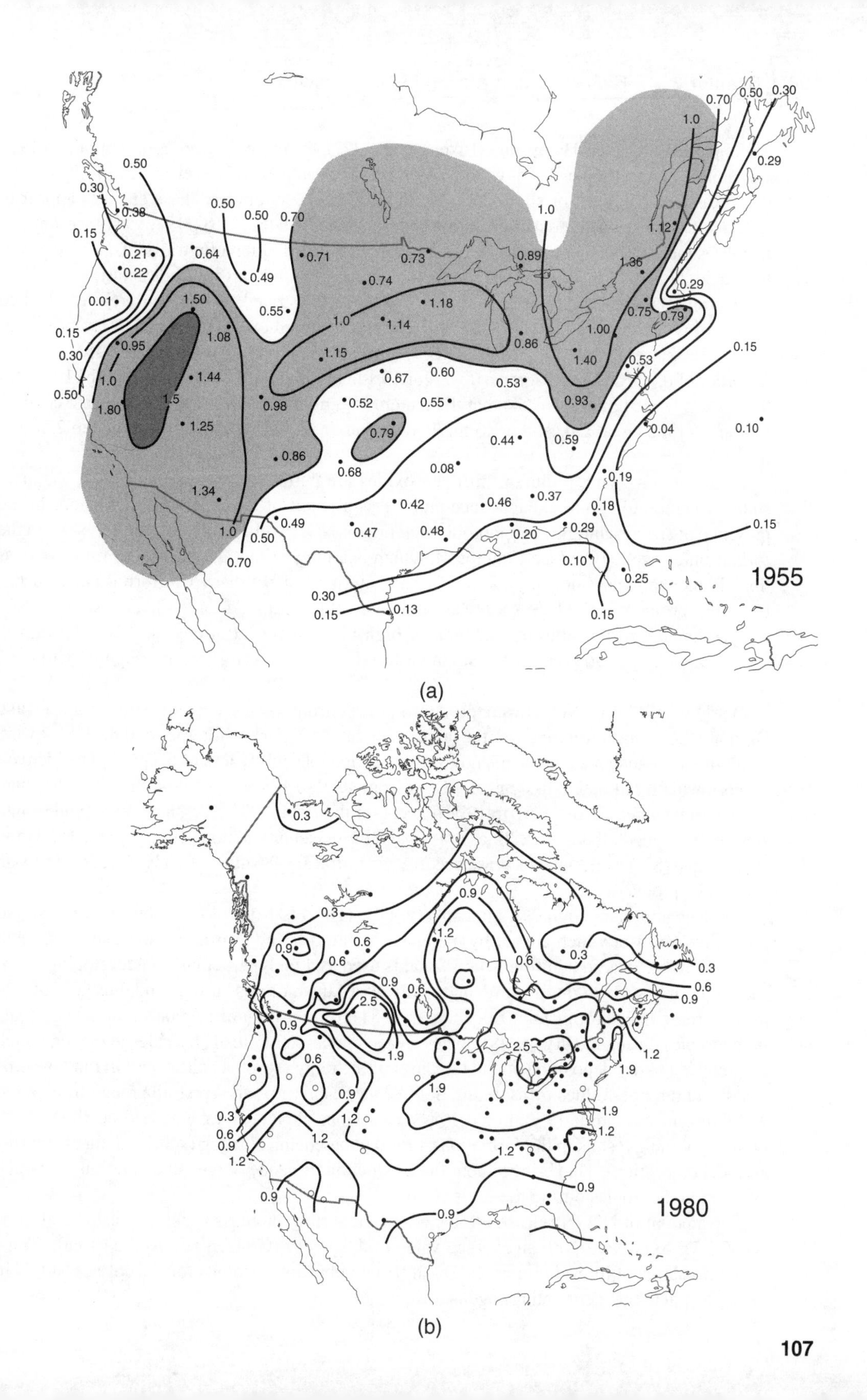
1955
0.50
0.30
0.38
0.15
0.21
0.22
0.01
0.15
0.30
0.95
1.0
0.50
1.80
1.50
1.08
1.44
1.5
1.25
1.34
1.0
0.64
0.50
0.50
0.70
0.49
0.71
0.55
0.98
0.86
0.50
0.70
0.49
0.74
1.0
1.18
1.14
1.15
0.67
0.60
0.52
0.55
0.79
0.68
0.08
0.42
0.47
0.48
0.73
0.89
1.0
0.86
0.53
0.44
0.46
0.37
0.20
0.29
0.10
0.13
0.30
0.15
1.12
1.36
1.00
1.40
0.93
0.59
0.75
0.79
0.29
0.53
0.04
0.19
0.18
0.25
0.15
0.15
0.10
0.15
1.0
0.70
0.50
0.30
0.29
(a)
1980
0.3
0.3
0.9
0.6
0.6
0.9
2.5
0.6
0.9
1.2
0.9
0.6
1.9
0.9
0.6
0.3
0.3
0.6
0.9
2.5
1.2
1.9
1.9
1.9
1.2
1.2
0.3
0.6
0.9
1.2
1.2
1.2
0.9
0.9
0.9
(b)

at Hubbard Brook, New Hampshire (Likens et al. 1977), the nitrate concentration of rain in 1970 was double that in 1964 or 1955, which were similar to each other. However, from 1970 to 1981 there was a levelling off or slight decrease in nitrate concentrations. These changes appear to reflect NO_x emission trends in the northeastern United States (NRC 1983; and see above). Increases in nitrate concentrations resulted in a greater contribution of nitric acid, relative to sulfuric acid, to acidity in rain (see section below on acid rain for further discussion).

Overall increases in nitrate in rain can also be demonstrated for the entire eastern United States from 1955 to 1980. In 1978–1979, Pack (1980) found that the average NO_3^- concentration for the eastern United States was 1.6–1.7 mg/l NO_3^- versus an average for 1955–1956 on the order of 0.7 mg/l NO_3^-. Comparing Figure 3.14a (1955–1956) with Figure 3.14b (1980) for the eastern United States, we see that the pattern is similar but that the area of highest concentration over the Great Lakes, the 1.0 mg/l NO_3 contour in 1955–1956, corresponds to 2.5 mg/l in 1980.

Reactions involving pollutant nitrogen oxides are particularly obvious in photochemical smog, where automobile exhausts accumulate as a result of a temperature inversion. In the presence of strong sunlight, the photodissociation of NO_2 by ultraviolet radiation results in the formation of ozone (O_3) (see Chapter 2), which is a lung irritant and also harmful to vegetation. Photochemical smog production is characteristic of southern California but also has become a problem in the U.S. southwest and many other areas. Automobile emissions of NO_x are presumably responsible for the area of high nitrate concentrations in California rain in 1955–1956 (Junge 1963) and in the southwestern United States in 1980 (see Figures 3.14a and 3.14b).

Worldwide, 24.3 Tg NO_x-N were produced in 1986 from fossil fuel combustion (Hameed and Dignon 1992). North America was the largest source (7.5 Tg/yr), while Asia and the USSR were next with more than 5 Tg/yr each, and Europe contibuted slightly less (4.5 Tg/yr). The increase in worldwide fossil fuel emissions of NO_x-N from 1970 to 1986 is shown in Figure 3.15. There was a steady increase from 1970 to 1978 (except for the 1974–1975 recession). Worldwide emissions then levelled off due to declines in U.S. and European production beginning in 1979. However, a rapid (8%) worldwide increase occurred from 1983–1986 due to steady increases in Asian and USSR production.

Another important source of atmospheric NO_x (10%) is from biomass burning, forest and grass fires, most of which are set by humans for land clearing in tropical areas such as South America and Africa. In addition, some wood is used as a fuel, particularly in developing countries. Wildfires set by lightning amount to only 4% of the biomass burned, and thus 96% of biomass burning is anthropogenic (Logan et al. 1981). Combustion of vegetation on a large scale in the tropics during the dry season causes an accumulation of nitrogen oxides in the air, which are removed in dry deposition and at the onset of the rainy season as nitric acid in rain, in some cases acid rain; pH values measured in South America, Africa, and Australia range from 4.3 to 4.8 (Crutzen and Andrae 1990; Lewis 1981). Emissions of NO_x, along with hydrocarbons, from biomass burning results in the formation of rural photochemical smog (similar to that from fossil fuel combustion). This brings about the production of ozone, large concentrations of which build up in the tropics during the dry season.

The amount of NO_x-N released to the atmosphere from biomass burning is estimated to be 2.1–5.5 Tg NOx-N/yr, with an average value of 4 Tg N/yr (Crutzen and Andrae 1990). Thus, the combination of forest burning and fossil fuel combustion accounts for the 66% of NO_x-N in the atmosphere that is of anthropogenic origin.

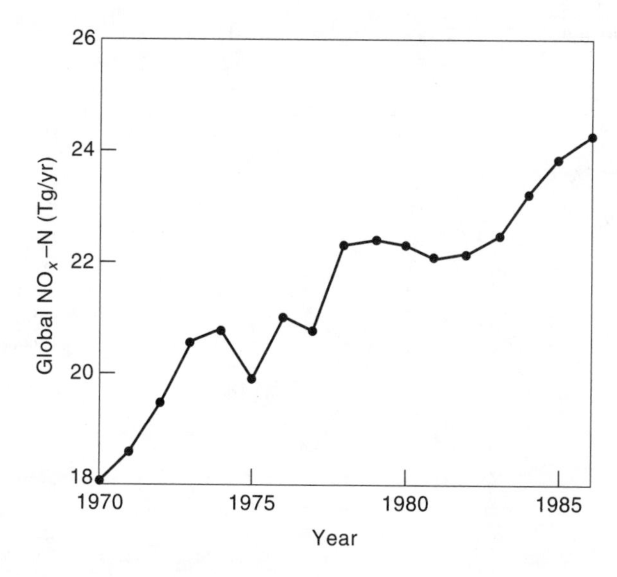

Figure 3.15. Worldwide production of NO_x-N by fossil fuel emissions from 1970 to 1986 in Tg N/yr. (Data from Hameed and Dignon 1992.)

Nitrate Deposition in Rain and the Nitrate–Nitrogen Cycle

Table 3.1, shown earlier, gives the concentration of nitrate in rain from various areas. (Nitrate concentrations in precipitation are less reliable and less numerous than those for other major ions because of analytical problems.) In estimating worldwide flux rates, however, there is a problem due to limited measurements of nitrate from remote continental areas that might be presumed to be unpolluted by humans. The minimum nitrate concentrations for inland continental rain (0.1–0.2 mg/l NO_3) are from Poker Flat, Alaska, Venezuela (Galloway et al. 1982), and the Amazon Basin (Stallard and Edmond 1981). The more usual unpolluted continental values probably average around 0.5–0.6 mg/l NO_3 (see Table 3.1). Europe and North America (particularly the United States) generally have much higher nitrate concentrations in rain, reflecting the high contribution from fossil fuel burning.

Estimates of the total wet deposition of NO_3-N over land based on careful analysis of precipitation data are quite similar: Ehhalt and Drummond (1982) give 10–24 Tg N/yr (17 Tg N/yr), and Logan (1983) estimates 8–30 Tg N/yr (19 Tg N/yr). Berner and Berner (1987) estimate 17 Tg N/yr, using continental areas, and average annual N flux in precipitation for each continent.

In addition to NO_3-N deposition in rain, there is also dry deposition of NO_3-N, primarily through uptake of NO_x gas by plants and water and from deposition of nitrate aerosol. Logan (1983) estimates a total dry deposition of 12-22 Tg N/yr. Since oceanic dry deposition is 4 Tg N/yr (Duce et al. 1991), this leaves 8–18 Tg N/yr or a mean of 13 Tg N/yr as dry deposition over the land. In summary, total land wet and dry deposition of NO_3-N (Table 3.14) amounts to 30 Tg N/yr.

Deposition of NO_3-N in rain over the oceans is estimated as 10 Tg N/yr (Duce et al. 1991). There is also an additional dry deposition of nitrate aerosol of 4 Tg N/yr. Duce et al. estimate that 50–70% of the oceanic deposition of NO_3^- in the Northern Hemisphere is anthropogenic

TABLE 3.14 Fluxes of NO_x–N: The Atmospheric Nitrate Cycle (in Tg N/yr)[a]

Process	Flux (Tg N/yr)	Percent of Total	Reference
Input sources:			
Lightning	3	7	Borucki & Chameides 1984
Conversion from N_2O—stratosphere	0.5	1	Logan 1983
Conversion of NH_3	1	2	Warneck 1988
Soil production of NO	10	24	See text
Biomass burning	4	10	Crutzen & Andrae 1990
Fossil fuel combustion (1986)	24	56	Hameed & Dignon 1992
Total input	42.5	100	
Removal:			
Land—rainfall	17		See text
Land—dry deposition	13		See text
Total land	30		
Ocean—rainfall	10		Duce et al. 1991
Ocean—dry deposition	4		Duce et al. 1991
Total ocean	14		
Total removal	44		

[a] $Tg = 10^{12} g$.

from fossil fuel combustion on the continents. This is feasible considering that NO_x has a 6-day atmospheric residence time. Pollutive nitrate transported from North America is particularly important in the North Atlantic. In the north Pacific ocean, the Asian continent is the major source of nitrate (Prospero and Savoie 1989). Between 25% and 30% of the anthropogenic fossil fuel NO_x emissions reach the oceans (Levy and Moxim 1989; Duce et al. 1991), and fossil fuel NO_x is spread around the globe in the northern hemisphere in winter (Penner et al. 1991).

Now that we have derived values for N deposition and for fluxes of NO_x to the atmosphere (discussed in the previous section), it is possible to construct an overall budget for NO_3^- in the atmosphere. This is given in Table 3.14. The total production of NO_x-N from various sources amounts to about 43 Tg NO_3^- N/yr. This estimate is somewhat smaller than the value of 50 Tg N/yr stated by Logan (1983). Lyons et al. (1990), based on fluxes of N in snow and rain in remote areas, estimate that the total natural NO_x-N flux is 19 Tg/yr; this agrees quite well with our value in Table 3.14 of 15 Tg N/yr.

Because of the relatively short residence time of NO_x in the atmosphere (6 days; Warneck 1988), the atmospheric nitrate cycle should balance. Independently derived input and output fluxes shown in Table 3.14, in fact, do balance with production and deposition of NO_x-N being nearly equal (43 versus 44 Tg N/yr). This agreement suggests that, even though there are uncertainties in all of the NO_x-N fluxes, particularly those for dry deposition, lightning, and soil production, the nitrogen cycle is better known than many other cycles.

Looking at Table 3.14, it would seem that fossil fuel combustion, which has been increasing over the last few decades, contributes more than half of the global NO_x-N production and, when combined with forest burning anthropogenic sources, amounts to about two thirds of global NO_x-N production. Fossil fuel combustion is most important in the Northern Hemisphere and biomass burning in the tropical Southern Hemisphere (South America and Africa). By contrast, natural sources, predominantly soil production and lightning, contribute only one third of the total. Thus, again we see how the activities of humans have made a major modification in the geochemical cycle of a major element.

Ammonium in Rain: The Atmospheric Ammonium–Nitrogen Cycle

Ammonium (NH_4^+) is the other major nitrogen-containing ion found in rain. Ammonium ion results from the partial reaction of ammonia gas (NH_3) with water:

$$NH_{3\,gas} + H_2O \leftrightarrows NH_4^+ + OH^-$$

This reaction tends, because of the production of hydroxyl ions (OH^-), to raise the pH of rainwater and partly counteract the acid effects of CO_2, NO_x, and SO_2. Ammonia is the only atmospheric gas that can do this. Junge (1963) has calculated that, in the absence of other gases, the average concentration of NH_3 in the air of 3 $\mu g/m^3$ would result in a high pH for rainwater, about 8.5.

However, since most rainwater has a pH of 4 to 6, ammonia obviously does not control its pH; the acidic gases CO_2, NO_x, and SO_2 are more important. In fact, since most rainwater is acid, NH_3 should be very soluble in rain and converted mainly to NH_4^+. (The above reaction also shows why alkaline soils tend to give off NH_3 and acid soils to take it up; see below.) NH_3 gas remains in the atmosphere for around 6 days before being converted to NH_4^+ (Warneck 1988), and once NH_4^+ is formed, it is removed in rain in about 5 days.

The measurement of NH_4^+ in rain is often subject to serious errors. For instance, contamination by bird droppings may be large if precautions are not taken to keep birds away from rain collectors. Thus, one should exert caution in interpreting NH_4^+ data (particularly old data), unless one has some knowledge of how the rain was collected.

There are five continental sources of atmospheric ammonia: (1) bacterial decomposition of domestic animal and human excreta; (2) bacterial decomposition of natural nitrogenous organic matter in soils; (3) release from fertilizers; (4) burning of coal (which contains organic nitrogen compounds); and (5) biomass burning. Concentrations of NH_4^+ in rain over the continents range from 0.01 to 1.0 mg/l NH_4. (See Table 3.1.) Agricultural sources are seasonal, being at a maximum in the spring and summer. This helps to explain the observation of increases in the NH_4^+ concentration of rain (and NH_3 in air) during warmer weather, as noted by several authors (Junge 1963; Freyer 1978; Lenhard and Gravenhorst 1980).

Decomposition of domestic animal excrement is considered an important source of atmospheric ammonia. In fact, urea from animal urine is the principal NH_3 source in air over the United Kingdom (Healy et al. 1970) and in Europe (Buijsman et al. 1987). The global flux from domestic animal waste has been estimated by Warneck (1988) to be 22 Tg NH_4-N/yr (see Table

3.16). Warneck assumes that animals in developing countries are more poorly fed and therefore, generate half the waste of those in developed countries. Schlesinger and Hartley (1992) calculates a flux 1.5 times this size, largely because they use European waste-generation rates for all animals. Both these values are within Söderlund and Svensson's (1976) flux: 20–35 Tg NH_3-N/yr from animal plus human wastes.

Release of NH_3 from soils as a result of the bacterial decomposition of natural organic matter is another source of atmospheric ammonia, particularly in remote areas. Measured soil fluxes of NH_3 are very variable and are greater from dry, alkaline soils (because of the conversion of NH_4^+ to NH_3 at higher pH). Because soil production of ammonia is dependent on bacterial decomposition, it is greater at higher temperatures (Junge 1963; Schlesinger and Hartley 1992). Many of the NH_3 soil flux measurements are made by placing chambers to trap gases at the soil surface. This misses any subsequent reabsorption of NH_3 by plants (Denmead et al. 1976). Plants apparently attempt to maintain a constant atmospheric NH_3 concentration, referred to as the *compensation point* (Farquhar et al. 1980), due to their internal chemistry. Thus, it is only the net loss of NH_3 above the plant canopy that results in ammonia in the atmosphere and in rain. Schlesinger and Hartley (1992) reviewed estimates of soil NH_3 fluxes from different undisturbed natural ecosystems in determining a global estimate. Their preferred estimate is 10 Tg N/yr, but soil losses are poorly known.

Extra ammonia can be released to the atmosphere by the addition of urea and ammonium fertilizers to the soil. Soil nitrogen, if not in the form of NH_3 is reduced to it by microorganisms. The release of fertilizer ammonia is influenced by the same factors as is the release of natural soil NH_3; that is, it is favored by dry, warm, high-pH soils. The percentage of applied fertilizer nitrogen that is volatized as ammonia varies with the fertilizer type. Buijsman et al. (1987) estimate that for Europe 5% of fertilizer N is lost as NH_3. Warneck (1988) estimates 5–10% loss of NH_3-N from fertilizers, and Schlesinger and Hartley estimate 10%. We adopt a value of 5% loss of annual fertilizer production (79 Tg/yr; FAO 1989), which gives 4 Tg N/yr (Table 3.16). The use of nitrogen fertilizers is increasing and doubled from 1976 to 1989. This could result in greater atmospheric NH_3 production in the future.

Ammonia is a very minor product of coal combustion. In fact, well-burned coal releases no NH_3 at all (Stedman and Shetter 1983), with all of the nitrogen appearing instead as NO_x. Using the estimate of NH_3-N released by the burning of hard coal (11.8 g NH_3-N per ton of coal) and brown coal (7.4 g NH_3-N per ton of coal) from Freyer (1978) and worldwide coal consumption of 1600 Tg of hard coal (less coke) and 750 Tg of brown coal (for 1978, Warneck 1988), only 0.03 Tg NH_3-N/yr are produced. This is considerably less than many estimates (Robinson and Robbins 1975; Söderlund and Svensson 1976; Warneck 1988). Low NH_3 release from fuel combustion tends to be confirmed by studies of NH_4^+ concentrations in rain and NH_3 concentrations in the atmosphere, which show little areal correlation with other fuel combustion products such as SO_2 (Healy et al. 1970; Junge 1963).

Some NH_3 is released by biomass burning, paricularly in the tropics, where 80% of the burning occurs. Crutzen and Andrae (1990) estimate a flux of O.5–2.0 Tg NH_3-N from biomass burning; the value 1.3 Tg N/yr (the mean) is adopted in Table 3.16.

A sea-surface source of NH_3 has been suggested by Quinn et al. (1988, 1990), based on low concentrations of NH_3 found in remote marine atmospheres (away from a continental source) and simultaneous seawater concentration measurements. The mean flux calculated is 7 μmol/m²/day (0.036 g NH_4-N/m²/yr) over the open oceans (Quinn et al. 1990), or 12 Tg NH_4-

N over the total ocean area (335 × 10^{12} m^2). (The lifetime of NH_3 gas formed in marine atmospheres is very short—6 hours—so little marine-derived NH_3 would be expected in continental air.) A source of this size is necessary to account for the deposition of 16 Tg N/yr NH_4^+ and NH_3 over the oceans as estimated by Duce et al. (1991), who state that much of this deposition is probably recycled from the ocean. [Schlesinger and Hartley (1992) include a similar sea-surface NH_3 source flux.]

Junge (1963) found that ammonium (NH_4^+) in U.S. precipitation in 1955–1956 (Figure 3.16a) was generally 0.1–0.2 mg/l. By 1972–1973 Miller (1974) found greatly increased concentrations, and the relative increases in NH_4^+ were greater than those for either sulfate or nitrate. The NH_4^+ concentrations in North American rain for 1980 (Figure 3.16b) show a similar pattern to that observed for NH_4^+ in 1955–1956 (Figure 3.16a), but a different pattern to that observed for H^+, NO_3^-, and SO_4^{2-} in rain. The ammonium concentrations are greatest (> 0.72 mg/l) in the Northern Plains area of the United States, where there are a lot of livestock feed lots (a large ammonium source), and downwind from this area (Barrie and Hales 1984). The maximum contour (0.72 mg/l) is more than double that in 1955. In addition, the concentrations in the area covered by the 0.36-mg/l contour are more than three times those in 1955–1956 (0.1 mg/l). These increases of NH_4^+ in rain seem to correlate with a tripling of U.S. NH_3 production (used mainly in fertilizer) from 1962 to 1975 (NRC 1979).

Using mean annual precipitation and the average NH_4^+ concentration in rain over each of the various continents (based on data from Table 3.1, Meybeck 1982, and Bottger et al. 1978), we have calculated the annual amount of nitrogen delivered as NH_4^+ in rainfall over all continents as 23 Tg NH_4-N/yr (Table 3.15). This is somewhat less than some other estimates (Söderlund and Svensson 1976; Warneck 1988). Adding this to the marine value, we obtain the global (continents plus oceans) rain removal of NH_4-N of 37 Tg N/yr.

In calculating ammonia deposition we have not included a dry deposition flux of ammonia gas (direct uptake of NH_3 by vegetation and soil). However, NH_3 reacts extensively with H_2SO_4 to form aerosol $(NH_4)_2SO_4$ (see following discussion), which is removed primarily in rain. In forested rural Massachusetts, where the concentration of particulate NH_4^+ is fairly high due to transport from other areas, dry deposition of particulate NH_4^+-N on leaf surfaces is a significant flux in contrast to dry deposition of NH_3, which is unimportant (Tjepkema et al. 1981). We have included a dry deposition flux of particulate NH_4^+ of 8 Tg NH_4-N/yr based on one third of the wet continental deposition Warneck (1988). The total of continental rain and dry deposition of NH_4^+-N thus amounts to 31 Tg N/yr. Reaction in the atmosphere with OH removes another 1 Tg N (Warneck 1988).

Our ammonia nitrogen budget, including both inputs and outputs, is summarized in Table 3.16. Although there are considerable uncertainties involved in all ammonia fluxes, we find that the atmospheric ammonia cycle is essentially in balance, as it should be because of the short turnover time of NH_3 in the atmosphere (6 days). Production and removal are both estimated at about 45–50 Tg N/yr. As with NO_xN, since production is mostly over land, there is net transport to the oceanic atmosphere, probably of about 4 Tg NH_4-N/yr (the difference between oceanic deposition and sea-surface production). Table 3.16 again illustrates the importance of anthropogenic influences on the cycle of a major rainwater constituent. Animal wastes (the largest NH_3 source), fertilizer release, and biomass burning are almost entirely due to humans. Thus, globally about 54% of total NH_3 production is due to humans, but over the continents that rises to 73%.

$(NH_4)_2SO_4$ Aerosol Formation: Interaction of the N and S Cycles

As noted earlier, NH_3 gas in the atmosphere reacts very readily with aqueous H_2SO_4 aerosols to form $(NH_4)_2SO_4$ aerosols. This reaction partially neutralizes the acid H_2SO_4 and converts gaseous NH_3 to a solid/liquid aerosol, which can be transported and removed in rain or dry deposition. It also provides a link between the atmospheric cycles of N and S (Galbally et al. 1982). In the atmospheric NH_3-N cycle, as pointed out by Galbally et al., there are essentially two competing processes regulating the atmospheric concentration of NH_3 gas (as opposed to NH_4^+ aerosol): (1) the tendency of plants to take up or give off atmospheric NH_3, maintaining a concentration of around 1 $\mu g/m^3$ (see earlier discussion); and (2) the reaction of NH_3 with H_2SO_4, which converts gaseous NH_3 to NH_4^+ in aerosols and prevents NH_3 from being recycled by plants. Thus, conversion of gaseous NH_3 to NH_4^+ in sulfate aerosol essentially removes ammonia gas from the biogenic N cycle.

NH_4^+ in aerosol form can be transported long distances (up to 5000 km) before removal in rain and dry deposition. Thus, the net effect of the formation of $(NH_4)_2SO_4$ aerosols is to spread NH_4^+-N over a large area and probably increase nutrient NH_4^+-N concentrations in rain in areas far from sources of NH_3 gas. Sulfate aerosols act as a carrier of nutrient N for remote marine and terrestrial areas, and the more sulfate aerosols present, the greater is the N deposition (Liss and Galloway 1993).

Strong interaction between N and S cycles in the marine atmosphere and ocean, far from sources of pollution, was found by Quinn et al. (1990). In a remote northern Pacific area there is a considerable natural flux of DMS (which converts to SO_2 and then to acid H_2SO_4 droplets in the atmosphere) along with NH_3 gas. The NH_3 and H_2SO_4 react in the atmosphere to form ammonium sulfate aerosol, which then is removed in marine rain (and about 5% by dry deposition). In the western North Atlantic Ocean off Bermuda, the situation is different. Here there is an appreciable anthropogenic input of both NH_3 and H_2SO_4 with two roughly equal sources of $NH_3 + NH_4^+$: pollutive transport eastward from North America and NH_3 emissions from the sea surface (Liss and Galloway 1993).

ACID RAIN

Besides bringing about changes in the concentrations of NO_3^-, SO_4^{2-}, and Cl^- in rain, the gases NO, NO_2, SO_2, and HCl also produce hydrogen ions, H^+. The result is called *acid rain.* Acid rain is rain that is more acid than it would be in the absence of these gases, and much of it is pollutive in origin. In this section we shall discuss the factors that affect the acidity of rain and the processes by which acid rain is formed. Throughout the discussion we shall express acidity in

Figure 3.16. (a) Ammonium (NH_4^+) concentration in rain over the continental United States (values in mg NH_4^+/l) for July–September 1955. (After C. E. Junge, "The Distribution of Ammonium and Nitrate in Rain Water over the United States," *Transactions of the American Geophysical Union* 39: 242, © 1958 by the American Geophysical Union.) (b) Annual mean precipitation-weighted ammonium concentration (NH_4^+) in North American precipitation in 1980; individual points are in μmol/l; heavy contours are in mg/l. (After Barrie and Hales 1984.)

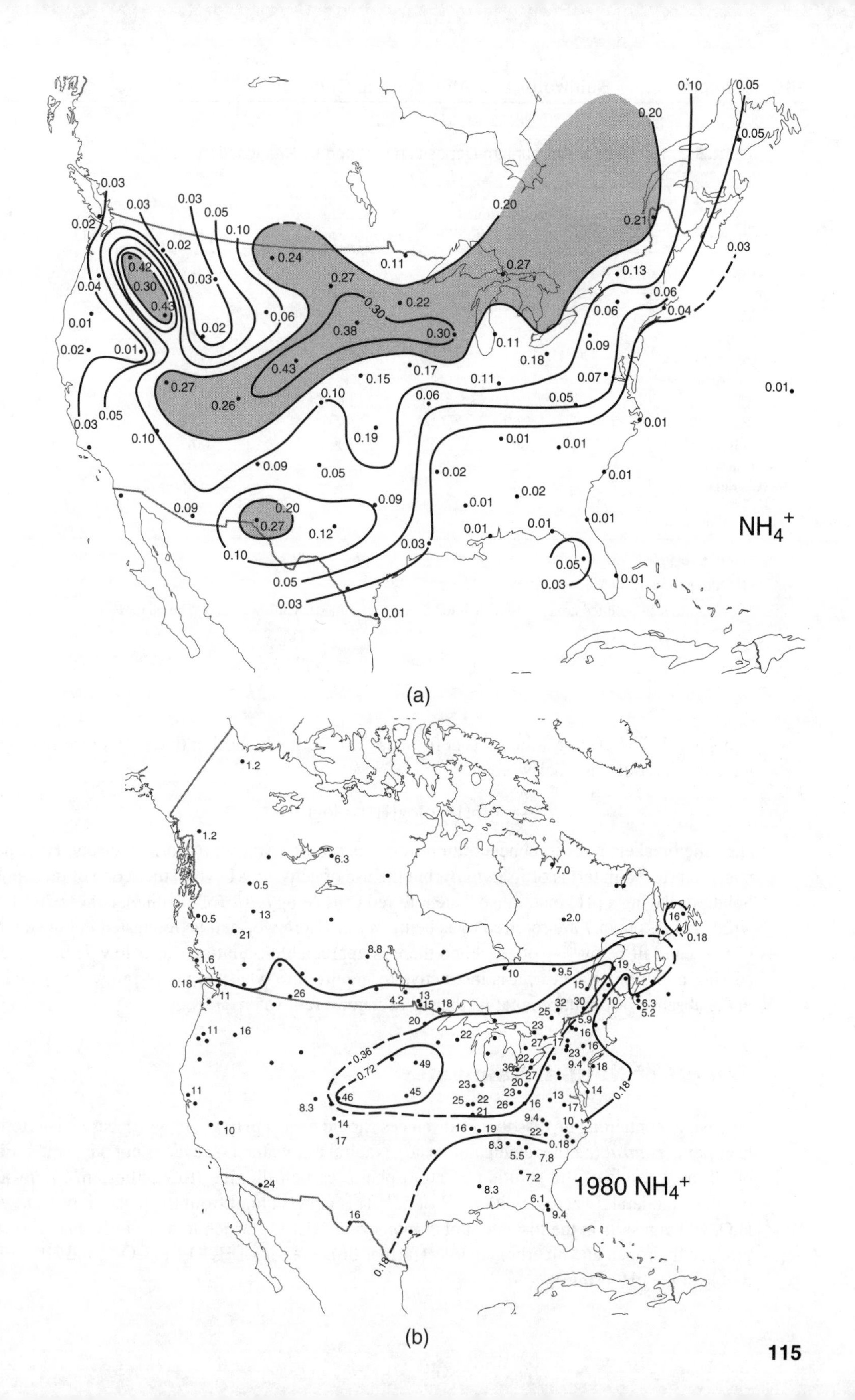

(a)

(b)

TABLE 3.15 Rate of Ammonium Deposition on Land by Precipitation

Continent	Mean Annual Precipitation[a] (cm)	Nondesert Area[b] (10^6 km^2)	[NH_4^+] (mg/l)	Deposition Rate (Tg N/yr)
Africa	100	21.8	0.3	5.1
Asia:	73			
Clean		24.8	0.3	4.2
Polluted		9.0	0.4	2.0
Europe: Polluted	73	10.0	0.65	3.7
North America:				
Clean	67	18.0	0.2	1.9
Polluted[c]	100	5.2	0.3	1.2
South America	165	17.1	0.2	4.4
Antarctica	17	12.2	0.01	0.02
Australia	44	4.9	0.05	0.1
Total on land				22.6

[a] From Lvovitch 1973.

[b] Nondesert area from Meybeck 1979.

[c] North American polluted area = eastern United States and Canada (Galloway and Whelpdale 1980).

terms of pH. The pH of a solution is defined as the negative logarithm (base 10) of the hydrogen ion concentration. In other words,

$$pH = -\log[H^+] = \log(1/H^+)$$

Here the brackets refer to concentration in moles per liter (mol/l). (More precisely, H^+ is normally expressed in terms of its *activity,* but the use of activity is beyond the scope of this book). Solutions having a pH greater than 7 are referred to as being *basic* (or alkaline); conversely, those with a pH less than 7 are referred to as being *acidic.* Here we shall be interested in how the pH of rain can fall below 7—often, when there is appreciable pollution, far below 7. Before discussing pollution, however, it is instructive to inquire into what the pH of rainwater would be in the absence of pollution, that is, the pH of natural rainwater.

The pH of Natural Rainwater

Pure water containing no dissolved substances should have a pH of 7, in which case it is referred to as being *neutral* (neither acidic nor basic). Natural rainwater, however, is not pure water. First of all, as a result of the solution of atmospheric carbon dioxide (to equilibrium), rainwater becomes moderately acidic, with a pH of 5.7. This comes about from the reaction of CO_2 with H_2O, which results in the formation of carbonic acid, H_2CO_3, which in turn partly dissociates to produce hydrogen and bicarbonate ions (further dissociation of HCO_3^- to CO_3^{2-} and H^+ is negligible at the pH of rain):

TABLE 3.16 Fluxes of Ammonia: The Atmospheric Ammonia Cycle

Process	Flux ($Tg\ NH_4$–N/yr)	Percent of Total	Anthropogenic as Percent of Total
Input sources:			
Domestic animal waste decomposition	22	45	45
Soil loss from organic matter (excluding fertilizer)	10	20	
Fertilizer release	4	8	8
Coal combustion	0.03	<<1	<<1
Biomass burning	1.3	3	3
Sea surface release	12	24	
Total input	49.3	100	56
Removal:			
Reaction in atmosphere with OH	1		
Land—rainfall	23		
Land—dry part. NH_4 deposition (1/3 wet)	8		
Ocean—rainfall (Duce et al. 1991)	14		
Ocean—dry deposition—particulate NH_4	2		
Total removal	48		

Sources: Warneck 1988; Schlesinger and Hartley 1992; Crutzen and Andrae 1990; Duce et al. 1991.

$$CO_{2\ gas} \rightleftharpoons CO_{2\ soln}$$
$$CO_{2\ soln} + H_2O \rightleftharpoons H_2CO_3$$
$$H_2CO_3 \rightleftharpoons H^+ + HCO_3^-$$

Here the double-headed arrows refer to partial reaction to chemical equilibrium. From the latter two reactions, it can be seen that CO_2 reacting with H_2O results in the formation of H^+ and HCO_3^- in equal amounts. By using equilibrium expressions for the above reactions and the fact that the concentrations of H^+ and HCO_3^- are equal, one can readily calculate (see Garrels and Christ 1965) that, at equilibrium,

$$[H^+] = 2.1 \times 10^{-6}\ mol/l = 10^{-5.67}$$

This is equivalent to a pH of about 5.7. Since the concentration of carbon dioxide in the atmosphere is everywhere about the same, one would therefore expect that natural rainwater, if no other reactions were involved, would exhibit a pH close to 5.7; that is, it would be moderately acidic. (Note that small deviations, on the scale of tenths of a pH unit, from this value can result from the uptake of CO_2 at different temperatures and pressures, and that, due to fossil fuel burning, atmospheric CO_2 has been increasing with time.)

In many cases the pH of natural (unpolluted) rainwater is either higher or lower than 5.7. Natural rainwater usually has a pH less than 5.7, in which case it falls in the category of acid rain. This can be due to the presence of naturally occurring H_2SO_4 from the oxidation of biogenic reduced-sulfur gases. Charlson and Rodhe (1982) suggest that, theoretically, an average pH of about 5.0 might occur in pristine areas (which lack neutralizing substances such as NH_3 and $CaCO_3$—see below) because of the emission of natural sulfur gases. However, since the biogenic sulfur-gas source is not uniform, considerable variation should occur in natural pH values. In addition, other acids such as HNO_3 affect the pH. Because the natural gaseous biogenic sulfur release is primarily over the oceans (17 TgS/yr; Bates et al. 1992), low natural pH values might be expected to occur over remote areas of the oceans because of the conversion of these gases to sulfuric acid. The continental biogenic sulfur source appears to be small (0.35 Tg S/yr; Bates et al. 1992), and 60% of this is in the tropics. Thus this area is the most likely place for naturally acid continental rain from biogenic sulfur. Volcanic SO_2 is also a possible natural source of sulfuric acid rain.

Weak organic acids, such as acetic or formic acid, can also supply an additional source of natural acidity in some local areas (Galloway et al. 1982; Keene and Galloway 1986, 1988; Talbott et al. 1988). It has been suggested that these acids could come either from natural terrestrial biogenic emissions or biogenic emissions from the sea surface. Organic acids apparently have a short lifetime of a day or less in the atmosphere. However, organic acids can also be of anthropogenic origin, from automotive emissions, and from biomass burning, which is an important source of organic acidity in tropical areas (Crutzen and Andrae 1990). In these latter cases the organic acids would be considered as *pollutive* in origin. However, formic and acetic acid have less impact on the environment than strong acids such as sulfuric and nitric, because the organic acids are rapidly oxidized to carbonic acid by microbes (Andrae et al. 1988).

Hydrochloric acid can also contribute to acidity, particularly in marine rain, where it comes from the reaction of sea-salt aerosol with strong acids or alternatively from reaction of O_3 with sea-salt aerosol. Combustion products from fossil fuel or biomass burning appear to enhance the volatilization (Keene et al. 1990).

Natural rain with a pH greater than 5.7 is a much less common situation on a worldwide basis and comes about mainly in arid regions (where air pollution is absent) as the result of the dissolution of wind-blown dust, which contains high concentrations of $CaCO_3$ (Kramer 1978). The reaction is

$$CaCO_3 + H^+ \rightarrow Ca^{2+} + HCO_3^-$$

This reaction not only results in the neutralization of acidity via the consumption of hydrogen ions but also in the production of Ca^{2+} and HCO_3^- ions. [A similar reaction to that with $CaCO_3$ but involving FeOOH dust—"brown dust" (Kramer 1978)—can also raise the pH of rain.] Calcium ion in rain in excess of sea-salt concentrations often indicates that the rain has reacted with $CaCO_3$ dust, as stated earlier. In most western U.S. rain, the concentration of Ca^{2+} varies with that of SO_4^{2-}, as would be expected for the neutralization of H_2SO_4 by $CaCO_3$ dust (Gillette et al. 1992). Neutralization to the point of producing "alkaline rain," however, is apparently rare in the United States because of the ubiquity of acid rain. In 1988 the highest pH of rain in the continental United States, even in dusty western areas, was found to be only 5.7 (see Figure 13.19c). In eastern U.S. rain, where abundant $CaCO_3$ dust is not available, H^+ concentrations vary with SO_4^{2+}.

An additional factor in pH of rain in the eastern United States has been the steep decline in base cations (Ca, Mg, K, Na), but especially Ca, in rain over the past 25 years. This means that less base cations are available to neutralize natural acidity than were available in the past (Hedin et al. 1994).

Neutralization of natural acidity in unpolluted rain over land can also take place by reaction with ammonia gas (NH_3). In regions where ammonia is emitted to the atmosphere from biological decay, agricultural activity, and so forth, there may be enough to bring about a slight rise in pH via the reaction

$$NH_3 + H^+ \rightarrow NH_4^+$$

For example, Charlson and Rodhe (1982) calculate that, in the absence of sulfate aerosol, NH_3, at the lowest concentrations found in continental areas (0.13 μg/m³), could raise the pH of CO_2-containing rain from 5.7 to 6.2. By contrast, in the presence of sulfate aerosol, there is usually insufficient NH_3 to neutralize acidity effectively. In the remote north Pacific Ocean (Quinn et al. 1990), the ocean emission fluxes of NH_3-N and DMS-S are in a mole ratio of 1.2:1, which is lower than the 2:1 ratio needed to neutralize all of the H_2SO_4 acidity derived from the DMS. Also, when pollutive NH_3 and H_2SO_4 are involved, there is normally insufficient ammonia for complete neutralization. The average concentration of NH_4^+ in (polluted) eastern U.S. rain is 0.3 mg/l (Pack 1980), and this would represent neutralization of 0.8 mg/l H_2SO_4 in rain. However, since the average eastern U.S. rain concentration of H_2SO_4 is 2.7 mg/l, it is not possible for available NH_3 to neutralize all the H_2SO_4 acidity, and consequently acid rain of average pH = 4.2 results.

The relative importance of NH_3 and soil dust in neutralizing acidity varies from area to area, but on the average in the United States about one third of the acid neutralization is due to NH_3 (Munger and Eisenreich 1983). However, because NH_3 gas or fine aerosol NH_4^+ can travel farther than coarse soil dust containing $CaCO_3$, the neutralization effect of NH_3 may affect areas farther from the source. In Beijing, China, where there is a lot of $CaCO_3$ dust and NH_3 along with a high sulfate concentration (4.39 mg/l), the pH is 6.2 rather than about 3.5 estimated if there were no neutralization (Galloway et al. 1987). One would expect such relatively high-pH rains in other arid areas where there are alkaline soils with greater NH_3 release and more $CaCO_3$ dust.

Over the oceans, sea-salt aerosol, which is alkaline (from bicarbonate and borate), may neutralize acids in marine precipitation to a small extent. When the concentration of sea-salt aerosol in rain is large (Na > 3.0 mg/l), it is possible to raise the original pH by about 0.05 pH units (Galloway et al. 1983; Pszenny et al. 1982).

Acid Rain from Pollution

Acid rain is defined here as that having a pH less than 5.7 due to reactions with acidic gases other than CO_2. The acidic gases are SO_2, NO_2, NO_x, and (to a lesser extent) HCl, and they result in the formation in the atmosphere and in rain clouds of sulfuric, nitric, and hydrochloric acids, respectively. (See also sections on sulfate and nitrate in rain.) Overall reactions are

$$SO_2 + OH \rightarrow \ldots \rightarrow H_2SO_4 \quad \text{(sulfuric acid)}$$

$$SO_2 + H_2O_2 \rightarrow H_2SO_4 \quad \text{(sulfuric acid)}$$

$$NO_2 + OH \rightarrow HNO_3 \quad \text{(nitric acid)}$$

$$HCl_{gas} \rightarrow HCl \quad \text{(hydrochloric acid)}$$

followed by the dissociation of these acids in rain water to form H^+ ions:

$$H_2SO_4 \rightarrow 2H^+ + SO_4^{2-}$$

$$HNO_3 \rightarrow H^+ + NO_3^-$$

$$HCl \rightarrow H^+ + Cl$$

Thus, as more and more of the precursor gases are added to the atmosphere by human activities, more and more hydrogen ions are produced and the pH of rainwater drops. This helps explain why the pH of rainfall at many locations has been decreasing over time.

Acid rain was first noted in northwest Europe in the early 1950s. Barrett and Brodin (1955) found that precipitation in southern Sweden had a pH between 4 and 5 and that the pH was lowest in the winter when the air flow is from the south, bringing pollution from central and western Europe. In 1968, Oden (1968) found that the acidity of northern European rain had increased since 1956; rain in some parts of Scandinavia was 200 times more acid than in 1956. A region of high acidity (pH 4–4.5) that was centered in the Benelux countries in the late 1950s had spread by the late 1960s to Germany, northern France, the eastern British Isles, and southern

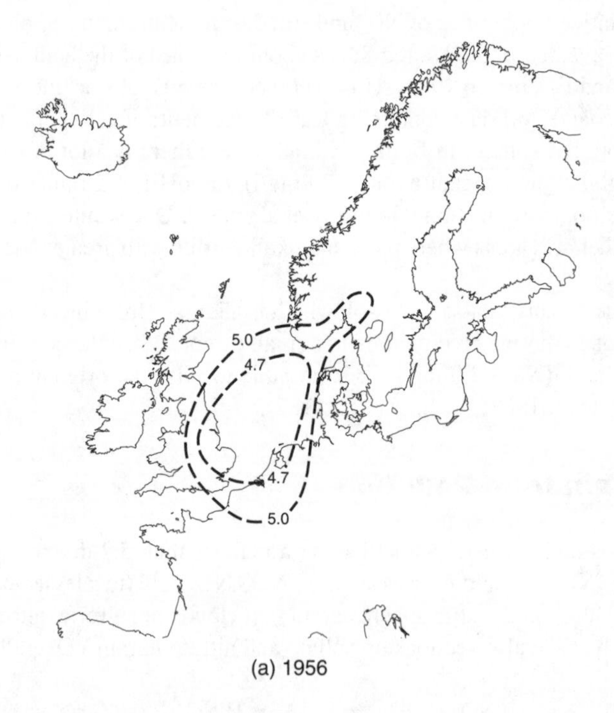

Figure 3.17. Acid rainfall in northwestern Europe and its change with time. Dashed lines represent contours of constant pH. (a) Situation in 1956. (b) Situation in 1974. (Modified from G. E. Likens et al., "Acid Rain." Copyright © October 1979 by Scientific American, Inc. All rights reserved.) (c) In 1985. [from *World Resources 1988–89*, Fig. 23.2, p. 337. *Source:* Co-operative Program for Monitoring and Evaluation of the Long-Range Transmission of Air Pollutants in Europe (EMEP); Summary Report from the Chemical Coordinating Centre for the Third Phase of EMEP (Norwegian Institute for Air Research, Lillestrom, *Norway,* 1987).]

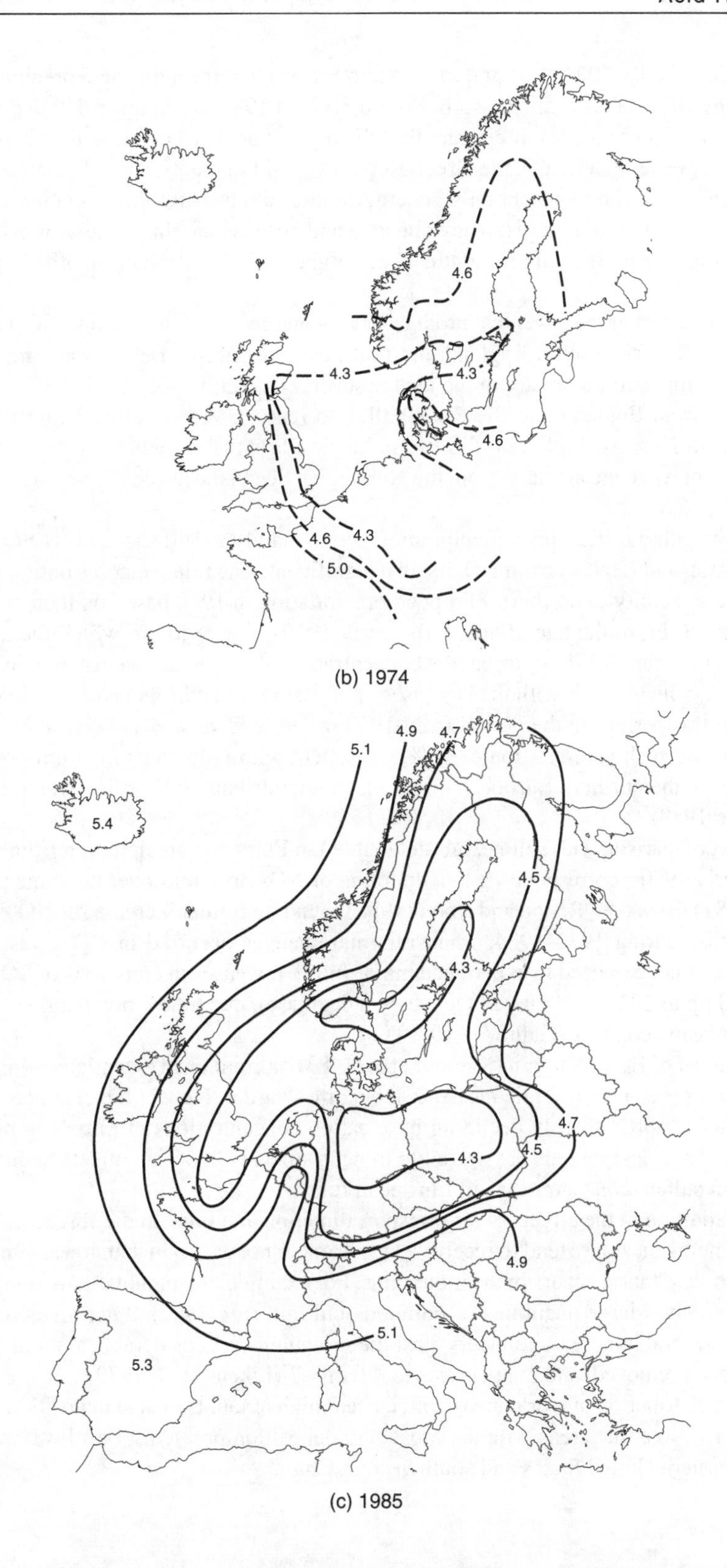

(b) 1974

(c) 1985

Scandinavia. By 1974 most of northwestern Europe was receiving acid precipitation (pH < 4.6) (Likens 1976). This is all shown in Figure 3.17. In 1985, when more detailed information was available (Figure 3.17c), it appears that the highest acidity (< 4.3) is found in southern Scandinavia, western Germany, the Czech Republic, and Poland. (It should be noted that maps such as Figure 3.17 can represent only general trends, both because there is considerable local geographic variation of rain pH in any one area and because monthly values of pH vary from yearly averages, with the winter months being higher and the summer months lower; see Kallend et al. 1983.)

A large part of the gases that produce acid Scandinavian rain (SO_2 and NO_2) are due to industrial activity far away in England and north central Europe. Because anthropogenic SO_2 and the resulting sulfate remain in the air for several days, it is possible for SO_2 and sulfate in air masses from England and the Ruhr Valley to reach Sweden before being rained out (Bolin 1971). In Norway, Forland (1973) found a similar situation when air masses came from the south; but when the air came from the Norwegian Sea to the north, the rain was much less acidic (pH 5.1–6.6).

The acidity of European precipitation is due mainly to sulfuric acid (H_2SO_4) from SO_2 gas, but nitric acid (HNO_3) from NO_x, is also significant. The relative contribution of nitrate versus sulfate to acidity in northern European precipitation in 1980 based on their relative emissions was 1:3.8. From the late 1950s to the early 1970s, the period in which the spread of acidity shown in Figure 3.17b occurred, the concentration of sulfate in Swedish precipitation increased by 50%, which agrees with the increases of emissions of anthropogenic SO_2 in northern Europe during this period. In the period from 1972 to 1986, there was a decrease in concentration of SO_4 in Swedish precipitation of 40%, consistent with reductions in anthropogenic emissions of SO_2 in the northern European countries that contribute to Swedish precipitation (Leck and Rodhe 1989).

By comparison with sulfate, nitrate doubled in European precipitation from the late 1950s to the early 1970s, corresponding to a doubling of NO_x emissions over the same period (Rodhe et al. 1981). However, Rodhe and Rood (1986) found no further increase for NO_3^- in Swedish precipitation during 1972–1984, when minimal changes occurred in NO_2 emissions in Europe. Europe is not expected to experience major future increases in emissions of SO_2 and NO_x in the period up to 2020 (and may in fact show a decrease), due to low population growth and possible emission controls (Galloway 1989).

The pH of European rain (Kallend et al. 1983) has remained essentially constant since 1965. Because pH is a logarithmic measure, it becomes harder to lower the pH once it is already low. Another factor is sharply declining base cation concentrations in European rain (Hedin et al. 1994). Thus, less cations are available to neutralize acidity. This offsets about 35–60% of the drop in sulfate concentrations in European rain.

In addition to the effects of air transport direction and time on the spread of acid rain, windward mountain slopes tend to receive a larger amount of acid rain than lower-lying areas because more precipitation occurs in the mountains. For example, the mountainous southern Norwegian coast receives large quantities of polluted acid rain (pH = 4.2). After air masses have moved inland several hundred kilometers, past the mountains, a considerable amount of the pollutants have been removed, and the pH rises to 4.6 or 4.7 (Likens et al. 1979).

The distribution of world areas with current high amounts of acid deposition is shown in Figure 3.18. Note that areas with the most acid deposition problems, besides Europe, include the northeastern United States and southwestern China.

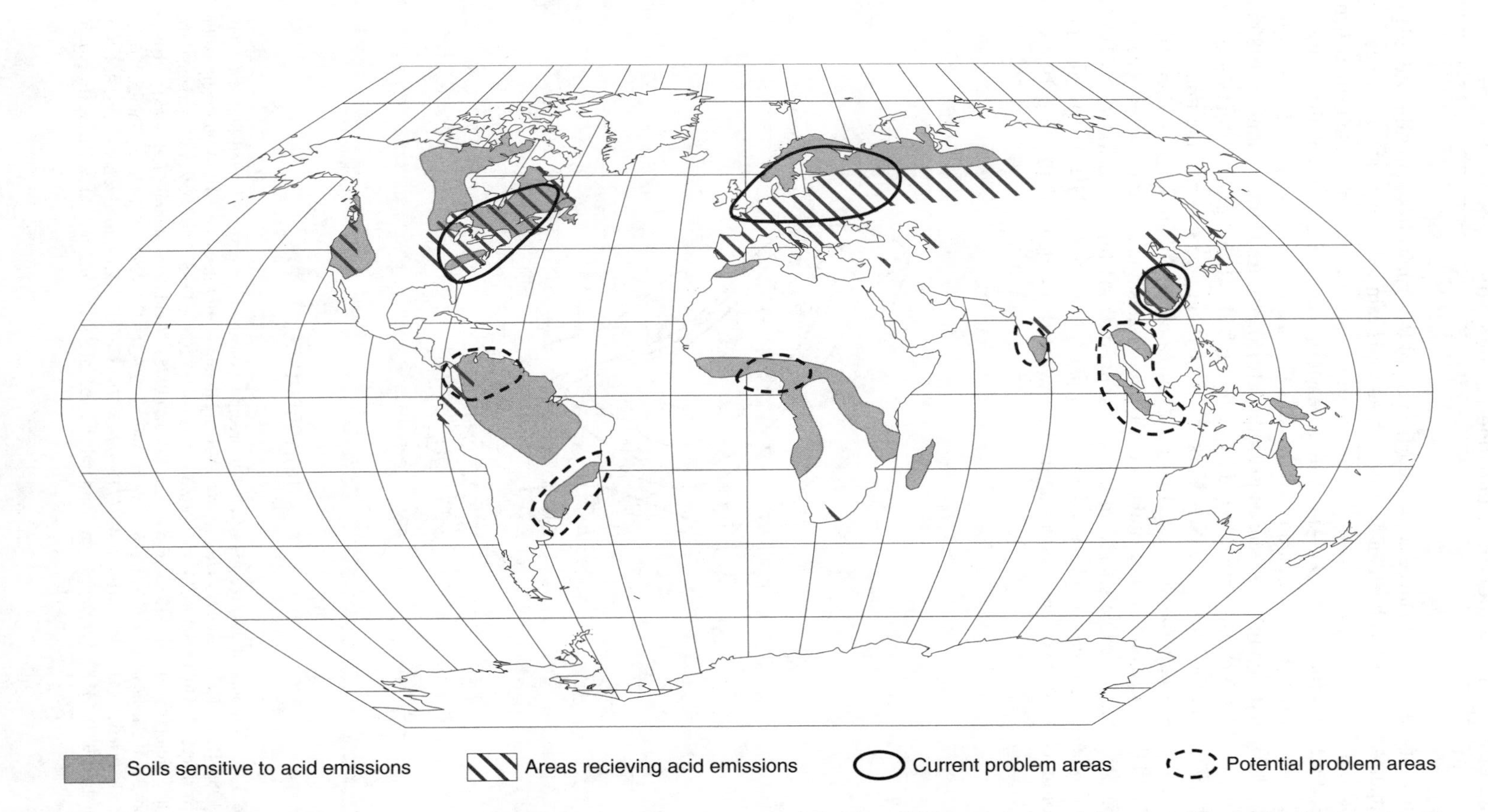

Figure 3.18. World areas with present or future acid deposition problems. (a) Areas, circled by solid lines, that *presently* have a problem with anthropogenic acid (sulfur and nitrogen) emissions due to sensitive soils and (b) *potential* problem areas (circled by dashed lines)—areas with sensitive soils that have or are expected to have rapidly growing acid emissions. (*Source:* Adapted from Rodhe 1989.)

Most of the United States east of the Mississippi River now has acid rain with pH values less than 4.6. There has been an area of very low-pH rain (pH < 4.5) at least since the mid-1950s, but the intensity and areal distribution of the acid rain has increased since then. Figure 3.19a shows the pH in 1955–1956 in the eastern United States as calculated by Cogbill and Likens (1974) from Junge and Werby's (1958) data. Figure 3.19b is Likens's map for 1972–1973, and Figure 3.19c shows pH values for the entire United States in 1985 *(World Resources 1988–89)*. The area covered by acid rain had become greater from 1955–1956 to 1972–1973, and the rain in the New York–New England area had become more acid. Data for 1985 show a larger area affected by acid rain, a pH less than 4.2 in the central "bull's-eye," and the entire midwestern and eastern United States and eastern Canada receiving rain with a pH < 5.0. The two sites with the lowest average pH for the period 1975–1985 were State College, Pennsylvania (pH $= 4.15$) and Ithaca, New York (pH $= 4.18$). Stensland and Semonin (1982) have suggested that the midwestern drought of the mid-1950s, which

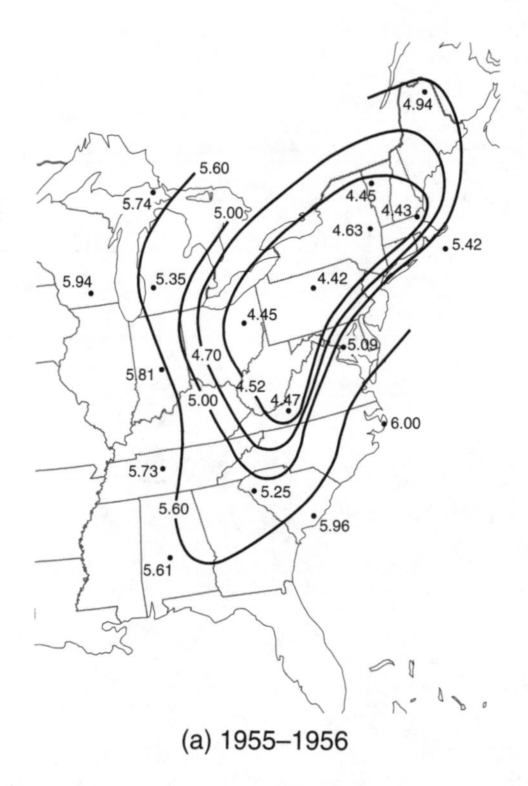

(a) 1955–1956

Figure 3.19. Contours of average pH of annual precipitation over the eastern United States. (a) For 1955–1956. [From G. E. Likens (1976), from C. V. Cogbill and G. E. Likens (1974), "Acid Precipitation in the Northeastern United States," *Water Resources Research* 10(6): 1135, © 1974 by the American Geophysical Union.] (b) For 1972–1973. [From G. E. Likens (1976), "Acid Precipitation." Reprinted with permission from *Chemical and Engineering News* 54(48): 29–44. Copyright © 1976 by the American Chemical Society.] (c) For 1985. (From *World Resources 1988–89,* Fig. 23.3, p. 337. *Source:* A. R. Olsen (September 1987), "1985 Wet Deposition Temporal and Spatial Patterns in North America, Pacific Northwest Laboratory, Rockland, Wash.")

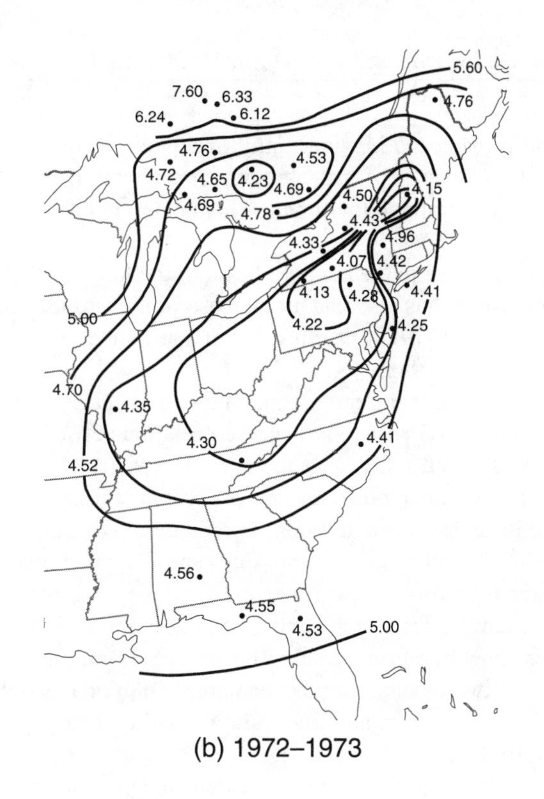

(b) 1972–1973

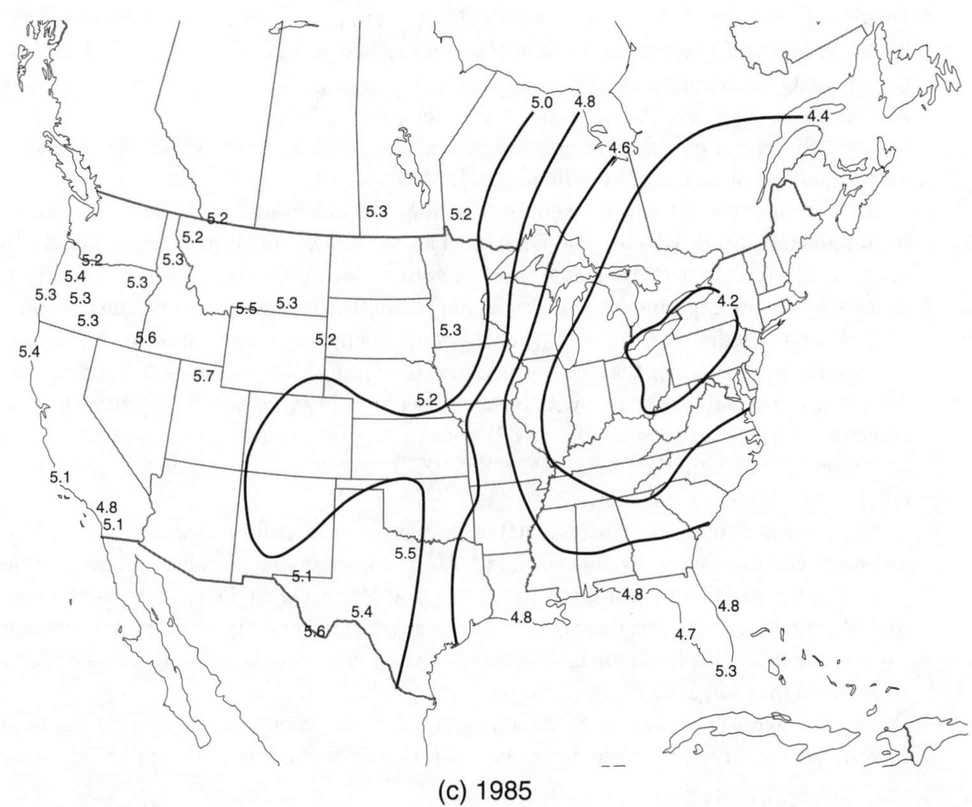

(c) 1985

increased dust-derived Ca and Mg concentrations in rain, may have resulted in decreased acidity; thus, some of the drop in pH of rain with time shown in Figure 3.19 over the midcontinent region may have been due to climatic change and not to an increase in the emissions of acidic gases.

The most rapid increase in the acidity of rain in the last 40 years has been in the southeastern United States, which has also had an increase in industrialization and urbanization during this period (Likens et al. 1979). (SO_2 emissions in the southeastern United States increased 33% from 1965 to 1978; NRC 1983.)

The acidity of U.S. rain in the eastern and midwestern states is due about two thirds to H_2SO_4 and about one third to HNO_3. There is a good correlation between U.S. rain acidity and the pollutive emission of SO_2 and NO_x. Over the period 1975–1987, in the eastern and midwestern United States there was a combined decline of 18% in emissions of SO_4 and NO_x. This resulted in an 18% decline in H^+ concentrations in precipitation overall in the area (Butler and Likens 1991). The correlation between sulfur emissions and SO_4 precipitation concentrations is unusually good at Hubbard Brook, New Hampshire, which is farthest downwind from emission sources, thereby allowing for better mixing and oxidation of SO_2 to SO_4. Other sites closer to emission sources show poorer correlation between SO_2 emissions and SO_4 concentrations in rain, possibly due to two factors: Limited availability of oxidants (OH, H_2O_2, O_3) may limit SO_4 concentrations in precipitation, and high rainfall acidity, below pH = 4.2, slows liquid-phase oxidation of SO_2 to SO_4.

The general correlation between sulfuric acid in precipitation and SO_2 emissions has also been noted in the eastern United States by NRC (1983) and in the western United States by Oppenheimer et al. (1985), where the sulfur is transported more than 1000 kilometers. The correlation between NO_x emissions and NO_3 concentrations in precipitation is much poorer in the eastern and midwestern United States (Butler and Likens 1991). Nitrate in precipitation shows highly variable concentrations, and faster atmospheric formation and removal of HNO_3 makes long-range transport of NO_3 less important than for SO_4.

Hedin et al. (1994) found a steep drop in basic cations in rain in the eastern United States, including Hubbard Brook (see also Driscoll et al. 1989), over the last 25 years. The decline was equal to from 50% to 100% of the drop in sulfate during the same period. This means that decreasing acidity expected from declining SO_4 was offset by rising acidity from less base cation neutralization. Hedin et al. did not find changes in either precipitation amounts or in wind speeds that would tend to affect the amount of wind-blown dust, so they concluded that the cation decline was probably correlated with observed declines in anthropogenic particulate emissions. However, Stensland and Semonin (1982) noted a drop in cations in the midwestern and western United States from the mid-1950s to the 1970s, which they attributed to a change to a wetter climate with less wind-blown dust.

Studies at remote areas such as Whiteface Mountain in the Adirondacks and Ithaca, New York, and south central Ontario (summarized in NRC 1983) show that the source of most of the acidity in the precipitation (and most of the H_2SO_4 and HNO_3) is the Ohio Valley and other industrial Midwest areas at a great distance to the southwest. In general, however, it is difficult in the eastern United States to distinguish between distant and local sources, because pollutants are well mixed over large areas (up to 1000 km on a straight line).

The average pH of rain in the western United States is about 5.2–5.3 (see Figure 3.19c). In general, the western United States has more alkaline dust than the eastern United States, and an average of 70% of the (SO_4 + NO_3) is neutralized by Ca (Gillette et al. 1992). However, dust production is episodic and not homogeneous; therefore, areas of more highly acidic precipitation are also found in parts of the western United States (see Figure 3.19c). For example, Lewis and Grant (1980), in a study of precipitation in a rural area in the Colorado Rockies, found increasingly acid precipitation, with the pH dropping from 5.43 in 1975 to 4.63 in 1978. Although sulfuric acid was the major cause of acidity, the increase in acidity correlated with an increase in nitric acid in the rain. Similarly, Byron et al. (1991), in a study on the crest of the Sierra Nevada mountains in California, downwind from Sacramento and San Francisco, found a pH drop from 5.4 to 4.9 in the period 1979–1986. The pH correlates with nitrate, which is slightly more important than sulfate as a cause of acidity. Nitric acid (much of it formed from automobile exhaust emissions) is also the dominant cause of acidity in southern California.

Acid rain that is definitely traceable to fossil fuel combustion is not confined to Europe and North America. There is also a large area in southern China that has very acid rain. Acid rain in China (Zhao and Sun 1986; Galloway et al. 1987) comes primarily from SO_2 from coal combustion, which accounts for 90% of the SO_2 in China and supplies 75% of the energy. Coal combustion in China occurs in small furnaces and household stoves, so the effects tend to be more localized than in the United States, although rural areas surrounding cities receive acid rain. Sulfate concentrations are very high in Chinese rain, both in north China and in south China, often reaching 13–20 mg/l (6–10 times higher than values in the eastern United States). Nitrate is usually somewhat lower than in the eastern United States due to the lower amounts of vehicle emissions. Despite the high concentrations of sulfate, northern Chinese rain (in Beijing, for example) has an average pH of 6.5 because of neutralization by NH_3 and $CaCO_3$, as discussed above. [Galloway et al. (1987) estimate that the pH in Beijing would be 3.5 without neutralization.] However, in southern China, which is more humid, there is less soil dust and NH_3 and the pH is 4–5. Galloway (1989) predicts that SO_2 emisssions could increase by a factor of 3 in China by 2020, which would greatly increase the rainfall acidity. Also, the acidity is not confined to southern China. In Machin, a remote small city in the eastern Tibetan Plateau in China, pH 2.25 rain, due mainly to nitric acid from coal burning, has been measured (Harte 1983).

One South American area noted by Rodhe (1989) as having potential problems with acid rain from fossil fuel combustion is the area around Sao Paulo in southern Brazil, which is known to have high concentrations of SO_2. Another potential problem area is northern Venezuela, from combined fossil fuel combustion and biomass burning (see Figure 3.18).

Problems in Distinguishing Naturally Acid Rain from That Due to Pollution

In areas such as the United States and Europe, where there are widespread anthropogenic acids in the atmosphere, it is difficult to see whether the "natural" pH of rain would in fact be about 5.7 as predicted from equilibrium with atmospheric CO_2, or whether rain might actually have a lower pH due to the presence of naturally occurring acids, as has been suggested by several workers (Kerr 1981; Charlson and Rodhe 1982; Galloway et al. 1982, 1984). Acid rain (pH < 5.7) has been found in a number of remote areas. Galloway et al. (1982) propose five possible causes of acid rain in such areas, and often several sources are involved: (1) local fossil fuel combustion; (2) very long range transport of sulfate and nitrate aerosol from distant anthropogenic sources; (3) natural emissions of reduced sulfur compounds, both from the ocean and from the land (as suggested by Charlson and Rodhe 1982); (4) agricultural (anthropogenic) and natural biomass burning; and (5) natural emissions of organic acids from vegetation and marine biogenic sources. The following discussion illustrates how many areas are affected by one or more of these sources in combination.

Acid rain can occur in remote areas from local fossil fuel combustion, as we have already mentioned for Tibetan China. The long-range transport of sulfate aerosols from distant anthropogenic sources has been implicated as a cause of acid rain in a number of other localities. For example, in Bermuda, in the North Atlantic Ocean, the average pH of rain is 4.9 in winter due to sulfuric acid from long-range transport of sulfate aerosol from the North American continent 1100 km away (Moody and Galloway 1988). However, rain from storms originating in the North Atlantic near the Bahamas is also acid (pH 4.84), presumably from natural oceanic biogenic sulfur (DMS), which has been observed in high concentrations in this area.

Acid rain (primarily from sulfuric acid) has also been found in Hawaii in the North Pacific (Miller and Yoshinaga 1981; Kerr 1981), where the pH decreases with increasing altitude from 5.2 at sea level to 4.3 at 2500 m. It is suggested that sulfuric acid from Asian sources thousands of kilometers eastward may be transported across the Pacific in the mid-troposphere at high altitudes (>2000 m), and rainout of this acid may explain the low pH found at high Hawaiian elevations. [At low Hawaiian elevations, 90% of the sulfate is marine biogenic in origin (Savoie and Prospero 1989).]

Rain from a marine air mass coming off the North Pacific Ocean was measured in a coastal location in Washington (Vong et al. 1988). The rain had a pH = 5.1, due mainly to excess sulfate, which presumably came mainly from marine biogenic sulfate produced over the oceans. According to Savoie and Prospero (1989), biogenic sulfur accounts for 80% of non-sea-salt sulfate over the mid-latitude Pacific Ocean, with the rest coming from anthropogenic Asian continental sources.

Amsterdam Island in the central Indian Ocean (Southern Hemisphere) receives slightly acid marine rain of pH 5.08 (Moody et al. 1991), whose acidity is mainly H_2SO_4, HCl, and organic acids (25% each), and HNO_3 (12%). The sulfuric acid is almost certainly from natural biogenic reduced-sulfur gases (DMS) from the ocean (Nguyen et al. 1992). Organic acids and NH_4 also appear to have a marine biogenic source. HCl comes from the reaction of either ozone or acids with sea-salt aerosol (Keene et al. 1990). The source of the nitrate is apparently biomass burning on Madagascar or the African continent, 3500–5000 km away.

A persistent Arctic haze occurs in the winter and spring and has been attributed to very long range transport, over 5000 km or 5–10 days' travel time, of sulfate aerosols from industrial areas in northern Eurasia (Barrie 1986). This occurs when circulation patterns favor long-distance transport and where there is little precipitation to remove pollutants. The pH of rain at Point Barrow, Alaska, is 5.1 (Dayan et al. 1985). Galloway et al. (1982) found an average pH of 5.0 in rain at an inland site at Poker Flat, Alaska, due mainly to sulfuric acid (Dayan et al. 1985). The sulfuric acid is thought to have been derived from a combination of Arctic haze aerosol, local sulfate pollution from Fairbanks 70 km away, Pacific Ocean biogenic sources, and long-distance anthropogenic transport from Japan and Russia. In Alaska there is also a considerable input from organic acidity in storms of local origin (Keene and Galloway 1988).

Anthropogenic biomass burning apparently produces acid rain in many continental tropical areas, including South America (Venezuela and Brazil), Africa (Congo and Ivory Coast), and Australia (Crutzen and Andrae 1990). The pH of these rains is from 4.3 to 4.8, and the acidity is due largely to organic acids (formic and acetic) and nitric acid. The effect is seasonal, with biomass burning during the dry season followed by particularly high acidity in rain at the beginning of the wet season (Lewis 1981; Galloway et al. 1982; Lacaux et al. 1987). Rains in monsoon areas tend to be very dilute, with few neutralizing basic cations. Rain from Katherine, Australia (Likens et al. 1987), had a pH of 4.73, with 64% of the acidity being due to organic acids. The authors attribute the acidity to terrestrial vegetative emissions, but there is a great deal of biomass burning in the area during the dry season and the organic acidity is highest early in the wet season.

Natural emissions of reduced sulfur compounds have been implicated in sulfuric acid rain (mean pH 5.05) in the Amazon River Basin, Brazil (Stallard and Edmond 1981). The pH, 4.7–5.7, is considerably higher than that of pollution-impacted rain in the northeastern United States where the pH range is 3.4–5.4 with most values being less than 4.0 (see Figure 3.20). This area of the Amazon Basin has low dust, no pollution sources (including no agricultural burning), and might be expected to have high emissions of reduced biogenic sulfur since it is a tropical rain forest (see section on sulfur). This seems to be the best-documented occurrence of naturally acid continental rain (presumably due to biogenic sulfur emissions).

Rain in Iceland, a remote North Atlantic location, has a pH of 5.5 (when there are no volcanic eruptions) (Gislason and Eugster 1987). This perhaps represents the closest approach to the pH 5.7 value for simple equilibration with atmospheric CO_2 and little input from other acidic gases.

In summary, judging from the above discussion, it appears that after inputs from human activities are removed, the mean pH of natural rain is probably greater than 5.0. This agrees with the conclusions of Galloway et al. (1982) and Charlson and Rodhe (1982) and applies best to marine and unpolluted coastal rain, which is low in neutralizing soil dust and often has considerable

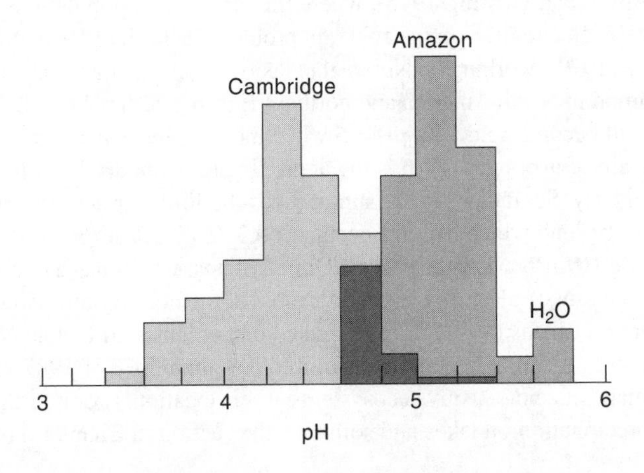

Figure 3.20. Histograms comparing the pH of Amazon rain (Brazil) with that from Cambridge, Massachusetts. The mean pH is 5.05 and 4.19, respectively. The value for water equilibrated with atmospheric CO_2, is labeled as H_2O. [After R. F. Stallard and J. M. Edmond (1981), "Chemistry of the Amazon, H_2O Precipitation Chemistry and the Marine Contribution to the Dissolved Load at the Time of Peak Discharge," *Journal of Geophysical Research* 86(C10):9852, © 1981 by the American Geophysical Union.]

marine biogenic sulfate input. There are very few examples of naturally acid continental rain other than that given above for the Amazon. Anthropogenic biomass burning, unfortunately, affects many supposedly pristine tropical areas that are remote from cities. It is certainly true that acid rain (pH < 5.7) of various origins is not uncommon in remote areas. However, as it appears that "natural" acid rain is less acidic than rain that has obvious pollutive influences, *in most continental areas and some marine ones, pollution is the major source of acid rain!*

Effects of Acid Rain

There are various serious effects of acid rain, ranging from increased destruction of structures, such as corrosion of metal and weathering of buildings, to the effect of acid rain on lakes, soils, and vegetation. There is evidence that acid rain causes increased leaching of nutrients from foliage and may disrupt leaf physiology and plant growth. Increased leaching of cations, particularly Ca^{2+}, Mg^{2+}, and Al^{3+} from the soil by acid rain has been noted (Overein 1972; Cronan and Schonfield 1979).

It has been suggested that acid deposition may contribute to the decline of high-elevation red spruce forests in areas such as the Adirondack Mountains in New York which are exposed to very acid fog (typically pH = 3.6; Mohnen 1988); leaf injury can occur at such pH values. Acid rain is one stress that may contribute to forest decline combined with other air pollutants from fossil fuel burning, particularly ozone, and toxic metals and natural stresses such as drought, cold weather, and insects (Likens 1989; Cowling 1989). In addition, excess nutrient substances, especially nitrogen from rain, can be taken up by plants, stimulating growth and bringing about cation deficiencies (Schultze 1989). Most of the evidence for forest decline is based on field observations, with limited experimental evidence.

Most important, acid rain can cause lakes and rivers to become acid (pH < 5). Examples include lakes in the Adirondack Mountains (Kramer et al 1986; Mohnen 1988; Schindler 1988) and in Scandinavia (Wright and Gessing 1976), where the loss of fish populations due to high acidity has focused increasing attention on the acid rain problem (Schofield 1976; NRC 1986; Schindler 1988). Henriksen (1979), working on Norwegian lakes, concludes that a calcium bicarbonate lake of the type common in North America and northern Europe, with pH 6.5, 2.0 mg/l Ca^{2+}, and 1.1 mEq/l HCO_3^-, will become acid with pH < 5 when the long-term average of precipitation is less than 4.6. Thus, rain of average pH 4.6 is the approximate boundary level for damage to aquatic ecosystems. Similarly, Schindler (1988) summarizes the limit for acid wet deposition on freshwater ecosystems beyond which damage occurs as 0.3–0.47 g S/m²/yr from H_2SO_4, which is the same as 19–31 mg H^+/m²/yr. For the eastern United States, which has average precipitation of 1 m/yr, this would be rain of pH 4.6–4.7. Clearly, most of the northeastern United States is beyond this limit. Butler and Likens (1991) have estimated that at Hubbard Brook, New Hampshire (pH = 4.28), additional reductions in acid deposition to less than 50% of 1987 values are needed to avoid future damage of acid-sensitive ecosystems at this location. (For a further discussion of the effects of acid precipitation on lakes and soils, see the section in Chapter 6 on acid lakes.)

REFERENCES

Anderson, I. C., and J. S. Levine. 1987. Simultaneous field measurements of biogenic emissions of nitric oxide and nitrous oxide, *J. Geophys. Res.* 92: 965–976.

Andrae, M. O., R. J. Charlson, F. Bruynseels, H. Storms, R. Van Grieken, and W. Maenhaut. 1986. Salts, silicates, and sulfates: Internal mixture in marine aerosols, *Science* 232: 1620–1623.

Andrae, M. O., and I. E. Galbally. 1984. Sulfur and nitrogen emissions in remote areas. Paper delivered at NATO Advanced Research Inst., Bermuda, October 1984.

Andrae, M. O., and H. Raemdonck. 1983. Dimethylsulfide in the surface ocean and the marine atmosphere: A global view, *Science* 221: 744–747.

Andrae, M. O., R. W. Talbot, T. W. Andrae, and R. C. Harriss. 1988. Formic and acetic acid over the central Amazon region, Brazil, 1. Dry season, *J. Geophys. Res.* 93: 1616–1624.

Barrett, E., and G. Brodin. 1955. The acidity of Scandinavian precipitation, *Tellus* 7: 251–257.

Barrie, L. A. 1986. Arctic air pollution: An overview of current knowledge, *Atmos. Environ.* 20: 643–663.

Barrie, L. A., and J. M. Hales. 1984. The spatial distribution of precipition acidity and major ion wet deposition in North America during 1980. *Tellus* 36B: 333–355.

Bates, T. S., J. D. Cline, R. H. Gammon, and S. R. Kelly-Hansen. 1987. Regional and seasonal flux of oceanic dimethylsulfide to the atmosphere, *J. Geophys. Res.* 92: 2930–2938.

Bates, T. S., B. K. Lamb, A. Guenther, J. Dignon, and R. E. Stoiber. 1992. Sulfur emissions to the atmosphere from natural sources, *J. Atmos. Chem.* 14: 315–337.

Benton, G. S., and M. A. Estoque. 1954. Water vapor transport over North American continent, *J. Meteorol.* 11: 462–477.

Berner, E. K., and Berner, R. A. 1987. *The Global Water Cycle: Geochemistry and Environment.* Englewood Cliffs, N.J.: Prentice-Hall.

Berresheim, H., and W. Jaeschke. 1983. The contribution of volcanoes to the global atmospheric sulfur budget, *J. Geophys. Res.* 88(C6): 3732–3740.

Bluth, G. J. S., C. C. Schnetzer, A. J. Krueger, and L. S. Walter. 1993. The contribution of explosive volcanism to global atmospheric sulphur dioxide concentrations, *Nature* 366: 327–329.

Bohn, H. L., B. L. McNeal, and G. A. O'Connor. 1979. *Soil Chemistry.* New York: John Wiley.

Bolin, B. (ed.). 1971. Air pollution across national boundaries: The impact on the environment. In *Report of the Swedish Preparatory Committee for the U.N. Conference on the Human Environment.* Stockholm: Royal Ministry of Agriculture.

Bonsang, B., B. C. Nguyen, A. Gandry, and G. Lambert. 1980. Sulfate enrichment in marine aerosols owing to biogenic gaseous sulfur compounds, *J. Geophys. Res.* 85: 7410–7416.

Borucki, W. J., and W. L. Chameides. 1984. Lightning: Estimates of the rates of energy dissipation and nitrogen fixation, *Rev. Geophys.* 22: 363–372.

Bottger, A. D., D. H. Ehhalt, and G. Gravenhorst. 1978. Atmospharische Kreislaufe von Stickoxiden und Ammoniak, *Ber. Kernforschungsanlange Julich,* Nr. 1558.

Bowden, W. B. 1986. Gaseous nitrogen emissions from undisturbed terrestrial ecosystems: an assesment of their impacts on local and global nitrogen budgets, *Biogeochemistry* 2: 249–279.

Bowen, H. J. M. 1966. *Trace Elements in Biochemistry.* New York: Academic Press.

Buijsman, E. H., F. M. Maas, and W. A. H. Asman. 1987. Anthropogenic NH_3 emissions in Europe, *Atmos. Environ.* 21: 1009–1022.

Burns, R. C., and R. W. F. Hardy. 1975. *Nitrogen Fixation in Bacteria and Higher Plants.* New York: Springer-Verlag.

Busenberg, E., and C. C. Langway, Jr. 1979. Levels of ammonium sulfate, chloride, calcium, and sodium in ice and snow from southern Greenland, *J. Geophys. Res.* 84(C4): 1705–1709.

Butler, T. J., C. V. Cogbill, and G. E. Likens. 1984. Effect of climatology on precipitation acidity, *Bull. Am. Meteorol. Soc.* 65: 639–640.

Butler, T. J., and G. E. Likens. 1991. The impact of changing regional emissions on precipitation chemistry in the eastern United States, *Atmos. Environ.* 25A: 305–315.

Byron, E. R., R. P. Axler, and C. R. Goldman. 1991. Increased precipitation acidity in the central Sierra Nevada, *Atmos. Environ.* 25A: 271–275.

Carroll, D. 1962. Rainwater as a chemical agent of geologic processes—A review, *USGS Water Supply Paper 1535-G.*

Chameides, W. L., D. D. Davis, J. Bradshaw, M. Rodgers, S. Sandholm, and D. B. Bai. 1987. An estimate of NO_x production rate in electrified clouds based on NO observations from the GTE/CITE fall 1983 field operation, *J. Geophys. Res.* 92: 2153–2156.

Charlson, R. J., T. L. Anderson, and R. E. McDuff. 1992. The sulfur cycle. In *Global Biogeochemical Cycles,* ed. S. S. Butcher, R. J. Charlson, G. H. Orians, and G. V. Wolfe, pp. 285–300. San Diego, Calif.: Academic Press.

Charlson, R. J., and H. Rodhe. 1982. Factors controlling the acidity of natural rainwater, *Nature* 295: 683–685.

Christensen, J. P., J. W. Murray, A. H. Devol, and L. A. Codispoti. 1987. Denitrification in continental shelf sediments has major impact on the oceanic nitrogen budget, *Global Biogeochem. Cycles* 1: 97–116.

Church, T. M., J. N. Galloway, T. D. Jickells, and A. H. Knap. 1982. The chemistry of western Atlantic precipitation at the mid-Atlantic coast and on Bermuda, *J. Geophys. Res.* 87(C13): 11013–11018.

Codispoti, L. A., and J. P. Christensen. 1985. Nitrification, denitification and nitrous oxide cycling in the eastern tropical south Pacific Ocean, *Marine. Chem.* 16: 277–300.

Cogbill, C. V., and G. E. Likens. 1974. Acid precipitation in the northeastern United States, *Water Resources Res.* 10(6): 1133–1137.

Connor, J. J., and H. T. Shacklette. 1975. Background geochemistry of some rocks, soils, plants and vegetables in the conterminous United States, *USGS Prof. Paper 574-F.*

Cowling, E. B. 1989. Recent changes in chemical climate and related effects on forests in North America and Europe, *Ambio* 18: 167–171.

Cronan, C. S., and C. C. Schofield. 1979. Aluminum leaching in response to acid precipitation: Effects on high-elevation watersheds in the northeast, *Science* 204: 304–306.

Crozat, G. 1979. Sur l'emission d'un aerosol riche en potassium par la foret tropical, *Tellus* 31: 52–57.

Crutzen, P. J., and M. O. Andrae. 1990. Biomass burning in the tropics: Impact on atmospheric chemistry and biogeochemical cycles, *Science* 250: 1669–1678.

Dawson, G. A. 1980. Nitrogen fixation by lightning, *J. Atmos. Sci.* 37: 174–178.

Dayan, U., J. M. Miller, W. C. Keene, and J. N. Galloway. 1985. An analysis of precipitation chemistry data from Alaska, *Atmos. Environ.* 19: 651–657.

Denmead, O. T., J. R. Freney, and J. R. Simpson. 1976. A closed ammonia cycle within a plant canopy, *Soil Biol. Biochem.* 8: 161–164.

Driscoll, C. T., G. E. Likens, L. O. Hedin, J. S. Eaton, and F. H. Bormann. 1989. Changes in the chemistry of surface waters, 25-year results at the Hubbard Brook Experimental Forest, NH, *Environ. Sci. Technol.* 23: 137–143.

Duce, R., P. S. Liss, J. T. Merrill, E. L. Atlans, P. Buat-Menard, B. B. Hicks, J. M. Miller, J. M. Prospero, R. Atimoto, T. M. Church, W. Ellis, J. N. Galloway, L. Hansen, T. D. Jickells, A. H. Knap, K. H. Reinhardt, B. Schneider, A. Soudine, J. J. Tokos, S. Tsunogai, R. Wollast, and M. Zhou. 1991. The atmospheric input of trace species to the world ocean, *Global Biochem. Cycles* 5: 193–259.

Duce, R. A., F. MacIntyre, and B. Bonsang. 1982. Discussion of "Enrichment of sulfate in maritime aerosols" (Garland 1981), *Atmos. Environ.* 16(8): 2025–2026.

Ehhalt, D., and J. W. Drummond. 1982. The tropospheric cycle of NO_x. In *Chemistry of Unpolluted and Polluted Troposphere,* ed. H. Georgii and W. Jaesche, pp. 219–251. Dordrecht: D. Reidel.

Erickson, D. J., S. J. Ghan, and J. E. Penner. 1990. Global ocean-to-atmosphere dimethyl sulfide flux, *J. Geophys. Res.* 95: 7543–7552.

Eriksson, E. 1957. The chemical composition of Hawaiian rainfall, *Tellus* 9: 509–520.

Eriksson, E. 1960. Yearly circulation of chloride and sulfur in nature, meteorological, geochemical and pedological implications, Part 2, *Tellus* 12: 63–109.

Fanning, K. A. 1989. Influence of atmospheric pollution on nutrients in the ocean, *Nature* 339: 460–463.

Farquhar, G. D., P. M. Firth, R. Wetselaar and B. Weir. 1980. On the gaseous exchange of ammonia between leaves and the environment: Determination of the ammonia compensation point, *Plant Physiol.* 66: 710–714.

Feller, M. C., and J. P. Kimmins. 1979. Chemical characteristics of small streams near Haney in southwestern British Columbia, *Water Resources Res.* 15(2): 247–258.

Food and Agriculture Organization (FAO). 1989. *FAO Fertilizer Yearbook, Volume 39.* Rome: FAO.

Forland, E. J. 1973. A study of the acidity in the precipitation of southwestern Norway, *Tellus* 25: 291–299.

Freyer, H. D. 1978. Seasonal trends of NH_4^+ and NO_3^- nitrogen isotope composition in rain collected in Jülich, Germany, *Tellus* 30: 83–92.

Friend, J. P. 1973. The global sulfur cycle. In *Chemistry of the Lower Atmosphere,* ed. S. I. Rasool, pp. 177–201. New York: Plenum Press.

Galbally, I. E., G. D. Farquhar, and G. P. Ayers. 1982. Interactions in the atmosphere of the biogeochemical cycles of carbon, nitrogen and sulfur. In *Cycling of Carbon, Nitrogen, Sulfur and Phosphorus in Terrestrial and Aquatic Ecosystems,* ed. J. R. Freney and I. E. Galbally, pp. 1–9. Berlin: Springer-Verlag.

Galbally, I. E., and C. R. Roy. 1978. Loss of fixed nitrogen from soils by nitric oxide exhalation, *Nature* 275: 734–735.

Galloway, J. N. 1985. The deposition of sulfur and nitrogen from the remote atmosphere. In *The Biogeochemical Cycling of Sulfur and Nitrogen in the Remote Atmosphere,* ed. J. N. Galloway, R. J. Charlson, M. O. Andrae, and H. Rodhe, pp. 143–175. Dordrecht: Reidel.

Galloway, J. N. 1989. Atmospheric acidification: Projections for the future, *Ambio* 18: 161–166.

Galloway, J. N., A. H. Knap, and T. M. Church. 1983. The composition of Western Atlantic precipitation using shipboard collectors, *J. Geophys. Res.* 88(C15): 10859–10864.

Galloway, J. N., and G. E. Likens. 1978. The collection of precipitation for chemical analysis, *Tellus* 30: 71–82.

Galloway, J. N., G. E. Likens, and M. E. Hawley. 1984. Acid precipitation: Natural versus anthropogenic components, *Science* 236: 829–831.

Galloway, J. N., G. E. Likens, W. C. Keene, and J. M. Miller. 1982. The composition of precipitation in remote areas of the world, *J. Geophys. Res.* 87(11): 8771–8786.

Galloway, J. N., and D. M. Whelpdale. 1980. An atmospheric sulfur budget for eastern North America, *Atmos. Environ.* 14: 409–417.

Galloway, J. N., D. Zhao, J. Xiong, and G. E. Likens. 1987. Acid rain: China, United States, and a remote area, *Science* 236: 1559–1562.

Gambell, A. W., Jr. 1962. Indirect evidence of the importance of water-soluble continentally derived aerosols, *Tellus* 14: 91–95.

Gambell, A. W., and D. W. Fisher. 1964. Occurrence of sulfate and nitrate in rainfall, *J. Geophys. Res.* 69(20): 4203–4210.

Gambell, A. W., and D. W. Fisher. 1966. Chemical composition of rainfall, western North Carolina and southeastern Virginia, *USGS Water Supply Paper 1535-K.*

Garland, J. A. 1981. Enrichment of sulfate in maritime aerosols, *Atmos. Environ.* 15: 787–791.

Garrels, R. M., and C. L. Christ. 1965. *Solutions, Minerals, and Equilibria.* New York: Harper & Row.

Garrels, R. M., and F. T. Mackenzie. 1971. *Evolution of Sedimentary Rocks.* New York: W. W. Norton.

Gatz, D. F. 1975. Relative contributions of different sources of urban aerosols: Application of a new estimation method to multiple sites in Chicago, *Atmos. Environ.* 9: 1–18.

Gillette, D. A., G. J. Stensland, A. L. Williams, W. Barnard, D. Gatz, P. C. Sinclair, and T. C. Johnson. 1992. Emissions of alkaline elements calcium, magnesium, potassium, and sodium from open sources in the contiguous United States, *Global Biogeochem. Cycles* 6: 437–457.

Gislason, S. R., and H. P. Eugster. 1987. Meteoric water-basalt interactions. II: A field study in N.E. Iceland, *Geochim. Cosmochim. Acta* 51: 2841–2855.

Gordeev, V. V. and I. S. Siderov. 1993. Concentrations of major elements and their outflow into the Laptev Sea by the Lena River, *Marine Chem.* 43: 33–45.

Granat, L. 1972. On the relation between pH and the chemical composition in atmospheric precipitation, *Tellus* 24: 550–560.

Granat, L. 1978. Sulfate in precipitation as observed by the European atmospheric chemistry network, *Atmos. Environ.* 12: 413–424.

Granat, L., H. Rodhe, and R. O. Hallberg. 1976. The global sulfur cycle. In *Nitrogen, Phosphorus, and Sulfur Global Cycles,* ed. B. H. Svensson and R. Soderlund, pp. 89–134. SCOPE Report no. 7. Ecol. Bull. (Stockholm) 22.

Graustein, W. C. 1981. The effects of forest vegetation on solute acquisition and chemical weathering: A study of the Tesuque watersheds near Santa Fe, New Mexico. Ph.D. dissertation, Yale University, New Haven, Conn.

Gschwandtner, G., K. Gschwandtner, K. Eldridge, C. Mann, and D. Mobley. 1986. Historic emissions of sulfur and nitrogen oxides in the United States from 1900 to 1980, *J. Air Pollut. Control Assoc.* 36: 139–149.

Guenther, A., B. Lamb, and H. Westberg. 1989. U.S. national biogenic sulfur inventory. In *Biogenic Sulfur in the Environment,* ed. E. J. Salzman and W. C. Cooper, ACS Symp. Ser. 393, pp. 14–30. Washington, D.C.: American Chemical Society.

Hameed, S., and Dignon, J. 1992. Global emissions of nitrogen and sulfur oxides in fossil fuel combustion 1970–1986, *J. Air Waste Management Assoc.* 42: 159–163.

Handa, B. K. 1971. Chemical composition of monsoon rain water over Bankipur, *Indian J. Meteorol. Geophys.* 22: 603.

Harte, J. 1983. An investigation of acid precipitation in Quinghai Province, China, *Atmos. Environ.* 17(2): 403–408.

Healy, T. V., A. C. McKay, A. Pilbeam, and D. Scargill. 1970. Ammonia and ammonium sulfate in the troposphere over the United Kingdom, *J. Geophys. Res.* 75: 2317–2321.

Hedin, L. O., L. Granat, G. E. Likens, T. A. Buishand, J. N. Galloway, T. J. Butler, and H. Rodhe. 1994. Steep declines in atmospheric base cations in regions of Europe and North America, *Nature* 367: 351–354.

Henriksen, A. 1979. A simple approach for identifying and measuring acidification of freshwater, *Nature* 278: 542–545.

Hill, R. D., R. G. Rinker, and W. D. Wilson. 1980. Atmospheric nitrogen fixation by lightning, *J. Atmos. Sci.* 37: 179–192.

Hobbs, P. V, 1993. Aerosol-cloud interactions. In *Aerosol-Cloud-Climate Interactions,* ed. P. V. Hobbs, pp. 33–73. San Diego, Calif.: Academic Press.

Holland, H. D. 1978. *The Chemistry of the Atmosphere and Oceans.* New York: John Wiley.

Husar, R. B. 1986. Emissions of sulfur dixide and nitrogen oxides and trends for eastern North America. In *Acid Deposition: Long-Term Trends,* ed. G. H. Gibson. Washington, D.C.: National Academy Press.

Hutton, J. T., and T. I. Leslie. 1958. Accession of nonnitrogenous ions dissolved in rainwater to soils in Victoria, *Australian J. Agr. Res.* 9: 492–507.

Johansson, C., and L. Granat. 1984. Emission of nitric oxide from arable land, *Tellus* 36B: 25–37.

Johansson, C., H. Rodhe, and E. Sanhueza. 1988. Emission of NO in a tropical savanna and a cloud forest during the dry season, *J. Geophys. Res.* 93: 7180–7192.

Junge, C. E. 1958. The distribution of ammonium and nitrate in rain water over the United States, *Trans. Am. Geophys. Union* 39: 241–248.

Junge, C. E. 1963. *Air Chemistry and Radioactivity.* New York: Academic Press.

Junge, C. E. 1972. Our knowledge of the physico-chemistry of aerosols in the undisturbed marine environment, *J. Geophys. Res.* 77: 5183–5200.

Junge, C. E., and R. T. Werby. 1958. The concentration of chloride, sodium, potassium, calcium and sulfate in rainwater over the United States, *J. Meteorol.* 15: 417–425.

Kallend, A. S., A. R. Marsh, J. H. Pickles, and M. V. Proctor. 1983. Acidity of rain in Europe, *Atmos. Environ.* 17: 127–137.

Kaplan, W. A., S. C. Wofsy, M. Keller, and J. M. Dacosta. 1988. Emission of NO and deposition of O_3 in a tropical forest system, *J. Geophys. Res.* 93: 1389–1395.

Keene, W. C., and J. N. Galloway. 1986. Considerations regarding sources for formic and acetic acids in the troposphere, *J. Geophys. Res.* 91: 14466–14474.

Keene, W. C., and J. N. Galloway. 1988. The biogeochemical cycling of formic and acetic acids through the troposphere: An overview of current understanding, *Tellus* 40B: 322–334.

Keene, W. C., A. A. A. Pszenny, J. N. Galloway, and M. E. Hawley. 1986. Sea-salt corrections and interpretation of constituent ratios in marine precipitation, *J. Geophys. Res.* 91: 6647–6658.

Keene, W. C., A. A. P. Pszenny, D. J. Jacob, R. A. Duce, J. N. Galloway, J. J. Schultz-Tokos, H. Sievring, and J. F. Boatman. 1990. The geochemical cycling of reactive chlorine in the marine troposphere, *Global Biogeochem. Cycles* 4: 407–430.

Kellogg, W. W., R. D. Cadle, E. R. Allen, A. L. Lazrus, and E. A. Martell. 1972. The sulfur cycle, *Science* 175: 587.

Kerr, R. A. 1981. Is all acid rain polluted? *Science* 212(29): 1014.

Kramer, J. R. 1978. Acid precipitation. In *Sulfur in the environment, Part 1: The Atmospheric Cycle,* ed. J. O. Nriagu, pp. 325–370. New York: John Wiley.

Kramer, J. R., A. W. Andren, R. A. Smith, A. H. Johnson, R. B. Alexander, and G. Oehlert. 1986. Streams and lakes. In *Acid Deposition: Long-Term Trends,* ed. G. H. Gibson, pp. 231–299. Washington, D.C.: National Academy Press.

Lacaux, J. P., J. Servant, and J. G. R. Baudet. 1987. Acid rain in the tropical forests of the Ivory Coast, *Atmos. Environ.* 21: 2643–2647.

Langner, J., and Rodhe, H. 1991. A global three-dimensional model of the tropospheric sulfur cycle, *J. Atmos. Chem.* 13: 225–263.

Lawson, D. R., and J. W. Winchester. 1978. Sulfur and trace element relationships in aerosols from the South American continent, *Geophys. Res. Lett.* 5: 195–198.

Lawson, D. R., and J. W. Winchester. 1979. Sulfur, potassium, and phosphorus associations in aerosols from South American tropical rain forests, *J. Geophys. Res.* 84(C7): 3723–3727.

Leck, C., and Rodhe, H. 1989. On the relation between anthropogenic SO_2 emissions and concentration of sulfate in air and precipitation, *Atmos. Environ.* 23: 959–966.

Lelieveld, J. 1993. Multi-phase processes in the atmospheric sulfur cycle. In *Interactions of C, N, P and S Biogeochemical Cycles and Global Change,* ed. R. Wollast, F. T. Mackenzie, and L. Chou, pp. 305–331. Berlin-Heidelberg: Springer-Verlag.

Lenhard, U., and G. Gravenhorst. 1980. Evaluation of ammonia fluxes into the free atmosphere over Western Germany, *Tellus* 32: 48–55.

Levine, J. S., T. R. Augustsson, I. C. Anderson, and J. M. Hoell, Jr. 1984. Tropospheric sources of NO_x: Lightning and biology, *Atmos. Environ.* 18: 1797–1804.

Levine, J. S., W. S. Cofer III, D. I. Sebacher, E. L. Winstead, S. Seibacher, and P. J. Boston. 1988. The effects of fire on biogenic soil emissons of nitric oxide and nitrous oxide, *Global Biogeochem. Cycles* 2: 445–449.

Levy, H. II, and W. J. Moxim. 1989. NO_x emissions to ocean simulated global distribution and deposition of reactive nitrogen emitted by fossil fuel combustion, *Tellus* 41B: 256–271.

Lewis, W. M., Jr. 1981. Precipitation chemistry and nutrient loading by precipitation in a tropical watershed, *Water Resources Res.* 17(1): 161–181.

Lewis, W. M., Jr., and M. C. Grant. 1980. Acid precipitation in the western United States, *Science* 207: 176–177.

Likens, G. E. 1976. Acid precipitation, *Chem. Eng. News* 54(48): 29–44.

Likens, G. E. 1989. Some aspects of air pollution effects on terrestrial ecosystems and prospects for the future, *Ambio* 18: 172–178.

Likens, G. E., F. H. Bormann, L. O. Hedin, C. T. Driscoll, and J. E. Eaton. 1990. Dry deposition of sulfur: A 23 year record for the Hubbard Brook Forest ecosystem, *Tellus* 42B: 319–329.

Likens, G. E., F. H. Bormann, R. S. Pierce, and J. S. Eaton. 1984. The Hubbard Brook valley: Biogeochemistry. In *An Ecosystem Approach to Aquatic Ecology: Mirror Lake and Its Environment,* ed. G. E. Likens, chap. 2. New York: Springer-Verlag.

Likens, G. E., W. C. Keene, J. M. Miller, and J. N. Galloway. 1987. Chemistry of precipitation from a remote terrestrial site in Australia, *J. Geophys. Res.* 92: 13299–13314.

Likens, G. E., R. F., Wright, J. N. Galloway, and T. J. Butler. 1979. Acid rain, *Sci. Am.* 241(4): 43–51.

Lipschultz, F., O. C. Zafiriou, S. C. Wofsy, M. B. McElroy, F. W. Valois, and S. W. Watson. 1981. Production of NO and N_2O by soil nitrifying bacteria, *Nature* 294: 641–643.

Liss, P. S., and J. N. Galloway. 1993. Air-sea exchange of sulphur and nitrogen and their interaction in the marine atmosphere. In *Interactions of C, N, and P and S Biogeochemical Cycles and Global Change,* ed. R. Wollast, F. T. Mackenzie, and L. Chou, pp. 259–281. Berlin-Heidelberg: Springer-Verlag.

Lodge, J. P., Jr., J. B. Pake, W. Basbasill, G. S. Swanson, K. C. Hill, A. L. Lorange, and A. L. Lazrus. 1968. *Chemistry of U.S. Precipitation.* Report of Natl. Precipitation Sampling Network, Natl. Center for Atmospheric Research, Boulder, Colo.

Logan, J. A. 1983. Nitrogen oxides in the troposphere: Global and regional budgets, *J. Geophys. Res.* 88(Cl5): 10785–10807.

Logan, J. A., M. J. Prather, S. C. Wofsy, and M. B. McElroy. 1981. Tropospheric chemistry: A global perspective, *J. Geophys. Res.* 86(C8): 7210–7254.

Lvovitch, M. I. 1973. The global water balance, *Trans. Amer. Geophys. Union* 227(3): 60–70.

Lyons, W. B., P. A. Mayewski, M. J. Spencer, and M. S. Twickler. 1990. Nitrate concentrations in snow from remote areas: Implication for the global NO_x flux, *Biogeochemistry* 9: 211–222.

Mason, B. 1966. *Principles of Geochemistry,* 3rd ed. New York: John Wiley.

Mason, B. J. 1971. *Physics of Clouds,* 2nd ed. New York: Oxford University Press.

Means, J. L., R. F. Yuretich, D. A. Crerar, D. J. J. Kinsman, and M. P. Borcsik. 1981. *Hydrogeochemistry of the New Jersey Pine Barrens.* Dept. of Environmental Protection, N.J., Geol. Survey Bull. 76, Trenton, N.J.

Mészáros, E. 1982. On the atmospheric input of sulfur into the ocean, *Tellus* 34: 277–282.

Meybeck, M. 1979. Concentration des eaux fluviales en elements majeurs et apports en solution aux octans, *Rev. Geol. Dyn. Geogr. Phys.* 21: 215–246.

Meybeck, M. 1982. Carbon, nitrogen and phosphorus transport by world rivers, *Am. J. Sci.* 282: 401–450.

Meybeck, M. 1983. Atmospheric inputs and river transport of dissolved substances. In *Dissolved Loads of Rivers and Surface Water Quantity/Quality Relationships,* Proceedings of the Hamburg Symposium, August 1983, pp. 173–192. IAHS Publ. 141.

Miller, A. C., J. C. Thompson, R. E. Peterson, and D. R. Haragan. 1983. *Elements of Meteorology,* 4th ed. Columbus, Ohio: Chas. E. Merrill.

Miller, D. H. 1977. *Water at the surface of the Earth.* New York: Academic Press.

Miller, J. M. 1974. A statistical evaluation of the U.S. precipitation chemistry network. In *Precipitation Scavenging,* ed. R. G. Semonin and R. W. Beadle, pp. 639–661. ERDA Symp. ser. 41.

Miller, J. M., and A. M. Yoshinaga. 1981. The pH of Hawaiian precipitation, a preliminary report, *Geophys. Res. Lett.* 8: 779–782.

Mohnen, V. A. 1988. The challenge of acid rain, *Sci. Am.* 259: 30–38.

Möller, D. 1984. Estimation of global man-made sulfur emissions, *Atmos. Environ.* 18(1): 19–27.

Moody, J. L., and J. N. Galloway. 1988. Quantifying the relationship between atmospheric transport and the chemical composition of precipitation on Bermuda, *Tellus* 40B: 463–479.

Moody, J. L., A. A. P. Pszenny, A. Gaudry, W. C. Keene, J. N. Galloway, and G. Polian. 1991. Precipitation composition and its variability in the Southern Indian Ocean: Amsterdam Island, 1980–1987, *J. Geophys. Res.* 96: 20769–20786.

Munger, J. W. 1982. Chemistry of atmospheric precipitation in the north-central United States: Influence of sulfate, nitrate, ammonia and calcareous soil particulates, *Atmos. Environ.,* 16(7): 1633–1645.

Munger, J. W., and S. J. Eisenreich. 1983. Continental-scale variations in precipitation chemistry, *Environ. Sci. Technol.* 17(1): 32A–42A.

Nader, J. S. 1980. Primary sulfate emissions from stationary industrial sources. In *Atmospheric Sulfur Deposition—Environmental and Health Effects,* ed. D. S. Shriner, C. R. Richmond, and S. Lindberg, pp. 121–130. Ann Arbor, Mich.: Ann Arbor Science.

Neiburger, M., J. G. Edinger, and W. D. Bonner. 1973. *Understanding Our Atmospheric Environment.* San Francisco: W. H. Freeman.

Newell, R. E. 1971. The global circulation of atmospheric pollutants, *Sci. Am.* 224(1): 32–42.

Nguyen, B. C., N. Mihalopoulos, J. P. Putaud, A. Gaudry, B. Gallet, W. C. Keene, and J. N. Galloway. 1992. Covariations in oceanic dimethyl sulfide, its oxidations products and rain acidity at Amsterdam Island in the Southern Indian Ocean, *J. Atmos. Chem.* 15: 39–53.

NRC (National Research Council). 1979. *Ammonia.* Washington, D.C.: National Academy of Sciences.

NRC (National Research Council). 1983. *Acid Deposition: Atmospheric Processes in Eastern North America.* Washington, D.C.: National Academy Press.

NRC (National Research Council). 1986. *Acid Deposition: Long-Term Trends,* ed. G. H. Gibson. Washington, D.C.: National Academy Press.

Oden, S. 1968. The acidification of air and precipitation and its consequences on the natural environment (In Swedish), *Statens Naturvetenskapliga Forskningsrid, Stockholm. Bull.* 1: 1–86.

Oppenheimer, M., C. E. Epstein, and R. E. Yuhnke. 1985. Acid deposition, smelter emissions, and the linearity issue in the western United States, *Science* 229: 859–862.

Overein, L. N. 1972. Sulfur pollution pattern observed: Leaching of calcium in forest soil determined, *Ambio* 1: 145–147.

Paciga, J. J., and R. E. Jervis. 1976. Multielement size characterization of urban aerosols, *Environ. Sci. Technol.* 10(12): 1124–1128.

Pack, D. H. 1980. Precipitation chemistry patterns: A two-network data set, *Science* 208: 1143–1145.

Pearson, F. J., Jr., and D. W. Fisher. 1971. Chemical composition of atmospheric precipitation in the northeastern United States. *USGS Water Supply Paper 1535-P.*

Peixoto, J. P., and M. A. Kettani. 1973. The control of the water cycle, *Sci. Am.* 228(4): 46–61.

Penner, J. E., and C. S. Atherton, J. Dignon, S. J. Ghan, and J. J. Walton, and S. Hameed. 1991. Tropospheric nitrogen: A three-dimensional study of sources, distributions and deposition, *J. Geophys. Res.* 96: 959–990.

Petrenchuk, O. P. 1980. On the budget of sea salts and sulfur in the atmosphere, *J, Geophys. Res.* 85(Cl2): 7439–7444.

Petrenchuk, O. P., and E. S. Selezneva. 1970. Chemical composition of precipitation in regions of the Soviet Union, *J. Geophys. Res.* 75(18): 3629–3634.

Post, D., and H. A. Bridgman. 1991. Fog and rainwater composition in rural SE Australia, *J. Atmos. Chem.* 13: 83–95.

Prospero, J. M., and D. L. Savoie. 1989. Effect of continental sources on nitrate concentrations over the Pacific Ocean, *Nature* 339: 687–689.

Pruppacher, H. R. 1973. The role of natural and anthropogenic pollutants in clouds and precipitation formation. In *Chemistry of the Lower Atmosphere,* ed. S. I. Rasool, pp. 1–62. New York: Plenum Press.

Pszenny, A. A. P., F. MacIntyre, and R. A. Duce. 1982. Sea salt and the acidity of marine rain on the windward coast of Samoa, *Geophys. Res. Lett.* 9: 751–754.

Quinn, P. K., T. S. Bates, J. E. Johnson, D. S. Covert, and R. J. Charlson. 1990. Interactions between the sulfur and reduced nitrogen cycles over the central Pacific ocean, *J. Geophys. Res.* 95: 16405–16416.

Quinn, P. K., R. J. Charlson, and T. S. Bates. 1988. Simultaneous observations of ammonia in the atmosphere and ocean. *Nature* 335: 336–338.

Robinson, E., and R. C. Robbins. 1975. Gaseous atmospheric pollutants from urban and natural sources. In *The Changing Global Environment,* ed. S. F. Singer, pp. 111–123. Dordrecht: D. Reidel.

Rodhe, H. 1989. Acidification in a global perspective, *Ambio* 18: 155–160.

Rodhe, H., P. Crutzen, and A. Vanderpol. 1981. Formation of sulfuric and nitric acid during long-range transport, *Tellus* 33: 132–141.

Rodhe, H., and Isaksen, 1980. Global distribution of sulfur compounds in the troposphere estimated in a height/latitude transport model, *J. Geophys. Res.* 85 (CR): 7401–7409.

Rodhe, H., and Rood, M. J. 1986. Temporal evolution of nitrogen compounds in Swedish precipitation since 1955, *Nature* 321: 762–764.

Ryaboshapko, A. G. 1983. The atmospheric sulfur cycle. In *The Global Biogeochemical Sulphur Cycle,* ed. M. V. Ivanov and J. R. Freney, SCOPE 19, pp. 203–296. Chichester, U.K.: John Wiley.

Saltzman, E. S., and D. J. Cooper. 1988. Shipboard measurements of atmospheric dimethylsulfide and hydrogen sulfide in the Caribbean and Gulf of Mexico. *J. Atmos. Chem.* 7: 191–209.

Savoie, D. L., and J. M. Prospero. 1980. Water-soluble K, Ca and Mg in the aerosols over the tropical North Atlantic, *J. Geophys. Res.* 85(Cl): 385–392.

Savoie, D. L., and J. M. Prospero. 1989. Comparison of oceanic and continental sources of non-sea-salt sulphate over the Pacific Ocean, *Nature* 339: 685–687.

Savoie, D. L., J. M. Prospero, and E. S. Saltzman. 1989. Non-sea-salt sulfate and nitrate in trade wind aerosols at Barbados: Evidence for long-range transport, *J. Geophys. Res.* 94: 5069–5080.

Schindler, D. W. 1988. Effects of acid rain on freshwater ecosystems, *Science* 239: 149–157.

Schindler, D. W., R. W. Newbury, K. G. Beatty, and P. Campbell. 1976. Natural water and chemical budgets for a small Precambrian lake basin in central Canada, *J. Fish. Res. Bd. Can.* 33: 2526–2543.

Schlesinger, W. H., and A. E. Hartley. 1992. A global budget for atmospheric NH_3, *Biogeochemistry* 15: 191–211.

Schofield, C. L. 1976. Acid precipitation: Effects on fish, *Ambio* 5(5–6): 228–230.

Schultze, E.-D. 1989. Air pollution and forest decline in a spruce (*Picea abies*) forest, *Science* 24: 776–783.

Sellers, W. D. 1965. *Physical Climatology,* 2nd ed. Chicago: University of Chicago Press.

Sequeira, R. 1976. Monsoonal deposition of sea salt and air pollutants over Bombay, *Tellus* 28(3): 275–281.

Shacklette, H. T., J. C. Hamilton, J. G. Boerngen, and J. M. Bowles. 1971. Elemental composition of surficial materials in the conterminous United States, *USGS Prof. Paper 574-D.*

Slemr, F. and W. Seiler. 1984. Field measurements of NO and NO_2 emissions from fertilized and unfertilized soils, *J. Atmos. Chem.* 2: 1–24.

Smith, R. A., R. B. Alexander, and M. G. Wolman. 1987. Water quality trends in the nation's rivers, *Science* 235: 1607–1615.

Söderlund, R., and B. H. Svensson. 1976. The global nitrogen cycle. In *Nitrogen, Phosphorus and Sulphur—Global Cycles,* SCOPE Report 7. ed. B. H. Svensson and R. Soderlund, pp. 23–73. Stockholm: Ecol. Bull. 22.

Stallard, R. F. 1980. Major element geochemistry of the Amazon river system. Ph.D. dissertation, MIT/Woods Hole Oceanographic Inst., WHO I-80-29.

Stallard, R. F., and J. M. Edmond. 1981. Chemistry of the Amazon, precipitation chemistry and the marine contribution to the dissolved load at the time of peak discharge, *J. Geophys. Res.* 86(C10): 9844–9858.

Stedman, D. H., and R. E. Shetter. 1983. The global budget of atmospheric nitrogen species. In *Trace Atmospheric Constituents,* ed. S. E. Schwarz, pp. 411–454. New York: John Wiley.

Stensland, G. J., and R. G. Semonin. 1982. Another interpretation of the pH trend in the United States, *Bull. Am. Meteorol. Soc.* 63: 1277–1284.

Stensland, G. J., and R. G. Semonin. 1984. Response to comment of T. S. Butler, C. V. Cogbill, and G. E. Likens, *Bull. Am. Meteorol. Soc.* 63: 640–643.

Stoiber, R. E., S. N. Williams, and B. Huebert. 1987. Annual contribution of sulfur dioxide to the atmosphere by volcanoes, *J. Volcanol. Geotherm. Res.* 33: 1–8.

Sugawara, K. 1967. Migration of elements through phases of the hydrosphere and atmosphere. In *Chemistry of the Earth's Crust,* vol. 2, ed. A. P. Vinogradov, pp. 501–510. Israel Program for Scientific Translation Ltd., Jerusalem. Reprinted in *Geochemistry of Water,* ed. Y. Kitano, pp. 227–237. New York: Halsted Press, 1975.

Talbot, R. W., K. M. Beecher, R. C. Harriss, and W. R. Cofer III. 1988. Atmospheric geochemistry of formic and acetic acids at a mid-latitude temperate site, *J. Geophys. Res.* 93: 1638–1652.

Tanaka, N. and K. K. Turekian. 1991. Use of cosmogenic ^{35}S to determine the rates of removal of atmospheric SO_2, *Nature* 352: 226–228.

Tanaka, N., and K. K. Turekian. 1995. The determination of the dry deposition flux of SO_2 using cosmogenic ^{35}S and ^{7}Be measurements, *J. Geophys. Res.* 100 (D2): 2841–2848.

Tanaka, S., M. Darzi, and J. W. Winchester. 1980. Sulfur and associated elements and acidity in continental and marine rain from north Florida, *J. Geophys. Res.* 85(C8): 4519–4526.

Thornton, J. D., and S. J. Eisenreich. 1982. Impact of land use on the acid and trace element composition of precipitation, *Atmos. Environ.* 16: 1945–1955.

Tjepkema, J. D., R. J. Cartica, and H. F. Hemond. 1981. Atmospheric concentration of ammonia in Massachusetts and deposition on vegetation, *Nature* 294: 445–446.

Varhelyi, G., and G. Gravenhorst. 1981. Production rate of airborne sea salt sulfur deduced from chemical analysis of marine aerosols and precipitation. Paper presented at *IAMAP-ROAC Symposium,* Hamburg, Germany, August. 24–27, 1981.

Visser, S. 1961. Chemical composition of rainwater in Kampala, Uganda, and its relation to meteorological and topographical conditions, *J. Geophys. Res.* 66: 3759–3766.

Vong, R. J., H. C. Hansson, D. S. Covert, and R. J. Charlson. 1988. Acid rain: Simultaneous observations of a natual marine background and its acidic sulfate aerosol precursor, *Geophys. Res. Lett.* 15: 338–341.

Warneck, P. 1988. *Chemistry of the Natural Atmosphere,* Geophys. Ser. vol. 41. London: Academic Press.

Whitehead, H. C., and J. H. Feth. 1964. Chemical composition of rain, dry fallout and bulk precipitation, Menlo Park, Calif., 1957–1959. *J. Geophys. Res.* 69: 3319–3333.

Williams, E. J., D. D. Parish, and F. C. Fesenfeld. 1987. Determination of nitrogen emissions from soils: Results from a grassland site in Colorado, United States, *J. Geophys. Res.* 92: 2173–2179.

World Resources 1988–89. 1988. World Resources Institute and International Institute for Environment and Development in Collaboration with the United Nations Environment Programme. New York and Oxford: Basic Books.

World Resources 1992–93. 1992. World Resources Institute and International Institute for Environment and Development in Collaboration with the United Nations Environment Programme. New York and Oxford: Oxford University Press.

Wright, R. F., and E. T. Gjessing. 1976. Changes in the chemical composition of lakes, *Ambio* 5(5–6): 219–223.

Zhao, D., and B. Sun. 1986. Air pollution and acid rain in China, *Ambio* 15: 2–5.

Zobrist, J., and W. Stumm. 1980. Chemical dynamics of the Rhine catchment area in Switzerland: Extrapolation to the "pristine" Rhine river input into the ocean. In *River Inputs to Ocean Systems,* ed. J.-M. Martin, J. D. Burton, and D. Eisma, pp. 52–63. Rome: SCOR/UNEPIUNESCO Review and Workshop, FAO.

Zverev, V. P., and V. Z. Rubeikin. 1973. The role of atmospheric precipitation in circulation of chemical elements between atmosphere, hydrosphere, and lithosphere, *Hydrogeochemistry* 1: 613–620.

CHAPTER 4

CHEMICAL WEATHERING AND WATER CHEMISTRY

INTRODUCTION AND HYDROLOGIC TERMINOLOGY

Water falling to the surface of the continents as rain, upon striking the surface, undergoes major modification of both its mode of transport and its chemical composition. The water may infiltrate the soil, or it may immediately run off the surface. It may first be intercepted by vegetation and then "drip" to the ground. The water may be lost back to the atmosphere by evaporation from the ground or from trees. Finally, it may pass deep underground, only to emerge at a much later date in a river or lake. In all cases it comes into contact with substances that react with it and, as a result, alter its composition.

An idea of the paths that water may take once it strikes the ground is shown diagramatically in Figure 4.1. Water that has been intercepted by vegetation and then drips off it is termed *throughfall.* (Water running down plant stems or tree trunks is called stemflow). Water infiltrating the soil is called *soil water,* and that passing directly into the nearest stream is referred to as *surface runoff.* Once in the soil, the water either passes downward or is taken up by plant and tree roots. In the latter case, the water is transported up through the tree and eventually evaporated from leaf surfaces. In this way the water is returned to the atmosphere; the overall process is known as *transpiration.* Water trickling downward through the soil eventually encounters a level in the soil or underlying bedrock where all pore space is filled with water. At this point the water becomes *groundwater,* the rock or soil is said to be saturated with water, and the level where this occurs is known as the *water table* (Figure 4.1). (Above the water table, pore space is filled by a mixture of air and water, and this region is referred to as the unsaturated zone or zone of aeration.)

Groundwater flows underground until the water table intersects the land surface and the flowing water becomes surface water in the form of springs, rivers, swamps, and lakes. The continual contribution of groundwater to rivers, important between rainstorms, is known as *base flow.* Groundwater continues to flow due to a hydrostatic head built up by the recharge of new rainwater at the source. Because of day-to-day, seasonal, and longer-term climatic changes, the

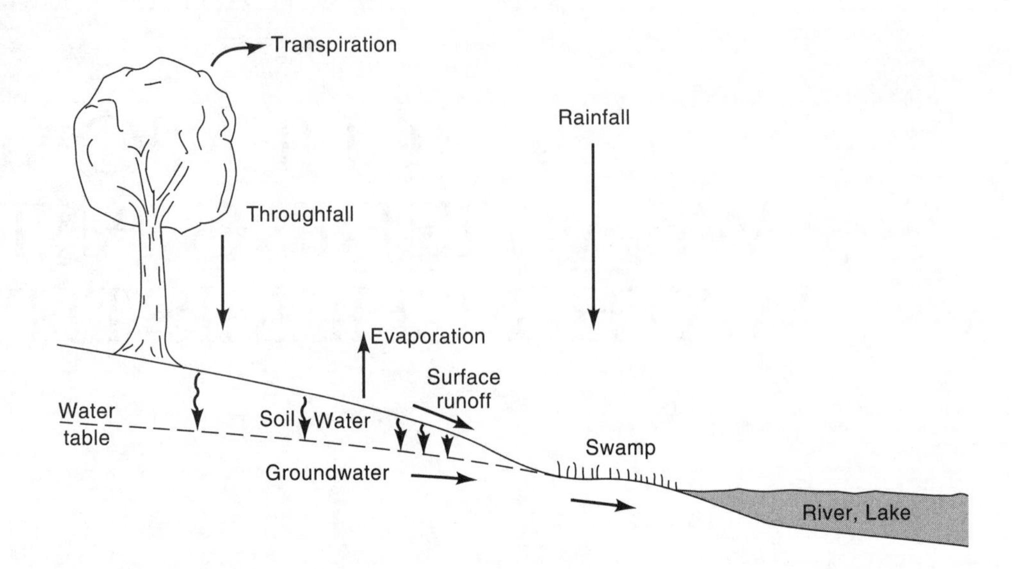

Figure 4.1. Pathways of water near the land surface.

rate of rainfall input and, consequently, the position of the water table can fluctuate, but the degree of fluctuation and its timing can be considerably damped and delayed depending on the capacity of the subsurface rocks to store groundwater. [For further details on groundwater hydrology consult, e.g., Freeze and Cherry (1979).]

Water coming into contact with rocks (and derived soils) reacts with primary minerals contained in them. The minerals dissolve to varying extents, and some of the dissolved constituents react with one another to form new, or secondary, minerals. Dissolution is brought about mainly by acids provided by plant activity and by bacterial metabolism (and, in areas of pollution, by acid rain); the overall process is called *chemical weathering.* Besides biological factors, chemical weathering is also aided by physical processes that act to break up rocks and expose additional mineral surface area to weathering solutions. This is known as *physical weathering,* and the dominant process is the fracturing of rocks by expansion accompanying the freezing of water in cracks. Thus, physical weathering is most important at higher latitudes and elevations. Together, chemical, biological, and physical weathering result in the breakdown of rock and the formation of soil.

Besides soil formation, chemical weathering also results in a radical change in the composition of soil water and groundwater. These changes reflect both the composition of the primary minerals and the degree of biological activity acting to bring about mineral dissolution. Although limited dissolution can occur by reactions of primary minerals with pure rainwater, it is safe to say that most weathering and consequent change in water composition is brought about, directly or indirectly, by biological activity, and it is this intimate interplay between rocks, water, and life that we shall emphasize in this chapter. In fact, we shall begin our coverage of weathering with a discussion of biological cycling or, in simple terms, a "tree's-eye view of weathering."

BIOGEOCHEMICAL CYCLING IN FORESTS

In forested areas, natural water chemistry is influenced by the uptake, storage, and release of nutrients by the vegetation. (Nutrients are elements needed by living things.) A forest can be considered a reservoir of nutrient elements between their input and their release in streamwater runoff. Thus, biogeochemical cycles exist that illustrate interaction of the hydrologic (water) cycle with plant nutrient cycles. The degree of interaction varies, and, in some cases, the biological cycle may be in large part independent of the weathering process.

The major nutrient elements needed by plants are shown in Table 4.1 in the order of concentrations required for growth. Life cannot exist without water and carbon dioxide, which are the principal sources of carbon, hydrogen, and oxygen. In addition, essential life components, such as proteins, nucleic acids, and ATP, also require nitrogen, phosphorus, and sulfur. Together these elements combine to produce an overall average composition for land plants (on a mole basis) of $C_{1200}H_{1900}O_{900}N_{35}P_2S_1$ (based on data of Table 4.1). Other major elements include magnesium and potassium, which are essential components in the chlorophyll used by plants for photosynthesis. Besides major elements, trace nutrients (Table 4.1) are also required but are taken up in much lower concentrations.

Nutrient elements are taken up by plants from four immediate sources (Likens et al. 1977). They are the atmosphere, dead organic matter, minerals, and soil solutions. Atmospheric nutrients include gases, which can be added directly to the plants, and aerosols, which are carried by the atmosphere and trapped on foliage. Dead organic matter in the soil contains nutrients that are made available to plants upon microbial decomposition. (In tropical rain forests, which are commonly developed on heavily leached soils, dead organic matter constitutes the principal source of nutrients.) Minerals, both primary and secondary, are the principal ultimate source of many nutrients but require solubilization, via weathering, before the nutrients can become available. Finally, soil solutions carry nutrients, derived from minerals, rainwater, and organic matter, directly to the plant roots.

An intermediate state of nutrient availability also exists as adsorbed ions and molecules on the surfaces of organic matter and secondary minerals such as clays. Adsorption represents a reservoir of nutrients that is in constant exchange equilibrium with the soil solution. Many factors, such as crystal chemistry, the presence of organic functional groups (e.g., —COOH),

TABLE 4.1 Elements Essential for Nutrition of Plants

Element	Adequate Concentration (% dry wt. of tissue)
Carbon	45
Oxygen	45
Hydrogen	6
Nitrogen	1.5
Potassium	1.0
Calcium	0.5
Phosphorus	0.2
Magnesium	0.2
Sulfur	0.1
Chlorine	0.01
Iron	0.01
Manganese	0.005
Zinc	0.002
Boron	0.002
Copper	0.0006
Molybdenum	0.00001

Source: Zinke 1977.

pH, temperature, etc., affect adsorption. [For a proper treatment of adsorption in soils, as well as a discussion of the allied process, ion exchange, consult Sposito (1989).]

Nutrients can also be provided by the trees themselves. Within the overall biogeochemical cycle of a forested watershed, there is an internal cycle that is at least partly independent of weathering (Likens et al. 1977; Graustein 1981). An element is taken up by a tree in solution through its roots and carried upward through the tree. A certain amount (which varies from element to element) is stored in living and dead vegetation (biomass storage) and the rest is released to be recycled as (1) litterfall, dead leaves lost to the forest floor (important for Ca and Mg); (2) throughfall, the washoff and leaching of an element from the leaves by precipitation (important for K); and (3) root exudates, loss of nonutilized elements in solution from living and dead roots (important for Na). By these processes the litterfall, throughfall, and exudates all add nutrients to the soil that are readily taken up again by the tree through its roots.

A generalized biogeochemical cycle of an element in a forested area, including the internal cycle just discussed, is shown in Figure 4.2. The trees take up nutrients from the various sources discussed above. The primary external inputs are atmospheric gases, rainfall, trapped aerosol dust, and underlying rocks. Atmospheric gases and rain provide some of the sulfur, nitrogen, and phosphorus, whereas aerosols and rocks provide most of the potassium, magnesium, calcium, sodium, and silicon. Losses from the forest take place via groundwater flow and streams.

If a forest is at steady state, that is, if it is not growing and storing nutrients, inputs equal outputs and the change in water composition, in passing from rainwater to streamwater, is due solely to the release of elements by rock weathering (and aerosol dust dissolution). For large forested regions the assumption of steady state is probably valid, but only in the absence of human deforestation (which has become a major global problem). For single, small watersheds, year-to-year variations in biological activity can have a major effect on runoff composition, and corrections for biomass storage (or release) should be made before any attempt is undertaken to estimate the degree of rock weathering from the composition of streamwaters or groundwaters.

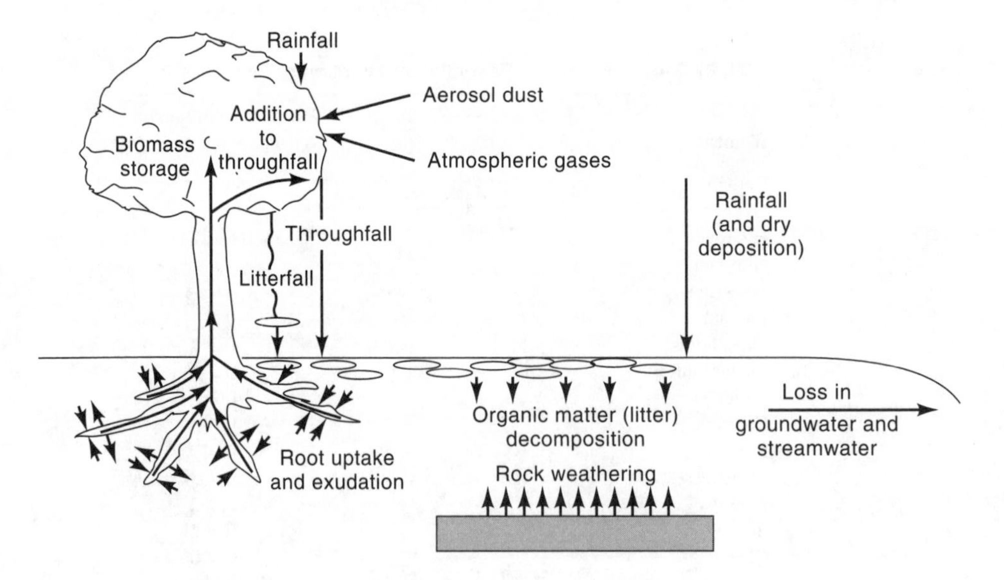

Figure 4.2. Biogeochemical cycling in forests.

An example of estimating rates of weathering by correcting for biomass storage is provided by the study of Taylor and Velbel (1991) of weathering in an area of the southern Appalachians (see also Velbel 1985). By constructing a series of mass balance expressions for the principal cations (Ca, Na, Mg, and K), they were able to combine stream and rainwater chemistry with determinations of mineralogy and the composition of biomass to calculate rates of weathering of the principal minerals in the rocks of this area. They found that by ignoring biomass storage they calculated mineral weathering rates that were too low by as much as a factor of four.

The relative importance of the various inputs and outputs, and resulting storage or release of elements, varies both with the area considered and with the element. For example, Likens et al. (1977), in a forested New Hampshire area (Hubbard Brook Experimental Watershed), found that 80–90% of the streamwater output of calcium, magnesium, potassium, and sodium came from rock weathering, whereas nitrogen and sulfur were derived almost entirely from rain and atmospheric gases with little or none from weathering. (In this area, aerosol dust input was believed to be small because of the relatively humid climate.) Likens et al. also showed that the amount of total weathering, calculated after correction for biomass storage, was considerably greater than that indicated simply by concentrations in runoff minus those in rainfall.

Graustein (1981), working in a mountainous area in New Mexico surrounded by arid lowlands, found a much greater input from aerosols. Spruce and fir trees trapped considerable quantities of wind-blown dust, which was washed off by precipitation and became part of the throughfall (in addition to elements recycled by the tree and released from the foliage). Based partly on strontium isotope studies, he concluded that tree-trapped dust accounted for about two thirds of the calcium added to throughfall as opposed to one third from biologic recycling. (The relative importance of dust versus botanical recycling was greater for sodium and less for magnesium and potassium.) The amount of calcium, sodium, magnesium, and potassium added by tree-trapped dust was about two to three times the amount originally present in precipitation. However, the type of tree was important: The trapping of dust by nearby aspen trees, by contrast, was found to be negligible.

In Graustein's (1981) study, the flux of spruce- and fir-trapped dust dissolved in throughfall was a significant portion of the flux of calcium and potassium in runoff but a minor part of sodium and magnesium losses. Thus, calculations of rates of chemical weathering for Ca^{2+} and K^+ based on streamwater output minus precipitation input of these elements without considering tree-trapped dust input would have resulted in too large an estimate for the amount of bedrock weathering.

Graustein (1981) also points out the importance of biological internal cycling on soil-water composition. Most of the dissolved K^+ in soil waters of his watersheds was derived from throughfall additions as a result of leaf exudation (not dust). This increased K^+ should affect the rate and type of weathering of potassium minerals as well as estimates of the importance of weathering on streamwater runoff composition.

If one measures biological activity as the ratio of an element stored annually in living and dead vegetation to the annual loss from the area in streamflow, the elements in order of decreasing biological activity are

$$P > N > K > Ca > S > Mg > Na$$

The most biologically affected element is thus P, with N coming second. K and Ca are considerably affected by vegetation, while Na is little affected. Thus, in New Hampshire (Likens et al. 1977), the annual increase in biomass storage for potassium was much greater than streamwater losses, while that for calcium was somewhat less than streamwater runoff. Further, the flux of potassium cycled internally by the vegetation was around 25 times the flux lost in streamflow, whereas that for Ca and Mg was about three to six times the loss via runoff.

The rate of turnover of nutrients via the decomposition of litter varies with the type of vegetation, which in turn is a strong function of climate. Schlesinger (1991) lists the following mean residence times for forest litter:

Boreal (high latitude) forest	353 yr
Temperate coniferous forest	17 yr
Temperate deciduous forest	4 yr
Mediterannean scrub	3.8 yr
Tropical rain forest	0.4 yr

The overall effect here appears to be mean annual temperature as it affects the rate of decomposition of soil organic matter and the liberation of nutrients. Although litter on a global basis (55,000 Tg) is considerably less abundant than total soil organic matter (1,500,000 Tg) (Schlesinger 1991), it is more effective in nutrient regeneration because of its much faster turnover rate. (Total organic matter has a mean residence time on the scale of a few thousand years).

In temperate climates with deciduous forests, the biological control over elements such as potassium and nitrogen is strongly seasonal. During the forest growing season in summer, potassium input in precipitation is greater than potassium loss in streamflow; apparently the vegetation takes up potassium from both the soil and from rainwater. Only during the dormant season, when leaves are on the ground, is the streamflow output greater than the precipitation input. This is in contrast to Ca, Mg, Na, and Al, which show greater output than input regardless of the season.

Phosphorus input in precipitation in many forests is greater than output in streamflow because tremendous amounts of phosphate are stored in vegetation. Only in agricultural areas, where large amounts of waste and fertilizer phosphate enter the runoff, does stream runoff exceed precipitation input. Weathering is not as important a source for phosphate as it is for Ca, Mg, K, and Na.

Although the principal immediate source of nitrogen by far in forest biomass is nitrogen recycled from soil organic matter and soil solution (plus adorbed NH_4^+ on clays), the ultimate source is almost entirely the atmosphere. Nitrogen is added from the atmosphere by way of nitrate and ammonium dissolved in rain and in certain plants (e.g., legumes) via the direct fixation of N_2 by symbiotic root microbes. (For a detailed discussion of the terrestrial cycle of nitrogen, see Chapter 5.) Very little nitrogen is added from rock weathering.

Besides nutrient cycling, trees exert indirect control on water composition. This is brought about by transpiration. During transpiration, pure water is lost via evaporation from leaf surfaces to the atmosphere. As a result the dissolved ions, such as Cl^-, that are not taken up by the tree become concentrated in soil water. Likens et al. (1977) have shown that at the New Hampshire location, where transpiration (plus evaporation) accounts for about 40% of the yearly water loss, the concentration of dissolved substances in runoff may be increased by as much as 60% over that which would have prevailed in the absence of this process.

SOIL WATER AND MICROORGANISMS: ACID PRODUCTION

The chemical composition of soil water is affected by inputs from throughfall and rainfall, additions from rock weathering, exchange with soil colloids, and inputs and removals due to biological activity. In addition to the nutrient cycling discussed in the previous section, important biologically induced changes are brought about by soil microorganisms, most notably the production of soil acids. These acids constitute the principal agents of rock weathering (Carroll 1970; Alexander 1961). Acids provide hydrogen ions, which replace cations on mineral surfaces, thus bringing about disintegration of the minerals. Also, some organic acids react with specific elements contained in minerals to form chelates or soluble, multiply bonded, metal-organic complexes. Acid production is accomplished by a variety of organisms including bacteria, fungi, actinomycetes, and algae. [For a detailed discussion of soil microbiology, consult Alexander (1961) or Richards (1987).]

The principal acids produced by soil microorganisms are carbonic and sulfuric acids and a whole host of various organic acids. Carbonic acid (H_2CO_3) is produced by the oxidation of organic matter by microbes to CO_2. In other words, representing organic matter as CH_2O,

$$CH_2O + O_2 \rightarrow CO_2 + H_2O$$

This CO_2 then combines with water to form carbonic acid, which further partly dissociates (it is a weak acid) to H^+ and HCO_3^-:

$$CO_2 + H_2O \rightarrow H_2CO_3$$

$$H_2CO_3 \rightarrow H^+ + HCO_3^-$$

Sulfuric acid (H_2SO_4) is produced by the bacterially catalyzed oxidation of sulfide minerals and, in soils developed on rocks with high sulfide mineral contents, high concentrations of this strong acid and, consequently, low pH values can result. This is especially true where mining has brought about enhanced oxidation of sulfides. (Further details on sulfuric acid weathering are provided later in this chapter.)

Organic acids are formed by the partial breakdown of organic matter. Chief among these is a variety of high-molecular-weight, poorly defined compounds collectively referred to as humic and fulvic acids (see, e.g., Schnitzer and Khan 1972). These common acids, which impart characteristic brown and yellow colors to soil solutions, attack minerals and chelate and solubilize several metallic elements (e.g., Fe, Al). Precipitation of the acids brings about the formation of humus. In addition, certain organisms secrete specific, well-defined acids. For example, fungi living in the litter layer and upper portions of the soil exude oxalic acid ($H_2C_2O_4$), which reacts with iron and aluminum minerals to form dissolved iron and aluminum oxalate chelates (Graustein 1981). Iron and aluminum, which are otherwise highly insoluble, are carried downward in the soil by oxalate until the oxalate itself is broken down to CO_2 and HCO_3^- by additional microorganisms. As a result, the iron and aluminum are precipitated in lower portions of the soil. The net effect of the oxalate is to move the iron and aluminum downward. This movement, which is also accomplished by a number of other soil acids, is one of the main processes that bring about the differentiation of soils into specific horizons (see section on soil formation).

The importance of organic acids in soil chemistry is demonstrated by the results of a study of volcanic ash weathering in Wyoming by Antweiler and Drever (1988). They found an excellent positive correlation between dissolved Al (or Fe) and dissolved organic matter and a good negative correlation between pH and dissolved organic matter in soil solutions extracted using porous-cup lysimeters. This is just the result one would be expect if organic chelation of Al and Fe were important and if soil water pH was affected by organic acid dissociation.

As a result of the attack on minerals by soil acids, the input from throughfall, the removal by plant roots, and vertical transport by organic chelates, concentrations of principal elements dissolved in soil water can vary from place to place and with depth in the soil. Some idea of depth variation, taken from the work of Graustein (1981), is shown in Table 4.2. Note that K^+ added at the top by throughfall is greatly depleted with depth due to root uptake. Continual increase of sodium with depth is due to rock weathering. High dissolved Al concentration in the shallow soil water is due to organic chelation, and decrease in Al with depth is due to precipitation upon downward transport and chelate decomposition. Although data of this sort are sparse (see also Drever 1988), these profiles are probably typical of many forest soils.

Before leaving the topic of soil acids, it should be noted that human activities have resulted in the addition of excess acid to soil. This includes H_2SO_4 and HNO_3 from acid rain and H_2SO_4 from the mining of coal or metallic sulfides. Discussion of these anthropogenic sources is given in Chapter 3 (rain) and Chapter 5 (rivers).

CHEMICAL WEATHERING

Minerals Involved in Weathering

The chemical weathering of rocks and minerals has been alluded to throughout the previous discussion without detailing exactly what rocks and minerals are involved. As a guide we present in Tables 4.3 and 4.4 lists of the most common minerals involved in weathering. Primary minerals (Table 4.3) are those undergoing destruction by weathering, while secondary minerals (Table 4.4) are those formed by weathering. (Technically speaking, all minerals can undergo destruction by weathering, but the secondary minerals are the most resistant.) In interpreting Table 4.3, those with no geological background may find it helpful to visualize rocks simply as aggregates of minerals. There are three basic types of rock: igneous, sedimentary, and metamorphic. Igneous rocks are formed by crystallization from a melt at high temperature and include the common rock types, granite (Na plagioclase feldspar, K-feldspar, quartz, biotite) and basalt (Ca-plagioclase feldspar, pyroxenes, olivine). Sedimentary rocks are deposited in water at the

TABLE 4.2 Concentrations of Na^+, K, Ca^{++}, and Al in Soil Waters of a Watershed Forested with Aspen Trees, Sangre de Cristo Mountains, New Mexico, and in Rainfall and Throughfall in the Same Area

		Concentration (μg/l)			
Water	Depth (cm)	Na^+	K^+	Ca^{++}	Al
Rainfall	—	67	120	360	5
Throughfall	—	85	2800	780	10
Soil water	30	710	2200	4400	350
Soil water	100	1600	430	1300	30
Soil water	150	3100	350	1050	38
Soil water	200	3800	510	2450	11

Source: Graustein 1981.

TABLE 4.3 Common Primary Minerals That Undergo Weathering

Mineral	Generalized Composition	Weathering Rock Type(s)	Main Reaction
Olivine	$(Mg,Fe)_2SiO_4$	Igneous	Oxid. of Fe Cong. diss. by acids
Pyroxenes	$Ca(Mg,Fe)Si_2O_6$ or $(Mg,Fe)SiO_3$	Igneous	Oxid. of Fe Cong. diss. by acids
Amphiboles	$Ca_2(Mg,Fe)_5Si_8O_{22}(OH)_2$ (also some Na and Al)	Igneous Metamorphic	Oxid. of Fe Cong. diss. by acids
Plagioclase feldspar	Solid solution between $NaAlSi_3O_8$ (albite) and $CaAl_2Si_2O_8$ (anorthite)	Igneous Metamorphic	Incong. diss. by acids
K-feldspar	$KAlSi_3O_8$	Igneous Metamorphic Sedimentary	Incong. diss. by acids
Biotite	$K(Mg,Fe)_3(AlSi_3O_{10})(OH)_2$	Metamorphic Igneous	Incong. diss. by acids Oxid. of Fe
Muscovite	$KAl_2(AlSi_3O_{10})(OH)_2$	Metamorphic	Incong. diss. by acids
Volcanic glass (not a mineral)	Ca,Mg,Na,K,Al,Fe-silicate	Igneous	Incong. diss. by acids and H_2O
Quartz	SiO_2	Igneous Metamorphic Sedimentary	Resistant to diss.
Calcite	$CaCO_3$	Sedimentary	Cong. diss. by acids
Dolomite	$CaMg(CO_3)_2$	Sedimentary	Cong. diss. by acids
Pyrite	FeS_2	Sedimentary	Oxid. of Fe and S
Gypsum	$CaSO_4 \cdot 2H_2O$	Sedimentary	Cong. diss. by H_2O
Anhydrite	$CaSO_4$	Sedimentary	Cong. diss. by H_2O
Halite	NaCl	Sedimentary	Cong. diss. by H_2O

Note: cong. = congruent; incong. = incongruent; diss. = dissolution; oxid. = oxidation.

earth's surface and include eroded debris from preexisting rocks (e.g., quartz and feldspar), as in sandstones; fine-grained secondary minerals formed by weathering (e.g., iron oxides and clays), as in shales; the skeletal remains of organisms (mainly $CaCO_3$), as in limestones; and precipitates from seawater (gypsum and halite), as in evaporites. Metamorphic rocks form by the recrystallization and alteration of sedimentary and igneous rocks at elevated temperatures and pressures (but without melting) and contain many different minerals, including amphiboles, muscovite, biotite, quartz and feldspar.

Also included in Table 4.3 are the principal weathering reactions that each primary mineral undergoes. Weathering reactions are classified here according to the nature of the attacking substance and whether the primary mineral simply dissolves or whether a portion of it reprecipitates to form a secondary mineral or minerals. Simple dissolution is referred to as *congruent dissolution,* and dissolution with reprecipitation of some of the components of the mineral is called *incongruent dissolution.* Attacking substances are separated into soil acids, dissolved oxygen, and water itself. Dissolved oxygen attacks only those minerals that contain reduced forms of elements, principally iron and sulfur, and that undergo oxidation to form new minerals.

TABLE 4.4 Common Secondary Minerals Formed by Weathering in Soils

Mineral	Composition
Hematite	Fe_2O_3
Goethite	$HFeO_2$
Gibbsite	$Al(OH)_3$
Kaolinite	$Al_2Si_2O_5(OH)_4$
Smectite	$(^1/_2\ Ca, Na)\ Al_3MgSi_8O_{20}(OH)_4 \cdot nH_2O$ (average composition)
Vermiculite	Basically biotite or muscovite composition with K^+ replaced by hydrated cations
Calcite	$CaCO_3$
Opaline silica (not a mineral)	$SiO_2 \cdot nH_2O$
Gypsum	$CaSO_4 \cdot 2H_2O$

Although most minerals are attacked mainly by soil acids, a few very soluble ones simply dissolve in water. This is shown in Table 4.3. In addition, these soluble minerals may also reprecipitate under arid conditions. This is why gypsum, for example, appears in Tables 4.3 and 4.4 as both a primary and a secondary mineral.

Minerals can be listed in order of their degree of resistance to weathering. In other words, if two minerals are present in the same soil and are attacked by the same acids for the same length of time, one will be destroyed faster than the other. On the basis of observations of partly weathered rocks and soils and of responses to different climatic conditions, a table has been prepared (Table 4.5) of minerals listed in order of increasing resistance to weathering. This table is similar to those prepared by others (e.g., Goldich 1938; Loughnan 1969; Carroll 1970) and is based largely on this older work. Although some reordering can occur in different soils, overall the order shown is considered to be well established. Goldich (1938) noted that the order shown for igneous silicate minerals parallels their temperature of formation from molten magma. In other words, the silicate minerals that weather fastest (e.g., olivine) are those that originally formed at the highest temperatures. The reason for this correlation is not clear, but a common explanation (e.g., Goldich 1938) is that minerals formed under conditions more distantly removed from those at the earth's surface are less stable there and thus weather faster. This explanation agrees with the position of the common secondary minerals at the bottom of the list but does not account for the high weatherability of nonsilicates, such as halite, gypsum, calcite, and pyrite, which also form under earth surface conditions.

Mechanism of Silicate Dissolution

Of all mineral groups, silicates have received the most attention in weathering studies because they make up the most abundant rock types. How primary silicates dissolve during weathering, however, is not well known. One theory is that silicate dissolution occurs by means of the formation of a protective surface layer of altered composition on each mineral grain (e.g., Luce et al. 1972; Paces 1973; Busenberg and Clemency 1976; Chou and Wollast 1984; Muir and Nesbitt

TABLE 4.5 Mineral Weatherability (Decreasing from Top to Bottom)

Halite
Gypsum-anhydrite
Pyrite
Calcite
Dolomite
Volcanic glass
Olivine
Ca-plagioclase
Pyroxenes
Ca-Na plagioclase
Amphiboles
Na-plagioclase
Biotite
K-feldspar
Muscovite
Vermiculite, smectite
Quartz
Kaolinite
Gibbsite, hematite, goethite

Note: Minerals are listed in order of increasing resistance to weathering. (Exact positions for some minerals can change one or two places due to effects of grain size, climate, etc.) See also Goldich 1938; Loughnan 1969; and Carroll 1970.

1992). This layer is assumed to be so tight that it severely inhibits the migration of dissolved species to and from the surface of the primary mineral, and in this way is protective. The layer forms from components of the underlying primary mineral and, as weathering proceeds, it increases in thickness. It was invoked originally to explain the results of laboratory dissolution experiments (simulating weathering), where rates of dissolution were seen to decrease with time, due presumably to the thickening of a protective surface layer. However, when applied to most mineral grains taken from actual soils (e.g., Berner and Holdren 1979; Berner and Schott 1982; Blum et al. 1991) or to the surfaces of mineral grains from experiments at neutral pH (e.g., Holdren and Berner 1979; Casey et al. 1989), attempts to prove the existence of a layer with any appreciable thickness greater than a few nanometers, using both electron microscopic and surface chemical techniques, have proven to be negative. Much recent experimental and theoretical work (e.g., Casey et al. 1989) has emphasized layer formation, but it is based almost entirely on results for low-pH experiments outside the normal pH range found in most soils.

What really appears to happen during weathering of most silicate minerals (at least for feldspars, pyroxenes, and amphiboles) is that soil solutions penetrate through permeable (nonprotective) clay layers right to the bare surfaces of the primary mineral grains and there react with them. Dissolution does not occur at all places on the surface so as to produce general rounding of the grains, as predicted by the protective surface layer theory, but instead affects only those portions of the surface where there is excess energy, such as at outcrops of dislocations. (Dislocations are rows of atoms in a crystal that are slightly out of place and therefore more energetic.) As a result of selective etching, distinct crystallographically controlled etch pits form on the mineral surface and, upon growth and coalescence, form interesting features. Some examples taken from our studies of soil feldspars and pyroxenes are shown in Figure 4.3. These etch pits reflect the crystal structure of the underlying mineral and therefore are regular in shape and aligned in certain directions. [For a general discussion of the microscopic study of soil minerals, consult Nahon (1991).]

For some minerals, principally the garnets, a protective surface layer dissolution mechanism may be operative. Velbel (1993b) has shown that the ratio of the molar volume of secondary weathering products to the molar volume of the primary mineral undergoing dissolution is a key factor in whether dissolution via a protective layer is possible. For most primary minerals, all reasonable weathering products (e.g., clay minerals and hydrous aluminum and iron oxides) have insufficient volume, during mole-for-mole weathering, to cover the primary mineral surfaces and thus provide a protective layer. Only the garnets, among common minerals, seem to fulfill the necessary citerion for total covering of the surface. In agreement with this prediction is the finding by Velbel of a lack of etch pitting and a tight covering of gibbsite and goethite on the surfaces of almandine garnet grains in soils from the southern Appalachian Mountains.

Dissolution of primary minerals, therefore, occurs via etch pit formation and growth of the pits. If the pits are located primarily at outcrops of dislocations, then a fundamental controlling factor on dissolution of a given mineral during weathering is the density of dislocations. (Dissolution of different minerals is still controlled by differences in chemical composition.) This can help to explain different rates of dissolution of minerals of similar composition under the same soil conditions. For example, the work of Holdren and Berner (1979) has shown that adularia, $KAlSi_3O_8$, reacts in the laboratory with hydrofluoric acid (as a simulator of soil acids) much more slowly than does microcline, which has essentially the same chemical composition, $KAlSi_3O_8$. The major difference between the two minerals is the presence of numerous twinning dislocations in microcline and very few in adularia. In addition, electron microscope studies (Berner and Schott 1982) show that coexisting hypersthene and augite (two pyroxene minerals in the same soil) weather at different rates, mainly because of differences in dislocation density. This is illustrated in Figure 4.3.

Silicate Weathering Reactions: Secondary Mineral Formation

Because silicate minerals constitute the fundamental components of most major rock types, it is important to inquire in detail into how they weather and to what they weather. As noted in Table 4.3, some weathering reactions involve simple congruent dissolution by water or acids. In the case of silicate minerals, congruent dissolution is rare and confined only to olivine, amphiboles, and pyroxenes that are relatively free of iron. (Congruent quartz dissolution is also rare.) In this case we have the following reactions, assuming that attack is by carbonic acid:

$$Mg_2SiO_4 + 4H_2CO_3 \rightarrow 2Mg^{2+} + 4HCO_3^- + H_4SiO_4 \quad (4.1)$$

$$2H_2O + CaMgSi_2O_6 + 4H_2CO_3 \rightarrow Ca^{2+} + Mg^{2+} + 4HCO_3^- + 2H_4SiO_4 \quad (4.2)$$

[Note that dissolved silica is represented here as H_4SiO_4, which closely represents the actual form found in solution. It is sometimes also represented by the alternative formula $Si(OH)_4$. We shall not represent silica in solution by SiO_2, as is commonly done, because it can be confused with quartz.]

Most other silicate minerals, especially those containing aluminum, dissolve incongruently with the consequent formation of iron oxides and/or clay minerals. ("Clay minerals" is a common term applied to fine-grained aluminosilicates and includes kaolinite, smectite, and vermiculite as well as other minerals, such as chlorite, that are not discussed here). The most abundant silicate mineral in the earth's crust, which also readily undergoes weathering, is plagioclase feldspar. On this basis we begin our discussion of weathering reactions using Na-plagioclase or albite.

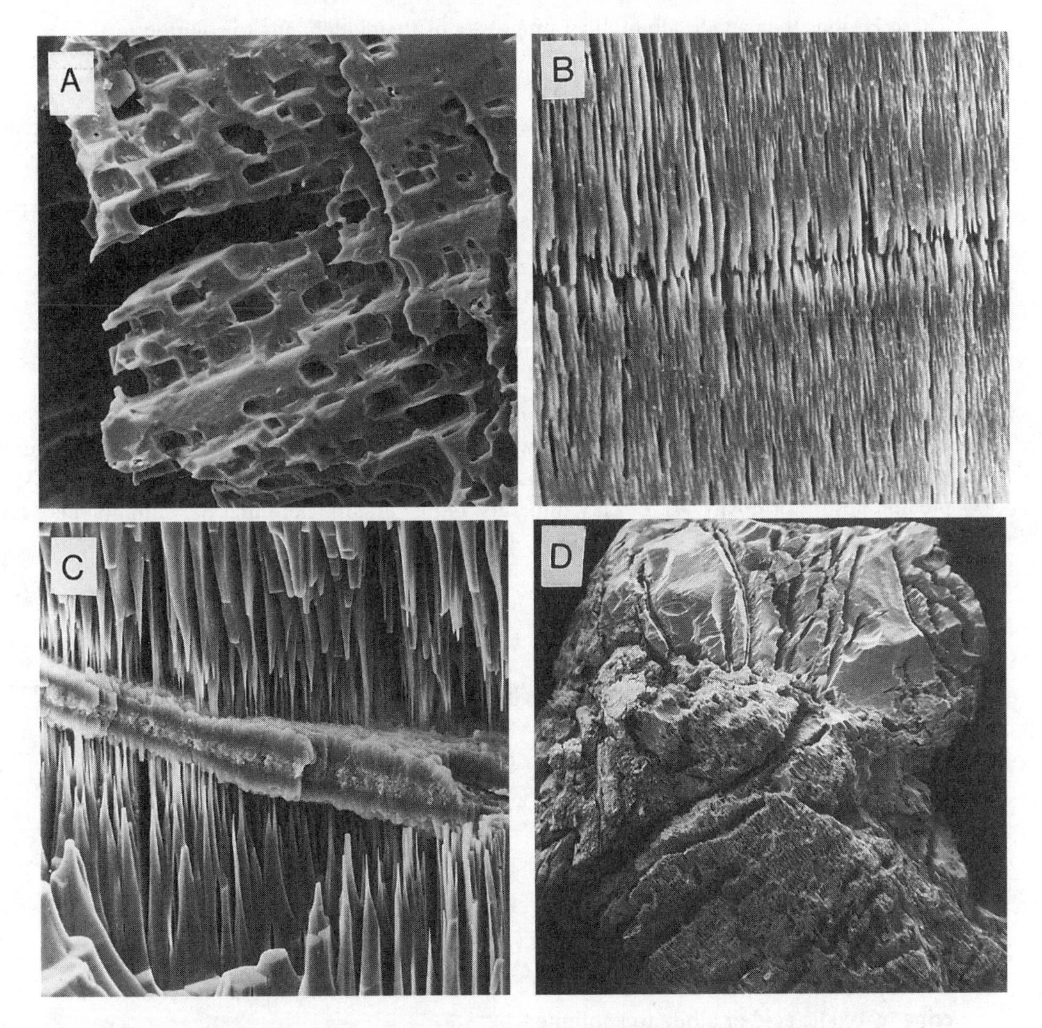

Figure 4.3. Scanning electron photomicrographs of partly weathered primary silicate mineral grains taken from soils. (A) Square-shaped ("prismatic") etch pits developed on dislocations in plagioclase feldspar (oligoclase from Piedmont, North Carolina). (× 3000) (After R. A. Berner and G. R. Holdren. "Mechanism of Feldspar Weathering: Some Observational Evidence," *Geology,* 5. p. 372. © 1977 by The Geological Society of America. All rights reserved.) (B) Lens-shaped etch pits developed on dislocations in amphibole (hornblende from Ashe County, North Carolina). (× 3000) (C) Clay-filled crack and surrounding "teeth" in hornblende formed by the coalescence of lens-shaped etch pits. (×1000) (D) Compound double pyroxene grain consisting of a single crystal of augite, $Ca(Mg,Fe)Si_2O_6$, in lower portion of photo and a single crystal of hypersthene. $(Mg,Fe)SiO_3$, in upper portion. Note greater degree of pitting of augite showing that it weathers faster because of a higher dislocation density. (×50) (B, C, D after R. A. Berner and J. Schott. "Mechanism of Pyroxene and Amphibole Weathering II: Observations of Soil Grains," *American Journal of Science,* 282, pp. 1219, 1222, 1223. © 1982 by the American Journal of Science, reprinted by permission of the publisher.)

Let us suppose that albite, $NaAlSi_3O_8$, is attacked by an organic acid, here represented as oxalic acid, $H_2C_2O_4$. First of all the oxalic acid dissociates to form H^+ ions:

$$H_2C_2O_4 \rightarrow 2H^+ + C_2O_4^{2-} \quad (4.3)$$

These H^+ ions then attack albite, liberating its constituent elements to solution:

$$4H^+ + 4H_2O + NaAlSi_3O_8 \rightarrow Al^{3+} + Na^+ + 3H_4SiO_4 \quad (4.4)$$

Since oxalate ion, $C_2O_4^{2-}$ readily reacts with Al^{3+} to form a chelate (Graustein 1981), we also have

$$Al^{3+} + C_2O_4^{2-} \rightarrow Al(C_2O_4)^+ \quad (4.5)$$

If we combine reactions (4.3)–(4.5) so as to cancel H^+ ion, we obtain

$$2H_2C_2O_4 + 4H_2O + NaAlSi_3O_8 \rightarrow Al(C_2O_4)^+ + Na^+ + C_2O_4^{2-} + 3H_4SiO_4 \quad (4.6)$$

This is a dissolution reaction characteristic of the uppermost, highly acid portion of most temperate soils.

Aluminum oxalate and oxalate ion in solution are not stable, however, due to bacterial decomposition (Graustein 1981). On passing downward in migrating soil water, the oxalate is oxidized by bacteria, and the liberated Al^{3+}, which is unstable in solution at most pH values found in soils, precipitates to form $Al(OH)_3$ or clay minerals. Let us assume here that the common clay mineral, kaolinite, $A1_2Si_2O_5(OH)_4$, is formed. The reactions are

$$2C_2O_4^{2-} + O_2 + 2H_2O \rightarrow 4HCO_3^- \quad (4.7)$$

$$2Al\,(C_2O_4)^+ + O_2 + 2H_4SiO_4 \rightarrow Al_2Si_2O_5\,(OH)_4 + 4CO_2 + H_2O + 2H^+ \quad (4.8)$$

Multiplying reaction (4.6) by 2, and adding to reactions (4.7) and (4.8), we obtain the overall reaction:

$$4H_2C_2O_4 + 2O_2 + 9H_2O + 2NaAlSi_3O_8 \rightarrow$$
$$Al_2Si_2O_5(OH)_4 + 2Na^+ + 4HCO_3^- + 2H^+ + 4CO_2 + 4H_4SiO_4 \quad (4.9)$$

As written, this reaction is not quite complete. Since H^+ and HCO_3^- rapidly react with one another, we must take into account

$$H^+ + HCO_3^- \rightarrow CO_2 + H_2O \quad (4.10)$$

Doubling reaction (4.10) and adding it to (4.9), we obtain the final overall reaction for the weathering by oxalic acid of albite to kaolinite:

$$4H_2C_2O_4 + 2O_2 + 7H_2O + 2NaAlSi_3O_8 \rightarrow$$
$$Al_2Si_2O_5(OH)_4 + 2Na^+ + 2HCO_3^- + 4H_4SiO_4 + 6CO_2 \quad (4.11)$$

Note that reaction (4.11), even though the attacking acid is oxalic acid, ends up with the production of only Na^+, HCO_3^-, and H_4SiO_4 in solution. (The CO_2 is readily lost as a gas from the soil.) These are the dissolved species that would be found if the water were to pass out of the soil zone to become groundwater and eventually stream and river water. There is no memory of the oxalate, and it is as though the albite had actually reacted with carbonic acid:

$$2H_2CO_3 + 9H_2O + 2NaAlSi_3O_8 \rightarrow$$
$$Al_2Si_2O_5(OH)_4 + 2Na^+ + 2HCO_3^- + 4H_4SiO_4 \quad (4.12)$$

In fact, recalling that H_2CO_3 forms by the reaction

$$H_2O + CO_2 \rightarrow H_2CO_3$$

the only difference between reactions (4.11) and (4.12) is

$$4H_2C_2O_4 + 2O_2 \rightarrow 8CO_2 + 4H_2O \quad (4.13)$$

which is the reaction for the oxidative decomposition of oxalic acid.

What has been said above for oxalic acid and albite is true of any organic acid and silicate (and carbonate) mineral. Thus, even though the actual acid attacking a mineral is organic, the overall reaction, as far as most groundwater composition is concerned, can be represented as though the only attacking acid were H_2CO_3. In other words, we find HCO_3^- and not $C_2O_4^{2-}$ in most ground and river waters. (In some rivers draining swampy areas of heavy organic matter production, organic acids and their anions can resist bacterial oxidation and survive to become carried some distance in the rivers; see Chapter 5.) The simplification provided by this reasoning enables the prediction of the origin of ions in groundwaters without concern for the type of organic acid actually attacking the primary minerals. Thus, the assumption that silicate weathering consists solely of attack by carbonic (and sulfuric) acids (Garrels 1967) is justified even though, when looked at in detail, it is not correct. The organic acids really do much of the attacking, but they disappear. From now on in our discussion of weathering we shall adopt the convention of writing reactions in terms only of H_2CO_3, but it should be remembered that this is only a convention and a shorthand way of representing a series of chemical weathering reactions that is far more complex.

In the above example, we assumed that all aluminum liberated by feldspar dissolution was precipitated to form a secondary mineral (in this case, kaolinite). This is a reasonable assumption for most soils (e.g., Loughnan 1969). Except for localized redistribution accompanying chelate transport, aluminum does not migrate any appreciable distance in solution and normally accumulates in soils as weathering proceeds. (However, in unusually acid soils containing appreciable H_2SO_4, Al can be removed in solution; see Chapter 6.) In fact, the change in the ratio of other elements to aluminum in soils is often used as a measure of the degree of removal by weathering (e.g., Goldich 1938). Iron, because of its insolubility in the presence of dissolved O_2, also accumulates in soils, as ferric oxides. Because of their lack of removal in solution, we shall continue in all overall weathering reactions, as is the custom, to assign Al and Fe only to secondary minerals and allow none to appear in solution.

Although all Al is reprecipitated to form a secondary mineral, it need not always be kaolinite. Two other common aluminous weathering products are gibbsite, $Al(OH)_3$, and smectite, a complex cation Al-silicate (see Table 4.4). (Vermiculite forms by the loss of K^+ from biotite and muscovite and constitutes a special case of structural inheritance which will not be discussed here.) Conditions under which each of these minerals would be expected to form can be deduced on the basis of some rather simple reasoning. Weathering of primary aluminosilicates to gibbsite, kaolinite, or smectite should occur under different conditions because of fundamental differences in the composition of the three phases. Smectite contains Al, Si, and various cations; kaolinite contains only Al and Si; and gibbsite, only Al (see Table 4.4). In addition, the ratio of Si/Al is higher in smectite than it is in kaolinite. Thus, we would expect that increase of H_4SiO_4

in solution would favor the formation of kaolinite over gibbsite and, at higher values, smectite over kaolinite. In addition, increase in cations, for example, Na^+, should favor smectite. Quantitative representation of these ideas is shown in Figure 4.4. Here a plot of log $[Na^+]/[H^+]$ versus log $[H_4SiO_4]$, where brackets represent molar concentrations in solution, shows the regions of stability for the various secondary minerals as well as that of albite in aqueous solution. If a natural water has concentrations of $[Na^+]$, $[H^+]$, and $[H_4SiO_4]$, that fall in, let us say, the field for kaolinite, we would expect to find kaolinite forming from this water, and so forth for the other minerals. Note that as both $[H_4SiO_4]$ and $[Na^+]$ are increased (for constant pH), kaolinite and smectite, respectively, become the favored phases. By use of this diagram the compositional evolution of a water reacting with albite can be predicted (for details consult Helgeson et al. 1969).

With reference to Figure 4.4, following the reasoning of Helgeson et al. (1969), imagine the sequence of events as albite (again used to represent a typical primary aluminosilicate) undergoes dissolution during weathering at constant pH. If we start with a very dilute solution (low

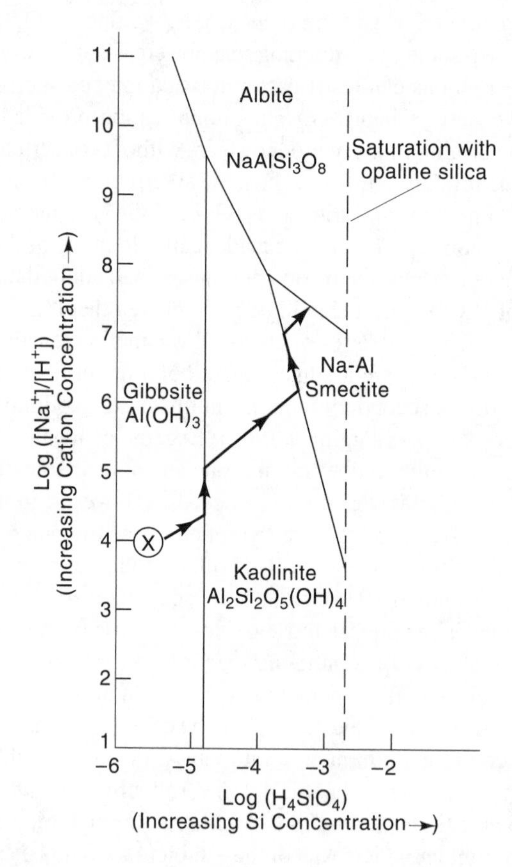

Figure 4.4. Stability fields of gibbsite, kaolinite, smectite, and albite as a function of log $[Na^+]/[H^+]$ and log $[H_4SiO_4]$ in solution. (Brackets denote concentration in moles per liter.) Smectite is represented by its pure Na-Al end member, Na-beidellite, and analcite (and other zeolite) formation is ignored. (Albite is metastable with respect to analcite). A typical weathering reaction path taken for a closed system is shown by the heavy line with arrows. (Modified and recalculated from Bricker and Garrels 1965; Helgeson et al. 1969.)

Na^+; low H_4SiO_4) its composition will fall in the field of stability of gibbsite (marked X on Figure 4.4). The weathering reaction in this case is

$$7H_2O + H_2CO_3 + NaAlSi_3O_8 \rightarrow Al(OH)_3 + Na^+ + HCO_3^- + 3H_4SiO_4 \quad (4.14)$$

If the water does not leave the rock, concentrations of Na^+, HCO_3^-, and H_4SiO_4 will build up and the solution composition will move to the northeast on the diagram following the arrow-marked path. When the boundary between gibbsite and kaolinite is reached, gibbsite begins to convert to kaolinite by the addition of silica:

$$2H_4SiO_4 + 2Al(OH)_3 \rightarrow Al_2Si_2O_5(OH)_4 + 5H_2O \quad (4.15)$$

During this reaction all silica released by albite dissolution is used to convert gibbsite to kaolinite, and this is why the solution path on the diagram turns abruptly northward. After all gibbsite is converted to kaolinite, we proceed again along a northeasterly trend until the boundary between kaolinite and smectite is reached. Again we follow the kaolinite–smectite boundary until all kaolinite is converted to smectite by the addition of silica and cations. We then proceed again to the northeast as smectite forms from albite until we reach the albite–smectite boundary. At this point, the solution is saturated with albite; it cannot continue to dissolve, and weathering therefore ceases.

The scenario described above is what would be expected if the water always stayed in contact with the albite (and there were no kinetic problems with the precipitation of secondary minerals). In other words, this is analogous to adding water and albite to a beaker and letting them react until albite solubility is reached. It is a closed system. Soils, on the other hand, are open to flow of water through them. Thus, concentrations will build up during contact of the water with albite (or other primary minerals), but further buildup will cease when the water leaves the rocks. The faster the rock is flushed with water, the shorter will be the time of contact with the primary minerals, and the lower will be the dissolved concentrations in the exiting waters. A steady state is attained between rate of addition by dissolution and rate of water flow, so that the water composition for any given soil containing albite may fall anywhere along the reaction path of Figure 4.4 depending on the relative magnitude of these two rates. Gibbsite formation should represent a high degree of flushing with removal of both cations and silica; kaolinite should represent less rapid flushing where less silica is removed; and smectite should represent rather stagnant conditions of water flow so that appreciable buildup of both silica and cations can occur. Also, for a given rate of flushing, more rapidly reacting minerals providing more silica and cations to solution should favor the formation of smectite or kaolinite over gibbsite.

These predictions are borne out when actual soils are examined. In general, gibbsite forms only in areas where there is very rapid flushing due to a combination of high rainfall and good drainage due to high relief. An example is the island of Jamaica, where valuable deposits of bauxite (an ore of aluminum consisting largely of gibbsitelike minerals) are formed as a result of intense weathering accompanying high rainfall in a mountainous terrain. Less strong flushing, but still enough to remove all cations, is found in most tropical and subtropical soils and here, as predicted, the characteristic secondary mineral is kaolinite.

Smectite is the characteristic mineral of soils of semiarid regions and where volcanic glass is the primary material undergoing weathering. In semiarid regions, rainfall is low and water adheres to soil grains for long periods before being replaced by new water. Volcanic glass is the most reactive silicate known and is weathered very rapidly. Again, both observations are in agreement with our predictions.

A nice demonstration of the effects of flushing by water on the weathering of a single rock type is shown by the studies of Sherman (1952) and Mohr and van Baren (1954). Sherman found that clay mineralogy of the soils developed on the basalt of the island of Hawaii are correlated very well with mean annual rainfall. (Rainfall varies considerably on Hawaii because of the effects of rain shadowing by mountains.) This is shown in Figure 4.5. Mohr and van Baren found that for the same rock type and same rainfall, soils on islands of Indonesia showed different clay minerals depending on the degree of drainage. Upland soils, where drainage was good, consisted of kaolinite, whereas those in poorly drained or swampy depressions consisted of smectite. This is again what would be expected from our predictions.

Further illustration of the importance of flushing is shown by the study of Velbel (1984). On hill slopes of the southern Appalachian Mountains of the United States, Velbel found that differences in water flow path resulted in the formation of different clay minerals from the same plagioclase-rich rock subjected to the same climate and same relief. Gibbsite formed in surficial zones where the water residence time was short because of a small flow path. At depth, both gibbsite and kaolinite were found where the water travel distance was much greater. The entrapment of water in the slightly weathered and deeply buried underlying rock resulted in long residence times and the formation of smectite.

Rate of Silicate Weathering

There has been much interest in determining the rate of chemical weathering of regions ranging in size from small watersheds to continents in order to calculate rates of chemical lowering of the regions as well as to explain natural water chemistry. Continental and global weathering rates are covered in the present book in Chapter 5, where the chemistry of large rivers is discussed. The purpose here is to briefly cover the topic of weathering rates on the small-watershed scale and how they are obtained. Further information on weathering rates can be found in the book by Colman and Dethier (1986).

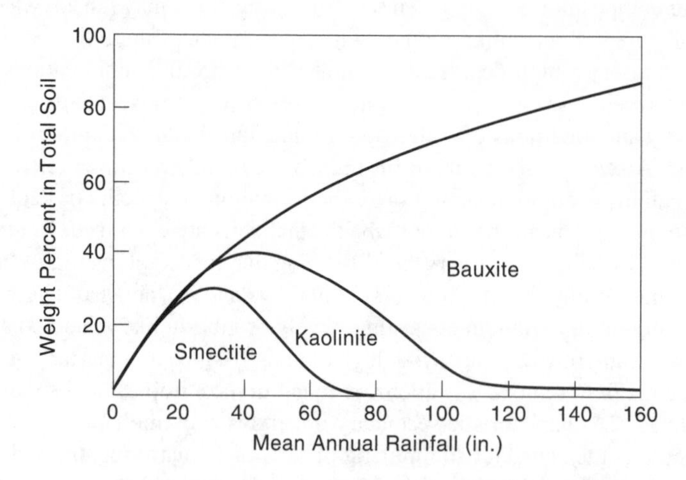

Figure 4.5. Weathering products of basalt in Hawaii. Note the excellent correlation of clay mineral type with rainfall in agreement with predictions based on the degree of flushing of the soil with water. After Sherman (1952).

The normal procedure for measuring rates of weathering on the watershed scale is to compute an input–output budget from mean annual flow and chemical composition of the water at the stream outlet of the watershed combined with input fluxes from rain and snow and corrections for biological and soil storage or release (Likens et al. 1977). The difference between stream output and rainfall input, when added to biological and soil storage, is assumed to represent mineral dissolution. An example of this type of study is that of Velbel (1985) of an area in the southern Appalachian Mountains of the United States. By constructing a series of mass balance expressions for the principal cations (Ca, Na, Mg, and K), Velbel was able to combine stream and rainwater chemistry with determinations of mineralogy and the composition of biomass to calculate rates of weathering of the principal minerals in the rocks of this area.

In addition to field studies, models have been constructed to predict the rates of liberation of ions to solution during weathering. For example, Sverdrup and Warfvinge (1988) have calculated rates of dissolution of silicate minerals for a Swedish field area from quantitative rate data based on silicate mineral laboratory dissolution experiments. Experimental rate constants plus functional relations to important variables (principally pH, temperature, and concentrations of dissolved reacting species including organic acids) were combined with field data on soil thickness, pH, temperature, surface areas of dissolving minerals, and the degree of wetting of the surfaces of these minerals (as revealed by average soil moisture), to calculate rates of weathering. Results were then compared to rates calculated from the chemical input–output budget, and good agreement between results was found.

The work of Sverdrup and Warfvinge highlights the importance of estimating correctly how minerals actually dissolve in the field. Extrapolating laboratory results to soils is fraught with many problems, as is indicated by common major disagreements between predictions based on laboratory mineral dissolution rates and rates based on field watershed studies (see Velbel 1993a for a review). In agreement with Velbel, we believe that much of this disagreement is due to a failure to correct for major differences in the exposure of minerals to solutions in the lab and in the field. In soils, the situation of variable flow paths and the wetting and drying of mineral surfaces is very different from laboratory studies, where exposure of suspended particles to stirred solution is continuous. Obviously, wetting in soils is important because without it dissolution does not take place. Flow paths in soils are complicated by inhomogeneities such as pedogenic clay horizons, worm and rodent burrows, root molds, dessication cracks, and primary layering. This, combined with spatial and temporal variations in soil moisture due to variations in rainfall intensity, make it very difficult to estimate the degree and duration of wetting of mineral surfaces in order to use laboratory dissolution data to predict rates of weathering in the field.

Many factors affect the rate of chemical weathering. These include (1) rainfall, relief, and permeability as they affect flushing of the rock (see previous section); (2) physical weathering as it brings about increased interfacial area between primary minerals and soil solutions due to the disintegration of bedrock; (3) mean annual temperature as it affects the rate of mineral dissolution and the rate of microbial activity; and (4) vegetation as it brings about organic acid production and enhanced contact of water with primary minerals due to both the protection by roots against soil erosion and the recycling of moisture via transpiration.

A good example of the effect of higher plants on chemical weathering is shown by the study of Drever and Zobrist (1993) in the southern Swiss Alps. Here they determined the stream flux of dissolved bicarbonate from silicate weathering as a function of elevation on similar hill slopes developed on the same rock type with essentially the same mean annual precipitation. Sampling sites varied from deciduous forest at the lowest elevations, through coniferous forest at intermediate elevations, to barren or lichen-encrusted rocks above treeline. Mean annual temperature also decreased with elevation. They found that the HCO_3^- weathering flux was about 25 times higher at the lowest elevations than at the highest. This represents the combined effects of both temperature and vegetation. If the temperature effect is assumed to be similar to that found in laboratory silicate mineral dissolution experiments, which for this area would be a factor of 3, then the residual effect due to the presence of deciduous forest versus its absence is a factor of 8. In other words, in this area deciduous vegetation caused silicate weathering to increase by eight times.

In general agreement with the results of Drever and Zobrist are the unpublished results by one of us (R.A.B.) and M. F. Cochran of Yale University. Cochran and Berner studied the chemistry of solutions draining a series of experimental plots at the Hubbard Brook Experimental Station in New Hampshire [see Bormann et al. (1987) for a detailed description of the plots]. Each small (about 60 m^2) plot was set up to have the same rock type (feldspar-rich glacial sand), the same rainfall input, and the same aspect, relief, etc., and each was underlain by impermeable material so that water percolating through each plot could be collected at an exit pipe. One plot was planted with pine trees, another with grass, and a third was left fallow. After 10 years of tree growth, results by Cochran and Berner indicate that the enhancement of sodium release from the weathering of plagioclase feldspar is about 3 to 10 times greater for the pine tree-covered plot than for the fallow one.

Silicate Weathering: Soil Formation

Soil can be defined as "a complex system of air, water, decomposing organic matter, living plants and animals, in addition to the residues of rock weathering, organized into definite structural patterns as dictated by the environmental conditions" (Loughnan 1969: 115). The dominant control of soil formation, on a worldwide basis, is climate. In fact, most classifications of soils (zonal soils) are based primarily on climatic differences. A common recent classification is that shown in Table 4.6. [For a more detailed discussion of weathering as it relates to soil formation, consult Loughnan (1969), Paton (1978), Birkeland (1984), and Retallack (1990).] In addition to the role of rainfall as it affects flushing of the rock (as discussed above), climate exerts an important influence on weathering via its control on vegetation and the organic content of soils. High rainfall, which enhances plant growth, and low temperature, which retards bacterial destruction, both favor the accumulation of organic matter in soils. Organic matter is important, as mentioned earlier, in that it is the source of carbonic and organic acids and chelating compounds that react with silicate minerals. It is because of the accumulation of organic matter that the most acid soils in Table 4.6 are those formed under cool, humid conditions, in other words, the spodisols. (Histosols are even richer in organic matter, but, due to lack of water circulation, little weathering takes place in them.) By contrast, less organic matter accumulates in the warm-climate oxysols (latosols and laterites), due to almost complete destruction of organic matter by microorganisms, and in desert and semiarid soils (aridosols), due to a lack of appreciable plant growth.

Since our concern is mainly with water chemistry, we need not go into great detail about various soil types and classifications. However, a few additional comments concerning some of the soils of Table 4.6 are in order. The characteristic soils of humid temperate regions are the spodisols and alfisols. Spodisols (formerly podzols) are characterized by a strong vertical zonation, an example of which is shown in Figure 4.6. The top of the soil (O horizon) consists of leaf litter,

TABLE 4.6 Soil Classification

Entisol	Negligible degree of soil formation or zonation, due to short time available or location on slopes subject to constant removal by erosion
Inceptisol	Intermediate stage of development between entisols and other soil types; immature soils
Histosol	Organic-rich soils with thick, peaty horizons; formed in low/lying permanently water-logged areas
Vertisol	Uniform, thick, clay-rich profiles with deep cracks and hummocky topography produced by intense seasonal drying of expandable clay minerals such as smectites; high exchangeable cations (base-rich)
Mollisol	Well-developed base-rich surface horizon of mixed clay and organic matter plus a subsurface clayey, calcareous, or gypsiferous horizon; found under grassland vegetation in subhumid to semiarid climates
Aridosol	Soils of arid regions with shallow calcareous, gypsiferous, or saline horizons; often contain wind-blown dust
Spodisol	Strong zonation showing a base-poor and amorphous Al- and Fe-rich (B) horizon normally overlain by a quartz-rich, strongly leached (E) horizon in turn overlain by a highly acidic surface organic layer; typically under coniferous forest in temperate climates; (formerly called podzols)
Alfisol	Forest soils with a base-rich (smectitic) clayey subsurface layer (B) horizon overlain by a well-defined, light-colored surface horizon (A); high content of primary weatherable minerals (e.g., feldspars)
Ultisol	Forest soils (mainly deciduous) with base-poor (e.g. kaolinitic) clayey subsurface (B) horizon and low content of primary minerals due to extensive weathering in humid, warm climates
Oxysol	Deeply weathered, reddish, kaolinitic, clay-rich and iron-rich soils of tropical humid climates with almost no remaining primary minerals; (formerly called laterites)

Source: After Retallack 1990.

which becomes decayed to structureless humus upon burial (A horizon). The downward flow of water, which becomes highly acidic upon passing through the organic layers, results in intense leaching and removal of the immediately underlying material, resulting in a residue consisting mainly of resistant quartz (E horizon). Cations, silica, organic matter, and even Al and Fe are removed. The Al and Fe, however, as well as some silica and dissolved organic matter (humic and fulvic acids), do not leave the rock and are precipitated at depth in the so-called B horizon. Underlying the B horizon is the C horizon (Figure 4.6) or zone, where the rock is only slightly weathered and is little affected by biological activity.

In other soil types one also encounters vertical zonation, but often different from that exhibited by spodisols. For example, in mollisols, the characteristic soils of subhumid temperate grasslands, as well as in aridosols, one often finds the precipitation of $CaCO_3$ at depth within the B horizon due to the dominance, during dry periods, of evaporation over downward percolation and leaching.

In tropical and subtropical soils there is little burial of organic matter but there is still intense leaching of cations and silica, due to heavy rainfall, resulting in the buildup of residual iron and aluminum within the B horizon (ultisols and oxysols). In desert soils (aridosols) or in soils that have been subjected to weathering for only short periods (entisols), vertical zonation is often poorly developed due to a relative lack of weathering. Examples of the latter soils are those developed on young glacial deposits or on relatively recent volcanic flows.

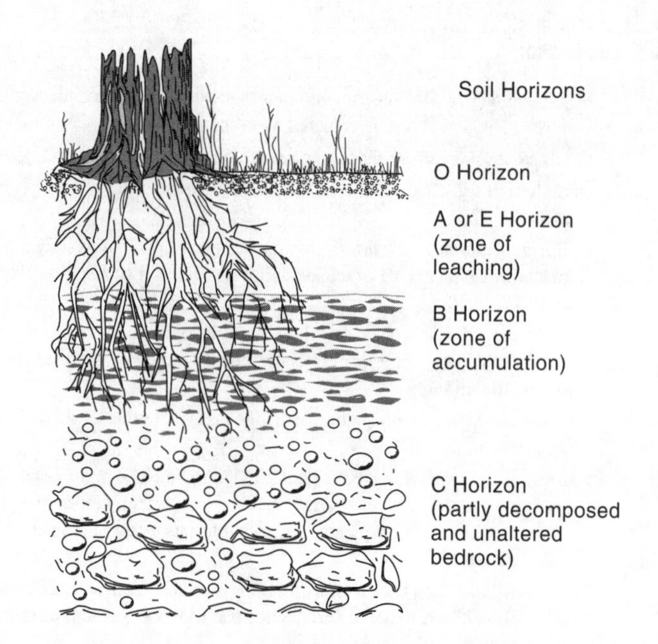

Figure 4.6. Soil horizons formed in a temperate humid climate (James I. Drever, *The Geochemistry of Natural Waters,* 2e, © 1992, p. 145. Reprinted by permission of Prentice Hall, Englewood Cliffs, New Jersey.)

Carbonate Weathering

Compared to silicates, carbonate minerals weather far more simply. Dissolution is normally congruent, and the overall reaction, accomplished by carbonic acid (H_2CO_3) attack (or organic acid attack, which can be represented as H_2CO_3; see above) is for calcite,

$$H_2CO_3 + CaCO_3 \rightarrow Ca^{2+} + 2HCO_3^- \quad (4.16)$$

and for dolomite,

$$2H_2CO_3 + CaMg(CO_3)_2 \rightarrow Ca^{2+} + Mg^{2+} + 4HCO_3^- \quad (4.17)$$

Carbonate minerals are not as abundant as silicate minerals, but they exert a dominant influence on groundwater and river water composition. Most of the dissolved Ca^{2+} and HCO_3^- in river water, on a worldwide basis, arises from carbonate dissolution via reactions (4.16) and (4.17).

Calcium carbonate dissolution, under certain circumstances, can be followed by reprecipitation. In soil water and groundwater where carbonic (and organic) acid concentrations are high, the resulting concentrations of Ca^{2+} and HCO_3^- can build up to high values. If the waters then encounter conditions where degassing of dissolved CO_2 occurs, the waters may become supersaturated with respect to $CaCO_3$. Carbonic acid constantly maintains equilibrium with dissolved CO_2:

$$H_2CO_3 \leftrightarrows H_2O + CO_2 \quad (4.18)$$

and if CO_2 is lost from solution, reaction (4.18) shifts to the right, using up H_2CO_3, and consequently, reaction (4.16) shifts to the left to replace the H_2CO_3. If Ca^{2+} and HCO_3^- are present in high enough concentrations, $CaCO_3$ precipitation may then take place.

An outstanding example of the reprecipitaton of $CaCO_3$ occurs in limestone caves. Limestone is a rock consisting largely of calcite. It undergoes congruent dissolution during weathering along joints, cracks, and other avenues of water flow by carbonic acid derived from organic matter decomposition in the overlying soils. The Ca^{2+} and HCO_3^- are normally removed from the rock and, as a result, enlargement of cracks and ultimately the formation of caves takes place. If a cave is sufficiently connected to the atmosphere, via cracks, a low, atmospheric value of CO_2 gas is maintained in the cave air. In this case, water emerging in the cave from above and containing high levels of H_2CO_3, CO_2, and HCO_3^- can lose CO_2 to the cave atmosphere. As a result, supersaturation with $CaCO_3$ is attained and calcite is precipitated to form stalactites, stalagmites, and other cave deposits. Cave deposits also form in dolostones (dolomite rocks), but in this case dolomite is not reprecipitated. There are severe problems in precipitating dolomite so that, in solutions supersaturated with respect to dolomite and calcite, calcite is invariably the precipitate. Thus, in dolostones, we find caves that contain deposits of calcite, not dolomite (see, e.g., Holland et al. 1964).

Calcite can also precipitate in soils. (The Ca^{2+} and HCO_3^- may come from the weathering of silicates rather than carbonates.) This comes about by a similar process of degassing, but the precipitation occurs much closer to the ground surface. In arid climates, downward-percolating soil waters undergo degassing and loss of CO_2 at a depth of only a few tens of centimeters in the soil. (The CO_2 is picked up from organic decay at even shallower depths.) As a result, calcite precipitation takes place and the resulting deposit (Loughnan 1969) is referred to as *caliche.* [For a discussion of soil carbonate, see also Retallack 1990).]

Sulfide Weathering

Sulfide minerals occur in minor quantities in many different rock types and locally in major quantities in ore deposits. The most abundant mineral, by far, is pyrite (FeS_2), which is found mainly in organic-rich, fine-grained sedimentary rocks known as black shales, and in coal. Upon exposure to dissolved oxygen during weathering, pyrite (and other sulfides) are chemically unstable and rapidly undergo oxidative decomposition. This decomposition is important in that it results in the production of sulfuric acid, which can be used, in turn, to bring about further weathering of silicate and carbonate minerals. Sulfuric acid forms as follows:

$$4FeS_2 + 15O_2 + 8H_2O \rightarrow 2Fe_2O_3 + 8H_2SO_4 \quad (4.19)$$

$$H_2SO_4 \rightarrow 2H^+ + SO_4^{2-} \quad (4.20)$$

(Various other reactions can be written involving Fe^{2+} or Fe^{3+} in solution, etc., but in all cases there is a distinct lowering of pH due to the formation of H_2SO_4.) Almost always the oxidation is catalyzed by bacteria (Stumm and Morgan 1981). The acidity of water draining rocks where sulfide oxidation is taking place depends on the content of sulfides and the presence of other minerals, especially carbonates, which can readily neutralize acid. During coal mining, large expanses of pyrite-bearing coal are suddenly exposed to oxygenated water, and because associated rocks generally contain little carbonate, the waters draining coal-mining localities are very acid. Values of pH of less than 3, resulting from bacterial catalysis of pyrite oxidation, are common for such mine drainage waters (Stumm and Morgan 1981; Kleinmann and Crerar 1979).

By contrast, calcareous rocks that also contain pyrite do not weather to produce highly acid water. Instead the sulfuric acid is neutralized by $CaCO_3$. Reactions are

$$4FeS_2 + 15O_2 + 8H_2O \rightarrow 2Fe_2O_3 + 16H^+ + 8SO_4^{2-} \quad (4.21)$$

$$16H^+ + 16CaCO_3 \rightarrow 16Ca^{2+} + 16HCO_3^- \quad (4.22)$$

which, added together, give

$$4FeS_2 + 15O_2 + 8H_2O + 16CaCO_3 \rightarrow 2Fe_2O_3 + 16Ca^{2+} + 8SO_4^{2-} + 16HCO_3^- \quad (4.23)$$

In this way Ca^{2+}-SO_4^{2-}-HCO_3^- groundwaters and streamwaters can arise. In arid and semiarid regions, where evaporative concentration of soil waters is common, the Ca^{2+} and SO_4^{2-} concentrations often reach the point where the precipitation of gypsum, $CaSO_4 \cdot 2H_2O$ takes place. The common occurrence of gypsum crystals in weathered outcrops of calcareous black shales in the western interior of the United States is an example of this process.

Sulfuric acid is also partially neutralized by silicate minerals. It is this neutralization, in fact, that constitutes one of the more important weathering reactions for silicates. For example, for albite,

$$H_2SO_4 + 9H_2O + 2NaAlSi_3O_8 \rightarrow Al_2Si_2O_5(OH)_4 + 2Na^+ + SO_4^{2-} + 4H_4SiO_4 \quad (4.24)$$

Sometimes, if there is sufficient pyrite (or other sulfides), acidity can be so high that common secondary minerals, such as kaolinite and iron oxides, become unstable and dissolve:

$$6H^+ + Al_2Si_2O_5(OH)_4 \rightarrow 2Al^{3+} + 2H_4SiO_4 + H_2O \quad (4.25)$$

$$6H^+ + Fe_2O_3 \rightarrow 2Fe^{3+} + 3H_2O \quad (4.26)$$

In this case, nonsilicate minerals such as alunite, $KAl_3(SO_4)_2(OH)_6$, or jarosite, $KFe_3(SO_4)_2(OH)_6$, may form. Finding these minerals in soils is a good indication that the soils are highly acidic (van Breemen 1976).

GROUNDWATERS AND WEATHERING

Upon percolating downward below the water table, soil waters become groundwaters. Once the zone dominated by biological activity is left, weathering and rock dissolution are much slower but by no means negligible. Given sufficient time, silicate rock decomposition at depth can be extensive, giving rise to the formation of thick accumulations of clays, iron oxides, etc., known as *saprolites*. In contrast to soils saprolites retain the texture (e.g., layering) of the rock they are replacing, whereas in soils original texture is destroyed by roots and the burrowing activity of soil-dwelling organisms.

An example of extensive saprolitization is provided by the deep weathering profiles found in the southeastern United States. Such saprolitization takes place by means of the attack on primary minerals of carbonic acid and by water itself, and little direct biological activity is involved (Carroll 1970). (However, the carbonic acid is derived indirectly from microbiological organic matter decomposition in the overlying soil.) This deep weathering takes place and affects water composition both above and below the water table (Cleaves 1974; Velbel 1984). Below the water table the groundwater is out of contact with the atmosphere and, as a result of continuing reactions, may become anoxic (O_2-free). In this case, dissolved Fe^{2+} and Mn^{2+} appear in solution because of a lack of dissolved O_2, which would otherwise remove them by oxidation plus precipitation. Such groundwater, when pumped to the surface, undergoes rapid O_2 uptake from the atmosphere and consequent ferric hydroxide and manganese hydroxide precipitation. This explains the "rusty color" often encountered when using well water (see, e.g., Crerar et al. 1979).

The results of chemical reaction taking place at depth, such as Fe^{2+} liberation, as well as the many weathering reactions occurring in overlying soils, are manifested together in the composition of groundwater. In many circumstances, these reactions can be deduced from the chemical composition of the groundwater, and because of the abundance of chemical data (see, e.g., White et al. 1963), various schemes can be constructed for deducing the origin of groundwater composition. This will be pursued here (see also Drever 1988). Throughout the discussion it should always be kept in mind, however, that we are assuming that differences in water composition between groundwater and rainwater are due only to rock weathering. This may not be true if there are large effects on groundwater composition of soil-water alterations due to biomass increases or decreases or the dissolution of wind-blown dust as discussed at the beginning of this chapter. Fortunately, for several elements such effects can often be ignored, as shown by good agreement between predictions based on groundwater composition and actual observations of soils.

Studies of groundwater and springwater draining silicate rocks have been used to construct a mobility series for elements reminiscent of the mineral weatherability series shown in Table 4.5. One can define element mobility as the weight fraction of total dissolved matter that is made up by a given element divided by the weight fraction of the same element in the primary rock undergoing weathering. Examination of much groundwater data by Feth et al. (1964) has led to the following order of mobility:

$$\mathrm{Ca > Na > Mg > Si > K > Al = Fe}$$

In other words, Ca, Na, and Mg are the most mobile and most easily liberated by weathering; Si and K are intermediate; and Al and Fe are essentially immobile and remain in the soil. This order is in agreement with results of mineralogical studies in that the most rapidly weathered silicate minerals are Na-Ca silicates (plagioclase feldspars) and Mg-containing silicates (e.g., pyroxenes, amphiboles), whereas Al, Fe, and Si (the latter to a lesser degree) form secondary minerals and remain in the soils, and K is contained in less rapidly weathered minerals, specifically biotite, muscovite, and K-feldspar.

Garrels's Model for the Composition of Groundwaters from Igneous Rocks

Since plagioclase is the most abundant mineral in the earth's crust, and is weathered rapidly, it is reasonable to assume that much of the composition of natural groundwaters in silicate rocks can be explained in terms of plagioclase weathering. Based on his earlier research with F. T. Mackenzie, this has been done by Garrels (1967), and his discussion of the origin of groundwater composition will be described here. [For an extensive coverage of the ideas discussed in this section, the reader is referred to the book by Garrels and Mackenzie (1971).]

In examining the composition of groundwaters draining through igneous rocks, Garrels concluded that the principal dissolved species deserving explanation are Ca^{2+}, Na^+, HCO_3^-, and H_4SiO_4. (Lesser concentrations of K^+ and Mg^{2+} were explained in terms of biotite and amphibole weathering and will not be discussed here.) To explain the relative concentrations of these species, he made the following assumptions.

1. Water compositions (after correcting for rainfall input) result from the attack by carbonic acid on primary silicate minerals.
2. Plagioclase feldspar is the sole mineral source of Na^+ and Ca^{2+} because it is abundant and readily weathered. (Also, it is assumed that there is no $CaCO_3$ present.)
3. Dissolved silica is derived almost entirely from the weathering of plagioclase (a small amount comes from Fe-Mg minerals).
4. Rainfall correction for Na^+ is made by subtracting an amount equivalent to the measured Cl^- concentration from the total measured Na^+ concentration.

First, Garrels constructed a series of weathering reactions of plagioclase reacting to form gibbsite, kaolinite, or smectite. For example, for weathering to kaolinite of a plagioclase with 50% $NaAlSi_3O_8$ and 50% $CaAl_2Si_2O_8$,

$$4Na_{0.5}Ca_{0.5}Al_{1.5}Si_{2.5}O_8 + 6H_2CO_3 + 11H_2O \rightarrow$$
$$3Al_2Si_2O_5(OH)_4 + 2Na^+ + 2Ca^{2+} + 6HCO_3^- + 4H_4SiO_4 \qquad (4.27)$$

According to this reaction, one would expect to find a molar ratio in solution of $Na:Ca:HCO_3^-:H_4SiO_4$ of 1:1:3:2. For the weathering of plagioclase of different Na-Ca compositions to kaolinite, different ratios result, and likewise for the weathering of the same or different compositions of plagioclase to gibbsite, or to smectite. As a result of writing many reactions of this type, Garrels was able to construct a diagram, reproduced here as Figure 4.7, which shows solution compositions, in terms of $[HCO_3^-]/[H_4SiO_4]$ versus $[Na^+]/[Ca^{2+}]$, predicted for the weathering of plagioclase separately to gibbsite, kaolinite, and smectite.

By plotting natural groundwater compositions on the plagioclase weathering diagram, Garrels was able to deduce the major plagioclase weathering reactions taking place in nature. The data (Figure 4.7) indicate that the dominant reactions are those involving the formation of either kaolinite, smectite, or a mixture of both. This prediction is in accord with mineralogical observations of most soils and indicates the basic validity of the Garrels model.

Garrels also constructed plots of various parameters versus HCO_3^- concentration. Bicarbonate concentration is a measure of the degree of ion buildup in solution and, thus, the contact time of water with plagioclase. Garrels found that at low HCO_3^- concentrations, water appeared to have HCO_3^-/H_4SiO_4 ratios expected for the weathering of plagioclase (and other minerals) to kaolinite, whereas at high concentrations of HCO_3^-, the HCO_3^-/H_4SiO_4 ratios suggested weathering to smectite. In addition, at low HCO_3^- concentrations, solutions behaved as though they were undersaturated with respect to smectite; at high concentrations they appeared to attain a constant solubility product expected if smectite were present. These observations led Garrels to conclude that at low contact times of water with plagioclase ("initial weathering"), the primary product is kaolinite, and at high contact times both kaolinite and smectite form. This is in full agreement with what has been said earlier in this chapter. Moderately rapid flushing prevents

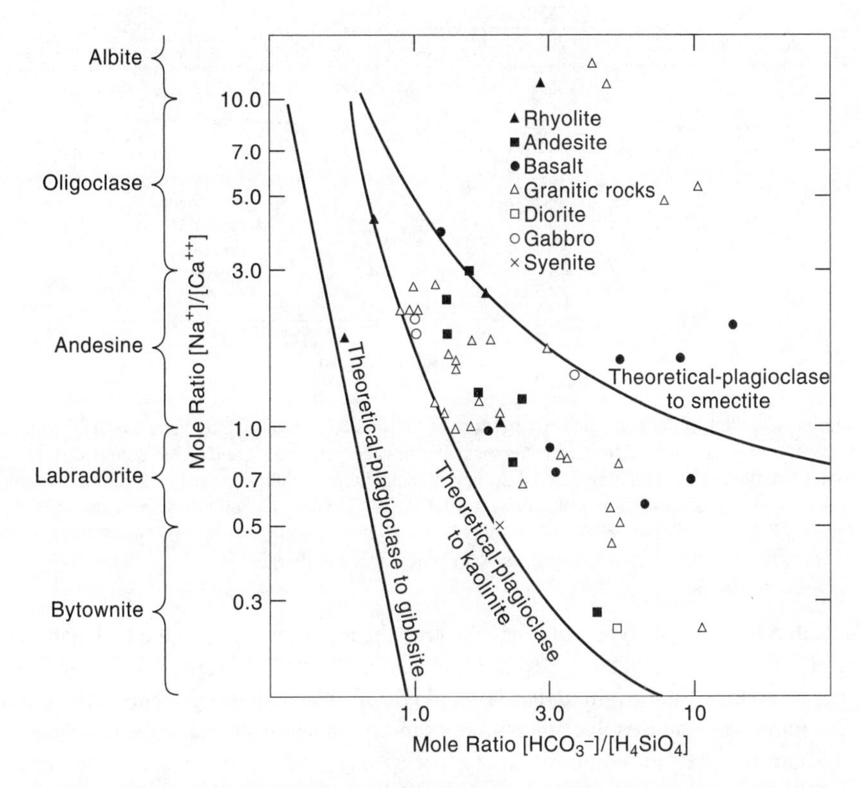

Figure 4.7. Mole ratios of $[HCO_3^-]/[H_4SiO_4]$ and $[Na^+]/[Ca^{2+}]$ in solution for groundwaters from various types of igneous rocks. (Brackets denote concentration in moles per liter.) Rock types are shown by different symbols. Nomenclature for plagioclase feldspar, composition from which the solution Na^+/Ca^{2+} ratio is derived, is shown on the left. Note that most of the water compositions plot between the theoretical curves for the weathering of plagioclase to kaolinite and to smectite. (After R. M. Garrels. "Genesis of Some Ground Waters from Igneous Rocks." In *Researches in Geochemistry,* ed. P. H. Abelson, p. 412. Copyright © 1967, John Wiley & Sons, Inc. Reprinted by permission of John Wiley & Sons, Inc.)

cation buildup and consequent smectite formation, which explains its relative absence in well-drained soils in regions of moderate to heavy rainfall.

Garrels also found that there was an interesting correlation of the concentration product $[Ca^{2+}][CO_3^{2-}]$ versus HCO_3^- concentration. This is shown in Figure 4.8. At high HCO_3^- levels, the product approaches that (10^{-8}) expected for calcite saturation, in other words, the solubility product for $CaCO_3$. This means that high water/plagioclase contact time, as reflected by high HCO_3^- concentrations, should lead to calcite precipitation. Again, this prediction is borne out. In semiarid regions, where water contacts soil grains for long periods before being replaced by new rainwater, concentrations in solution build up, due to prolonged weathering and evaporation, and often $CaCO_3$ precipitation in the soil results, sometimes in the form of caliche (see section on carbonate weathering).

These calculations show that chemical modelling of groundwater composition can prove to be a useful tool for studying chemical weathering. However, it must be remembered that in the Garrels model it is assumed that all Ca^{2+} arises from the weathering of plagioclase and all HCO_3^- from the weathering of silicates. This is reasonable only if the groundwater associated with a

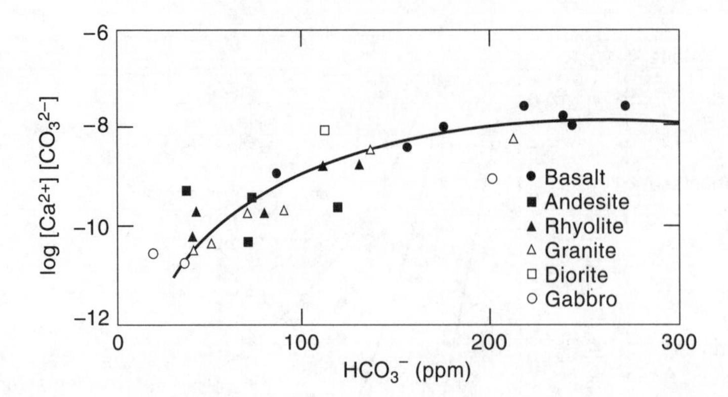

Figure 4.8. Plot of the concentration product of dissolved Ca^{2+} and CO_3^{2-} versus HCO_3^- concentration for groundwaters from different igneous rocks. (Brackets around Ca^{2+} and CO_3^{2-} denote concentration in moles per liter.) Note that above about 200 ppm HCO_3^- the product approaches a constant value, which is that expected for saturation with calcite. (After R. M. Garrels. "Genesis of Some Ground Waters from Igneous Rocks." In *Researches in Geochemistry,* ed. P. H. Abelson, p. 415. Copyright © 1967, John Wiley & Sons, Inc. Reprinted by permission of John Wiley & Sons, Inc.)

specific igneous rock type has contacted only the major minerals of the rock and has not encountered any carbonate minerals. Even if trace amounts of $CaCO_3$ are present—say, as fracture fillings in a basalt—the origin of the Ca^{2+} and HCO_3^- in an igneous groundwater becomes doubtful. Calcium carbonate dissolves much more rapidly than silicate minerals, and small amounts in a rock can dominate the composition of associated ground water. Thus, caution must be employed in selecting groundwaters for use in chemical modelling of silicate weathering reactions to be sure that problems associated with carbonate dissolution can be excluded.

Origin of Major Constituents of Groundwater: A Summary

By way of summary, the minerals and types of reactions involved in the production of the various common components of groundwater are shown in Table 4.7. Included also are inputs due to rainfall and biological effects where pertinent. Quantitative estimation of the relative importance of the various sources for each element is done in Chapter 5 when discussing riverwater and will not be repeated here. Table 4.7 is offered merely as a summary of what has been said in the present chapter. As can be seen, there are a variety of sources, and each element exhibits its own idiosyncrasy.

TABLE 4.7 Origin of Major Components of Groundwater (Major Processes Only)

Component	Origin
Na^+	NaCl dissolution (some pollutive)[a]
	Plagioclase weathering
	Rainwater addition
K^+	Biotite weathering
	K-feldspar weathering
	Biomass decreases
	Dissolution of trapped aerosols
Mg^{++}	Amphibole and pyroxene weathering
	Biotite(and chlorite) weathering
	Dolomite weathering
	Olivine weathering
	Rainwater addition
Ca^{++}	Calcite weathering
	Plagioclase weathering
	Dolomite weathering
	Dissolution of trapped aerosols
	Biomass decreases
HCO_3^-	Calcite and dolomite weathering
	Silicate weathering
SO_4^{--}	Pyrite weathering (some pollutive)[a]
	$CaSO_4$ dissolution
	Rainwater addition
Cl^-	NaCl dissolution (some pollutive)[a]
	Rainwater addition
H_4SiO_4	Silicate weathering

Note: Order presented is approximate order of decreasing importance. For further information consult Chapter 5.

[a] For a discussion of pollutive contributions, see Chapter 5.

REFERENCES

Alexander, M. 1961. *Introduction to soil microbiology.* New York: John Wiley.

Antweiler, R. C., and J. I. Drever. 1988. The weathering of a late Tertiary volcanic ash: Implication of organic solutes, *Geochim. Cosmochim. Acta* 47: 623–629.

Berner, R. A., and G. R. Holdren. 1977. Mechanism of feldspar weathering: Some observational evidence, *Geology* 5: 369–372.

Berner, R. A, and G. R. Holdren. 1979. Mechanism of feldspar weathering II: Observations of feldspars from soils, *Geochim. Cosmochim. Acta* 43: 1173–1186.

Berner, R. A., and J. Schott. 1982. Mechanism of pyroxene and amphibole weathering II: Observations of soil grains, *Am. J. Sci.* 282: 1214–1231.

Birkeland, P. W. 1984. *Soils and Geomophology.* Oxford: Oxford University Press.

Blum, A. E., M. F. Hochella, A. F. White, and J. Harden. 1991. A comparison between models of mineral dissolution and growth kinetics and morphologic evidence of weathering, *Abstr. Ann. Meeting Geol. Soc. Am.,* 105.

Bormann, F. H., W. B. Bowden, R. S. Pierce, S. P. Hamburg, G. K. Voight, R. C. Ingersoll, and G. E. Likens. 1987. The Hubbard Brook sandbox experiment. In *Restoration Ecology,* ed. R. Jordan, M. E. Gilpin, and J. D. Aber, pp. 251–256. Cambridge: Cambridge University Press.

Bricker, O. P., and R. M. Garrels. 1965. Mineralogic factors in natural water equilibria. *In Principles and applications of water chemistry,* ed. S. Faust and J. V. Hunter, pp. 449–469. New York: John Wiley.

Busenberg, E., and C. V. Clemency. 1976. The dissolution kinetics of feldspars at 25°C and 1 atm. CO_2 partial pressure, *Geochim. Cosmochim. Acta* 40: 41–50.

Carroll, D. 1970. *Rock Weathering.* New York: Plenum Press.

Casey, W. H., H. R. Westrich, G. W. Arnold, and J. F. Banfield. 1989. The surface chemistry of dissolving labradorite feldspar, *Geochim. Cosmochim. Acta* 53: 821–832.

Chou, L., and R. Wollast. 1984. Study of the weathering of albite at room temperature and pressure with a fluidized bed reactor, *Geochim. Cosmochim. Acta* 48: 2205–2218.

Cleaves, E. T. 1974. Petrologic and chemical investigations of chemical weathering in mafic rocks, eastern piedmont of Maryland, *Maryland Geol. Surv., Rep. Invest. 25.*

Colman, S. M., and D. P. Dethier. 1986. *Rates of Chemical Weathering of Rocks and Minerals.* Orlando, Fla.: Academic Press.

Crerar, D. A., G. W. Knox, and J. L. Means. 1979. Biogeochemistry of bog iron in the New Jersey Pine Barrens, *Chem. Geol.* 24: 111–135.

Drever, J. I. 1988. *The Geochemistry of Natural Waters,* 2nd ed. Englewood Cliffs, N.J.: Prentice-Hall.

Drever, J. I., and J. Zobrist. 1993. Chemical weathering of silicate rocks as a function of elevation in the southern Swiss Alps, *Geochim. Cosmochim. Acta* 56: 3209–3216.

Feth, J. H., C. E. Roberson, and W. L. Polzer. 1964. Sources of mineral constituents in water from granitic rock, Sierra Nevada, California and Nevada, *USGS Water Supply Paper 1535-I.*

Freeze, R. A., and J. A. Cherry. 1979. *Groundwater.* Englewood Cliffs, N.J.: Prentice Hall.

Garrels, R. M. 1967. Genesis of some ground waters from igneous rocks. In *Researches in Geochemistry,* ed. P. H. Abelson, pp. 405–420. New York: John Wiley.

Garrels, R. M., and F. T. Mackenzie. 1971. *Evolution of Sedimentary Rocks.* New York: W. W. Norton.

Goldich, S. S. 1938. A study on rock weathering, *J. Geol.* 46: 17–58.

Graustein, W. C. 1981. The effect of forest vegetation on solute aquisition and chemical weathering: A study of the Tesuque watersheds near Santa Fe, New Mexico. Ph.D. dissertation, Yale University, New Haven, Conn.

Helgeson, H. C., R. M. Garrels, and F. T. Mackenzie. 1969. Evaluation of irreversible reactions in geochemical processes involving minerals and aqueous solutions 11: Applications, *Geochim. Cosmochim. Acta* 33: 455–482.

Holdren, G. R., and R. A. Berner 1979. Mechanism of feldspar weathering I. Experimental studies, *Geochim. Cosmochim. Acta* 43: 1161–1171.

Holland, H. D., T. V. Kirsipu, J. S. Huebner, and U. M. Oxburgh. 1964. On some aspects of the chemical evolution of cave waters, *J. Geol.* 72: 36–67.

Kleinmann, R. L. P., and D. A. Crerar. 1979. *Thiobacillus ferrooxidans* and the formation of acidity in simulated coal mine environments, *Geomicrobiol. J.* 1: 373–388.

Likens, G. E., F. H. Bormann, R. S. Pierce, J. S. Eaton, and N. M. Johnson. 1977. *Biogeochemistry of a Forested Ecosystem.* New York: Springer-Verlag.

Loughnan, F. C. 1969. *Chemical Weathering of the Silicate Minerals.* New York: American Elsevier.

Luce, R. W., R. W. Bartlett, and G. A. Parks. 1972. Dissolution kinetics of magnesium silicates, *Geochim. Cosmochim. Acta* 36: 35–50.

Mohr, E. J. C., and F. A. van Baren. 1954. *Tropical soils.* New York: Interscience.

Muir, I. J., and Nesbitt, H. W. 1992. Controls on differential leaching of calcium and aluminum from labradorite in dilute electrolyte solutions, *Geochim. Cosmochim. Acta* 56: 3979–3985.

Nahon, D. B. 1991. *Introduction to the Petrology of Soils and Chemical Weathering.* New York: John Wiley.

Paces, T. 1973. Steady-state kinetics and equilibrium between ground water and granitic rock, *Geochim. Cosmochim. Acta* 37: 2641–2663.

Paton, T. R. 1978. *The Formation of Soil Material.* London: Allen & Unwin.

Retallack, G. J. 1990. *Soils of the Past.* London: Harper-Collins.

Richards, B. N. 1987. *The Microbiology of Terrestrial Ecosystems.* New York: John Wiley.

Schlesinger, W. H. 1991. *Biogeochemistry: An analysis of Global Change.* New York: Academic Press.

Schnitzer, M., and S. U. Khan. 1972. *Humic Substances in the Environment.* New York: Marcel Dekker.

Sherman, G. D. 1952. The genesis and morphology of the alumina-rich laterite clays. In *Problems in Clay and Laterite Genesis.* pp. 154–161. American Institute of Mining and Metallurgical Engineers.

Sposito, G. 1989. *The Chemistry of Soils.* Oxford: Oxford University Press.

Stumm, W., and J. J. Morgan. 1981. *Aquatic Chemistry.* New York: John Wiley.

Sverdrup. H., and P. Warfvinge. 1988. Weathering of primary silicate minerals in the natural soil environment in relation to a chemical weathering model, *Water Air and Soil Pollution* 38: 387–408.

Taylor, A. B., and M. A. Velbel. 1991. Geochemical mass balances and weathering rates in forested watersheds of the southern Blue Ridge II. Effects of botanical uptake terms, *Geoderma* 51: 29–50.

Van Breeman, N. 1976. Genesis and solution chemistry of acid sulfate soils in Thailand, Agricultural Research Report 848. Wageningen, the Netherlands: Centre for Agricultural Publishing and Documentation.

Velbel, M. A. 1984. Mineral transformations during rock weathering and geochemical mass-balances in forested watersheds of the southern Appalachians. Ph.D. dissertation, Yale University, New Haven, Conn.

Velbel, M. A. 1985. Geochemical mass balances and weathering rates in forested watersheds of the southern Blue Ridge, *Am. J. Sci.* 285: 904–930.

Velbel, M. A. 1993a. Constancy of silicate-mineral weathering-rate ratios between natural and experimental weathering: Implication for hydrologic control of differences in absolute rates, *Chem. Geol.* 105: 89–99.

Velbel, M. A. 1993b. Formation of protective surface layers during silicate mineral weathering under well-leached, oxidizing conditions, *Am. Mineral.* 78: 405–414.

White, D. E., J. D. Hem, and G. A. Waring. 1963. Chemical composition of subsurface waters: Data of geochemistry, 6th ed., *USGS Prof. Paper no. 440-F.*

Zinke, P. J. 1977. Man's activities and their effect upon the limiting nutrients of primary productivity in marine and terrestrial ecosystems. In *Global Cycles and Their Alterations by Man,* ed. W. Stumm, pp. 89–98. Berlin: Dahlem Konferenzen.

CHAPTER 5

RIVERS

INTRODUCTION

Having considered water vapor in the atmosphere, rain, and the reactions of rainwater with rocks and soil (weathering), we come next in the water cycle to rivers, which serve as the major routes by which continental rain and the products of continental weathering reach the oceans. Rivers do not contain (on the scale of the water cycle) a very large percentage of the total water on the earth (see Chapter 1). Their importance derives from the major role they play in the transport of water as well as dissolved and suspended solids. In this they dwarf the other transport agents from the continents to the oceans, the atmosphere, and groundwaters.

Rivers have played an important role in human development. Historically, settlement along rivers occurred because the rivers provided both water supply and transportation. As a result of human proximity, rivers have been considerably affected by activities ranging from agriculture and flood control to the input of human and industrial wastes. These effects, which are not only of recent origin, have a considerable impact on the transport of water and on dissolved and suspended matter.

River Water Components

River water transported to the oceans can be thought of as being made up of a number of components:

1. Water.
2. Suspended inorganic matter. Major elements include Al, Fe, Si, Ca, K, Mg, Na, and P.
3. Dissolved major species: HCO_3^-, Ca^{2+}, SO_4^{2-}, H_4SiO_4, Cl^-, Na^+, Mg^{2+}, and K^+. These can be further divided into:

a. Elements with no gaseous phases in the atmosphere (for which an input and output balance within the water cycle can be more easily done). These include Ca^{2+}, Cl^-, H_4SiO_4, Na^+, Mg^{2+}, and K^+.

b. Elements with gasous phases, SO_4^{2-} and HCO_3^-, which are derived from atmospheric gases (e.g., SO_2 and CO_2, respectively) as well as from rocks.

4. Dissolved nutrient elements, N and P (and to a certain extent Si), which are used biologically and whose concentrations vary due to this.
5. Suspended and dissolved organic matter.
6. Trace metals, both dissolved and suspended, which will not be discussed in any detail here (however, see, e.g., Nriagu and Pacyna 1988; Martin and Whitfield 1981).

In this chapter we attempt to quantify fluxes of these substances to the oceans by individual rivers as well as for the entire earth. In examining results, the reader should be forewarned that flux calulations are susceptible to a number of errors. This includes [see Meybeck (1988) for further discussion] inadequate knowledge of discharge (water flux), especially of seasonal variations; inadequate sampling of rivers for concentrations of dissolved and suspended components; extrapolation of results on known rivers to a large number of unsampled rivers; inadequate knowledge of the chemical composition of rainfall falling on river basins; inadequate estimates of human effects; and lack of sampling right at the river mouth (sampling upstream of the mouth can give erroneous results because of processes occurring before reaching the sea, e.g., deposition of riverine suspended matter on coastal alluvial plains).

River Runoff

Rivers result from the *runoff* of water from the continents. River water itself comes ultimately from precipitation, some of which is evaporated from the land, some of which passes through the ground to the river at shallow depths (surface flow), and some of which remains in the ground much longer and reaches greater depths before entering the river (groundwater). On a global scale, continental runoff, which includes predominantly river runoff and a small amount of direct groundwater discharge to the oceans (Meybeck 1984), can be thought of as being equal to the excess of oceanic evaporation over precipitation. This excess ocean evaporation results in water vapor transport to the continents, where water vapor is converted to rain and the nonevaporated portion is ultimately lost as river runoff, thereby balancing the water cycle (see Chapter 1). Thus, the initial amount of rainfall on land is determined by the evaporation rate over the oceans (where 85% of total evaporation occurs), and by the global circulation of water vapor, which is driven by meridional differences in solar heating (and temperature).

On a continental scale, in order to have river runoff, there must be *net* precipitation on land (i.e., precipitation must exceed continental evaporation). Average continental evaporation rates are temperature dependent and decrease rapidly with increasing latitude (see Chapter 1). Rainfall is not uniformly distributed because it is dependent on the atmospheric circulation of water vapor. The result is that there are two major belts where precipitation exceeds evaporation (i.e., where there is net positive precipitation), and these are the areas where most of the large rivers occur. One belt is near the equator (10°N–10°S) where there is both high rainfall (due to high concentrations of water vapor) and high evaporation (due to high temperatures) but with rainfall exceeding evaporation. This results in large rivers such as the Amazon and Zaire. The second belt is in the temperate zone (30°–60°N and S). Here there is generally adequate rainfall (and a good supply of water vapor) as well as lower evaporation rates (due to lower average surface temperature). Two major rivers originating in this area are the Mississippi and the Yangtze. Between these two belts there is a region of low river runoff in the subtropics (15°–30°N and S), where evaporation exceeds precipitation and deserts form. This area has fairly high surface temperatures and high evaporation. Subarctic cold regions from about 60°N to 70°N also have relatively low river runoff (10–20 cm/yr) and include rivers such as the Mackenzie and Lena. Here precipitation is low but evaporation is also very low due to the cold temperatures.

Another type of variation in river runoff which occurs on a subcontinental scale is caused by cyclic fluctuations with time in the general atmospheric circulation (Probst and Tardy 1989). High and low pressure anomalies move across continents and are correlated with low and high runoff, respectively. This is often related to the Southern Pacific Oscillation which produces El Nino.

Superimposed on the continental-scale factors influencing the amount of river runoff are more local effects resulting from the distribution of rainfall in both space and time. Geographic differences in rainfall distribution are due primarily to relief, with the windward sides of mountains receiving large amounts of rain and the leeward sides very little. Geographic heterogeneity can increase the runoff by as much as 20% (Holland 1978). For two continents having the same average annual rainfall, the one with greater geographically induced variation in the rainfall rate will have less loss of rainfall due to evaporation and thus more runoff. Seasonality of rainfall also increases runoff because, again, there is less annual loss by evaporation. An extreme example of this is the monsoonal climate of southern Asia, where an "average" monthly rainfall of 17 cm varies seasonally from about 1 cm to 69 cm per month (Miller et al. 1983). This gives rise to such rivers as the Ganges and Brahmaputra, which undergo large seasonal fluctuations in flow rates and which flood often.

When we discussed atmospheric CO_2 in Chapter 2, we noted that an increase in the average temperature of the earth as a whole would greatly affect runoff. Overall, worldwide temperature increases should speed up the hydrologic cycle and produce more runoff. The oceans would be warmer and there would be more evaporation and more water vapor to produce more rain and runoff. However, the geographic distribution of belts of excess precipitation would be different, with the result that certain areas, such as the interior of the United States, might have less runoff.

The *runoff ratio* is the ratio of average river runoff (per unit area) to average rainfall (per unit area). The world average value is about 0.46, suggesting that about 50% of rainwater that reaches the ground is returned directly to the atmosphere by evaporation and never reaches rivers. However, there is considerable variation in the runoff ratio by continent due to the factors discussed above, from a high of 0.54 in Asia (with its areas of high mountains and monsoon climate) to a low of 0.28 in Africa (which has large areas of desert and where most of the rainfall occurs in low-lying areas). South America (0.41), Europe (0.42), and North America (0.38) have intermediate runoff ratios (See Table 5.6).

Major World Rivers

Table 5.1 lists some major world rivers, in approximate order of water discharge, that empty directly into the sea. The first 13 rivers (with a total discharge of 14,000 km^3 water per year) account for about 38% of the total water discharge to the oceans (37,400 km^3/yr). Most remarkably, one river, the Amazon, accounts for 17% of the total water discharge to the oceans, and

TABLE 5.1 Major Rivers that Flow to the Sea, Listed in Order of Discharge

River	Location	Annual Discharge: Water (km^3/yr)	Dissolved Solids (Tg/yr)	Suspended Solids (Tg/yr)	Dissolved/ Suspended Ratio	Drainage Area (10^6 km^2)
1. Amazon	S. America	6300	275	1200	0.23	6.15
2. Zaire (Congo)	Africa	1250	41	43	0.95	3.82
3. Orinoco	S. America	1100	32	150	0.21	0.99
4. Yangtze (Chiang)	Asia (China)	900	247	478	0.53	1.94
5. Brahmaputra	Asia	603	61	540	0.11	0.58
6. Mississippi	N. America	580	125	210 (400)	0.60	3.27
7. Yenisei	Asia (Russia)	560	68	13	5.2	2.58
8. Lena	Asia (Russia)	525	49	18	2.7	2.49
9. Mekong	Asia (Vietnam)	470	57	160	0.36	0.79
10. Ganges	Asia	450	75	520	0.14	0.975
11. St. Lawrence	N. America	447	45	4	11.3	1.03
12. Parana	S. America	429	16	79	0.20	2.6
13. Irrawaddy	Asia (Burma)	428	92	265	0.35	0.43
15. Mackenzie	N. America	306	64	42	1.5	1.81
17. Columbia	N. America	251	35	10 (15)	3.5	0.67
20. Indus	Asia (India)	238	79	59 (250)	1.3	0.975
Red (Hungho)	Asia (Vietnam)	123	?	130	?	0.12
Huanghe (Yellow)	Asia (China)	59	22	1100	0.02	0.77

Note: Tributaries are excluded. Tg = 10^6 tons = 10^{12} g.

Sources: Water and suspended solids from Milliman and Meade (1983) and Milliman and Syvitski (1992). Dissolved solids calculated from Table 5.7 and Pinet and Souriau (1988) (for the Irrawaddy). Suspended load values in parentheses indicate pre-dam values.

has over 10 times the discharge of the Mississippi River. Also, most of the world's largest rivers are in underdeveloped countries and as a result have not been studied very well.

SUSPENDED MATTER IN RIVERS

Total Suspended Matter

The total suspended load carried by rivers to the oceans was about 20×10^9 metric tons/yr before there was increased dam construction starting about 1950 (Milliman and Syvitski 1992). The previous estimate of suspended load, 13.5×10^9 tons/yr (Milliman and Meade 1983), underestimated the considerable contribution of small mountain rivers to suspended sediment transport. Other estimates of the suspended load include: 15.5×10^9 tons/yr (Martin and Meybeck 1979) and 18.4×10^9 tons/yr (Holeman 1968). The total suspended sediment load can be used to calculate the mechanical denudation rate for the continents, which is about 8.3 cm of elevation reduction in 1000 years (assuming an average rock density of 2.7 and ignoring isostatic uplift, which occurs over longer geologic time periods). However, as pointed out by Milliman and

Meade (1983), the sediment discharge rate from the continents to the oceans by rivers *is not the same* as the total rate of soil erosion. This is because much sediment is eroded from upland areas and deposited in lowland areas, stream valleys, floodplains, and reservoirs without reaching the sea. The sediment discharge rate from the conterminous United States to the oceans is only about 8% of the total erosion rate over the same area. [This number is based on an erosion rate of 5.3 $\times 10^9$ tons/yr from Holeman (1980), and a sediment discharge rate of 0.445×10^9 tons/yr, from Curtis et al. (1973).] Similarly, Meade and Parker (1985) state that only 10% of the sediment eroded in the United States reaches the oceans.

One third of the total world suspended load carried to the sea is from southern Asia, and another 45% is from the large Pacific and Indian Ocean islands (particularly Taiwan, New Guinea, and New Zealand) (Milliman and Syvitski 1992). This is despite the fact that no river from the large Pacific islands ranks among the top 25 rivers in water discharge. For example, approximately the same amount of suspended sediment is removed from Taiwan as from a major drainage area, 575 times as large, on the African continent. It is the sediment *yield* (yield = load/drainage area) that sets these oceanic island rivers apart. Yield is a measure of the erodability of a drainage basin. Figure 5.1 shows the total suspended sediment load and suspended sediment yield from the various world river basins, and Table 5.2 gives the sediment yield and discharge and drainage area for the various continents. Globally, the average sediment yield is 226 tons/km^2/yr, based on the land area of external drainage of 88.6×10^6 km^2.

The individual rivers carrying the largest sediment load shown in Table 5.1 are, in order, the Amazon River, the Huanghe (Yellow) River, the Brahmaputra River, the Ganges River, and the Yangtze River. The Huanghe River, whose sediment load is 6% of the world total, carries only

TABLE 5.2 Suspended Sediment Carried by Rivers to the Ocean (in Metric Tons)

Continent	Drainage Area Contributing Sediment to Ocean (10^6 km^2)	Sediment Discharge (10^6 tons/yr)	Sediment Yield (tons/km^2/yr)	Mean Continental Elevation (km)
North America	15.4	1020	66	0.72
Central America[a]	2.1	442	210	—
South America	17.9	1788	97	0.59
Europe	4.61	230	50	0.34
Eurasian Arctic	11.17	84	8	~0.2
Asia	16.88	6349	380	0.96
Africa	15.34	530	35	0.75
Australia	2.2	62	28	0.34
Pacific & Indian Ocean Islands[b]	3.0	9000[c]	3000[c]	~1.0
World total	88.6	20,000[d]	226[d]	

[a] Includes Mexico.

[b] Japan, Indonesia, Taiwan, Phillipines, New Guinea, and New Zealand (Oceania).

[c] From Milliman and Syvitski (1992).

[d] From Milliman and Syvitski (1992). Data reflect greater sediment discharge from South America, the Alps-Caucusus Mountains, and northwest Africa, in addition to Oceania.

Sources: After Milliman and Meade (1983) and Milliman and Syvitski (1992), elevations from Fairbridge (1968).

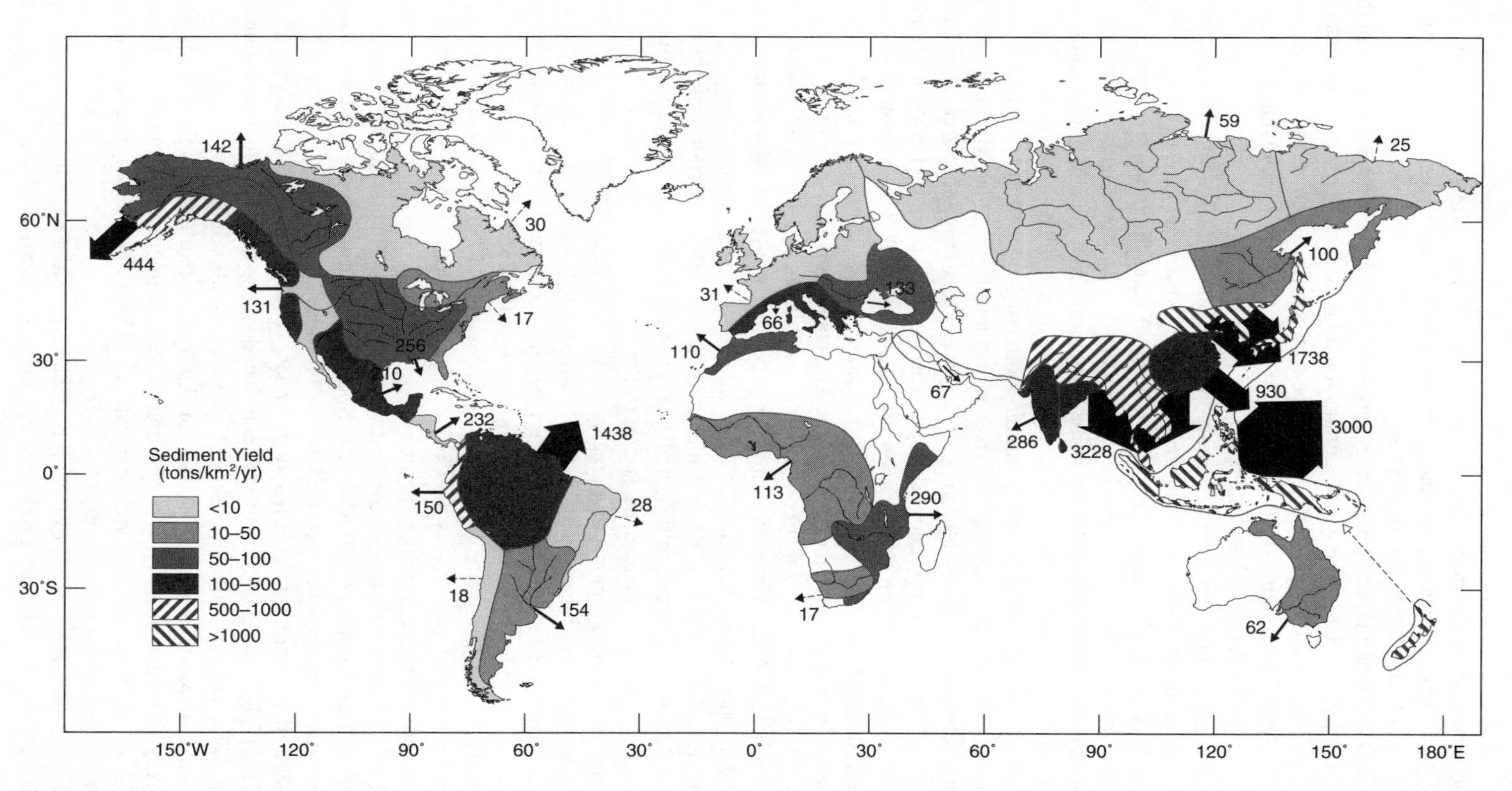

Figure 5.1. Discharge of suspended sediment from world drainage basins (in 10^6 tons/yr) as indicated by arrows. Sediment yield (tons/km^2/yr) for various drainage basins is also shown by appropriate pattern (see legend). Open pattern indicates essentially no sediment discharges to the oceans. [After J. D. Milliman and R. H. Meade. "World-Wide Delivery of River Sediment to the Oceans," *Journal of Geology* 91(1): 16. Copyright © 1983 by The University of Chicago Press, reprinted by permission of the publisher.]

0.1% of the total river water discharge. The reason for this lies in the nature of the sediment source; for the Huanghe River it is the easily eroded and heavily farmed yellow loess of north central China (Holeman 1968; Milliman et al. 1987).

The Ganges and Brahmaputra rivers carry large sediment loads because they drain the readily erodible foothills of the Himalaya Mountains, the highest in the world, in an area of heavy seasonal (monsoonal) rainfall and large runoff. The Amazon River gets most of its sediment load only from the Andes Mountains (Gibbs 1967), not from the Brazilian lowlands; as a result, it does not have a particularly high sediment yield.

A number of natural factors control the suspended sediment load of rivers (Milliman 1980): (1) relief of the drainage basin, (2) drainage basin area, (3) amount of water discharge, (4) climate, (5) geology of the river basin, and (6) presence of lakes along the river length. The suspended load of most rivers is influenced by more than one of these factors in combination. In addition, human activities, such as deforestation, agriculture, and the building of dams, have had a very large effect.

The most important factors in determining sediment load and sediment yield are relief (expressed as maximum elevation of the river basin) and basin area (Milliman and Syvitski 1992). Relief (elevation) is especially important. Greater relief produces greater water velocity and consequently greater competence (ability to carry coarser particles). Milliman and Syvitski classify rivers into elevation categories: high mountains (>3000 m), mountains (1000–3000 m), uplands (500–1000 m), lowlands (100–500 m), and coastal plains (<100 m). The largest sediment yields come from the high mountain rivers. Mountain rivers have greater yields than upland rivers, which have greater yields than lowland rivers, which have greater yields than coastal plain rivers. The southern Asian rivers draining the very high Himalaya Mountains, such as the Ganges-Brahmaputra, Indus, Yangtze, Mekong, and the Irawaddy, have especially high sediment yields. Climate and runoff are usually less important factors. However, mountain rivers in southern Asia and Oceania (the large Pacific islands) have much greater sediment yields (two- to threefold) than other mountain areas. This is due to the combined influences of human activity (deforestation and farming), monsoon climate, and geology (loess).

Small mountain rivers located a short distance from the ocean, with steep slopes and little area for sediment storage in flood plains, have very high sediment yields. In South America, the rivers on the steep Andes Mountain slopes draining into the Pacific exemplify this effect (Figure 5.1). The U.S. river with the largest known sediment yield (1750 tons/km^2/yr) is the Eel River of northern California, which drains the Coast Ranges (reaching elevations of more than 2000 m) located a short distance from the Pacific.

For large river basins, suspended sediment yield can be correlated with the geological age of the terrain (Pinet and Souriau 1988). Pinet and Souriau stress the tectonic control of basin relief, pointing out that rivers draining mountainous areas whose last orogeny was less than 250 million years ago (those draining the Himalayas, Andes, and Alps) have much higher mean basin elevations and higher suspended sediment yields than do those draining older mountains. In fact, the areas with the largest sediment yield in Figure 5.1 are all young, tectonically active areas. Taiwan, with a suspended yield of 10,000 tons/km^2/yr, probably the highest in the world, is presently very active tectonically, with the transported sediment being rapidly subducted and recycled. This suspended sediment yield translates to a lowering of elevation of about 1.5 mm/yr out of an actual uplift rate of 5.5 mm/yr (Li 1976).

In large rivers the presence or absence of high elevations at a large distance from the mouth can exert an effect on sediment discharge. For example, erosion of the Andean foothills is believed to be the dominant control on both the amount and composition of suspended material in the Amazon River (Gibbs 1967), even though the foothills are thousands of kilometers away from the river mouth. By contrast, rivers in areas of consistently low relief have very low sediment yields. The Eurasian Arctic rivers (the Yenisei and Lena, the seventh and eighth largest rivers) are located in a vast low-lying area and have a very low sediment yield (5 tons/km²/yr).

Garrels and Mackenzie (1971), Hay and Southam (1977), and Holland (1978) point out that suspended sediment yield seems to increase exponentially with mean continental elevation, despite the fact that the elevation of a whole continent is highly variable. This is shown in Figure 5.2, in which Milliman and Meade's (1983) data for suspended load per unit area of *external* drainage from the continents are plotted against mean continental elevation (Fairbridge 1968). [Hay and Southam (1977) point out that external drainage area, i.e., area draining to the oceans, should be more relevant than total drainage area, since suspended load is based on rivers that enter the ocean.] Also included in Figure 5.2 are the Pacific islands and the Eurasian Arctic (with a rough estimate of their elevation), which tend to confirm the correlation. Africa has a somewhat lower sediment yield than might be expected based on mean elevation. Hay and Southam (1977) suggest that this is due to the fact that the continental slope of Africa starts 400 m higher than that of most continents; in other words, Africa is a high plateau. If the mean elevation of Africa is measured as being that above 400 m, then the sediment yield plots near Europe and Australia in Figure 5.2 and the empirical relation provides a better fit to the data.

Milliman and Syvitski (1992) found that sediment load increases (and sediment yield decreases) with increasing basin area, but the scatter is very large. The correlation between load and area is much better when basins are subdivided into categories by relief. The greater sediment yield in small basins is due to their inability to store sediment because of steep slopes and the absence of flood plains. The decrease in sediment yield with increasing drainage area of streams

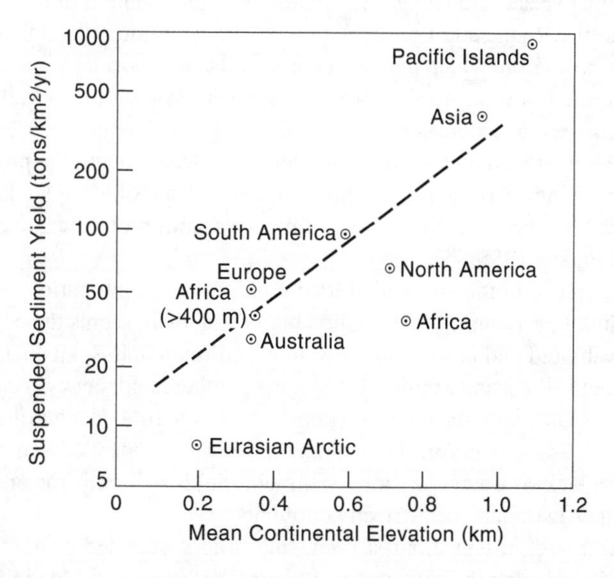

Figure 5.2. Suspended sediment yield (tons/km²/yr) versus mean continental elevation (km). [Suspended sediment data from Milliman and Meade (1983); mean continental elevation from Fairbridge (1968), except for the Pacific Islands and Eurasian Arctic, which are our rough estimates.]

is attributed by Trimble (1977) at least partly to human activities. The inability of large-drainage-area streams to carry artificially increased solid erosion loads due to human activities may result in the deposition of the excess sediment in channels and flood plains.

The amount of water discharged by a river (either total annual discharge or runoff per unit area) seems to have less effect than other factors on the suspended load, either on a local basis (Milliman and Meade 1983; Holland 1978) or on a continental basis (Hay and Southam 1977). Milliman and Syvitski (1992) found only a random relationship betweeen sediment load and runoff. Although the Amazon (the world's largest river in terms of discharge) does carry a large sediment load, its sediment yield per unit area is only about average for world rivers (150 tons/km^2/yr). Certainly, some of the largest rivers in terms of discharge (the Congo, Yenisei, and Lena) have abnormally low sediment yields (less than 10 tons/km^2/yr), and others (the Ganges-Brahmaputra) have abnormally high yields. Most major rivers have an average suspended sediment concentration between 100 mg/l and 1000 mg/l (Milliman 1980). Dividing the total suspended load (20×10^9 tons/yr) by the total water discharge (37,400 km^3/yr) gives a global average suspended sediment concentration for all rivers of 535 mg/l.

The geology of the river basin can have a very important effect on the sediment load. The most obvious case is the Huanghe (Yellow) River, with one of the largest of all sediment loads, which flows through vast areas of loess (wind-blown glacial dust), a material that is extremely easily eroded and is heavily farmed. Similarly, rivers in parts of the U.S. Midwest that are underlain by loess (Iowa, Illinois, etc.) have unusually high suspended sediment concentrations (Meade and Parker 1985). The presence of active glaciers also increases the sediment load of rivers (Milliman and Meade 1983) because of the production of easily eroded rock debris (moraine, drift, etc.) from glacial grinding. Examples of this are Alaskan rivers draining from glaciers, which average suspended sediment yields of 1000 tons/km^2/yr. Three rivers (the Copper, Yukon, and Susitna) draining the Alaska Range have the largest sediment loads in the United States after the Mississippi (Meade and Parker 1985). Similarly, the southern European rivers draining Alpine glacial areas (the Rhone and Po) carry a much larger sediment load than most European rivers. (By contrast, areas from which the rock debris has been *removed* by glaciers—northeastern North America and the Eurasian Arctic—tend to have very low sediment yields.)

Tectonism, besides increasing relief, also tends to create erodable materials leading to increased sediment load and yield. Examples are volcanism and the production of easily erodible volcanic ash, and earthquakes leading to the formation of loose landslide debris. The Columbia River load has been greatly increased by the eruption of volcanic ash from Mt. St. Helens (Meade and Parker 1985).

Climatic effects on the suspended load are due to a combination of mean annual temperature, total rainfall, and rainfall seasonality. In general, there seems to be little correlation between sediment load/yield and *mean annual runoff* (Milliman and Syvitski 1992; Walling and Webb 1983) or rainfall (Pinet and Souriau 1988). On the other hand, heavy *seasonal* rainfall and runoff, such as is associated with the monsoon climate in southern Asia, contributes to a large suspended sediment load. Also, rivers in desert areas have high suspended sediment concentrations but (because of a lack of water) low total sediment yields, which is the situation for the large arid regions of the Australian and African continents.

When lakes are present along a river, they trap suspended sediment before it reaches the ocean, greatly reducing the river sediment load (Milliman and Meade 1983). Examples are the effects of the Great Lakes on the St. Lawrence River; Lake Constance, which traps most of the Rhine River sediment; and lakes along the Zaire River. On a larger scale, the Black Sea traps more than half of the sediment draining off nonarctic Europe, and the Mediterranean Sea traps much of the rest, thus preventing the sediment from reaching the Atlantic Ocean.

To summarize, the greatest suspended sediment loads/yields are from mountainous areas of high relief such as the mountainous Pacific islands and southern Asia. Particularly high loads are found in southern Asia because its monsoon climate and extensive human activity are combined with high mountainous relief. Higher sediment yields are also found in areas with active glaciers, small mountainous basins near coasts, and areas draining loess (Figure 5.1).

Human Influence

Human activities may increase or decrease the suspended sediment load. Increases are due to (1) deforestation and cultivation of the land, (2) overgrazing, and (3) construction. Decreases in the suspended load result from (1) building of dams and reservoirs, which trap sediment, (2) bank stabilization of rivers, and (3) soil conservation practices. The suspended load of most world rivers, except Arctic rivers, has been changed by human activities (Milliman and Syvitski 1992).

The initial effect of humans was to increase the suspended sediment load through deforestation and conversion of the land into cropland. These effects can be seen as far back as Roman times, when human activity resulted in large increases in sediment deposition in an Italian lake (Judson 1968). Similar effects have been noted in the eastern United States (Trimble 1975) corresponding to the arrival of European settlers. However, there is a difference between the increased *erosion* rate (the rate at which soil is removed from the upland land surfaces) due to human activities and the stream *transport* rate of suspended sediment to the oceans. In the southeastern United States, apparently streams were unable to handle the increased erosional loads due to European settlement, and only 5% of the material eroded from upland slopes has been exported to the oceans as suspended sediment. Most of the eroded sediment has been deposited instead as colluvium and alluvium in lowland valleys and floodplains (Trimble 1975). Trimble (1983), in a study of Coon Creek Basin, Wisconsin, found that $< 10\%$ of the sediment eroded after 1850, when erosion was accelerated by farming, was exported from the basin.

Many major rivers that transport large amounts of sediment, such as those in India and China, have been affected by agriculture for centuries (Milliman and Meade 1983). Milliman et al. (1987) estimate that the Yellow River sediment load prior to 200 B.C. was one tenth of the present load. Overgrazing by animals can also cause unusually large suspended sediment yields. One such case noted by Milliman and Meade (1983) is the Tana River in Kenya.

More recently, humans have been building huge dams that trap large amounts of riverine suspended sediment. Milliman and Meade (1983) point out that the suspended sediment loads normally transported to the oceans by the Nile and Colorado rivers have been reduced to almost nothing by dams, and the load of the Mississippi River has been reduced by one third. The loads of the Zambezi River, the Indus River, and the Rio Grande have been similarly reduced. The loss in suspended sediment transported to the ocean due to dams on large rivers is, according to Milliman and Meade (1983), about 0.5×10^9 tons of sediment a year or around 3% of the total river suspended load. (Pearce 1991, has estimated that 13% of river sediment is dammed.) Most U.S. reservoirs trap at least half of the river sediment that enters them (Meade and Parker 1985). Meade (1982) estimated that present southeastern U.S. sediment loads may be nearly equal to pre-European values due to the effect of dams offsetting anthropogenically increased erosion. Similarly, deforestation and increased erosion in Nepal has caused a larger sediment load in the Brahmaputra River, which is counterbalanced by dams after entering the Ganges River (Hossain

1991). Soil conservation plus bank stabilization on the Orange River in South Africa have also greatly reduced the suspended sediment load there.

Smith el al. (1987) found that increases in the suspended load in U.S. rivers from 1974 to 1981 (which occurred in the Mississippi and Arkansas-Red basins) were significantly correlated with cropland soil erosion. Conversely, decreasing U.S. trends in suspended load, where they occurred, did not seem to be due to reservoir construction during this period.

What is the net effect of the various human influences on the suspended load in both U.S. and world rivers? In the United States, deforestation and the conversion of land to cropland and pasture since the arrival of European settlers has greatly increased the soil erosion rate (Judson 1968). From a knowledge of the present total U.S. erosion rate, and the percentage of it contributed by various land types (forests, rangeland, cropland, pasture) combined with the areal extent of these land types, we have estimated the pre-European suspended sediment load.

First, we note from Table 5.3 that the present erosion rate from cropland (1371 tons/km^2/yr) is about eight times the rate from forested land, that from pasture about four times, and that from rangeland about three times the forest rate. The rates given in Table 5.3 are rates for *erosion* of soil from the land surface, but what we want is the *transport* (yield) of sediment from drainage basins to the oceans as river suspended sediment load, which, as pointed out above, can be a considerably lower number. If we assume that annual river suspended sediment yield for each land type and for the total continent are in the same proportion as is found for relative erosion rates (given above and in Table 5.3), then we can calculate the present sediment yield rate for forested land only. Let n equal this rate; from Table 5.3, the yield rate, on the above assumption, is then about $8n$ for cropland, $4n$ for pasture, and $3n$ for rangeland. Multiplying these rates by their respective areas, we obtain the following equation by summing all four types and equating the result to the total U.S. suspended load (Curtis et al. 1973):

$$1.43 \times 10^6 \text{ km}^2(8n) + 0.47 \times 10^6 \text{ km}^2(4n) + 2.61 \times 10^6 \text{ km}^2(3n)$$

$$+ 2.38 \times 10^6 \text{ km}^2(n) = 445 \times 10^6 \text{ tons/yr}$$

Solution of this equation results in $n = 18.9$ tons/km^2/yr, the value for the present forested land sediment yield rate.

Before the arrival of Europeans, the United States was originally one half forest and one third grassland (*Encyclopedia Americana* 1959, vol. 27: 311) with the latter, by analogy with rangeland, assumed to have a natural erosion rate three times that of forests. (The remaining one sixth of the United States is and was desert, which presumably has few rivers and contributes little sediment to the sea.) From this we calculate that the pre-European total erosion rate for the United States was roughly equal to 1.5 times the present forest erosion rate [$\frac{1}{2}(n) + \frac{1}{3}(3n) = 1.5n$; see Table 5.3]. We assume that the total pre-European suspended sediment yield rate was also 1.5 times the the present forest yield rate, so that multiplying the value of n above by 1.5, we obtain 28 tons/km^2/yr for the pre-European sediment yield rate for all land. By comparison the present value is 65 tons/km^2/yr. Thus, the effect of European settlement of the United States has been to increase the sediment loss to the oceans by a factor of about 2.3.

By a similar procedure, one can calculate a prehuman sediment yield for the world. We start with the present-day sources of sediment. We assume that the sediment yield rates for different types of land for the world have approximately the same relative proportions as those for the United States. In other words, if the forest sediment yield rate is n, then the rate for grassland is $3n$ and that for cropland is $10n$. (The latter value is raised from $8n$ to account for the poorer

TABLE 5.3 U.S. (Excluding Alaska) Soil Erosion Rates by Land Type

Land Type	U.S. Area (10^6 km^2)[a]	Percent of Total U.S. External Drainage Area[b]	Calculated Percent of Total Erosion Load[c]	Relative Erosion Load (10^9 tons/yr)[c]	Erosion Rate (tons/km^2/yr)	Erosion Rate[d]
Arable (cropland plus pasture)	1.90	28	—	—	—	—
Cropland	1.43	21	37	1.96	1371	$8n$
Pasture	0.47[e]	7	7	0.37	787	$4n$
Forest	2.38	34	8	0.42	176	$\equiv n$
Rangeland	2.61[f]	38	26	1.38	529	$3n$

[a] From *World Almanac* 1982, New York: Pharos Books.

[b] Total U.S. external drainage area is 6.89×16^6 km^2 (Curtis et al 1973).

[c] Total soil erosion load (5.3×10^9 tons/yr) and percentages of this load from Holeman (1980).

[d] See text for calculation.

[e] Calculated as difference between arable and cropland.

[f] Equal to 6.89 – (1.90 + 2.38).

worldwide soil conservation on agricultural land.) The worldwide areas of various land types are given in Table 5.4 along with the total world exterior drainage area (area over which rivers flow to the ocean rather than dry up by evaporation), and the world suspended sediment load of 20×10^9 tons/yr (Milliman and Syvitski 1992). The equation for the world analogous to that for the United States is

$$51 \times 10^6 \text{ km}^2(n) + 14 \times 10^6 \text{ km}^2(10n) + 24 \times 10^6 \text{ km}^2(3n) = 20 \times 10^9 \text{ tons/yr}$$

Solving for *n*, we get a forested land suspended sediment yield rate of 76 tons/km^2/yr. Assuming that the prehuman land distribution for the world was the same as for the pre-European United States, we multiply, as above, this value by 1.5 to obtain a prehuman yield rate of 114 tons/km^2/yr. Over a total world exterior drainage area of 89×10^6 km^2, this equals a prehuman suspended load of 10.15×10^9 tons/yr. The present world suspended sediment load is 20×10^9 tons/yr, so the effect of agriculture worldwide has caused an increase in sediment yield by a factor of about 2. Thus, our results for both the United States and the world indicate that human activity has had a profound influence on the suspended sediment yield. This is in overall agreement with the results of Judson (1968), who used a somewhat different method of calculation and obtained 9.3×10^9 tons/yr for the prehuman load, and with those of Wold and Hay (1990), who estimated 10.9×10^9 tons/yr. McLennan (1993) estimated a prehuman load of 12.6×10^9 tons/yr.

Chemical Composition of Suspended Matter

Table 5.5 shows the average concentrations of the major elements in river suspended (particulate) material and dissolved material compared with the concentration in average surficial rocks and soils. Because of chemical weathering and the dissolution of soluble elements combined with reprecipitation of insoluble elements in secondary weathering products, the river suspended matter is enriched in the less soluble elements (such as Al and Fe) relative to the parent rock and strongly depleted in the most soluble elements (Na and Ca). The ratio of concentrations of various elements in river suspended matter to average surficial rock is also shown in Table 5.5. A ratio greater than 1.0 indicates enrichment in river suspended matter, as is the case for Al and Fe. Although Si has about the same concentration in river suspended matter as in surficial rocks,

TABLE 5.4 World Areas of Different Land Types, and Estimated Sediment Yield

Land Type	Area[a] (10^6 km^2)	Relative Suspended Sediment Yield[b] (tons/km^2/yr)
Cropland	14	$10n$
Grassland	24	$3n$
Forests	51	n

[a] After Zehnder and Zinder 1980; Logan 1983.

[b] Total world exterior drainage area = 89×10^6 km^2; total suspended sediment load = $20{,}000 \times 10^6$ tons/yr (Milliman and Syvitski 1992).

TABLE 5.5 Concentrations of Major Elements in Continental Rocks and Soils and in River Dissolved and Particulate Matter

	Continents		Rivers				Element Weight Ratio	
Element	Surficial Rock Concentration (mg/g)	Soil Concentration (mg/g)	Particulate Concentration (mg/g)	Dissolved Concentration (mg/l)	Particulate Load (10^6 tons/yr)	Dissolved Load (10^6 tons/yr)	River Particulate/ Rock	Particulate/ (Particulate + Dissolved)
Al	69.3	71.0	94.0	0.05	1457	2	1.35	0.999
Ca	45.0	35.0	21.5	13.40	333	501	0.48	0.40
Fe	35.9	40.0	48.0	0.04	744	1.5	1.33	0.998
K	24.4	14.0	20.0	1.30	310	49	0.82	0.86
Mg	16.4	5.0	11.8	3.35	183	125	0.72	0.59
Na	14.2	5.0	7.1	5.15	110	193	0.50	0.36
Si	275.0	330.0	285.0	4.85	4418	181	1.04	0.96
P	0.61	0.8	1.15	0.025	18	1.0	1.89	0.96

Note: Elements with no gaseous phase only. Particulate and dissolved loads based, respectively, on the total loads, 15.5×10^9 tons solids/yr and 37,400 km^3 water/yr.

Sources: After Martin and Meybeck 1979; Martin and Whitfield 1981; Meybeck 1979, 1982.

it is somewhat enriched and left behind in soils; this is due partly to the resistance of quartz (SiO_2) to weathering. Sodium and calcium in river suspended material are strongly depleted to about half of their concentrations in surficial rocks, and Mg and K are also fairly strongly depleted. Phosphorus, which is a nutrient, is enriched greatly over surficial rocks, but this is probably a biological and possibly a pollutional effect.

The relative size of the suspended particulate load to the total load (dissolved plus suspended) is also shown in Table 5.5. As expected from their enrichment in the suspended load, Al and Fe are carried almost entirely in the suspended load, and Si is largely carried there. At the other extreme, Na and Ca, which are the most depleted in the river particulate load, have only 40% of their total load in the particulate form. Potassium and magnesium are interesting in that they are depleted similarly in the river particulate load over rock concentrations (enrichment = 0.7), but a larger part of the Mg (40%) appears in the dissolved load than does K (14%). This may have to do with the fact that K is retained in clay minerals in soils while Mg, although it is found in clay minerals, also occurs in soluble carbonate rocks. In addition, more K is stored in the biosphere.

Martin and Meybeck (1979) believe that the geographic variations in major element concentrations in river particulate matter can be explained by differences in climate and resultant weathering regimes between river basins. They distinguish between intensely weathered tropical river basins, and temperate and arctic river basins, which have less intense weathering. The tropical rivers have high Al and Fe concentrations because their particulates originate from soil material enriched in insoluble elements left behind when soluble elements were leached away. Temperate and arctic rivers, on the other hand, have lower concentrations of Al and Fe in the suspended matter relative to soluble elements, because smaller amounts of soluble elements have been removed. Their suspended load comes from rock debris and poorly weathered material, especially in mountainous areas, and the composition of the suspended matter is closer to the average composition of surficial rocks. Martin and Meybeck also state that Ca and Na concentrations are lower in tropical river particulate matter than in temperate and arctic river particulate matter. Based on their data (for 15 rivers), particulate Al does seem generally to be higher and Ca lower in tropical rivers, but there are a number of exceptions and the effect of variations in the original lithology of the rocks and relief of the river basin certainly must be considered. In fact, those rivers that have higher suspended matter Ca concentrations also have high dissolved Ca concentrations, suggesting that the rocks in their river basins contain more Ca than other basins, probably in the form of limestone. Also, in general, sedimentary rocks are more abundant in temperate areas than in tropical areas (Meybeck 1987). Tropical rivers with high particulate Al concentrations also tend to have high dissolved SiO_2 concentrations, which is suggestive of both a siliceous and aluminous lithology in the source regions and higher weathering rates associated with higher temperatures (Meybeck 1987).

For most of the major elements listed in Table 5.5, nearly 90% of the total transport is in the particulate load. However, HCO_3^-, which is the dominant dissolved ion, is not included in the elements listed. In addition, as far as the availability of major elements for chemical reactions in the oceans is concerned, obviously the dissolved elements will be more important and the suspended load much less so. A large part of the suspended load is merely dumped and buried upon reaching the oceans. There are, nevertheless, some changes involving the suspended load, such as ion exchange, when it goes from river water to the ocean, and this can affect ocean water chemistry. (Discussion of this topic is deferred to Chapters 7 and 8 on estuaries and the oceans.)

CHEMICAL COMPOSITION OF RIVERS

World Average River Water

The chemical composition of world average river water according to Meybeck (1979) is given in Table 5.6. This represents river transport to the oceans of dissolved components and does not include transport to internal basins. "Natural" world river water is corrected for pollution and "actual" river water includes pollution. Meybeck's estimate for natural river water is similar to earlier estimates (Livingstone 1963). (Note that dissolved silica, in keeping with tradition, is here expressed in terms of SiO_2, and not the more correct chemical form H_4SiO_4 as used in Chapter 4.)

In estimating natural (prehuman) world river water composition, Meybeck attempted to avoid conspicuously polluted river data by using early (pre-1900) data for such rivers as the Mississippi, St. Lawrence, and Rhine. In addition, he made a further correction for pollution by estimating anthropogenic inputs from the changes with time in the composition of five large river basins in industrial areas and from direct measurements of river pollution. He corrected the various continents differently for pollution based on their population and stage of industrial development. The main constituents affected are Cl^-, SO_4^{2-}, and Na^+ and, to a lesser extent, Ca^{2+} and Mg^{2+}. As can be seen in Table 5.6, Meybeck estimated that about 30% of the Na^+, Cl^-, and SO_4^{2-} concentrations in actual river water can be considered as arising from pollution. (However, we estimate that sulfate pollution is higher; see Table 5.11 and our detailed discussion of river pollution later in this chapter.) Meybeck gives only the natural river water values for each continent. We here calculate the actual (polluted) values by continent based on his correction scheme.

Looking at world average river water, we can see first that the total concentration of dissolved major ions (total dissolved solids, TDS) in river water is about 100 mg/l or about 20 times greater than the concentration in rain. Additional ions are added to rainwater on the continents before it becomes river water. However, river water would be more concentrated than rain even if no additional ions were added, due to the loss of water, through evaporation, after it reaches the ground. Using the runoff ratio of 0.46 given in Table 5.6, we calculate that the concentration of TDS in river water due solely to evaporation should be 2.2 times greater than the concentration in rainwater. This is a considerably lower concentration factor than the value of 20 actually found, and the difference is due mainly to contributions from rocks during weathering. There is also considerable anthropogenic input, particularly for ions such as Na^+, Cl^-, and SO_4^{2-}.

Some examples are instructive. Figure 5.3 compares the North American average natural river water concentration of various ions from Table 5.6 to 2.6 times the average U.S. rainwater concentration from Chapter 3. (The value 2.6 is the inverse of the North American runoff ratio of 0.38.) The estimated pollutive input to North American rivers is also shown. On the order of 10–15% of the Ca, Na, and Cl in U.S. river water, one quarter of the K, and nearly half of the sulfate (most of which is pollutive) comes from rain. By comparison, SiO_2 and HCO_3^- are essentially entirely from rock weathering (0% from rain).

Considering the effect of the runoff ratio on river concentration by continent, one might expect on the basis of the runoff ratio that African surface water (runoff ratio = 0.28) would be more concentrated and Asian surface water (runoff ratio = 0.54) more dilute than other continents, but this is not the case. African and South American river waters (which have a greater influence of crystalline rocks) are both more dilute (TDS = 61 and 55, respectively) than Asian, North American, and European river waters, which are more influenced by sedimentary rocks.

TABLE 5.6 Chemical Composition of Average River Water

By Continent	River Water Concentration[a] (mg/l)									Water Discharge (10^3 km^3/yr)	Runoff Ratio[b]
	Ca^{++}	Mg^{++}	Na^+	K^+	Cl^-	SO_4^{--}	HCO_3^-	SiO_2	TDS		
Africa:											
Actual	5.7	2.2	4.4	1.4	4.1	4.2	26.9	12.0	60.5	3.41	0.28
Natural	5.3	2.2	3.8	1.4	3.4	3.2	26.7	12.0	57.8		
Asia:											
Actual	17.8	4.6	8.7	1.7	10.0	13.3	67.1	11.0	134.6	12.47	0.54
Natural	16.6	4.3	6.6	1.6	7.6	9.7	66.2	11.0	123.5		
S. America:											
Actual	6.3	1.4	3.3	1.0	4.1	3.8	24.4	10.3	54.6	11.04	0.41
Natural	6.3	1.4	3.3	1.0	4.1	3.5	24.4	10.3	54.3		
N. America:											
Actual	21.2	4.9	8.4	1.5	9.2	18.0	72.3	7.2	142.6	5.53	0.38
Natural	20.1	4.9	6.5	1.5	7.0	14.9	71.4	7.2	133.5		
Europe:											
Actual	31.7	6.7	16.5	1.8	20.0	35.5	86.0	6.8	212.8	2.56	0.42
Natural	24.2	5.2	3.2	1.1	4.7	15.1	80.1	6.8	140.3		
Oceania:											
Actual	15.2	3.8	7.6	1.1	6.8	7.7	65.6	16.3	125.3	2.40	—
Natural	15.0	3.8	7.0	1.1	5.9	6.5	65.1	16.3	120.3		
World average:											
Actual	14.7	3.7	7.2	1.4	8.3	11.5	53.0	10.4	110.1	37.4	0.46
Natural (unpolluted)	13.4	3.4	5.2	1.3	5.8	8.3 (5.3)[c]	52.0	10.4	99.6	37.4	0.46
Pollution	1.3	0.3	2.0	0.1	2.5	3.2 (6.2)[c]	1.0	0	10.5	—	—
World % pollutive	9%	8%	28%	7%	30%	28% (54%)[c]	2%	0%	—	—	—

[a] *Actual* concentrations include pollution. *Natural* concentrations are corrected for pollution.

[b] Runoff ratio = average runoff per unit area/average rainfall (calculated from Meybeck).

[c] We have raised pollutive contribution; see Table 5.11. (Our values are in parentheses.)

Source: All river water concentrations and discharge values from Meybeck (1979) except "actual" concentrations by continent, which were calculated from Meybeck's data. (M. Meybeck, "Concentrations des eaux fluviales en éléments majeurs et apports en solution aux oceans," *Rev. Géol. Dyn. Georgr. Phys.*, 21(3), 220, 227. Copyright © 1979. Reprinted by permission of the publisher.)

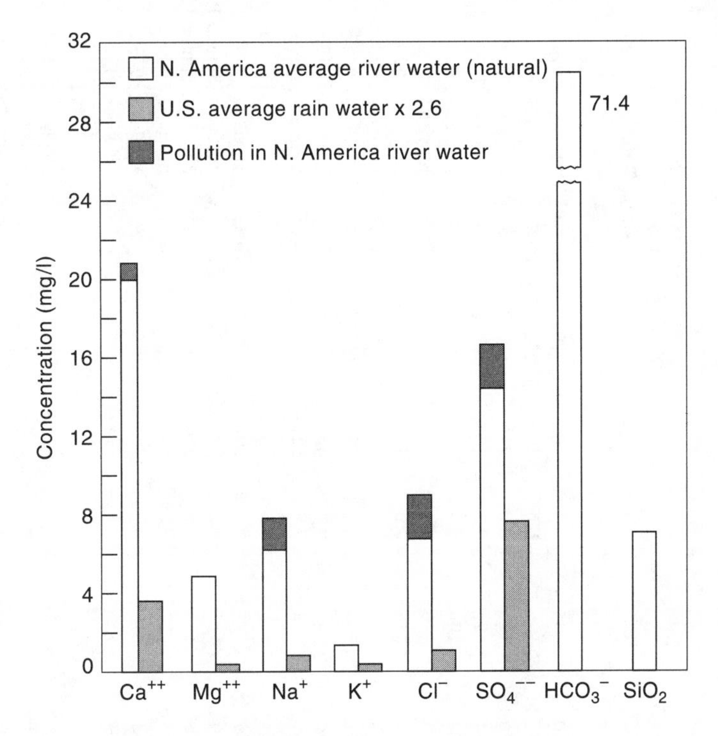

Figure 5.3. Comparison of dissolved composition of North American natural and polluted river water (data from Meybeck 1979) with U.S. rainwater (concentrations in mg/l). Rainwater concentrations are multiplied by 2.6 to correct for evaporation from the continents (see text).

Thus, for the continents as a whole, the runoff ratio does not seem to be the dominant influence on river concentrations.

The above discussion emphasizes the importance not only of total concentration of dissolved ions but also of chemical composition. World average river water composition is dominated by Ca^{2+} and HCO_3^-, both of which are derived predominantly from limestone weathering. Meybeck (1979) found that 98% of all river water is of the calcium carbonate type (i.e., had Ca^{2+} and HCO_3^- as the principal ions). Less than 2% of surface waters have Na^+ (linked with Cl^-, SO_4^{2-}, or HCO_3^-) as the principal ion. In the next section, we shall discuss the natural chemical composition of river water in an effort to show the origin of various major dissolved ions and how they are affected by (1) the amount and nature of rainfall and evaporation, (2) geology of the river basin and weathering history, (3) average temperature, (4) relief, and (5) vegetation and biologic uptake. The chemical composition of some major world rivers is given in Table 5.7.

Chemical Classification of Rivers

Several attempts have been made to classify rivers on the basis of their chemistry. The reason for classifying rivers is to determine which of a number of natural environmental factors that affect river water chemistry (i.e., those listed above) are most important. By studying well-known rivers, it is possible to extrapolate results to those whose environments are less well known. As we have seen in the previous section, there are also many human influences that tend to increase

TABLE 5.7 Major Ion (Dissolved Only) Chemical Composition of Principal World Rivers

River	Concentration (mg/l)									Discharge	Drainage Area	Reference
	Ca^{++}	Mg^{++}	Na^{+}	K^{+}	Cl^{-}	SO_4^{--}	HCO_3^{-}	SiO_2	TDS	(km^3/yr)	(10^6 km^2)	
North America												
Colorado 1960s	83	24	95	5.0	82	270	135	9.3	703	20	0.64	Meybeck 1979
Columbia	19	5.1	6.2	1.6	3.5	17.1	76	10.5	139	250	0.67	Meybeck 1979
Mackenzie	33	10.4	7.0	1.1	8.9	36.1	111	3.0	211	304	1.8	Meybeck 1979
St. Lawrence 1870	25	3.5	5.3	1.0	6.6	14.2	75	2.4	133	337	1.02	Meybeck 1979
Yukon	31	5.5	2.7	1.4	0.7	22	104	6.4	174	195	0.77	Meybeck 1979
Mississippi 1905	34	8.9	11.0	2.8	10.3	25.5	116	7.6	216	580	3.27	Meybeck 1979
Mississippi 1965–67	39	10.7	17	2.8	19.3	50.3	118	7.6	265	580	3.27	Meybeck 1979
Frazer	16	2.2	1.6	0.8	0.1	8.0	60	4.9	93	100	0.38	Meybeck 1979
Nelson	33	13.6	24	2.4	30.2	31.4	144	2.6	281	110	1.15	Meybeck 1979
Rio Grande: Laredo	109	24	117	6.7	171	238	183	30	881	2.4	0.67	Livingstone 1963
Ohio	33	7.7	15	3.6	19	69	63	7.9	221	—	—	Livingstone 1963
Europe												
Danube	49	9	(9)	(1)	19.5	24	190	5	307	203	0.8	Meybeck 1979
U. Rhine: unpolluted	41	7.2	1.4	1.2	1.1	36	114	3.7	307	—	—	Zobrist & Stumm 1980
L. Rhine: polluted	84	10.8	99	7.4	178	78	153	5.5	256	68.9	0.145	Zobrist & Stumm 1980
Norwegian rivers	3.6	0.9	2.8	0.7	4.2	3.6	12	(3.0)	31	383	0.34	Meybeck 1979
Black Sea rivers	43	8.6	17.1	1.3	16.5	42	136	—	265	158	1.32	Meybeck 1979
Icelandic rivers	3.9	1.5	8.8	0.5	4.4	4.8	35.5	14.2	73.4	110	0.1	Meybeck 1979
South America												
U. Amazon: Peru	19	2.3	6.4	1.1	6.5	7.0	68	11.1	122	1512	—	Stallard 1980
L. Amazon: Brazil	5.2	1.0	1.5	0.8	1.1	1.7	20	7.2	38	7245	6.3	Stallard 1980
L. Negro	0.2	0.1	0.4	0.3	0.3	0.2	0.7	4.1	6	1383	0.76	Stallard 1980
Madeira	5.6	0.2	2.6	1.6	0.8	5.6	28	9.4	53	1550	2.6	Stallard 1980
Parana	5.4	2.4	5.5	1.8	5.9	3.2	31	14.3	69	567	2.8	Meybeck 1979
Magdalene	15.0	3.3	8.3	1.9	(13.4)	14.4	49	12.6	118	235	0.24	Meybeck 1979
Guyana rivers	2.6	1.1	2.6	0.8	3.9	2.0	12	10.9	36	240	0.24	Meybeck 1979
Orinoco	3.3	1.0	(1.5)	(0.65)	2.9	3.4	11	11.5	34	946	0.95	Meybeck 1979

TABLE 5.7 *continued*

River	Concentration (mg/l)									Discharge	Drainage Area	Reference
	Ca^{++}	Mg^{++}	Na^{+}	K^{+}	Cl^{-}	SO_4^{--}	HCO_3^{-}	SiO_2	TDS	(km^3/yr)	(10^6 km^2)	
Africa												
Zambeze	9.7	2.2	4.0	1.2	1	3	25	12	58	224	1.34	Meybeck 1979
Congo (Zaire)	2.4	1.4	2.0	1.4	1.4	1.2	13.4	10.4	34	1215	3.7	Probst et al. 1992
Ubangui	3.3	1.4	2.1	1.6	0.8	0.8	19	13.2	43	90	0.5	Probst et al. 1992
Niger	4.1	2.6	3.5	2.4	1.3	(1)	36	15	66	190	1.12	Meybeck 1979
Nile	25	7.0	17	4.0	7.7	9	134	21	225	83	3.0	Meybeck 1979
Orange	18	7.8	13.4	2.3	10.6	7.2	107	16.3	183	10	0.8	Meybeck 1979
Asia												
Brahmaputra	14	3.8	2.1	1.9	1.1	10.2	58	7.8	99	609	0.58	Sarin et al. 1989
Ganges	25.4	6.9	10.1	2.7	5.	8.5	127	8.2	194	393	0.975	Sarin et al. 1989
Indus	26.4	5.6	9.0	2.0	7.1	26.4	90	5.1	171	238	0.97	Meybeck 1979
Mekong	14.2	3.2	3.6	2.0	(5.3)	3.8	58	8.9	99	577	0.795	Meybeck 1979
Japanese rivers	8.8	1.9	6.7	2.2	5.8	10.6	31	19	86	550	0.37	Meybeck 1979
Indonesian rivers	5.2	2.5	3.8	1.0	3.9	5.8	26	10.6	58	1734	1.23	Meybeck 1979
New Zealand rivers	8.2	4.6	5.6	0.7	5.8	6.2	50	7	88	400	0.27	Meybeck 1979
Yangtze (Changjiang)	30.2	7.4	7.6	1.5	9.1	11.5	120	6.9	194	928	1.95	Zhang et al. 1990
Yellow (Hwanghe)	42	17.7	55.6	2.9	46.9	71.7	182	5.1	424	43	0.745	Zhang et al. 1990
Ob	21.	5.0	4.0	3.0	10.	9.	79	4.2	135	433	2.99	Gordeev & Siderov 1993
Yenisei	21.	4.1	2.3	w.Na	9.0	8.6	74.	3.8	123	555	2.5	Telang et al. 1991
Lena	17.1	5.1	5.2	w.Na	12.0	13.6	53.1	2.9	109	525	2.49	Gordeev & Siderov 1993
Philippines	31	6.6	10.4	1.7	3.9	13.6	131	30.4	228	332	0.3	Meybeck 1979

the total dissolved solids (TDS) in many rivers and particularly increase Cl^-, SO_4^{2-}, and certain cations relative to HCO_3^- and SiO_2.

According to the classification of Gibbs (1970), the major natural mechanisms controlling world surface-water chemistry are (1) atmospheric precipitation, both composition and amount; (2) rock weathering; and (3) evaporation and fractional crystallization. A boomerang-shaped diagram resulted when he plotted the ratio of two major cations in world surface waters, Ca^{2+} and Na^+ plotted as Na/(Na + Ca), versus total dissolved salts (TDS) (see Figure 5.4). Rivers plot either in areas dominated by each of the three mechanisms, in other words in the three corners of the "boomerang," or in intermediate areas where more than one mechanism controls their chemistry. The TDS axis is roughly inversely proportional to rainfall or runoff. Atmospheric precipitation-controlled rivers are in areas of high rainfall, and evaporation-crystallization-controlled rivers are in arid areas, while rock-dominated rivers are in areas of intermediate rainfall. Thus, this classification is based to a considerable extent on the amount of rainfall and, thus, runoff. Gibbs also

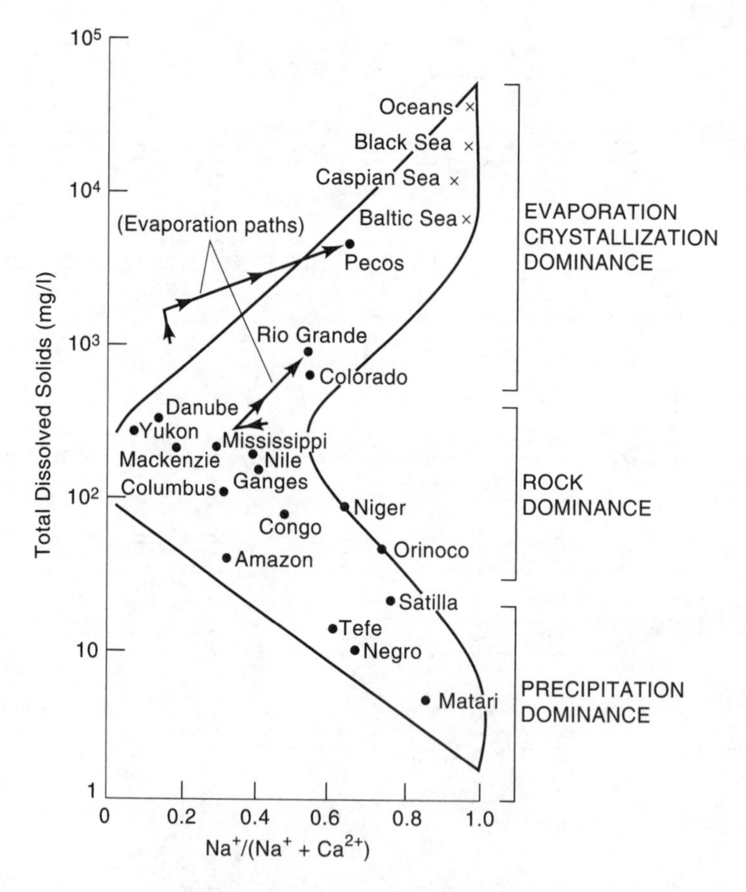

Figure 5.4. Variation of the weight ratio, $Na^+/(Na^+ + Ca^{2+})$, as a function of total dissolved solids, of world rivers (•) and other water bodies (×). The "evaporation paths" in a downstream sequence of samples from the Pecos and Rio Grande rivers are discussed in the text. (Redrawn after R. J. Gibbs, "Mechanisms Controlling World Water Chemistry," *Science* 170: 1088. Copyright © December 4, 1970, by The American Association for the Advancement of Science, reprinted by permission of the publisher.)

found that practically identical locations for almost all rivers were found if a similar diagram was plotted which used $Cl/(Cl + HCO_3)$ in place of $Na/(Na + Ca)$.

According to Gibbs, rivers whose composition is controlled mainly by atmospheric precipitation are those whose composition resembles that of rainfall; that is, they have low total dissolved solids and high Na relative to Ca (or high Cl relative to HCO_3). In Figure 5.4 they plot in the lower right corner of the boomerang. These are generally tropical rivers of South America and Africa and Atlantic Coastal Plain rivers of the United States in areas of high rainfall, low relief, and clay-rich, heavily weathered material, or rivers draining sandy rocks with a resulting low supply of dissolved salts. Gibbs states that the ratios of the major dissolved ions in atmospheric precipitation-controlled Amazon Basin rivers are similar to sea salt except for H_4SiO_4 and K, which are derived from rock weathering.

The middle portion of the boomerang, where values of TDS are intermediate and values of $Na/(Na + Ca)$ are low, is the location of rivers whose composition is controlled, according to Gibbs, by rock weathering. This includes most of the major world rivers (Mississippi, Ganges, Nile, etc.). For these rivers, rock weathering supplies most of the dissolved salts. Since sedimentary rocks occupy about 75% of the earth's surface and their weathering is dominated by the dissolution of $CaCO_3$ (see Chapter 4), one would expect that rivers controlled by rock weathering would consist mainly of $Ca^{2+} + HCO_3^-$ resulting from carbonate dissolution. This is why such rivers plot at low values of $Na/(Na + Ca)$ and low values of $Cl/(Cl + HCO_3)$.

The upper right corner of the boomerang in Figure 5.4 is the domain of rivers whose composition is controlled by evaporation and fractional crystallization. These rather atypical rivers have a high concentration of total dissolved salts and high Na (or Cl^-) relative to Ca^{2+} (or HCO_3^-). Two examples given by Gibbs are the Rio Grande and Pecos. The composition evolves in a sequence of river samples taken in a downstream direction from rock-weathering-type waters found in headwaters to more concentrated and sodic waters as a result of evaporation and concentration of salts accompanied by the crystallization of $CaCO_3$ (see arrows in Figure 5.4). As evaporation continues, the concentrations of Na and Cl increase but those of Ca and HCO_3 do not because the latter two are constantly removed by precipitation of $CaCO_3$. Much of the evaporation can be ascribed to the use of the water for irrigation, where a greater surface area of water is made available for evaporation before the remaining water is returned to the river. (For a further discussion of evaporation and fractional crystallization, see the section on saline lakes in Chapter 6.)

Gibbs' classification has proved to be controversial. The main points of contention have to do with the two high $Na/(Na + Ca)$ corners of the boomerang. Stallard and Edmond (1983) point out that several of the Amazon tributaries, used by Gibbs as prototypes of rainfall-controlled rivers, upon resampling and reanalysis of both rivers and nearby rainwater, prove to have compositions rather different from that of sea salt. In fact, Stallard and Edmond disagree with Gibbs's assumption that rivers in the "precipitation dominance" field, with low concentration of dissolved salts (TDS) and high Na^+ relative to Ca^{2+}, are dominated by atmospheric precipitation. When Stallard and Edmond made a correction for cyclic salts in their data for coastal and lowland Amazon rivers, they found that it made only a minor change in the position of these rivers on Gibbs's diagram (except for one river, the Matari). They reasoned that if these rivers were dominated by atmospheric precipitation of cyclic salts (i.e., marine sea salt), then correcting for cyclic salt should reduce their TDS and Na^+ concentrations and they should be shifted toward the less sodic part of the diagram. However, this was not the case. Only in the coastal Matari River was the cyclic contribution to TDS large (42%). The Negro River, which is about 1500 km inland and which falls in Gibbs's atmospheric precipitation-dominated field, contains only on the order of 10% cyclic salt. Gibbs (1970) states that 81% of the Na^+, K^+, Ca^{2+}, and Mg^{2+} in the coastal Rio Tefe (which also falls in his precipitation-dominated field) is from precipitation, but *using Stallard's (1980) data* we calculate that only 10% of the Na^+, K^+, Ca^{2+}, and Mg^{2+} and 3–5% of the TDS consists of cyclic salt.

Stallard and Edmond (1983) believe that the chemical composition of these dilute, high-Na/Ca rivers results from their geology and erosional regime rather than from atmospheric precipitation. Their high Na/Ca ratio arises because the rocks are mainly siliceous (as compared to calcareous), and the low TDS arises because the sediments and soils of the drainage basins have already undergone intense weathering and are, thus, relatively unreactive. Thus, they would consider these rivers to be rock dominated (in Gibbs's terminology).

A river that might better fit Gibbs's "atmospheric precipitation-dominated" category is the Satilla River in Georgia (Beck et al. 1974). The Satilla drains a swampy area of the U.S. coastal plain, is very dilute, has a low pH, and its organic constituents are greater (30 ppm C) than its inorganic constituents (15 ppm). Sodium and chloride dominate the inorganic ions in the Satilla River, and the river ratios of Na^+/Cl^- and SO_4^{2-}/Cl^- are very similar to sea salt. Beck et al. concluded that rainfall input is the main source of the major element ions in the river except for minor additions of K and Ca from the soil. If all the river Cl^- is assumed to be from sea salt, then we calculate that about 50% of the TDS in the Satilla River and most of the Na^+ and SO_4^{2-} (see Table 5.14) are also derived from sea salt. Pine Barrens rivers in the New Jersey coastal plain are also atmospheric controlled; that is, they receive a large part of their composition from precipitation (Yuretich et al. 1981).

The other major controversial points of Gibbs's classification concerns the saline rivers. Feth (1971) states that the evaporation-crystallization process is not the controlling mechanism for the composition and concentration for two of Gibbs's major examples, the Pecos and Rio Grande rivers. Feth agrees that the river water becomes progressively more concentrated downstream due to evapotranspiration in this arid area and by the addition of irrigation return water. However, he feels that the *main* increase in dissolved salts downstream in the Pecos River is from groundwater flow into the river of NaCl brines derived from the dissolution of near-surface halite deposits (see also Gibbs 1971 and Feth 1981).

Stallard (1980), based on Amazon Basin data, also feels that high-Na/Ca and high-TDS rivers are mainly the result of the weathering of evaporites (which have a high Na/Ca ratio) rather than the concentration and evolution of typical "rock-weathering" river water. Stallard found that three fourths of the chloride in the main Amazon River comes from evaporites (the rest is cyclic salt), with 90% of this evaporite-derived chloride coming from the weathering of halite-bearing salt domes in the Peruvian Andes. Buried evaporites there result in the formation of salt springs and saline rivers. Here the salt domes do not occur in an arid area (precipitation is about 150 cm/yr); thus, highly saline rivers can result from evaporite weathering in the *absence* of aridity.

The evaporation-crystallization process of Gibbs apparently has occurred in the Nile River since construction of the Aswan Dam. According to Kempe (1988), the Cl concentration in the Nile has risen by a factor of 3.6, while the concentrations of Ca, Mg and HCO_3 have risen by only a factor of 1.55. Irrigation water from the Nile selectively precipitates carbonates, leaving Cl and other ions behind in the return water; thus, any evaporative concentration increase should be indicated by Cl. About one third of the concentration increase is due to evaporation in Lake Nasser, behind the dam.

Another process that could change the relative amounts of Na and Ca in river water is ion exchange of Na on clays in marine shales for dissolved Ca during weathering, thus increasing Na concentraions and the Na/Na + Ca ratio in river water (Cerling et al. 1989).

Stallard and Edmond (1983) have created their own classification of Amazon Basin rivers, emphasizing the role of geology and erosional regime as the major control on river water composition. They group rivers in terms of total cation charge, which for the sake of simplification we have roughly converted to mg/l total dissolved solids (TDS). Stallard and Edmond's categories, and the Amazon Basin geology and erosional regime to which they correspond, are shown in Table 5.8.

Although the Stallard and Edmond classification is based on study of rivers of the Amazon Basin, where there is rapid tropical weathering, a lack of evaporative concentration, and little pollution, the categories can be expanded to cover other areas. Thus, we have also included here other world rivers that fit these categories. The first category (< 20 mg/l TDS) comprises rivers draining intensely weathered and cation-poor siliceous rocks, soils, and saprolites, and it includes Gibbs's "atmospheric precipitation control" rivers. The second category (20–40 mg/l TDS) represents rivers draining (cationic) siliceous terrains (igneous and metamorphic rocks and noncalcareous shales), and it falls between Gibbs's "atmospheric precipitation control" and "rock (weathering) dominance." The third category (40–250 mg/l TDS) includes rivers draining marine sedimentary rocks. Such rivers have high Ca^{2+} and HCO_3^- concentrations as a result of the weathering of carbonate minerals, and high sulfate from the weathering of gypsum and pyrite in shale. This corresponds to Gibbs's rock dominance field and, if extrapolated to world rivers, includes most major world rivers. The fourth category (TDS > 250 mg/l) represents rivers draining evaporites. As pointed out above, this corresponds to Gibbs' "evaporation-crystallization control" rivers.

As further evidence of the geologic or rock-weathering control of Amazon rivers, Stallard (1980) and Stallard and Edmond (1983) show that rivers in the first two categories (i.e., those draining cationic siliceous rocks and heavily weathered areas) generally have a molar ratio of 2:1 between Si and (Na + K) after (minor) cyclic salt corrections are made (see Table 5.8). This is the ratio predicted for water draining areas dominated by the weathering of primary silicate minerals (such as Na-plagioclase and K-feldspar) to kaolinite:

$$\underset{\text{plagioclase}}{2NaAlSi_3O_8} + 2H^+ + 9H_2O \rightarrow \underset{\text{kaolinite}}{Al_2Si_2O_5(OH)_4} + 2Na^+ + 4H_4SiO_4$$

or

$$\underset{\text{K-feldspar}}{2KAlSi_3O_8} + 2H^+ + 9H_2O \rightarrow \underset{\text{kaolinite}}{Al_2Si_2O_5(OH)_4} + 2K^+ + 4H_4SiO_4$$

The ion ratio of the weathering products of these reactions is 2Na:4Si and 2K:4Si or, in other words, 1(Na + K):2Si. (For further discussion, see Chapter 4.) Similarly, rivers in Stallard's third and fourth groups (draining carbonate and evaporite terrains) have molar ratios of 1:1 for Na:Cl and 1:1 for (Ca + Mg):(½ HCO_3^- + SO_4^{2-}). These are the ratios expected for waters draining areas dominated by carbonate and evaporite weathering (Table 5.9) (see also Chapter 4).

Drever (1988) and Garrels and Mackenzie (1971) also stress the importance of rock type in determining river composition. Drever's description of the river water produced by different rock types is similar to Stallard's. Drever characterizes water from shales as having variable TDS, and a lower ratio of Si to total cations than igneous rocks, because illite and quartz (the

TABLE 5.8 Stallard and Edmond's River Classification

Total Cationic Charge (μeq/l)	Approximate TDS (mg/l)	Predominant Source-Rock Type	Characteristic Water Chemistry[a]	Examples	Gibbs Category
<200	<20	Intensely weathered (cation-poor) siliceous rocks and soils (thick regolith)	Si-enriched; low pH; Si/(Na + K) = 2; high Na/(Na + Ca)	Amazon tributaries (Matari, Tefe, Negro)	Atmosphere-precipitation controlled
200–450	20–40	Siliceous (cation-rich); igneous rocks and shales (sedimentary silicates)	Si-enriched; (higher Si from igneous and metamorphic rocks); Si/(Na + K) = 2; intermediate Na/(Na + Ca)	L. Amazon, Orinoco, Zaire	Between atmosphere-precipitation controlled and rock-dominated
450–3000	40–250	Marine sediments; carbonates, pyrite; minor evaporites	Na/Cl = 1; (Ca + Mg)/($^1/_2HCO_3 + SO_4$) = 1; low Na/(Na + Ca)	Most major rivers	Rock-weathering dominated
>3000	>250	Evaporites; $CaSO_4$ and NaCl	Na/Cl = 1; (Ca + Mg)/($^1/_2HCO_3 + SO_4$) = 1; high Na/(Na + Ca)	Rio Grande, Colorado	Evaporation-crystallization

[a] Element ratios given refer to mole ratios.

Source: Data from Stallard 1980; Stallard and Emond 1983.

TABLE 5.9 Carbonate and Evaporite Weathering Reactions

Reaction	Molar Ratio of Products in solution
$NaCl \rightarrow Na^+ + Cl^-$	$Na^+/Cl^- = 1:1$
$CaSO_4 \rightarrow Ca^{++} + SO_4^{--}$	$(Ca^{++} + Mg^{++})/(^1/_2HCO_3^- + SO_4^{--}) = 1:1$
$CaCO_3 + H_2CO_3 \rightarrow Ca^{++} + 2HCO_3^-$	$(Ca^{++} + Mg^{++})/(^1/_2HCO_3^- + SO_4^{--}) = 1:1$
$2CaCO_3 + H_2SO_4 \rightarrow 2Ca^{++} + SO_4^{--} + 2HCO_3^-$	$(Ca^{++} + Mg^{++})/(^1/_2HCO_3^- + SO_4^{--}) = 1:1$
$CaMg(CO_3)_2 + 2H_2CO_3 \rightarrow Ca^{++} + Mg^{++} + 4HCO_3^-$	$(Ca^{++} + Mg^{++})/(^1/_2HCO_3^- + SO_4^{--}) = 1:1$
$2CaMg(CO_3)_2 + 2H_2SO_4 \rightarrow 2Ca^{++} + 2Mg^{++} + 2SO_4^{--} + 4HCO_3^-$	$(Ca^{++} + Mg^{++})/(^1/_2HCO_3^- + SO_4^{--}) = 1:1$

Note: Mineral names are: NaCl is halite; $CaSO_4$ is anhydrite, $CaCO_3$ is calcite, $CaMg(CO_3)_2$ is dolomite.

predominant mineral sources of Si in shales) do not weather readily. Sulfate from pyrite weathering and Cl^- from trapped NaCl (probably originally seawater) are the primary shale-derived anions. From these considerations, waters from shales should fall in Stallard's categories 2 and 3. (Because black shales contain pyrite, which readily weathers to H_2SO_4, they tend to produce waters that extend over more than one category.)

Meybeck (1984, 1987) has studied French rivers draining single rock types and from these rivers has derived representative river compositions for each rock type, again emphasizing the importance of rock type on river composition. Meybeck (1987) concludes that *rock type* rather than its *area* of surficial outcrop is most important in determining river dissolved loads. He found that crystalline igneous (plutonic and volcanic) and metamorphic rocks, which comprise 34% of world outcrop area, contribute only 12% of the dissolved river load. Evaporites (only 1.3% of outcrop area) contribute 17% of river load and carbonates (16% of outcrop area) contribute 50% of river load. This emphasizes the disproportionate contribution of evaporites and carbonates to the chemical composition of river water.

Reeder et al. (1972), in a study of the Mackenzie River drainage in Canada, have concluded that salinity is controlled largely by lithology, with higher salinities resulting from the weathering of carbonates and evaporites. In general, the total dissolved solids in rivers draining sedimentary rocks (including carbonates) tend to be at least twice the TDS in rivers draining only crystalline (igneous and metamorphic) rocks (Holland 1978).

Stallard plotted the composition of Amazon Basin rivers (corrected for cyclic salt) on a ternary diagram with the three vertices represented by siliceous rock weathering (Si), carbonate weathering (HCO_3^-), and evaporite (plus pyrite) weathering ($Cl^- + SO_4^{2-}$). He found that the total dissolved solids (represented by total cation charge) increases from the Si vertex to the HCO_3^- vertex and then from the HCO_3^- vertex to the $Cl^- + SO_4^{2-}$ vertex.

As an extension of Stallard's work (and also that of Hu et al. 1982), we have plotted world rivers on a similar ternary diagram (Figure 5.5). Here the rivers are not corrected for cyclic salt, so those few rivers with considerable cyclic salt are shifted toward the right as compared to Stallard's diagram. The data for Figure 5.5 are from Table 5.7, in the previous section. The diagram also includes world average river water and certain other rivers for comparison. "Natural world average river water" (Meybeck 1979), which has been corrected for pollution (denoted as World Average N), plots near a number of major world rivers (Orange, Columbia, Brahmaputra, and Upper Amazon). (The inclusion of the Orange River in this group is surprising because it is from

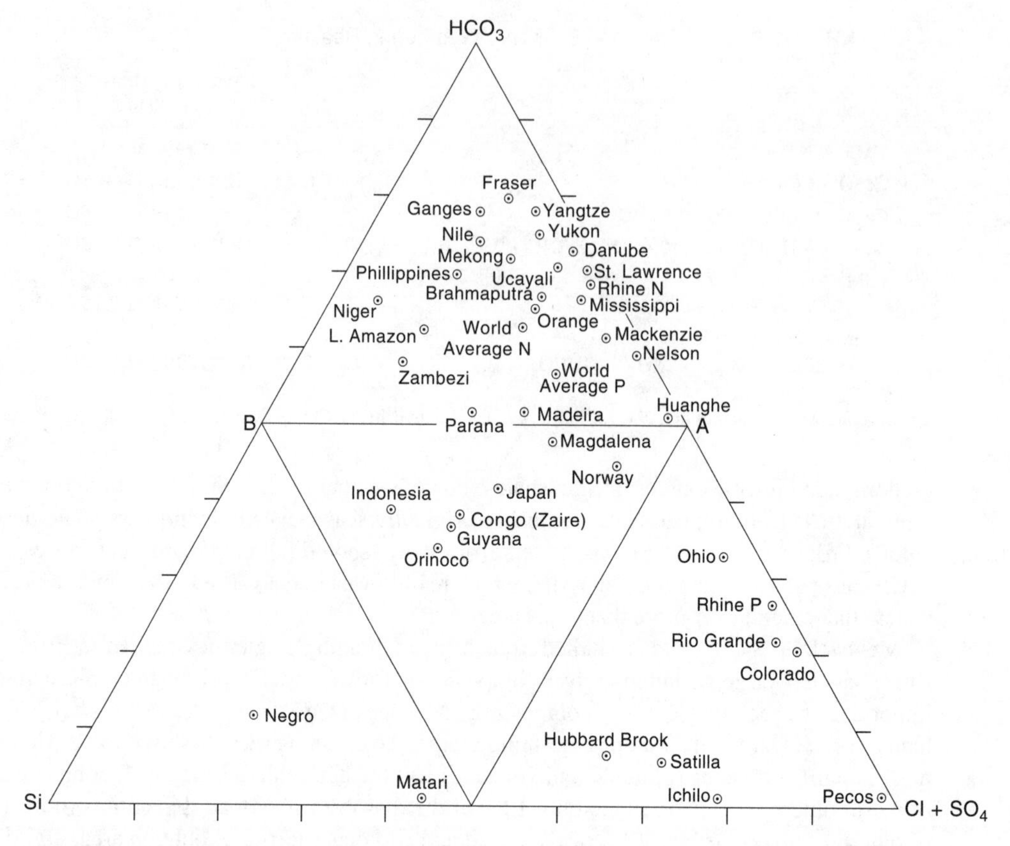

Figure 5.5. Major rivers: Percentage of Si (μmol/l); HCO_3^- (μEq/l); and sum of $Cl^- + SO_4^{2-}$ (μEq/l). For example, 100% HCO_3^- plots at the HCO_3^- vertex; 50% HCO_3^- and 50% Si plots at point *B;* and 50% HCO_3^- and 50% ($Cl^- + SO_4^{2-}$) plots at point *A*. River TDS increases from Si vertex to HCO_3^- vertex to ($Cl^- + SO_4^{2-}$) vertex. (Data from Tables 5.7 and 5.14.)

an arid area.) Uncorrected or polluted world average river water (denoted as World Average P) after Meybeck (1979) is also plotted and shows a shift toward the ($Cl^- + SO_4^{2-}$) vertex since these two ions are the major pollutants. Similarly, when the unpolluted Rhine River (Rhine N) is compared to the polluted Rhine River (Rhine P), there is a tremendous increase in Cl^- and SO_4^{2-} (data from Zobrist and Stumm 1980).

From Figure 5.5 one can see also that the vast majority of major world rivers have more than 50% HCO_3^- and from 10% to 30% $Cl^- + SO_4^{2-}$. Thus, most large rivers are dominated by sedimentary rock weathering and are mainly Ca^{2+} and HCO_3^- waters derived from carbonate minerals. [To a lesser extent, rivers draining areas of incomplete silicate weathering where cation-rich soils are forming will also plot in this area (Stallard and Edmond 1983).] Arid-area rivers (such as the Rio Grande and Colorado) plot along the axis between (Cl^- and SO_4^{2-}) and HCO_3^- toward the ($Cl^- + SO_4^{2-}$) vertex, and the Pecos plots virtually at the vertex. The Huanghe River is interesting in that its composition is predominantly due to the weathering of carbonates but is shifted toward the Cl-SO_4 vertex. There is extensive evaporation in the arid part of its course, and there is also probably considerable input from soil salts and even carbonate precipitation (as in Gibbs's evaporation-crystallization) in this area (Hu et al. 1982).

Some other rivers shown in Figure 5.5 represent special situations. The sulfuric acid weathering trend of pyritiferous black shales described by Stallard for the Madeira River (and in particular one tributary, the Ichilo) also includes Japanese rivers (which have thermal springs and volcanic input). Hubbard Brook, which is a very small river receiving sulfuric acid rain on a siliceous terrain, plots as an extension of the same general trend (see Figure 5.5). The Congo, Guayana, and Orinoco rivers, which are all tropical rivers flowing over highly weathered siliceous terrain, plot together. We have also included, on Figure 5.5, the Satilla and Matari rivers, which we mentioned earlier as having considerable atmospheric precipitation influences. The main reason they are included in Figure 5.5 is that both are quite acidic, and this results in a very low proportion of HCO_3^-. (In the case of the Satilla, this is due to organic acidity; see later discussion.)

Relief and River Water Composition

Relief is another factor in determining river composition that is not mentioned specifically in the classifications discussed above. Relief is an important factor in Stallard and Edmond's (1983) "erosional regime" which is used to produce their classification of Amazon Basin rivers using total cation concentration or TDS. Greater relief means greater physical erosion and faster exposure of fresh rock to chemical weathering. Because transport away of weathering material is rapid with high relief, chemical weathering in the soil is incomplete and the amount depends on how easily the rocks undergo decomposition. Thus, both lithology (carbonates versus silicates) and texture are important. For instance, massive silicates (shield rocks) weather more slowly than more porous silicates such as immature sediments and volcaniclastics. With low relief, chemical weathering is much more complete and thick soils and saprolites develop over bedrock, so variations in geology become less important. For example, where evaporites are exposed along the lower Amazon Basin, their contribution to rivers is fairly minor (more in proportion to their relative abundance—Stallard 1985), as compared to their conspicuous contributions to Andean rivers. (However, even in the Amazon Basin lowland areas, massive igneous and metamorphic silicates weather more slowly than porous sedimentary silicates, and rivers draining carbonates have a greater TDS concentration than those draining silicates.)

Relief in a particular area often correlates with rock type, temperature, and vegetation, and these factors are hard to separate (Drever 1988). For example, Gibbs (1967) points to relief as the dominant factor in controlling the chemical composition of the Amazon River system. In the Amazon River about 70% of the dissolved load is derived from the Andean highlands, which are underlain by easily weathered carbonates and evaporites (Stallard 1980). Here, high relief and easily weathered rock types tend to correlate. In North America and other areas, by contrast, the headwaters of rivers are often in crystalline rock, which is resistant to chemical weathering, and the headwater streams are very dilute. These rivers become more concentrated downstream as more easily weathered rocks are encountered (Livingstone 1963). The Mississippi River is an example of this effect.

MAJOR DISSOLVED COMPONENTS OF RIVER WATER

In this section we consider the sources of the major elements in river water via a detailed discussion of each element. Throughout the discussion we will refer repeatedly to Table 5.11, which is presented later in this section. At this point we should mention that Meybeck (1987), using water analyses of rivers draining major rock types and their outcrop proportion at the earth's surface, has independently determined weathering sources of the major elements, and his results agree well with our estimates.

Chloride and Cyclic Salt

Chloride in rocks is extremely mobile and very soluble. Once in solution, Cl^- has a very uneventful geochemistry. It reacts very little with other ions, does not form complexes, is not greatly adsorbed on mineral surfaces, and is not active in biogeochemical cycling (Feth 1981). Because it is so chemically unreactive, Cl is used as a tracer of natural water masses, particularly in the oceans, but also in ground and surface waters.

The main sources of chloride in river water are (according to Feth 1981): (1) sea salt, that is, cyclic salt from rain and dry fallout; (2) dissolution during weathering of halite (NaCl) in bedded evaporites or dispersed in shales; (3) thermal and mineral springs in volcanic areas; (4) redissolution of saline crusts in desert basins (not a primary source); and (5) pollution such as domestic and industrial sewage, oil well brines, mining, and road salt.

"Cyclic salt," or sea salt carried inland in the atmosphere and deposited in rain or snow, has been assumed by some to be the only or main source of Cl in river water. However, as pointed out by Feth (1981), this may be true only for some very dilute rivers, such as those draining nonsedimentary rocks, particularly those along coasts. In rivers with higher Cl concentrations, Feth (1981) states that Cl from rain probably makes up less than 20% of the total Cl concentration. Similarly, Zverev and Rubeikin (1973) estimate the average contribution of Cl of atmospheric origin in all USSR rivers to be 21%. In the section on chemical classification of rivers, we mentioned several exceptional coastal rivers whose Cl content is presumably dominated by cyclic salt, for example, the Matari River in the Amazon Basin and the Satilla River in Georgia. Some other notable, but small, river basins whose Cl input is mainly from sea salt include Hubbard Brook, New Hampshire (Likens et. al. 1977); Pond Branch, Maryland, (Cleaves et al. 1970); the Pine Barrens rivers, New Jersey (Yuretich et al. 1981); and the Mattole River basin in coastal northern California (Kennedy and Malcolm 1977). The Ganges River, which has a monsoon climate, is somewhat intermediate, with about 40% of the Cl being marine sea salt (calculated from Sarin et al. 1989).

Table 5.10 gives the percentage of Cl^- (and other ions) in world average river water that comes from cyclic salt as estimated by us and by other studies. Our values are based on Stallard and Edmond's (1981) estimate of cyclic salt in the relatively unpolluted Amazon River plus tributary basins. They determined the relationship between Cl^- concentrations in Amazon River basins draining terrains with no chloride-bearing rocks and the distance of the rivers from the Atlantic Ocean. The Cl^- concentration dropped rapidly from the coast, and leveled off at 2000 km inland; this inland Cl^- value was then used to estimate cyclic Cl^- in inland rivers. Overall, they found that 18% of the Cl^- in the Amazon River was cyclic and, as they suggest, this seems to be applicable to world average river water. However, this value is for unpolluted rivers, and since about 30% of riverine Cl worldwide is due to pollution (Meybeck 1979), we calculate that only 70% of Stallard and Edmond's value of 18%, or 13%, represents the fraction of total Cl^- carried globally by rivers that is due to cyclic salt. The cyclic Cl^- estimates of Holland (1978), and particularly those of Garrels and Mackenzie (1971) and Meybeck (1983, 1984), are much larger than our estimate of 13% (see Table 5.10). The primary reason for this is the underestimation by these

TABLE 5.10 Atmospheric Cyclic Salt (Sea Salt) as a Percentage of World Average River Water

Element	Present Study[a]	Holland 1978[b]	Garrels and Mackenzie 1971[b]	Meybeck 1983[c]
Cl^-	13 (18)	27	55	72
Na^+	8 (11)	19	35	53
SO_4^{--}	2 (2)	39[d]	6	19
Mg^{++}	2 (2)	<3	7	15
K^+	1 (1)	<14	15	14
Ca^{++}	0.1 (0.1)	1.3	0.7	2.5

[a] Values are given as a percentage of "actual" (including pollution) world average river water (Meybeck 1979) and, in parenthesis, of "natural" world river water (corrected for pollution). [Cyclic Cl is set equal to 18%, which is the Amazon value given by Stallard and Edmond (1981). Sea salt contributions to other ions are set in proportion to this based on the composition of average seawater.]

[b] Based on world average river water from Livingstone (1963), which is not corrected for pollution.

[c] Based on "natural" world average river water.

[d] Based on total atmospherically derived sulfur (natural + pollution).

authors of the large contribution of Cl from sedimentary halite (see below). Also, Meybeck has an unusually large sea-salt contribution because of his emphasis on the importance of coastal rivers (within 100 km of the coast) returning locally derived cyclic Na^+ and Cl^- back to the ocean. However, analyses of many major rivers are taken farther upstream than this, so it confuses the matter to emphasize coastal effects.

After accounting for cyclic salt (13%) and pollution (30%), the remaining Cl^- in river water (57% of total Cl) comes predominantly from the weathering of sodium chloride. (Contributions from saline groundwater and hydrothermal springs are included here, since the Cl^- comes ultimately from NaCl dissolution.) NaCl occurs in bedded evaporite rocks as the mineral halite, which is normally accompanied by $CaSO_4$ minerals, and as inclusions in shale formed by the trapping of interstitial seawater. Since NaCl is both widespread and very soluble, it provides an abundant source of Cl^- in both ground and surface waters. For example, according to Mattox (1968), 25% of continental areas are underlain by evaporites and 15% of these areas have salts containing Cl. (In the Northern Hemisphere, more than half of the land areas are underlain by evaporites.) Evaporite NaCl occurs most commonly in the subsurface, because surface outcrops are usually dissolved away. Likewise, shales generally contain more Cl^- in the subsurface because of removal or preferential dissolution near the earth's surface (see Feth 1981).

The halite sources of Cl in a river basin are most often quite localized. In the Amazon Basin, for example, Stallard (1980) found that 90% of the Cl (after correction for cyclic salt) comes from the Peruvian Andes, where Cl is added from salt domes, salt springs from buried evaporites, and saline water migrating up fault planes from depth. The Pecos River in Texas receives tremendous quantities of Cl^- from saline springs, which are leaching subsurface halite. Feth (1981) notes that two thirds of the continental United States is underlain by groundwater with high dissolved solid concentrations (>1000 mg/l), with considerable Cl. This saline groundwater is likely derived from NaCl dissolution and may constitute a source of Cl^- to rivers. Chloride is also added to surface water by redissolution of saline evaporative crusts in desert basins, but this is not a primary source—merely a secondary, concentrated one. For example, 60% of the chloride in the Ganges River is estimated to be from saline groundwater added to the river in its semiarid alluvial plain (Sarin et al. 1989).

Hydrothermal alteration of volcanic rocks (as in New Zealand) constitutes an additional source of Cl^-. In Japan, around 5% of the Cl in river water is estimated as being from thermal and mineral springs (Sugawara 1967). Meybeck (1984) estimates a hydrothermal Cl input to world average river water equal to 8%.

Pollution can be an important Cl source for many rivers. Worldwide, Meybeck estimates that about 30% of Cl in river water arises from pollution (see Table 5.11). Domestic sewage contains considerable Cl, due primarily to the consumption by humans of table salt. According to Feth (1981), chlorination of public water supplies for purification adds 0.5–2.0 mg/l Cl to the Cl^- concentration of water, and domestic uses add another 20–50 mg/l to sewage. In addition to direct sewage discharge, there is additional seepage from septic tanks. Other sources of Cl^- pollution include road salt, fertilizer, and industrial Cl-containing brines. Over the period 1974–1981, Cl (and Na) increases by more than 25% in U.S. rivers were moderately correlated with population increases and highly associated with road-salt use on highways, which has increased greatly especially in the Midwest (Smith et al. 1987). Increases in Cl did not correlate with irrigation increases. In some areas, mining of salt can add Cl locally to river water; and in the western United States, oil field brines are also a local source. Two notable examples of rivers bearing a large component of chloride pollution are the St. Lawrence and the lower Rhine. Three quarters of the Cl in the lower Rhine is estimated to be pollutive, from potash mining, coal mining, industry, and sewage (van der Weijden and Middelburg 1989).

When plotted against runoff, the concentration of Cl^- in river water from most rivers behaves in the manner expected for simple dilution of a limited number of highly concentrated sources (Holland 1978). In other words, there is an inverse proportionality between chloride concentration and runoff. This provides further evidence for the importance of localized deposits of halite and point-source pollution (i.e., cities) as major sources of this ion.

Sodium

Because sodium is a major component of seawater, it is a principal contributor to atmospheric cyclic salt, and, therefore, should be prominent in river water. However, cyclic salt is not the major source of Na in river water. Using the sea-salt value of Na/Cl, total concentrations of Na and Cl in world average river water (Table 5.6), and the proportion of Cl^- in average river water that is contributed by cyclic sea salt (13%; see Table 5.10), we can calculate that only about 8% of river water sodium owes its origin to cyclic salt. This is shown in Table 5.10. A much more important source of Na is halite in sedimentary rocks. Since Na^+ is the major ion accompanying Cl^- in halite, everything said in the previous section concerning the rock sources of Cl^- also refers to Na^+. In fact, by knowing the relative importance of each source of Cl^-, we can calculate from the stoichiometry of NaCl the relative importance of the same source for Na. The result (Table 5.11) is that about 42% of Na comes from the weathering of halite in evaporite beds and from NaCl occluded in shales.

The thing about Na^+ that sets it apart from Cl^- is that it is a major component of silicate rocks. As discussed previously in the chapter on weathering and groundwater, sodium in silicate rocks is present mainly as the albite component of plagioclase, with the formula $NaAlSi_3O_8$. Plagioclase is a major source of Na for groundwater and thus is also a major source for river water. If we sum up all the sources of Na from pollution (28%), sea salt (8%), and halite (42%), the

TABLE 5.11 Sources of Major Elements in World River Water (in Percent of Actual Concentrations)

Element	Atmos. Cyclic Salt	Weathering			Pollution[b]
		Carbonates	Silicates	Evaporites[a]	
Ca^{++}	0.1	65	18	8	9
HCO_3^-	<<1	61[c]	37[c]	0	2
Na^+	8	0	22	42	28
Cl^-	13	0	0	57	30
SO_4^{--}	2[d]	0	0	22[d]	54
Mg^{++}	2	36	54	<<1	8
K^+	1	0	87	5	7
H_4SiO_4	<<1	0	99+	0	0

[a] Also includes NaCl from shales and thermal springs.

[b] Values taken from Meybeck (1979) except sulfate, which is based on calculation given in the text.

[c] For carbonates, 34% from calcite and dolomite and 27% from soil CO_2; for silicates, all 37% from soil CO_2; thus, total HCO_3^- from soil (atmospheric) CO_2 = 64% (See also Table 5.13).

[d] Other sources of river SO_4^{--}; natural biogenic emissions to atmosphere delivered to land in rain, 3%; volcanism, 8%; pyrite weathering, 11%.

remaining (22%) must come from the weathering of plagioclase. This is all summarized in Table 5.11. The ways and means by which Na is added to soil water and groundwater, and eventually to river waters by silicate weathering, is summarized in Chapter 4. Our point here is merely to show that an appreciable fraction (22%) of Na in river water arises from silicate weathering.

Another possible source of Na in river water is from cation exchange of dissolved Ca^{2+} with Na^+ on detrital clay minerals during marine shale weathering (Cerling et al. 1989). Cation exchange would produce excess Na relative to Cl in river water similar to silicate weathering. However, cation exchange is probably of minor importance globally as a Na source in river water because of the lack on a global scale of enough clay minerals that contain exchangeable Na (Berner et al. 1990).

Pollution is an important source of sodium in river water (28% of the total; see Table 5.11). In fact, along with Cl^- and SO_4^{2-}, Na^+ is one of the ions most affected by pollution. Most of pollutive sodium is associated with Cl^- in NaCl, and thus, a number of the sources of Cl pollution mentioned previously are also Na sources. These include domestic sewage, mining of halite, industrial brines, and road salt. Other Na salts, such as Na_2CO_3, Na_2SO_4, and Na borate, are mined and used industrially for paper, soaps and detergents, and other products (Skinner 1969), and sodium is also used in water softeners as sodium zeolite and Na_2CO_3, replacing Ca^{2+} and Mg^{2+} in industrial and domestic water by Na^+. These provide an additional source of Na in sewage.

Potassium

Potassium in river water comes predominantly (nearly 90%; see Table 5.11) from the weathering of silicate minerals, particularly potassium feldspar, as orthoclase and microcline, and mica, as biotite. These silicate minerals are found in sedimentary, metamorphic, and igneous rocks. Meybeck (1984) estimates that about three fourths of silicate weathering K comes from silicates in sedimentary rocks and one fourth from silicates in igneous and metamorphic rocks. Potassium is not dissolved and released as quickly during weathering as most other cations, because the potassium-containing primary minerals weather more slowly than those containing Na, Ca, and Mg (see Chapter 4), and therefore, considerable amounts of potassium remain in the soil. Holland (1978) estimates that, on average, only about 50% of rock potassium is released to solution during silicate weathering. This would agree with the fact that the soil concentration of potassium is about half that of average surficial rocks (Table 5.5). From Table 5.5, we can also see that only 15% of the river transport of K is in the dissolved load, while the rest is particulate. Thus, in behavior, K is intermediate between rapidly weathering Ca, Na, and Mg and resistant Si and Al. (We have also assumed that 5% of weathering K comes from evaporates as KCl, after Meybeck 1984. This is minor compared to silicate weathering.)

Nonweathering sources of potassium include pollution (7%) and minor contributions from cyclic salt (1%). Rare evaporite deposits of KCl and similar salts are mined for K fertilizer, and river pollution of potassium results from such mining as well as from the use of the K fertilizers themselves. An example is the mining of KCl in Alsace (France), which is a contributor to K pollution in the Rhine River (as well as to NaCl pollution; Meybeck 1979). Some 23% of the Rhine K is pollutive (van der Weijden and Middelburg 1989). Since the world's consumption of K salts for fertilizers is increasing at the rate of 10% a year (Skinner 1969), potassium pollution of rivers should be expected to increase with time.

Although K in river water is ultimately predominantly from silicate weathering, it is a very biogenic element due to its utilization by growing vegetation. For example, Dion (1983) found that 60% of K^+ from groundwater in the Connecticut River basin was taken up by biological activity. In temperate drainage basins, there is a seasonal variation in stream K concentrations, with K being lower during summer, when plants are growing and taking up K, than during the winter, when plants are dormant. In Hubbard Brook, New Hampshire, for example, Likens et al. (1977) found K concentrations to vary by about ±33% due to this effect. Potassium is concentrated in leaves of trees, so when trees lose their leaves in the fall, there tends to be a rise of K concentration in streamwater due to the leaching of K from the leaf litter (Cleaves et al. 1970; Likens et al. 1977). Also, unlike most other elements, potassium tends to increase in concentration during increases in discharge (e.g., from flood flow, spring runoff, or heavy rains), due to the dissolution of soluble salts from trees, leaf litter, and the top of the soil (Cleaves et al. 1970; Miller and Drever 1977). Because there is a net accumulation of potassium in vegetation (the net yearly biomass accumulation in the Hubbard Brook area is three times the stream output (Likens et al. 1977; see also Chapter 4), forest cutting results in a large increase in stream K concentrations.

Overall, worldwide, there is little variation in potassium concentrations among major rivers, and it is always the least abundant of the four major cations (Meybeck 1984). (The average potassium concentration is 1.3 mg/l, and the range is 0.5–4.0 mg/l; Meybeck 1980). Highest concentrations are found in high-TDS rivers from arid areas (e.g., the Nile, Colorado, and Rio Grande), but, in general, potassium is the least variable of the major dissolved ions in river water (Meybeck 1980). A number of factors probably contribute to the low variability of potassium concentrations in major world rivers. First, there is not much difference between the average concentration of K in sedimentary rocks (2.0% K_2O; Holland 1978) and in crystalline or igneous rocks (3.2% K_2O; Holland 1978), so the lithology of various drainage basins should be relatively less important. Second, K is released much more slowly and less completely during weathering than many other major dissolved ions. Third, because K, released by silicate weathering, is taken up to a considerable degree by the biomass, its release to streamwater is partly controlled by organic decay, and over a year, stream release should be related to the balance between biological uptake and decay. In established vegetation, which presumably is adjusted for K uptake and release, this should help make K release more uniform on an annual basis.

Calcium and Magnesium

Calcium and magnesium are contributed to river water almost entirely from rock weathering. Only 9% of Ca and 8% of Mg arise from pollution (Table 5.11), and < 1% of Ca and 2% of Mg^{2+} from cyclic sea salt (Table 5.10). The sources of Ca consist mainly of carbonate rocks containing calcite, $CaCO_3$, and dolomite, $CaMg(CO_3)_2$, with a lesser proportion derived from Ca-silicate minerals, chiefly calcian plagioclase, and a minor amount from $CaSO_4$ minerals. Mg-silicate minerals, chiefly amphiboles, pyroxenes, olivine, and biotite (for compositions see Chapter 4), as well as dolomite, constitute the main sources of Mg.

The relative proportions of different minerals contributing Ca and Mg to river water have been calculated by Holland (1978) and more recently by Berner et al. (1983) and Meybeck (1984; 1987), and the results of the four studies are in general agreement. The results of the Berner et al. study are included in Table 5.12. As pointed out by Holland, these types of calculations rest on a variety of assumptions having to do with such things as the area of the continents underlain by sedimentary (as opposed to igneous and metamorphic) rocks, the contribution of Ca in river water from Ca in sedimentary rocks, the rate of weathering of carbonates relative to silicates, the average Mg/Ca ratio for both carbonates and silicate rocks, and the weathering flux of sulfate from $CaSO_4$ minerals. Thus, the calculation is open to a variety of potential errors; nonetheless, we feel that the values given in Table 5.12 are the best that can be obtained from presently available data. At least there is essential agreement among four independent studies.

The most important finding from the data of Table 5.12 is the predominance of carbonate minerals (calcite and dolomite), which occur almost entirely in sedimentary rocks, as the main source (65%) of Ca^{2+} in river water. Since the most abundant cation in world average river water is Ca^{2+}, this result further emphasizes the importance of the weathering of sedimentary carbonate rocks

TABLE 5.12 Sources of Ca and Mg in World Average River Water

Source	Percent of Total Ca	Percent of Total Mg
Weathering		
Calcite, $CaCO_3$	52	—
Dolomite, $CaMg(CO_3)_2$	13	36
$CaSO_4$ minerals	8	—
Ca-silicates	18	—
Mg-silicates	—	54
Cyclic sea salt	<<1	2
Pollution	9	8
Total	100	100

Source: Data for rock sources from Berner et al. 1983. Cyclic sea salt from Table 5.10 and pollution from Meybeck 1979.

to the composition of natural waters. As a crude, first-order approximation, average river water can be characterized as a $Ca(HCO_3)_2$ solution derived from the dissolution of limestone.

Bicarbonate (HCO_3^-)

Like Ca and Mg, almost all bicarbonate in average river water is derived from rock weathering. Pollution contributes only 2% (Meybeck 1979) and cyclic sea salt far less than 1%. (Where acid rain is important, pollution can even be considered as destroying HCO_3^-—see below under acid rivers.) The weathering-derived bicarbonate, as discussed in Chapter 4, comes from two sources. One source is the carbon in carbonate minerals, such as calcite and dolomite. The other arises as a result of the reaction of carbon dioxide dissolved in soil water and groundwater with carbonate and silicate minerals. The carbon dioxide is derived almost entirely from the bacterial decomposition of soil organic matter. Two representative weathering reactions (see Chapter 4) are

$$CO_2 + H_2O + CaCO_3 \rightarrow Ca^{2+} + 2HCO_3^-$$

$$2CO_2 + 11H_2O + 2NaAlSi_3O_8 \rightarrow 2Na^+ + 2HCO_3^- + Al_2Si_2O_5(OH)_4 + 4H_4SiO_4$$

As these reactions demonstrate, half of the HCO_3^- resulting from carbonate weathering and all of the HCO_3^- from silicate weathering is derived from soil CO_2. It should be noted that soil CO_2 from the decomposition of organic matter was originally atmospheric CO_2, since organisms photosynthetically fix CO_2. Thus, the ultimate source of much of the HCO_3^- in river water is the atmosphere.

If all weathering were accomplished only by dissolved CO_2 (which reacts by way of the intermediate formation of carbonic acid), then the amount of HCO_3^- added to river water from each weathering source could be obtained by calculating that accompanying the release each of Na^+, K^+, Ca^{2+}, and Mg^{2+} according to the stoichiometry of reactions like the two given above. However, some silicate and carbonate weathering is also brought about by sulfuric acid formed in soils by the oxidation of pyrite (and other sulfides). Typical reactions are

$$H_2SO_4 + 2CaCO_3 \rightarrow 2Ca^{2+} + 2HCO_3^- + SO_4^{2-}$$

$$H_2SO_4 + 9H_2O + NaAlSi_3O_8 \rightarrow 2Na^+ + SO_4^{2-} + Al_2Si_2O_5(OH)_4 + 4H_4SiO_4$$

Note that, in this case, HCO_3^- arises only from the carbon in carbonate minerals and none comes from soil CO_2. To take into account H_2SO_4 weathering, we calculate the percentage contribution of HCO_3^- from various weathering sources as follows: From the total measured concentration of HCO_3^- (Table 5.6) is subtracted the HCO_3^-, added by weathering, which was originally contained within carbonate minerals. The latter is equivalent to the amount of Ca and Mg added by carbonate weathering (Table 5.12). The remaining HCO_3^- must be derived from soil CO_2. This HCO_3^- is then subdivided and assigned to each of Na^+, K^+, Ca^{2+}, and Mg^{2+} in the same proportions as these cations are given off to river water from silicate plus carbonate weathering (see Tables 5.11 and 5.12). However, the assignment is not simply to balance the charge of each cation, according to CO_2-type weathering reactions, but rather the proportioning is done so that the total HCO_3^- added from all four cations equals the remaining CO_2-derived HCO_3^- calculated above. In this way correction is automatically made for H_2SO_4 weathering. (The relative proportioning of H_2SO_4 weathering to each cation is assumed to be the same as that for CO_2 weathering.)

Results of our calculations of the sources of HCO_3^- in river water are shown in Table 5.13. Note that most HCO_3^- (64%) is derived from soil CO_2, with roughly equivalent proportions coming from carbonate and silicate weathering. Only about half as much (34%) comes from carbon originally contained within carbonate minerals. [These proportions are in excellent agreement with similar calculations by Holland (1978) and with calculations by Meybeck (1987) based on average river composition and outcrop area.] For specific rivers, Probst et al. (1992) estimate the percentage of HCO_3^- from atmospheric CO_2 as 67% for the Amazon and 75% for the Congo (which has only 5% carbonate rocks). If it were not for the fact that HCO_3^- in river water can exchange carbon with dissolved CO_2 obtained from the atmosphere, the proportions calculated here could be checked using the stable isotopes of carbon. The $^{13}C/^{12}C$ ratio of soil CO_2 is distinctly different from that of carbonate minerals, and knowledge of the $^{13}C/^{12}C$ ratio of river water bicarbonate could, in principle, be used to calculate the relative importance of each carbon source. Unfortunately, because of isotopic exchange with the atmosphere, and consequent change in river water $^{13}C/^{12}C$, this sort of calculation is much more difficult than it at first appears. What limited isotopic work has been done along these lines suggests that our calculated proportions of soil CO_2 and carbonate-derived HCO_3^- are essentially correct.

Since the advent of acid rain (see Chapter 3), one might expect that HCO_3^- in rivers might have decreased. However, most major rivers have shown little change in HCO_3^- concentration during the last century (Meybeck 1979; Zobrist and Stumm 1980). Decreased HCO_3^- concentrations in rivers could arise by titration of HCO_3^- by H^+ to carbonic acid (H_2CO_3), and because sulfuric and nitric acid weathering tends to replace carbonic acid weathering in affected land areas. Apparently these effects are still to be noted on a worldwide basis. However, the Lower Rhine river shows a 10% loss of HCO_3^- from 1930 to 1975–1984 (van der Weijden and Middelburg 1989).

Carbonate weathering increases with increasing runoff. Due to the rapid dissolution reaction of carbonates, there is little dilution of the ion concentrations in rivers with greater runoff. In other words, carbonate dissolution is so rapid that waters remain close to saturation as flushing by groundwater increases (e.g., Stallard and Edmond 1987). This is not the case for silicate weathering, which undergoes dilution at high runoff (see section below on silica). Similarly, relief appears to be important for silicate weathering (see section on silica) but not for carbonate weathering. The latter is well demonstrated by extensive subsurface evidence for carbonate dissolution in flat, low-lying Florida.

TABLE 5.13 Sources of Rock Weathering-Derived HCO_3^- in World Average River Water

Weathering Source	Percent of Total HCO_3^- from Soil CO_2	Percent of Total HCO_3^- from Carbonate Minerals
Calcite + Dolomite	27	34
Ca-silicates	13	—
Mg-silicates	15	—
Na-silicates	6	—
K-silicates	3	—
Total	64	34

Note: For method of calculation, see text. (An additional 2% of total HCO_3^- is added by pollution; see Table 5.11.)

An increase in atmospheric CO_2 should ultimately increase the input of HCO_3^- to rivers, due to weathering of both silicates and carbonates. This is due to both greenhouse-induced higher temperatures and higher runoff, especially at higher latitudes. Minerals dissolve faster with an increase in temperature, and greater runoff brings about a greater weathering flux. With high-latitude warming, permafrost and ice-covered land becomes forested, which further accelerates the weathering rate. Further, greater atmospheric CO_2 should fertilize the growth of some plants, which means faster plant-mediated weathering. [For a further discussion of atmospheric CO_2 and weathering, see Berner (1994).]

Silica

Dissolved silica in river water comes almost entirely from silicate weathering. Silicate minerals are very plentiful, but because carbonate minerals weather so much more rapidly, silica is usually swamped by the ions from carbonate weathering. For example, in Figure 5.5, relatively few rivers have more than 20% dissolved Si relative to HCO_3^- and ($Cl^- + SO_4^{2-}$) expressed in equivalents. Meybeck (1987) suggests that a minor (8%) Si source is amorphous silica (chert) in carbonate rocks.

Meybeck (1980) states that the dissolved silica content of rivers is determined predominantly by the average temperature of the drainage basin and by its geology. The correlation between the mean temperature of the drainage basin and the dissolved silica concentration of rivers (expressed as milligrams of SiO_2 per liter) is shown in Figure 5.6. For nonvolcanic rivers, arctic rivers (average temperature < 4°C) have a silica concentration of about 3 ± 2 mg/l, temperate rivers (average temperature 4° to 19° C) have an average SiO_2 concentration of about 6 ± 3 mg/l, and tropical rivers (average temperature > 20°C) contain about 13 ± 4 mg/l SiO_2. Thus the increase from the arctic to the tropics is around four times for nonvolcanic rivers. Meybeck attributes the temperature effect on dissolved silica to the differences in silicate weathering products (clays) formed in different climatic zones. In the tropics, chemical weathering is more complete and silicate minerals tend to weather to kaolinite, which releases 1.5 times as much dissolved silica as weathering to smectite, which is more common in temperate areas. The formation of gibbsite, which is the result of even more intense tropical weathering, would release twice the

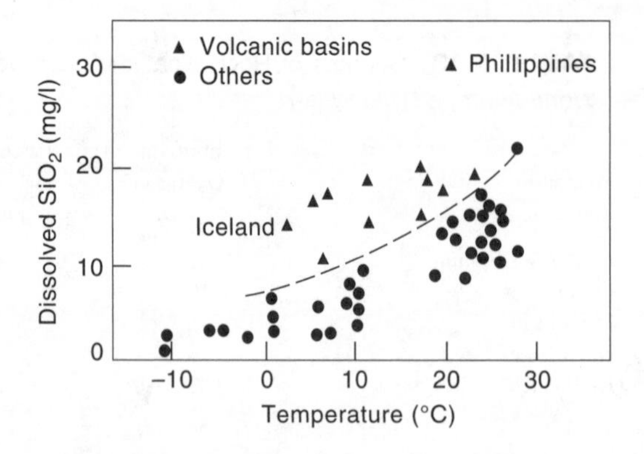

Figure 5.6. Variation in dissolved silica content of world rivers (mg/l SiO_2), with average temperature of drainage basin (°C) for nonvolcanic and volcanic basins. (After Meybeck 1980.)

amount of silica as weathering to smectite. (However, the type of clay formed is also controlled by the amount of runoff—see below).

Variations in the silica concentration in rivers with the average temperature of the basin could also be caused by the effect of temperature on silicate weathering rates. Laboratory studies (e.g., Brady 1991) and field studies (e.g., Velbel 1993) have shown that silicate minerals dissolve faster at higher temperatures. Meybeck (1984, 1987) also found that silica concentrations drop with increasing average altitude (and lower temperature) in rivers draining silicate terrains. However, this is also due to lower silicate weathering rates due to less vegetation and soil cover at higher elevations and/or the presence of snow and frozen soil, which restricts water circulation (Drever and Zobrist 1992; Berner 1994).

White and Blum (1995), studying small river watersheds with granitoid silicate weathering, found that fluxes of SiO_2 (and Na) (mass/area/yr) increase with precipitation (or runoff) and with temperature. For a constant amount of precipitation, the flux is an exponential function of temperature. These authors emphasize the relation between temperature and flux rather than concentration. This is because they found that in watersheds with very low precipitation, the effects of evapotranspiration tend to increase solute concentrations and thus make them a poor measure of weathering.

The geology of the river basin controls the differences in river silica concentration within any temperature zone. For the same temperature, the weathering of volcanic rocks results in twice the concentration of silica as weathering of crystalline plutonic and metamorphic rocks (see Figure 5.6), which give similar silica concentrations as weathering of sedimentary silicates (sandstone and shale) (Meybeck 1987). Stallard (1980), however, found greater silica concentrations in igneous and metamorphic rivers than in sedimentary silicate rivers of the Amazon Basin. The reason that sedimentary silicates (sandstone and shale) release less silica is that they consist largely of minerals that are basically resistant to weathering (quartz, micas, and clays). The large silica release from volcanic rocks is due to the presence in them of easily weathered minerals, such as pyroxenes and Ca-plagioclase, and to the fact that volcanic glass, a major constituent found only in volcanics, weathers more rapidly than any silicate mineral (see Chapter 4). In addition, volcanic rocks are often porous and permeable, which speeds up weathering by enhancing the area of contact between minerals and water.

Meybeck (1980) points to the Philippine rivers, which have a very high silica concentration (30 mg/l), as an example of the effect of volcanic rocks on silica release. Because these rivers are also influenced by carbonate weathering, they do not show a particularly high ratio of Si to HCO_3^- (Figure 5.6), but the concentration (and dissolved load) of silica is very high due to the weathering of volcanic rocks, which is nearly as rapid as carbonate weathering. Other high-silica-content volcanic rivers include Japanese rivers (19 mg/l SiO_2), New Guinea rivers (19 mg/l SiO_2), and those from Iceland, which average 14 mg/l SiO_2 despite their arctic location.

The weathering of igneous silicate rocks in basins presumed to be composed entirely of silicates has been studied by a number of authors (Dunne 1978; Meybeck 1984, 1987; Peters 1984). Rivers draining only igneous rocks show an increase in the chemical denudation rate (measured by the total concentration of cations plus silica released) that is proportional to the two-thirds power of runoff (Dunne 1978). This largely reflects dilution of the weathering signal at high runoff. However, as far as silica is concerned, it is complicated by the fact that with increasing runoff the ratio of silica to cations increases. A good example is the ratio of silica to silicate-derived Na (Dunne 1978). (Silicate-derived Na is obtained by subtracting NaCl-Na from total Na.) The increase in ratio is probably the result of changes in weathering reactions with runoff. Low runoff and a hot, dry climate with high evaporation favor smectite formation, which, for albite plagioclase weathering to smectite, ties up silica in the smectite and releases only one half of it to solution. Higher runoff favors kaolinite formation, which relases two thirds of the silica, and very high runoff favors gibbsite [$Al(OH)_3$] formation, which releases all of the silica to solution. Thus, changes in the type of secondary weathering products formed with increasing runoff result in increases in the ratio of silica to Na. This also results in little change of silica concentration with runoff as compared to cations and bicarbonate whose concentrations decrease (leading to the two-thirds-power dependence discussed above). These conclusions are in excellent agreement with the weathering studies discussed in Chapter 4, which show that the nature of silicate weathering products is strongly affected by the degree of flushing of the rock undergoing weathering.

The effect of relief on silicate weathering and silica concentrations in rivers can be seen in the Amazon River basin, which is a humid tropical area with fairly constant runoff (Stallard 1985; Stallard and Edmond 1987). The effect of high relief, as found in the Andean foothills, is to enhance the removal of protective soil coverings via physical erosion. Greater exposure of fresh silicate minerals via stripping away of the soil brings about faster silicate weathering. In this case, the removal of silicate rock is limited by the rate of chemical weathering, and the overall process is referred to as weathering-limited erosion (Stallard 1985). By contrast, in the flat-lying Amazon lowland areas, a thick regolith of chemically resistant secondary weathering products (e.g., clays) builds up due to a lack of erosive removal. The thick regolith restricts the contact of water with underlying fresh bedrock, and this results in slow weathering. In this case, the chemical removal of silicate rock is limited by the rate by which the thick clay covering can be removed; this situation is referred to as transport-limited erosion.

Even though silicate weathering rates are higher in the Andean foothills, the concentration of silica in rivers draining the foothills is about the same as in rivers draining the heavily weathered, thick soils of the lowlands (Stallard and Edmond 1987). In the foothills region, riverine silica concentrations are lower than expected because of the formation of smectite, which retains silica in the soil and limits its release to rivers. [The foothills region is high in volcanic ash (Stallard 1985), which, because of its rapid dissolution, often leads to smectite formation—see Chapter 4.] Thus, in the Amazon drainage basin, silica concentrations are capped at a maximum level (about 12 mg SiO_2/l) even as weathering rates, and resulting high concentration of silicate-derived Na, increase.

Silica is a biogenic element and is used by diatoms (tiny floating organisms) in the formation of their tests. Oceanic silica concentrations are dominated by biogenic controls, but the effect is much less strong in rivers. However, silica concentrations in rivers can be affected by the presence of diatoms in lakes along their course or in the river itself. Meybeck (1980) points out that the presence of diatoms in the Great Lakes may reduce the amount of silica that reaches the St. Lawrence River. Hu et al. (1982) suggest that diatoms in reservoirs along the Huangho River in China may lower the Si concentration, and a similar diatom influence on Pine Barrens rivers of New Jesey is suggested by Yuretich et al. (1980). Dion (1983) found a strong seasonal variation in silica concentrations in the Connecticut River, which is wide with a low gradient toward its mouth. Concentrations of dissolved silica in the winter were around 6–7 mg/l SiO_2 as opposed to concentrations of 4 mg/l in the summer, when diatom populations were high. However, because diatom tests dissolve when the organisms die, returning dissolved silica to the river, such silica removal is not permanent. The question is how much of the silica is not returned by diatom dissolution; failure to correct for this can result in incorrect river flux data. Meybeck (1984) estimates that silica removal in lakes might amount to as much as 8% of the global river flux of silica.

Sulfate

The sources of sulfate in river water are also shown in Table 5.11. The amount of sulfate from cyclic salt (Table 5.10) is quite low (2%). The other, more significant sources are rock weathering (33%) and pollution (54%), with a minor fraction (8%) coming from volcanic activity and natural biogenically derived sulfate (3%). Rock-weathering sources include the two major forms of sulfur in sedimentary rocks: sulfide sulfur in pyrite (FeS_2) and sulfate sulfur as gypsum ($CaSO_4 \cdot H_2O$) and anhydrite ($CaSO_4$).

Pyrite weathers rapidly by oxidation to sulfuric acid (H_2SO_4), which then reacts with silicate and carbonate minerals in the enclosing rock to release cations and sulfate into river water (see Chapter 4). Since pyrite occurs as a minor dispersed phase, in order to release SO_4^{2-} the rock must be continually broken down. With this in mind we have calculated the percentage of pyrite-derived sulfate in river water by the following method: The ratio of pyrite-S to Ca in sedimentary rocks (times three to convert to SO_4^{2-}) is multiplied by the amount of Ca in river water derived from the weathering of sedimentary rocks:

$$(SO_4^{2-})_{riv.(pyrite)} = [S_{pyrite}/Ca]_{sed.rocks} \times 3 \times (Ca^{2+})_{riv.(sed.rocks)}$$

In this expression we substitute $S_{pyrite} = 0.3\%$ and Ca in sedimentary rocks = 8.5% (Holland 1978). The amount of Ca^{2+} in river water from the weathering of all rocks (excluding Ca from $CaSO_4$) is 13.6 mg/l, of which 86%, or 11.7 mg/l, is derived from sedimentary rocks (Berner et al. 1983). Then

$$\begin{aligned}(SO_4^{2+})_{riv.(pyrite)} &= (0.3/8.5) \times 3 \times (11.7 \text{ mg/l}) \\ &= 1.24 \text{ mg/l}\end{aligned}$$

The amount of sulfate in river water from pyrite weathering thus amounts to 11% of the total concentration of river sulfate (total concentration = 11.5 mg/l; see Table 5.6).

Gypsum and anhydrite weathering are the other main rock-weathering sources of sulfate in river water. They occur mainly as beds within evaporite deposits associated sometimes with halite but more often with dolomite and calcite. Gypsum (or anhydrite) should weather faster than pyrite, because (1) it goes into solution faster (see Table 4.5) and (2) it occurs in discrete beds, which can be selectively removed during weathering, as compared with pyrite, which is disseminated and requires breakdown of the surrounding rock before sulfate is released. Assuming that gypsum weathers twice as fast as pyrite (Berner and Raiswell 1983) and that the abundance of sedimentary sulfate-sulfur (gypsum and anhydrite) is approximately equal to the abundance of sedimentary sulfide-sulfur (Holland 1978), the percentage of river sulfate from gypsum and anhydrite weathering should be about twice that from pyrite weathering, or 22%. (Note: This also gives the minor amount of Ca from $CaSO_4$ in river water used in the pyrite calculation above, so this calculation has been recycled.)

The influence of different rock-weathering sources of sulfate to rivers can be seen in some examples. Since gypsum and pyrite occur in sedimentary rocks, sedimentary rock weathering tends to produce much larger sulfate concentrations in unpolluted rivers than does the weathering of igneous and metamorphic rocks. An example for Canadian rivers is a comparison of the sulfate concentration of 36 mg/l in the Mackenzie River (mainly sedimentary rocks) versus the Northwest Territory rivers (crystalline rocks), where the average concentration of SO_4 is 2 mg/l.

Another natural source of sulfate in river water (Table 5.11) is volcanism, which yields 15 Tg SO_4/yr directly to rivers (Friend 1973) and another 18 Tg SO_4/ yr to rivers via the continental atmosphere (Table 3.12), with the total (33 Tg SO_4) representing 8% of total river sulfate (430 Tg SO_4/yr). The influence of volcanism on sulfate concentrations can be seen in the rivers of some volcanic islands such as New Guinea and New Zealand, where sulfate is fairly high (~6 mg/l SO_4) even in the absence of pollution (Meybeck 1982).

If our estimates above are correct, weathering (33%), volcanism (8%), and cyclic salt (2%) account for 43% of river sulfate. The remaining 57%, or 245 Tg SO_4, must come from pollution and natural biogenic sulfur emissions to the atmosphere. In Chapter 3 we estimate that land biogenic sulfur emissions are 1 Tg SO_4/yr plus another 12 Tg SO_4 from the transport of marine biogenic sulfate. Assuming that all of this biogenic sulfate deposited on land (13 Tg SO_4) ends up in rivers, 3% of river sulfate would be from natural biogenic sulfur emissions to the atmosphere.

Based on the above discussion, we have estimated the proportion of sulfate in rivers that is derived from pollution (see Table 5.11). Here we have assigned the remainder of river sulfate, after allowing for weathering, volcanism, cyclic salt, and natural biogenic sulfate, to pollution. This results in 54% pollutive sulfate or 232 Tg SO_4. This is within the range that Husar and Husar (1985) estimate for human-derived sulfate pollution in rivers of 138–255 Tg SO_4/yr. The apparently large value of 54% is not unexpected judging from such observations as those of van der Weijden and Middelburg (1989), who estimate that *at least* half of the sulfate in the lower Rhine River is pollutive.

Pollutive sulfate in river water has been estimated by other studies to amount to only about 28% (or 120 Tg SO_4/yr) of the total (Berner 1971; Meybeck 1979). This value can be checked by two independent calculations. The first is based on calculating fluxes for the known sources of sulfate pollution to rivers: fertilizers, industrial and municipal wastes, enhanced pyrite weathering from mining activities, and sulfate-rich acid rain from fossil fuel and forest burning. Nriagu (1978) gives a 1975 world anthropogenic sulfate consumption of 144 Tg/yr, with about half (72 Tg/yr) from the use of sulfuric acid in the production of fertilizer and half from other industrial uses. Esser and Kohlmaier (1991) estimate that 25% of fertilizer sulfate ends up in rivers or ($0.25 \times 72 =$) 18 Tg SO_4/yr, whereas Möller (1984) estimates that 80% of fertilizer sulfate goes into rivers. Thus, fertilizer sulfate in rivers amounts to 18–58 Tg SO_4/yr. Freney et al. (1983) estimate that 28% of the sulfate from fossil fuel burning and smelting ends up in rivers. Our estimate (Table 3.12) of sulfate from fossil fuel and biomass burning deposited on the continents is 173 Tg SO_4/yr; 28% of 173 Tg SO_4/yr = 48 Tg SO_4/yr. Thus, the total contribution to rivers of anthropogenic sulfate from industry (maximum value of 72 Tg SO_4/yr), from fossil fuel burning (48 Tg SO_4/yr), and from fertilizer sulfate (18–58 Tg SO_4/yr) is 138–178 Tg SO_4/yr. This is distinctly higher than Meybeck's (1979) estimate of 120 Tg of pollutive river sulfate.

Meybeck (1993) has recently estimated that the *median* concentration of sulfate in major unpolluted rivers is 4.7 mg/l (transport of 176 Tg SO_4/yr), as opposed to his *average* estimate of 8.3 mg/l (310 Tg SO_4/yr). Our estimate for the unpolluted (natural) load is 46% of the total load or 198 Tg SO_4/yr, which is between his median unpolluted value and his average unpolluted value. This provides an additional check on our calculations.

The rain input of sulfate to a river basin may be considerably larger than the river dissolved sulfate output. The difference is attributed to transport of sulfur in river organic matter in the case of the organic-rich Satilla River in Georgia (Beck et al. 1974) and to biomass sulfur storage on land for Pond Branch, Maryland (Cleaves et al. 1970). The release of biogenic sulfur gases (H_2S, DMS, etc.) has been suggested as a possible way in which sulfur is being lost in lowland Amazon Basin rivers (Stallard and Edmond 1981), which also have more rain sulfate input than river output. However, sulfur released to the atmosphere would be recycled by rain and should not change the overall sulfur balance on land.

Sulfate Pollution and Acidic Rivers

Sulfuric acid rain is a source of sulfate pollution in rivers, particularly in areas such as the northeastern United States, Canada, and northern Europe. Hubbard Brook, New Hampshire (Likens et al. 1977), an example of such a river, has an unusually high sulfate concentration relative to its TDS (see Table 5.14) and is, in fact, acidic (pH = 4.9). It is a very dilute river in an area of glaciated crystalline bedrock with thin glacial-till soil cover. There is a lack of carbonate rocks, and silicate weathering is insufficient to neutralize the sulfuric acid rain (pH 4.1). However, sulfuric acid silicate weathering does result in, in addition to sulfate, a relatively high concentration of dissolved silica and Ca^{2+} in the river water (see Table 5.14) and also dissolved Al. Pond Branch, Maryland (Cleaves et al. 1970), another dilute river in a crystalline basin, also receives sulfuric acid rain (pH 4.6), but here there is a thick soil cover and sufficient silicate weathering in the basin to neutralize the rain, and the river, as a result, has a pH of 6.7. This river has a high silica concentration and considerable Mg^{2+}, K^+, and excess SO_4^{2-} (over sea-salt contributions) as a result of weathering by the sulfuric acid rain (see Table 5.14).

On a larger scale, trends in concentrations of streamwater sulfate and alkalinity (HCO_3^-), and in pH for the period 1965–1980 have been studied at 47 U.S. Geological Survey hydrologic benchmark stations in undeveloped stream basins all over the United States (Smith and Alexander 1983). In the northeastern United States as a whole, small decreases in stream sulfate (and increases in alkalinity) have occurred at streams in areas where SO_2 emissions have been reduced over this period, and small increases in sulfate (and decreases in alkalinity) have occurred at southeastern and western sites where SO_2 emissions have increased. This tends to confirm the suspicion that fossil fuel SO_2 and the resulting sulfuric acid rain contribute to stream sulfate and acidity. Stream sulfate and bicarbonate concentrations tend to be inversely correlated because the acid neutralizes HCO_3^- and this relationship is strongest at low-alkalinity stations. The ratio of HCO_3^- to major cation concentration (sum of Na, K, Ca, and Mg) is a sensitive indicator of acidification, and it declines with increased acidity in the stream basin because it is a measure of the importance of natural weathering by H_2CO_3 relative to that by H_2SO_4 (and HNO_3). The ratio of HCO_3^- to major cation concentration has declined at most streams west of the Mississippi River, and increased at most northestern U.S. stations approximately inversely to stream sulfate changes. Changes in the pH of these streams do not always follow sulfate changes, however, because in some cases acidification is due to HNO_3 rather than H_2SO_4 (Lewis and Grant 1979).

In general, for a given input of acid deposition, the response of streams varies greatly depending on the local bedrock geology and the ability of the associated minerals to neutralize acidity. On this basis the acid sensitivity of streams on a regional basis correlates with the relative outcrop area of various rock types (Bricker and Rice 1989). [For a more general discussion of the effect of acid deposition on streams and lakes, see Schindler (1988), and Chapter 6 on acid lakes.]

Sulfur in European rivers is also strongly influenced by sulfuric acid rain and dry fallout. For example, Odèn and Ahl (1978), in a study of Swedish rivers, estimate that 65% of the sulfate-sulfur in Swedish rivers comes from anthropogenic sulfur in rain and dry fallout, with minor

TABLE 5.14 Composition of Some Low-pH Sulfate-Rich and Organic-Rich Rivers Mentioned in Text

River	pH	Concentration (mg/l)										Rain pH	Comments	Reference
		Ca^{++}	Mg^{++}	Na^{+}	K^{+}	Cl^{-}	SO_4^{--}	SiO_2	HCO_3^{-}	TDS	DOC			
Hubbard Brook, N.H.	4.9	1.7	0.4	0.9	0.3	0.55	6.3	4.5	0.9	19	1.0	4.1	Sulfuric acid rain weathering	Likens et al. 1977
Pond Br., Maryland	6.7	1.4	0.8	1.7	0.9	2.1	1.3	9.3	7.7	25	—	4.6	Sulfuric acid rain weathering	Cleaves et al. 1970
X-14, Elbe Basin, Czech.	4.9	17.2	6.0	5.2	1.6	3.7	66.8	16.4	0	108		4.2 (3.2)[b]	Sulfuric acid rain weathering	Paces 1985
Ichio R., Amazon Basin	5.28	4.4	2.0	2.4	0.9	0.2	23.7	8.3	0.6	44	—	4.8–5.0	Natural pyrite weathering	Stallard 1980
Moshannon R., Pa.	2.9	44	21	4	1	10	300	7	0	387	—	acid	Mine drainage pyrite weathering	Lewis 1976
Satilla R., Ga.	4.3	1.3	0.7	4.1	0.8	5.4	1.2	6.85	—	20	24	acid	Organic-dominated	Beck et al. 1974
Pine Barrens, N.J.	4.5	1.1	0.6	2.7	0.6	4.7	6.4	4.3	—	20	2.2	4.4	Organic + sulfuric acid	Yuretich et al. 1981
U. Negro (above Branco), Amazon Basin	4.64	0.4	0.06	0.3	0.26	0.25	0.19	3.4	—	5	12	4.8–5.0	Organic-dominated	Stallard 1980; DOC—Leenheer 1980
Negro R. (above Manaus), Amazon Basin)	5.36	0.17	0.16	0.4	0.24	0.24	0.15	4.3	0.55	6	10	4.8–5.0	Organic-dominated	Stallard 1980; DOC—Leenheer 1980
Matari R., Amazon Basin	4.7	0.14	0.13	0.67	0.15	1.14	0.125	2.4	—	5	(5)[a]	4.8–5.0	Organic and marine rain	Stallard 1980

[a] (DOC) is rough estimate based on color measurements.

[b] Effective pH including dry deposition.

pollution from fertilizer sulfur (7%). In the central European Elbe River Basin, Paces (1985) studied an acidic river (pH 4.9) draining a forested mountain basin where part of the conifer forest has died and which faces an industrial area. The mountain basin receives acid rain (pH 4.2), but more important, because the air concentration of SO_2 is very high, about 10 times as much acid comes from SO_2 dry deposition as from the rain. Thus, the combined effects of dry deposition and acid rain would be equal to a pH 3.2 rain. The river has very high sulfate concentrations and high Ca and silica in addition to being acidic.

In coal-mining areas, natural pyrite weathering is greatly accelerated by the exposure of relatively large amounts of pyrite contained in the coal and by increased water circulation due to mining. If the groundwater circulating through the pyrite-rich coal beds does not contain sufficient HCO_3^- (from previous carbonate weathering) to neutralize the sulfuric acid being produced, extremely acid water (pH < 3) results, which can seep into nearby streams. This acid production via pyrite oxidation is greatly accelerated by bacteria (e.g., Kleinmann and Crerar 1979). Moshannon Creek (see Table 5.14), a tributary of the West Branch of the Susquehanna River in a coal-mining area of Pennsylvania, is an example of a mine-drainage river with a very high sulfate concentration (300 ppm) and low pH (2.9) (Lewis 1976). Further evidence of river sulfate pollution from coal mining is shown by increases in SO_4^{2-} from 1974 to 1981 in a number of midwestern U.S. rivers which were associated with increased surface coal production via strip mining (Smith et al. 1987).

Natural weathering of pyrite-rich black shales can produce high sulfate concentrations and quite acid streamwater because the pyrite is oxidized to sulfuric acid (see Chapter 4). An extreme example is the Ichilo River in the Madeira River drainage of the Amazon Basin (Stallard 1980) (see Table 5.14 and Figure 5.5), which has a very high sulfate concentration (on a mole basis, more than double the Ca concentration), and a pH of 5.28.

Organic Matter (Organic Carbon)

Organic matter is present in rivers in both dissolved and particulate forms. The concentration of dissolved organic matter in rivers is usually expressed in terms of dissolved organic carbon (DOC). The average amount of dissolved organic carbon in rivers is about 5.3 mg C/l, equivalent to a global transport flux of 200 Tg DOC/yr (Meybeck 1993), but there is considerable variation depending on climatic conditions (Meybeck 1988). High median concentrations of DOC are found in taiga (subarctic) rivers, which average 7 mg/l DOC, and in humid tropical rivers, which have about 8 mg/l DOC. Low concentrations are found in arctic and alpine rivers (2 mg/l DOC). Rivers draining swampy areas (such as the Satilla River in Georgia, which we shall discuss below) have the highest concentrations (~25 mg/l DOC). Most rivers show a flushing effect, with DOC concentrations increasing with increasing discharge, and showing a dominant soil and plant organic matter source of the DOC (Spitzy and Leenheer 1991).

In addition to dissolved organic carbon, there is also considerable transport of organic matter in the form of particulate organic carbon (POC) associated with the suspended load. Meybeck (1993) estimates the POC river load at 172 Tg/yr, while Ittekot (1988) estimates 231 Tg POC/yr. As the river suspended sediment load increases, the POC content decreases as a percentage of the total sediment solids (Meybeck 1982, 1988). This is due to dilution by increased erosion of rock and less biological activity in turbid water. On average, about 1% of the suspended sediment load is POC. Meybeck (1993) divides his total river load of POC into 55% young eroded soil POC and 45% fossil rock POC. Of the river POC, Ittekhot (1988) estimates that only 35% is labile, or degradable, and likely to be metabolized in rivers, estuaries, or the ocean, with the rest being highly degraded and nonreactive. Degens et al. (1991) estimate the global river flux of total organic carbon or TOC (= DOC + POC) (excluding Australia) as 330 $\times 10^6$ t/yr, similar to Meybeck's (1993) estimate of a natural load of 370 Tg TOC/yr (198 Tg DOC + 172 Tg POC). Meybeck estimates an additional pollutional flux of 100 Tg TOC/yr.

Because the average ratio of dissolved organic carbon to total (inorganic) dissolved solids in rivers is low (DOC:TDS = 1:19), chemical associations between organic matter and major inorganic ions do not have a large relative effect on the overall chemistry of most rivers. (By contrast, the behavior of trace metal ions is strongly affected.) However, in rivers with low total dissolved solids and large DOC concentrations, organic matter can dominate the river chemistry (see Table 5.14). An example is the Coastal Plain rivers of southeastern Georgia, particularly the Satilla River, which drains a swampy area, has a low pH (4.3), and a ratio of dissolved organic carbon (24 mg/l) to total dissolved (inorganic) solids (20 mg/l) of roughly 1:1 (Beck et al. 1974). It is not the absolute amount of dissolved organic carbon in the Satilla (which, however, is very high) that is so important, but rather its very high ratio to dissolved inorganic solids.

The dissolved organic matter which dominates the chemistry of organic-rich rivers such as the Satilla consists mainly of a mixture of humic and fulvic acids. Both substances are mixtures of complex (and poorly understood) high-molecular-weight organic polymers, which contain carboxyl groups and phenolic groups. (Humic acids are defined as being insoluble in strong acid, while fulvic acids are acid soluble.) The low pH of the Satilla River (3.8–5.0) results from the dissociation of the humic and fulvic acidic carboxyl groups (—COOH). The organic acids (R—COOH) lose a hydrogen ion and become $(R—COO)^-$ with a net negative charge:

$$\underset{\text{organic acid}}{(R—COOH)} \rightarrow \underset{\text{organic anion}}{(R—COO)^-} + H^+$$

The river has little or no bicarbonate (HCO_3^-), because any bicarbonate is used up in neutralizing the H^+ from the organic acid dissociation:

$$HCO_3^- + H^+ \rightarrow H_2O + CO_2$$

Thus, the overall reaction of HCO_3^- with organic acids is

$$(R—COOH) + HCO_3^- \rightarrow (R—COO)^- + H_2O + CO_2$$

As more HCO_3^- ions are added by carbonate-draining tributaries downstream, the hydrogen ions produced by organic acid dissociation are neutralized and the river pH goes up, producing large concentrations of organic anions.

This leads to the other major characteristic of rivers such as the Satilla: The sum of the charge on the major inorganic cations (equivalents per liter of Na^+, Mg^{2+}, K^+, Ca^{2+}, H^+) is greater than the sum of the charge on the major inorganic anions (equivalents of Cl^-, SO_4^{2-}, HCO_3^-, NO_3^-), leading to an apparent deficiency of inorganic anion charge. However, since there must be electrical balance in the river, the excess inorganic cation charge is balanced by organic anions, which result, as discussed above, from the dissociation of organic acid carboxyl groups. Thus, the overall charge balance in an organic-dominated river is

$$\sum \text{inorg. cations} = \sum \text{inorg. anions} + \sum \text{org. anions}$$

Another characteristic of organic-rich rivers is their large concentration of "dissolved" iron and aluminum relative to other rivers. Fe and Al form dissolved organic complexes or colloidal oxyhydroxides mixed with organic matter, and this leads to the mobilization and transport of Fe and Al, which are otherwise immobile and insoluble.

The major characteristics of the organic-dominated Satilla River (Beck et al. 1974), which might be used as criteria in identifying other rivers whose chemistry is dominated by organic matter, are as follows: (1) The ratio of the concentrations of dissolved organic matter (DOC) to dissolved inorganic matter (TDS) is high (here about 1:1); (2) there is an excess of total inorganic cation charge relative to total inorganic anion charge that is presumably balanced by organic anion charge; and (3) the river tends to be acid (although many acid rivers are not organic—see Table 5.14).

Some of the Amazon tributaries are chemically dominated by organic matter. The Negro River in the Amazon Basin, named for its typically organic-rich black color, is an example. Stallard (1980) gives the concentrations of major dissolved inorganic ions for the Upper Negro River (pH 4.6–4.8) and the Negro River (pH 4.95–5.4) (see Table 5.14). Both of these rivers have considerably greater inorganic cation charge than inorganic anion charge. The DOC concentration in the Upper Negro (Leenheer 1980) is 12 mg/l, and TDS is 5 mg/l. Thus, the ratio of DOC:TDS in the Upper Negro River is 2.4:1, fitting the criteria for an organic-dominated river. Similarly, the DOC:TDS ratio for the Negro River is 1.7:1. The Matari River, another Amazon tributory, with low TDS (4.8) and low pH (4.7), also has considerably greater inorganic cation charge than inorganic anion charge, and its dark color (Stallard 1980) suggests a large concentration of organic matter.

There has been considerable discussion about whether some of the effects in rivers and lakes attributed to pollutive sulfuric (and nitric) acid rain might rather be natural due to high concentrations of organic matter (e.g., Krug and Frink 1983). The criteria set forward above for recognizing organic rivers can be used to test whether or not a given acidic river or lake is naturally acidic due to high DOC content. For example, Hubbard Brook (which we discussed above as being affected by sulfuric acid rain) has a DOC:TDS ratio of 1:20 (less than world average river water) and an essential balance between inorganic cation charge and inorganic anion charge, and thus would not seem to be organic controlled. This agrees with the general observation that in subalpine northeastern forests, 75% of the charge balance in soil and groundwater is accomplished by sulfate (from acid rain) and not organic anions (Cronan et al. 1978).

The Pine Barrens rivers in New Jersey (Yuretich et al. 1981; Crerar et al. 1981) have a low pH (4.5), brown water, a fairly high concentration of organic matter (2.2 mg/l), a low TDS (20 mg/l), and a high concentration of SO_4 (6.4 mg/ l). The ratio of DOC:TDS is 1:10, which is about twice as organic rich as world average river water but nowhere near the 1:1 ratio of organic-rich rivers discussed above, and there is charge balance. Thus the low pH of the Pine Barrens rivers is due more to very high sulfate concentrations than to the presence of natural dissolved organic matter. This agrees with the work of Johnson (1979), who finds an increase in the acidity of several Pine Barrens streams from 1963 to 1978 which is attributed to acid rain. However, there is reason to suspect that the groundwater, and possibly the river, were already moderately acid, due to organic acids, before the advent of acid rain, as attested to by the presence of bog iron ores. (The iron most likely was transported in association with organic matter.)

Lowering of the pH of a river can also occur due to increases in the concentration of CO_2 from excessive microbial breakdown of organic matter. This occurs in the lower Rhine, where pollution provides nutrients, particularly phosphate, that greatly increase organic matter production (Buhl et al. 1991; Kempe 1988).

CHEMICAL AND TOTAL DENUDATION OF THE CONTINENTS AS DEDUCED FROM RIVER WATER COMPOSITION

Using the total concentration of dissolved ions in river water, one can calculate the chemical denudation rate of a drainage basin, a continent, or even the whole world. Denudation rate is expressed as the mass of dissolved material removed per unit area per unit time, and is obtained by multiplying the total concentration of the ions times the water discharge and dividing this by the drainage area. This approach is the only one available for rocks that dissolve completely, such as limestones and evaporites. However, it should always be kept in mind that water chemistry is a record of processes occurring only on a very short time scale—in other words, the present. The chemistry of a river can change with time, as factors affecting weathering of the drainage basin, such as climate, land use by humans, etc., change. Over longer time scales, a more accurate way to measure the chemical denudation of silicate rocks (which do not dissolve completely) is to quantify the mass of weathering products, in the form of clays, etc., formed, eroded, and deposited over time, and to calculate the mass of material lost by conversion of the original rock into the weathering products. Unfortunately, this is often impossible, so the second-best method is normally used, the study of river water chemistry.

Chemical denudation rate is a function of geology and of climate, as the latter affects rainfall and runoff, vegetation, and temperature. Runoff is important. Looking at individual rivers, chemical denudation rate tends to increase with increasing runoff (Dunne 1978; Holland 1978; Meybeck 1980). This is shown in Figure 5.7 (after Meybeck 1980). Although the total concentration of dissolved ions decreases with increasing runoff due to dilution, this effect on the total load is more than compensated by the greater volume of water being carried.

The geology (rock composition and weathering history) of the river basin has an important and probably dominant effect on the chemical denudation rate. As shown in Figure 5.7, the total dissolved river load per unit area in sedimentary (carbonate and evaporite) basins is five times greater than that from crystalline (igneous and metamorphic) rocks and 2.5 times greater than that from recent volcanic rocks (Meybeck 1980). Mixed-source rivers (crystalline plus sedimentary) plot in between rivers from the single rock types, but generally show more sedimentary rock influence. The influence of sedimentary rocks is due mainly to the presence of carbonates and evaporites. This is shown by the observation that HCO_3^- from carbonates comprises one third of the dissolved load.

From the studies of Hu et al. (1982) and Meybeck (1980), one can see the important influence of lithology on chemical denudation rate. The highest chemical denudation rates are for rivers with high runoff and in areas of high relief, but more important, draining a dominantly carbonate and evaporite terrain (such as the Yangtze and Brahmaputra rivers, whose load is around 100 tons/km^2/yr), or rivers whose drainage includes recent volcanic rocks (such as the Philippine rivers and New Guinea rivers with denudation rates of 250 tons/km^2/yr, and Japanese rivers with 185 tons/km^2/yr). By contrast, rivers with a very low chemical denudation rate are those draining predominantly crystalline shield terrains such as the Zaire. These different rivers are plotted on Figure 5.7 and tend to follow the trend one might expect for their rock type and runoff.

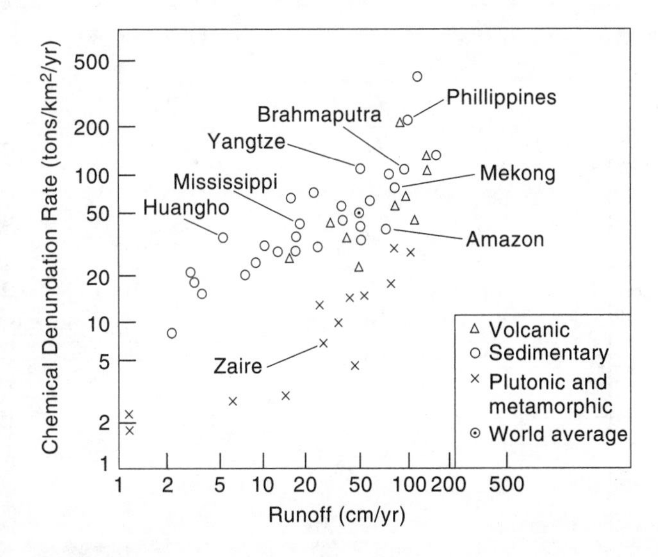

Figure 5.7. Influence of rock composition on total dissolved load per unit area (chemical denudation rate) versus runoff per unit area for major world rivers and some small basins. Certain major rivers discussed in the text are also included. (Adapted from Meybeck 1980; additional data from Hu et al. 1982.)

Meybeck (1987) has estimated the global average chemical denudation rate of different rock types from a study of the composition of individual rivers which drain a single rock type. Crystalline igneous and metamorphic rocks and sedimentary silicate rocks (sandstones and shales) have an average chemical denudation rate of 18–19 tons/km^2/yr, and volcanic igneous rocks have a rate 1.5 times higher. Both these rates are less than the world average of 42 tons/km^2/yr. This is because the denudation rate for carbonate rocks is 100 tons/km^2/yr, or three times the average rate, and that for evaporites is 423 tons/km^2/yr, or more than 10 times the average rate. This is all shown in Figure 5.8.

The effect of relief on denudation rate appears to vary with the geology of the river basin. Relief may be important in silicate weathering (see section on silica in rivers) but not particularly important for the weathering of carbonate rocks, which dissolve easily regardless of local topography (Holland, 1978; Berner 1994). For example, extensive limestone dissolution is found beneath southern Florida, where elevations are only a few meters above sea level. Because of the widespread distribution and rapid weathering rate of carbonates, relief does not correlate with total chemical denudation rate on a continental scale (Garrels and Mackenzie 1971; Hay and Southam 1977; Holland 1978) or for large drainage basins (Pinet and Souriau 1988) as it does with the suspended sediment load. Further, White and Blum (1995) recently have found no effect of hill slope steepness, or extent of recent glaciation, on chemical weathering fluxes in a large number of small granitoid watersheds, implying that physical erosion rates are *not* critical in influencing chemical weathering of silicates. Thus, worldwide, the dominant influences on chemical denudation rate are probably geology and climate rather than relief.

Vegetation has an important effect on chemical denudation. It increases chemical weathering by supplying CO_2 and organic acids to the soil, and it increases water contact time with minerals in the soil by retaining moisture and by locally accelerating water recycling via evapotranspiration-enhanced rainfall (Berner 1992). From the data of Drever and Zobrist (1992),

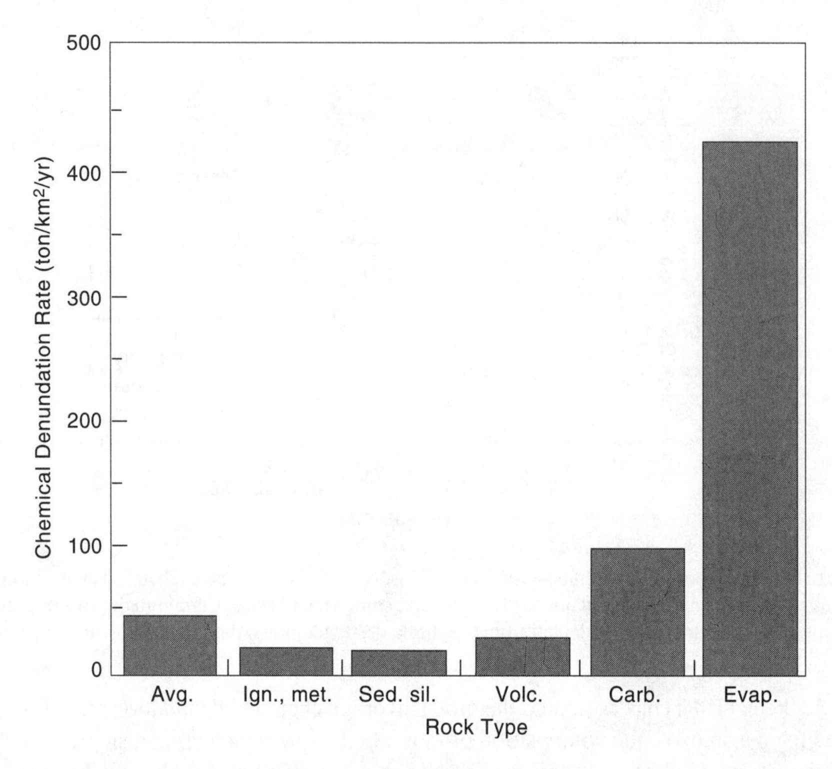

Figure 5.8. Relative chemical denudation rate in tons/km²/yr of various rock types compared to the average rate. Abbreviations for rock types are as follows: avg. = average; ign., met. = igneous and metamorphic; sed. sil. = sedimentary silicates; volc. = volcanics; carb. = carbonates; evap. = evaporites. (Data from Meybeck 1987.)

one can show that vegetation may increase the chemical weathering rate of silicates by as much as a factor of 8 (Berner 1994). Vegetation also affects weathering by stabilizing soil against erosion and thereby increasing the chemical weathering rate in areas of high physical erosion. Another effect of vegetation, particularly on slopes in the tropics, is to bring about higher dissolved concentrations in soil solutions via evapotranspiration. (For further discussion of the role of vegetation in weathering, consult Chapter 4.)

The role of temperature in chemical denudation by itself is not clear in that it is often tied up with other climatic factors such as rainfall (runoff) and vegetation. Nevertheless, one can calculate from experimental studies (see Chapter 4) that an increase of 10°C should result in an increase in silicate mineral dissolution rate by an average factor of 2–3 (Berner 1994). More fundamentally, the effect of temperature can be seen directly by the well-documented preponderance of physical over chemical weathering at high latitudes and high altitudes. The comparatively low content of the products of intense weathering, such as kaolinite, brought by rivers to high-latitude portions of the Atlantic Ocean (Biscaye 1965) further attests to the relative ineffectiveness of chemical weathering in cold climates.

The world average chemical denudation rate is 42 tons/km²/yr. This is based on the average "natural" concentration of total dissolved solids (TDS) in river water (100 mg/l) multiplied by the total world water discharge (37,400 km³/yr) and divided by the area of external drainage (89

$\times 10^6$ km^2) (i.e., drainage to the oceans) (after Meybeck 1979). However, this is not really the rate at which dissolved ions are being removed from the continents, because the value includes ions derived from rainfall and the atmosphere (Holland 1978; Meybeck 1979). Specifically, 64% of the total HCO_3 load in rivers comes from atmospheric CO_2, not carbonate rocks (see Table 5.13). This results in a reduction of 34% in the total dissolved load (or 14.3 tons/km^2/yr). Also, correction must be made for the ions in river water that come from atmospheric precipitation—about 4.5% of the total (1.9 tons/km^2). Thus, the corrected chemical denudation rate is about 26 tons/km^2/yr or about 60% of that derived from the total dissolved load. This is the rate at which continental ions are being removed to the oceans by river-dissolved transport.

The suspended sediment or mechanical denudation rate is about 226 tons/km^2/yr (see previous discussion). Thus, the combined mechanical and chemical denudation rate for the continents amounts to 252 tons/km^2/yr (or 22,314 Tg/yr), of which chemical denudation is only about 10%. The total denudation rate can be translated to a reduction of continental elevation of about 9.4 cm per 1000 years (assuming an average rock density of 2.7). This is less than the total physical plus chemical riverine erosion, as we noted earlier, because some eroded material is deposited on the continents.

The present dominance of mechanical denudation over chemical denudation may not always have been the case in the geologic past. First of all, as we have pointed out earlier in this chapter, the prehuman mechanical erosion rate was probably half that at present. Also, the suspended load is highly dependent on relief, while the dissolved load is not. Thus, at present, when the continents are fairly high, mechanical erosion dominates. With lower continental elevation (as may have been true at various times in the geological past), chemical erosion could have been more important (Holland 1978). A simple comparison between the percent of the total load that is dissolved for the Congo and Ganges-Brahmaputra rivers illustrates variations in the importance of the dissolved load. The combined Ganges and Brahmaputra rivers are in an area of high relief containing easily dissolved carbonate and evaporite rocks, but they have a dissolved load which is only about 10% of its total load. On the other hand, the Zaire River, draining an area of low relief and weathered crystalline rocks, has a dissolved load which is nearly 45% of its total load.

NUTRIENTS IN RIVER WATER

The two major nutrients in river water that will be discussed here are nitrogen and phosphorus. Since several aspects of the discussion of these elements overlap other water types, the reader is referred to further treatment of nitrogen and phosphorus under the subjects of rainwater and the atmosphere (Chapter 3), lakes (Chapter 6), estuaries (Chapter 7), and the oceans (Chapter 8).

Nitrogen in Rivers: The Terrestrial Nitrogen Cycle

The origin of nitrogen in river water is considerably more complex than it is for most other elements because nitrogen exists in solution in several different forms, is a major constituent of the atmosphere, and is intimately involved in biogeochemical cycling as an essential component of living tissue, both plant and animal. The subject of river water nitrogen involves study of a wide variety of nitrogen sources, which logically leads to study of the terrestrial nitrogen cycle. A qualitative diagramatic representation of this cycle is shown in Figure 5.9. We have discussed the atmosphere-soil part of the terrestrial nitrogen cycle already in detail in Chapter 3, so we include only a brief review of the relevant parts here. Nitrogen gas, N_2, which is a dominant part of the atmosphere, is not normally biologically available and must be "fixed," or combined with hydrogen, oxygen, and carbon, in order to be used by plants and organisms on land. Looking at the terrestrial nitrogen cycle in more detail (Table 5.15), we see that there are three major land inputs of fixed nitrogen in forms such as NO_3^- and NH_4^+. These include biological fixation, precipitation and dry deposition of previously fixed nitrogen, and the application of fertilizers (industrially fixed nitrogen).

Biological fixation is the dominant source of fixed nitrogen on land (~50%). This is accomplished by microorganisms living symbiotically in higher plants (particularly legumes) and lichens in trees. Overall, humans, by planting crops such as legumes and rice, are responsible for about 44 Tg N/yr out of the total biological fixation of 139 Tg N/yr (Burns and Hardy 1975).

Humans fix N_2 industrially as NO_3^- and NH_4^+ to produce about 85 Tg N/yr as fertilizer (FAO 1989), and fertilizer use has been growing. Precipitation (plus dry deposition) is another source

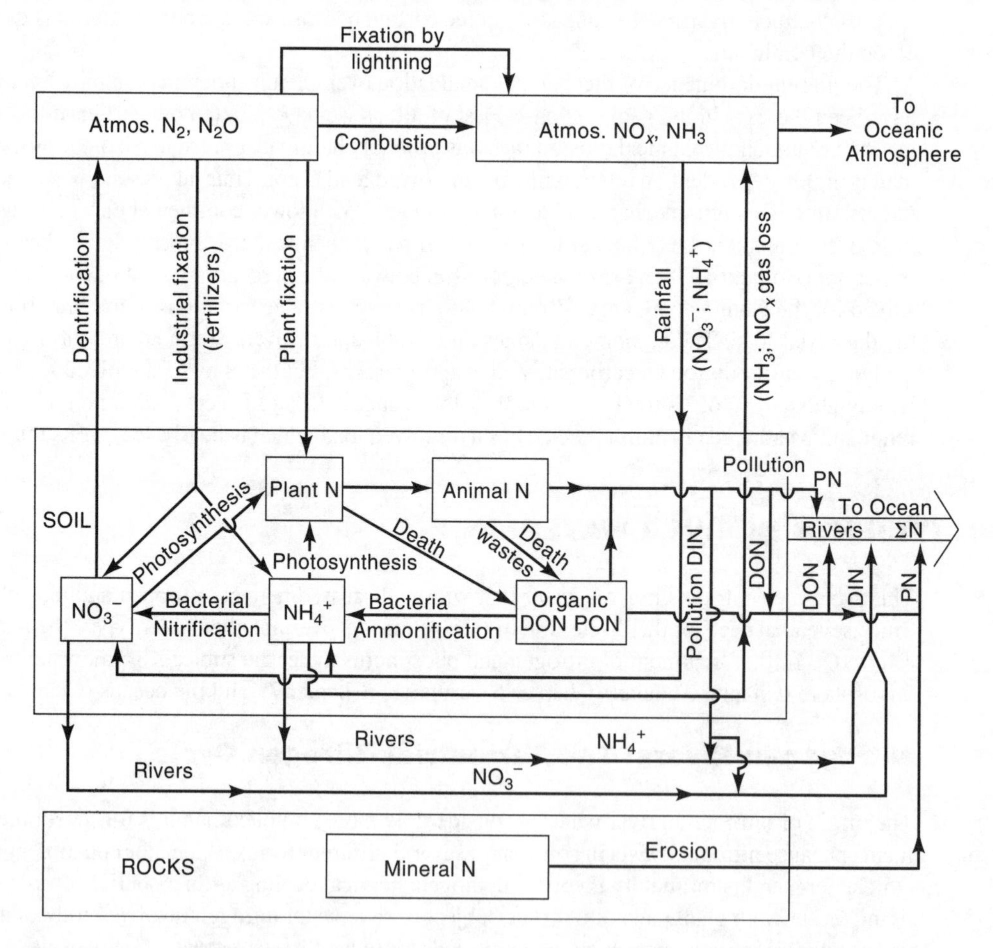

Figure 5.9. The terrestrial nitrogen cycle. DON = dissolved organic nitrogen; DIN = dissolved inorganic nitrogen; PN = particulate nitrogen; PON = particulate organic nitrogen; ΣN = total nitrogen. (See also Table 5.16 for details on river output).

TABLE 5.15 Terrestrial Nitrogen Cycle

Process	Total Flux (Tg N/yr)	Percent of Total Input or Output	Anthropogenic Flux (Tg N/yr)	Reference
Land input				
Biological fixation	139	49	44	Burns and Hardy 1975
Fertilizers & industry	85	30	85	FAO 1989
Precipitation and dry deposition	61	21	37	Table 3.15
Total input	285	100	166	
Land output				
River N	49–62	19	13–27	Table 5.16
Denitrification to N_2, N_2O	179	63	?	To balance (see text)
NH_3 gas loss	37	13	27	(See Chapter 3)
NO_x: soil gas loss and biomass burning	14	5	5	(See Chapter 3)
Total output	279–292	100	>45	

Note: Tg = 10^6 metric tons = 10^{12} g.

of terrestrial fixed nitrogen. Nitrogen in precipitation occurs as dissolved NO_3^- and NH_4^+. We estimate, of the total fixed nitrogen delivered to land in precipitation and dry deposition (61 Tg N), about 37 Tg N is due to human influences (see Chapter 3). Overall (Table 5.15), human influences are probably responsible for more than 50% of the total fixed nitrogen input to land in the form of polluted rain plus fertilizers.

Once it reaches the land, nitrogen is involved in various transformations in the terrestrial nitrogen cycle (Figure 5.9). As we have noted, some nitrogen is incorporated into organic matter directly from the atmosphere by fixation in plants. Plants also convert dissolved NO_3^- and NH_4^+ (from fertilizer, rain, or recycling of organic matter) into plant organic matter, some of which is eaten by animals and becomes animal organic nitrogen. The amount of nitrogen used in net primary production (1073 Tg N/yr; Melillo et al. 1993) is much larger than the yearly input of nitrogen (285 Tg N/yr); thus, a large amount of nitrogen is recycled within the biosphere.

When plants and animals die, their organic matter is broken down by bacteria into ammonia, some of which is dissolved in soil water as NH_4^+, and some of which escapes from the soil as NH_3 gas. Bacteria can also oxidize NH_4^+ to NO_2^- and NO_3^- in the soil (*nitrification*). Most of the NH_4^+ and NO_3^- from organic matter decay is recycled by plants. However, nitrogen can be lost from the land either directly to the atmosphere in the form of nitrogen-containing gases (as we have seen in Chapter 3) or in river water (our primary concern here). Some 80% of the total nitrogen output from land is gaseous (Table 5.15); denitrification breaks down soil nitrate to release N_2 (the major gaseous output) and N_2O, and lesser amounts of NH_3 and NO_x gas are also released at various stages of the nitrogen cycle.

River output of nitrogen is estimated to be 49–62 Tg N/yr (see Table 5.16). Because of denitrification and other gaseous emission processes, this amounts to only about one fifth of the total nitrogen loss from the land. This flux can be divided into different forms of nitrogen: (1) natural dissolved inorganic nitrogen (DIN), consisting mainly of ammonium and nitrate with a global riverine flux of 4.5 Tg N/yr (Meybeck 1982, 1993); (2) dissolved organic nitrogen (DON),

TABLE 5.16 River Nitrogen Transport (in Tg N/yr)

	Natural	Pollution	Total
Dissolved N			
DIN			
NO_3^--N	4.0		
NH_4^--N	0.5		
DON	10.0		
Total dissolved	14.5	7[a]–21[b]	22[a]–36[b]
Particulate N (PN)	21	6[b]	27–33[c]
Total N (TN)			49–63
Reactive N[d]			28–42

Note: [a] Meybeck 1993.

[b] Wollast 1993.

[c] Meybeck (1993), 21 Tg; Ittekkot and Zhang (1989),33 Tg; Wolast (1993), 27 Tg.

[d] Total dissolved N plus 22% of PN; see text.

Source: Meybeck 1982; 1993, except where noted.

with a flux of 10 Tg N/yr (Meybeck 1982, 1993); (3) dissolved pollutive nitrogen, with a flux estimated to lie between 7 Tg/yr (Meybeck 1993) and 21 Tg N/yr (Wollast 1993); and (4) Particulate nitrogen (PN), with a flux ranging from 27 to 33 Tg N/yr. Total PN is both natural and pollutional in origin and represents undecomposed organic N derived from soils and human activity along with NH_4^+ locked in K-containing silicate minerals. The natural PN flux is estimated by Meybeck (1982) to be 21 Tg N/yr and is based on the particulate organic carbon (POC) river load, and a weight ratio of POC/PN = 8.5. Ittekkot and Zhang (1989) get more PN (33 Tg/yr), partly because of a different estimate of the suspended load. Wollast estimates the pollutive PN flux to be 6 Tg N/yr. As much as 80% of the PN transport is in Asian rivers with a large suspended load (such as the Ganges, Bramaputra, Mekong, and Huanghe rivers), which are polluted from deforestation, nitrogen fertilizer application, and sewage.

Ittekkot and Zhang state that only 22% of the PN flux is labile (i.e., likely to be reactive in the oceans). The total *reactive* N riverine flux would thus be dissolved N (22–35 Tg N) plus 22% of particulate N (6–7 TgN), for a total reactive N flux of 28–42 Tg N/yr (Table 5.16).

The importance of biological processes to river-borne nitrogen is demonstrated by the fact that more of river dissolved nitrogen is organic than inorganic, and most of the rest (inorganic NO_3^- and NH_4^+) is derived from decomposition of organic matter (see Table 5.16). Further, the total river output of organic nitrogen amounts to only about 5% of the amount of nitrogen assimilated annually by the terrestrial biosphere through photosynthesis (1073 Tg N/yr; Melillo et al. 1993). This shows that biological recycling with the conservation of nitrogen is very efficient.

Human activities have influenced the river nitrogen load considerably. Meybeck (1982) estimates that, of the total dissolved nitrogen (DIN + DON) in river water, about one third is pollutive in origin, while Wollast estimates that two thirds of DIN + DON is pollutive (see Table 5.16). Since we estimate above that the total pollutional input of fixed nitrogen to the land is about 142 Tg N/yr, these values show that 15% or less of this pollutional input is lost as dissolved nitrogen in river water.

The nitrogen cycle shown in Table 5.15 simply assumes that the yearly river and gaseous output of nitrogen from land equals the input to the land. In other words, we *balance* the terrestrial cycle by assuming that denitrification (release of N_2 and N_2O gas) makes up the difference between input and the total known output in rivers plus NH_3 and NO_x gas loss; this gives 179 Tg N/yr for denitrification. Denitrification estimates vary widely: 107–161 Tg N/yr (Sönderlund and Svensson 1976); 43–390 Tg N/yr (Rosswall 1983); 13–233 Tg N/yr (Bowden 1986). Schlesinger and Hartley (1992) estimates denitrification as more than 150 Tg N/yr, which seems reasonable in light of our budget.

Any imbalance in N_2 due to changes in the amount of N_2 added to or subtracted from the atmosphere would not be apparent, because of the large mass and very slow turnover time of atmospheric N_2. It is possible, therefore, that the terrestrial nitrogen cycle is not balanced and that fixed nitrogen is building up in soils, groundwater, rivers, and lakes because of a large pollutional input (Delwiche 1970). The National Research Council (1972), based on an input–output budget, crudely estimated that nitrogen is being stored in U.S. soil and water (the amount being 20% of the U.S. fertilizer input at that time). However, the evidence is not good enough to say with any certainty whether the terrestrial nitrogen cycle is balanced by gas release (making up for pollutional nitrogen increases), or whether nitrogen is actually being stored in terrestrial soils and waters.

The type of nitrogen found in polluted waters varies. Nitrate is more important than ammonium in most polluted waters (see Table 5.16), but in poorly oxygenated rivers resulting from excessive organic matter loading, ammonium may reach 80% of the total dissolved inorganic nitrogen (nitrite, NO_2^-, is much less important). Urban wastes are higher in ammonium, and agricultural runoff is higher in nitrate, as are combustion products in rain. Europe and the United States are the biggest polluters, contributing 70% of the total dissolved pollutive nitrogen in world river water (Meybeck 1982).

The anthropogenic sources of nitrogen in rivers include (1) so-called point sources, which are discharged directly into surface waters, such as municipal and industrial sewage, septic tanks, refuse dumps, and animal feedlots; (2) diffuse sources, which result from runoff and leaching of rural and urban land; and (3) precipitation directly to lakes and streams (National Research Council 1972).

Municipal and industrial wastes, because they are discharged directly into the rivers, can produce large local increases of river nitrogen, particularly in urban areas. However, since sewage can be treated to remove around 40% of the nitrogen, this nitrogen source is easier to control than diffuse sources. Diffuse sources include natural leaching of nitrate from soil (a process which humans have accelerated by deforestation and cultivation), atmospheric deposition (see Chapter 3), and runoff from agricultural land. Nitrogen from agricultural land arises from application of nitrogen fertilizers, from animal wastes, and from the cultivation of plants such as legumes which are nitrogen fixers. Worldwide, the concentration of nitrate in a river correlates with the population density in the river basin (Peierls et al. 1991).

The concentration of nitrate increased in most U.S. rivers (on the average about 7%/yr) from 1974 to 1981 (Smith et al. 1987). Point-source N loads decreased due to better waste treatment, while nonpoint sources increased. Increased atmospheric deposition of nitrate, in particular, may have caused N increases in midwestern and eastern rivers (Figure 5.10). For example, Fisher et al. (1988) estimated that 16% of the N input to the rivers draining into Long Island Sound and 30% of the N input to the rivers draining into Chesapeake Bay was from atmospheric N deposition on each watershed. Apparently, in forests which have received large amounts of pollutive

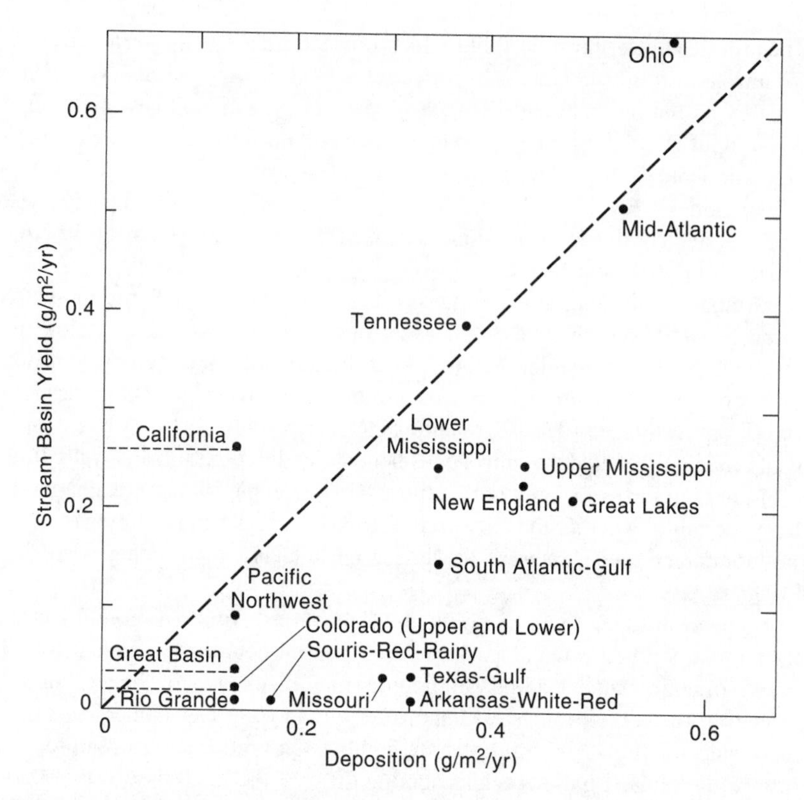

Figure 5.10. Median yield of nitrate ($g/m^2/yr$) in relation to the atmospheric deposition rate of nitrate ($g/m^2/yr$) in U.S. river basins grouped by region. (After R. A. Smith, R. B. Alexander, and M. G. Wolman, "Water Quality Trends in the Nation's Rivers," *Science* 235; 1612. Copyright © 1987 by the American Association for the Advancement of Science, reprinted by permission of the publisher.)

atmospheric nitrate deposition, the nitrate retention mechanism becomes saturated and a considerable part of atmospheric nitrate deposition moves straight through forests to rivers (Hedin 1994). The effect is worse in old-growth forests. Smith et al. (1987) also found that nitrogen increases in U.S. rivers were strongly associated with agricultural activity, including N fertilizer use (increasing by 6%/yr), livestock population increases, and an increasing number of animal feedlots.

Phosphorus in Rivers: The Terrestrial Phosphorus Cycle

Phosphorus, unlike nitrogen (or sulfur), has no stable gaseous phases in the atmosphere. For this reason, in contrast to nitrogen, most phosphorus is lost from land by way of river runoff, and a considerably smaller proportion, on average, of land input is provided by precipitation. The dominant feature of the terrestrial phosphorus cycle is the fact that phosphorus is an important and often limiting nutrient in fresh water. Also, the amount of phosphorus incorporated into organic matter each year is much greater than its production by weathering or its loss by rivers. The deficit is made up by biologically recycled material. Phosphorus tends to be strongly conserved by biological systems, so most phosphorus released by organic decay is rapidly converted into organic matter.

The major ultimate source of phosphorus is weathering, that is, the removal of phosphorus from rocks. However, phosphorus in rocks and sediments is mainly in a relatively insoluble form as the calcium phosphate mineral, apatite. Even when released as soluble phosphate by weathering, phosphorus is usually quickly tied up in the soil as iron, aluminum, and calcium phosphates or by clay minerals to produce insoluble forms that are not accessible to plants. Because of the relative insolubility of phosphorus, it is often a limiting nutrient in biological systems; that is, it is in short supply. Humans have intervened in the phosphorus cycle by deforestation, which increases erosion; by the use of phosphorus fertilizers; and through the production of industrial wastes, sewage, and detergents. (For more details on the phosphorus cycle, see Stumm 1972; Pierrou 1976; Richey 1983; Esser and Kohlmaier 1991; Meybeck 1982, 1993.) Particularly in lakes, the introduction by humans of greatly increased phosphorus has stimulated productivity and led to eutrophication (see Chapter 6). Here, however, we shall focus on river runoff from the land as it relates to the overall terrestrial phosphorus cycle.

Phosphorus concentrations in rainfall are small (0.01–0.04 mg/l total P) and difficult to measure due to contamination. The amount of phosphorus delivered to land in precipitation has been estimated at 1 Tg P/yr (Meybeck 1982, 1993), with total P deposition (including dry deposition) being 3.2 Tg P/yr (Graham and Duce 1979). This is of the same order of magnitude as the dissolved river transport (2 Tg P/yr) (see Table 5.17). In fact, in some remote areas, such as Hubbard Brook, New Hampshire (Likens et al. 1977), precipitation input appears to be greater than dissolved river output. The main source of phosphorus deposited on the continents, via precipitation and dry deposition, is soil dust (3.0 Tg P/yr or 94% of total deposition; Graham and Duce

TABLE 5.17 Phosphorus Fluxes in Rivers and Rain (in Tg P/yr)

Source	Total Flux	Polluted Part	Reference
P in river runoff			
Dissolved ortho-P	0.8	0.4	Meybeck 1982; 1993
Dissolved organic P [a]	1.2	0.6	
Total dissolved P	2.0	1.0	Meybeck 1982; 1993
Particulate organic-P	8.0	?	Meybeck 1982; 1993
Particulate inorganic-P [a]	12.	?	
Total particulate P	20.0	?	Meybeck 1982; 1993
Total output	22	>1	
Reactive P output[b]	5		See text
P in rain + dry deposition to land			
Soil particle origin	3.0	0.2	Graham and Duce 1979
Industry, combustion	0.21	0.21	Graham and Duce 1979
Sea salt	0.03		Graham and Duce 1979
Total rain and dry deposition	3.2	0.41	
Rain only to land	1.0	—	Meybeck 1982

[a] Calculated by difference from total; no data.

[b] Total dissolved P plus 15% of particulate P (after Berner and Rao 1994).

1979). Other, minor phosphorus sources in rain include industry and combustion, sea salt, and biogenic aerosols. (Table 5.17). However, atmospheric soil dust is derived from weathering on land and is thus recycled, and not a primary input to the terrestrial system.

The total amounts of suspended particulate phosphorus (inorganic and organic) transported by rivers at present is approximately 20 Tg P/yr (Meybeck 1982, 1993). (Similarly, Howarth et al. (1994) estimate 20 Tg P/y including pollution, and Richey (1983) estimates 17 Tg P/yr.) Part of this is from the erosion of phosphate minerals in rocks, primarily apatite, but much of the rest is from human effects which include increased soil erosion and transport due to deforestation and agriculture, increased organic matter removal due to agriculture, and fertilizer application. Froelich et al. (1982) estimate that the preagricultural river phosphorus flux due to weathering was about 10 Tg P/yr based on a preagricultural continental denudation rate of 10,000 Tg/yr [Judson 1968; Gregor 1970] and 0.1% P in the earth's crust. (Using our preagricultural denudation rate derived earlier gives essentially the same result.) Thus, roughly half of present-day river suspended phosphorus transport may be due to human activities. Whether pollutional in origin or not, overall suspended phosphorus transport includes organic P, and P adsorbed on hydrous ferric oxides and clay minerals, as well as detrital apatite (Berner and Rao 1994).

Land plant material contains considerable phosphorus, and plants convert 200 Tg P/yr into organic matter through photosynthesis (Richey 1983). A large proportion of this phosphorus is from recycled organic matter, but some is also new phosphorus from weathering. As a result of the biological involvement of phosphorus, more than half of the dissolved river phosphorus and 40% of the suspended load is in an organic form (Table 5.17). All of the dissolved phosphorus (organic and inorganic or 2 Tg P/yr) and perhaps 15% of the particulate phosphorus (another 3 Tg P/yr; Berner and Rao 1994) is reactive, that is, biologically active. Thus, the river "reactive phosphorus" flux, which can be considered as P lost by terrestrial organisms, is 5 Tg P/yr, or only about 3% of the phosphorus involved in the terrestrial biological cycle. This further demonstrates the efficiency of terrestrial biological recycling.

About 40% of the dissolved phosphorus transported by rivers is present as inorganic phosphate, generally orthophosphate anions (PO_4^{3-}, HPO_4^{2-}, $H_2PO_4^-$), which are the best known, and most commonly measured, type of phosphorus in rivers (see Table 5.17). Dissolved inorganic phosphate in rivers can come from natural weathering and solution of phosphate minerals, accelerated dissolution due to human-induced soil erosion and transport, natural and artificially enhanced (agricultural) release of phosphate from organic P, release of P from fertilizers, and soluble phosphate from detergents and domestic and industrial wastes (Stumm 1972). The rest of dissolved P is organic P. Meybeck estimates that 1 Tg P/yr or half of the total dissolved river P is anthropogenic.

Because the natural dissolved total phosphorus levels are low in rivers (0.025 mg P/l), the addition of pollutive phosphorus can result in large local increases in concentration. Although the world river dissolved phosphcrus load is estimated to have been doubled by human activities, in the polluted rivers of the United States and North America, the phosphorus concentration in many places is 10 times natural levels (Meybeck 1982). The median concentration in U.S. rivers in 1974–1981 was 0.13 mg/l P (Smith et al. 1987).

Esser and Kohlmaier (1991) estimate phosphate rock mining worldwide in 1985 as 24.4 Tg P, of which 90% was used to make fertilizer (22.2 Tg P/yr) and the rest was used for detergents and industry. They estimate that dissolved pollutive P input to rivers includes all of the detergent P (2.2 Tg P/yr), 5–10% of the fertilizer P (1.5 Tg P/yr), plus 1.6 Tg P/yr from food consumption for a total of 5 Tg P/yr. Richey (1983) estimated the total P input to waters as 4–7 Tg

P/yr including pollution. Since only about 2 Tg P/yr of total dissolved P is carried by rivers to the sea (Table 5.17), much of this added P must be converted to particulate P, e.g., adsorbed on sediment grains.

Smith et al. (1987) found that from 1974 to 1981, total dissolved P decreased in the Great Lakes and Upper Mississippi River basins due to reductions in pollutive point sources, such as municipal and industrial sewage. Much of this decrease probably was due to the removal of phosphorus from detergents. By contrast, there were increases in P in other U.S. rivers where pollutive inputs were from nonpoint sources, such as fertilizer use, increases in cattle population density, and P liberated from an elevated suspended load due to increased cropland soil erosion. These agricultural nonpoint sources have been much more difficult to control than the point sources. However, expected river increases in P concentrations from increased agricultural activity may have been delayed by sediment-bound P storage in stream channels.

Meybeck (1982, 1993) and Kempe (1984) have observed that dissolved N and P concentrations are correlated in polluted river water but with considerable scatter. Figure 5.11 summarizes the dissolved nitrate versus phosphate concentrations in unpolluted and polluted major rivers. Low nutrient concentrations and low N/P are found in tropical rivers such as the Amazon (N/P = 16) and Zaire, while arctic rivers such as the Mackenzie have low concentrations and high N/P ratios. Industrial and urban pollution, for example, in the Thames River, tends to produce higher P concentrations and lower N/P ratios than agricultural pollution, for example,

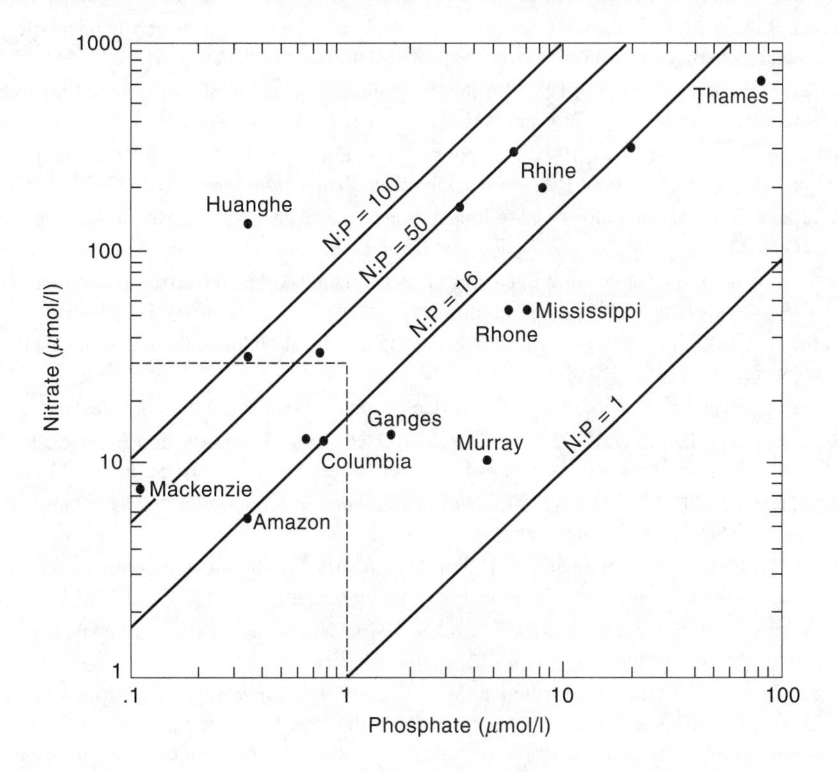

Figure 5.11. Mean nitrate concentration versus mean phosphate concentration for various world rivers. Pristine rivers are enclosed by the dashed line. The diagonals are lines of equal N/P ratio (in mol/mol). (Data from Kempe 1984 and Meybeck 1993.)

in the Huanghe and Yangtze Rivers, which results in high N/P ratios due to the use of manure as a fertilizer.

At high nutrient concentrations, sometimes the river atomic N/P ratio is less than 16. This is the Redfield ratio, which is the nutrient ratio used by marine phytoplankton in photosynthesis. In areas with high population densities, coastal waters receive excess P relative to N from polluted rivers, causing N limitation in both the rivers and the estuaries. Limitation of planktonic production by N is a common situation in estuaries (see Chapter 7).

REFERENCES

Beck, K. C., J. H. Reuter, and E. M. Perdue. 1974. Organic and inorganic geochemistry of some coastal plain rivers of the southeastern U.S., *Geochim. Cosmochim. Acta* 38: 341–364.

Berner, R. A. 1971. Worldwide sulfur pollution of rivers, *J. Geophys. Res.* 76: 6597–6600.

Berner, R. A. 1992. Weathering, plants and the long-term carbon cycle, *Geochim. Cosmochim. Acta* 56: 3225–3231.

Berner, R. A. 1994. GEOCARB II: A revised model of atmospheric CO_2 over Phanerozoic time, *Am. J. Sci.* 294: 56–91.

Berner, R. A., E. K. Berner, P. A. Schroeder, and T. W. Lyons. 1990. Comment on "Sodium-calcium exchange in the weathering of shales: Implications for global weathering," *Geology* 18(2): 190.

Berner, R. A., A. C. Lasaga, and R. M. Garrels. 1983. The carbonate-silicate geochemical cycle and its effect on atmospheric carbon dioxide over the past 100 million years, *Am. J. Sci.* 283: 641–683.

Berner, R. A., and R. Raiswell. 1983. Burial of organic carbon and pyrite sulfur in sediments over Phanerozoic time: A new theory, *Geochim. Cosmochim. Acta* 47(5): 855–862.

Berner, R. A., and J.-L. Rao. 1994. Phosphorus in sediments of the Amazon River and Estuary: Implications for the global flux of P to the sea, *Geochim. Cosmochim. Acta* 58: 2333–2339.

Biscaye, P. E. 1965. Mineralogy and sedimentation of recent deep-sea clay in the Atlantic Ocean and adjacent seas and oceans, *Bull. Geol. Soc. Am.* 76: 803–832.

Bowden, W. B. 1986. Gaseous nitrogen emissions from undisturbed terrestrial ecosystems: An assessment of their impacts on local and global nitrogen budgets, *Biogeochemistry* 2: 249–279.

Brady, P. V. 1991. The effect of silicate weathering on global temperature and atmospheric CO_2, *J. Geophys. Res.* 96: 18101–18106.

Bricker, O. P., and K. C. Rice. 1989. Acidic deposition to streams, *Environ. Sci. Technol.* 23(4): 379–385.

Buhl, D., R. D. Neuser, D. K. Richter, D. Reidel, B. Roberts, H. Strauss, and J. Veizer. 1991. Nature and nuture: Environmental isotope story of the river Rhine, *Naturwissenschaften* 78: 337–346.

Burns, R. C., and R. W. F. Hardy. 1975. *Nitrogen Fixation in Bacteria and Higher Plants.* Berlin, Heidelberg, and New York: Springer-Verlag.

Cerling, T. E., B. L. Pederson, and K. L. Von Damm. 1989. Sodium-calcium ion exchange in the weathering of shales: Implications for global weathering budgets, *Geology* 17: 552–554.

Cleaves, E. T., A. E. Godfrey, and O. P. Bricker. 1970. Geochemical balance of a small watershed and its geomorphic implications, *Geol. Soc. Am. Bull.* 81: 3015–3032.

Crerar, D. A., J. L. Means, R. F. Yuretich, M. P. Borcsik, J. L. Amster, D. W. Hastings, G. W. Knox, K. E. Lyon, and R. F. Quiett. 1981. Hydrochemistry of the New Jersey coastal plain, 2. Transport and deposition of iron, aluminum, dissolved organic matter, and selected trace elements in stream, ground, and estuary water, *Chem. Geol.* 33: 23–44.

Cronan, C. S., W. A. Reiners, R. C. Reynolds, and G. E. Lang. 1978. Forest floor leaching: Contributions from mineral, organic and carbonic acids in New Hampshire subalpine forests, *Science* 200: 309–311.

Curtis, W. F., J. K. Culbertson, and E. B. Chase. 1973. Fluvial-sediment discharge to the oceans from the conterminous United States, *USGS Circ. 670.*

Degens, E. T., S. Kempe, and J. E. Richey. 1991. Summary: Biogeochemistry of major world rivers. In *Biogeochemistry of Major World Rivers,* SCOPE, ed. E. T. Degens, S. Kempe, and J. E. Richey, pp. 323–347. Chichester, U.K.: John Wiley.

Delwiche, C. C. 1970. The nitrogen cycle, *Sci. Am.* 223: 137–146.

Dion, E. P. 1983. Trace elements and radionuclides in the Connecticut River and Amazon River estuary. Ph.D. dissertation, Dept. of Geology and Geophysics, Yale University, New Haven, Conn.

Drever, J. I. 1988. *The Geochemistry of Natural Water,* 2nd ed. Englewood Cliffs, N.J.: Prentice-Hall.

Drever, J. I., and J. Zobrist. 1992. Chemical weathering of silicate rocks as a function of elevation in the southern Swiss Alps, *Geochim. Cosmochim. Acta* 56: 3209–3216.

Dunne, T. 1978. Rates of chemical denudation of silicate rocks in tropical catchments, *Nature* 274: 244–246.

Esser, G., and G. H. Kohlmaier. 1991. Modelling terrestrial sources of nitrogen, phosphorus, sulphur and organic carbon to rivers. In *Biogeochemistry of Major World Rivers,* ed. E. T. Degens, S. Kempe, and J. E. Richey, pp. 297–322. Chichester, U.K.: John Wiley.

Fairbridge, R. W. 1968. *Encyclopedia of Geomorphology,* pp. 177–186. New York: Reinhold.

FAO. 1989. *FAO Fertilizer Yearbook 39.* Rome: Food and Agricultural Organization.

Feth, J. H. 1971. Mechanisms controlling world water chemistry: Evaporation-crystallization process, *Science* 172: 870–871.

Feth, J. H. 1981. Chloride in natural continental water—A review, *USGS Water Supply Paper 2176.*

Fisher, D., J. Ceraso, and M. Oppenheimer. 1988. *Polluted coastal waters: The role of acid rain.* New York: Environmental Defense Fund.

Freney, J. R., M. V. Ivanov, and H. Rodhe. 1983. The sulphur cycle. In *The Major Biogeochemical Cycles and Their Interactions,* SCOPE 21, ed. B. Bolin and R. Cook, pp. 56–61. Chichester, U.K.: John Wiley.

Friend, J. P. 1973. The global sulfur cycle. In *Chemistry of the Lower Atmosphere,* ed. S. I. Rasool, pp. 177–201. New York: Plenum Press.

Froehlich, P. N., M. L. Bender, N. A. Luedtke, G. R. Heath, and T. Devries. 1982. The marine phosphorus cycle, *Am. J. Sci.* 282: 474–511.

Garrels, R. M., and F. T. Mackenzie. 1971. *Evolution of Sedimentary Rocks.* New York: W. W. Norton.

Gibbs, R. J. 1967. Amazon River: Environmental factors that control its dissolved and suspended load, *Science* 156 (3783): 1734–1737.

Gibbs, R. J. 1970. Mechanisms controlling world water chemistry, *Science* 170: 1088–1090.

Gibbs, R. J. 1971. Mechanisms controlling world water chemistry: Evaporation-crystallization process, *Science* 172: 871–872.

Gordeev, V. V., and I. S. Siderov. 1993. Concentrations of major elements and their outflow into the Laptev Sea by the Lena River, *Marine Chem.* 43: 33–45.

Gorham, E., J. K. Underwood, F. B. Martin, and J. G. Ogden III. 1986. Natural and anthropogenic causes of lake acidification in Nova Scotia, *Nature* 324: 451–453.

Graham, W. F., and R. A. Duce. 1979. Atmospheric pathways of the phosphorus cycle, *Geochim. Cosmochim. Acta* 43: 1195–1208.

Gregor, B. 1970. Denudation of the continents, *Nature* 228: 273–275.

Hay, W. W., and J. R. Southam. 1977. Modulation of marine sedimentation by the continental shelves. In *The Fate of Fossil Fuel* CO_2 in the Oceans, ed. N. R. Andersen, and A. Malahoff, pp. 569–604. New York: Plenum Press.

Holeman, J. N. 1968. The sediment yield of major rivers of the world, *Water Res.* 4(4): 737–747.

Hedin, L. O. 1994. Stable isotopes, unstable forest, *Nature* 372: 725–726.

Holeman, J. N. 1980. Erosion rates in the U.S. estimated by the Soil Conservation Services inventory, *EOS* 61(46): 954.

Holland, H. D. 1978. *The Chemistry of the Atmosphere and Oceans.* New York: John Wiley.

Hossain, M. M. 1991. Total sediment load in the lower Ganges and Jumuna. Bangladesh University of Engineering and Technology.

Howarth, R. W., H. Jensen, R. Marino, and H. Postma. 1994. Transport and processing of phosphorus in estuaries and oceans. In *Phosphorus Cycling in Terrestrial and Aquatic Ecosystems,* SCOPE, ed. H. Tiessen (in press).

Hu, Ming-Hui, R. F. Stallard, and J. M. Edmond, 1982. Major ion chemistry of some large Chinese rivers, *Nature* 289(5): 550–553.

Husar, R. B., and J. D. Husar. 1985. Regional sulfur runoff, *J. Geophys. Res.* 90(C1): 1115–1125.

Ittekkot, V. 1988. Global trends in the nature of organic matter in river suspensions, *Nature* 332: 436–438.

Ittekkot, V., and S. Zhang. 1989. Pattern of particulate nitrogen transport in world rivers, *Global Biogeochem. Cycles* 3: 383–391.

Johnson, A. H. 1979. Evidence of acidification of headwater streams in the New Jersey pinelands, *Science* 206(16): 834–836.

Judson, S. 1968. Erosion of the land, *Am. Sci.* 56(4): 356–374.

Kempe, S. 1984. Sinks of the anthropogenically enhanced carbon cycle in surface fresh waters, *J. Geophys. Res.* 89: 4657–4676.

Kempe, S. 1988. Freshwater carbon and the weathering cycle. In *Physical and Chemical Weathering in Geochemical Cycles,* ed. A. Lerman and M. Meybeck, pp. 197–223. Dordrecht: Kluver.

Kennedy, V. C., and R. L. Malcolm. 1977. Geochemistry of the Mattole River of Northern California, *USGS Open-File Report 78–205.*

Kleinmann, R. L. P., and D. A. Crerar. 1979. *Thiobacillus ferrooxidans* and the formation of acidity in simulated coal mine environments, *Geomicrobiol. J.* 1: 373–388.

Krug, E. C., and C. R. Frink. 1983. Acid rain on acid soil: A new perspective, *Science* 221: 520–525.

Leenheer, J. A. 1980. Origin and nature of humic substances in the waters of the Amazon River Basin, *Acta Amazonica* 10(3): 513–526.

Lewis, D. M. 1976. The geochemistry of manganese, iron, uranium, lead-210, and major ions in the Susquehanna River. Ph.D. dissertation, Dept. of Geology and Geophysics, Yale University, New Haven, Conn.

Lewis, W. M., and M. C. Grant. 1979. Changes in the output of ions from a watershed as a result of the acidification of precipitation, *Ecology* 60(6): 1093–1097.

Li, Y.-H. 1976. Denudation of Taiwan Island since the Pliocene Epoch, *Geology* 4: 105–107.

Likens, G. E., F. H. Bormann, R. S. Pierce, J. S. Eaton, and N. M. Johnson. 1977. *Biogeochemistry of a Forested Ecosystem.* New York: Springer-Verlag.

Livingstone, D. A. 1963. Chemical composition of rivers and lakes, *USGS Prof. Paper 440G.*

Logan, J. A. 1983. Nitrogen oxides in the troposphere: Global and regional budgets, *J. Geophys. Res.* 88(C15): 10785–10807.

Martin, J. M., and M. Meybeck. 1979. Elemental mass-balance of material carried by major world rivers, *Marine Chem.* 7: 173–206.

Martin, J. M., and M. Whitfield. 1981. The significance of river input of chemical elements to the ocean. In *Trace Metals in Sea Water,* ed. C. S. Wong, E. Boyle, K. W. Bruland, J. D. Burton, and E. D. Goldberg, pp. 265–296. New York: Plenum Press.

McLennan, S. M. 1993. Weathering and global denudation, *J. Geol.* 101: 295–303.

Meade, R. H. 1982. Sources, sinks, and storage of river sediment in the Atlantic drainage of the United States, *J. Geol.* 90(3): 235–252.

Meade, R. H., and R. S. Parker. 1985. Sediment in rivers of the United States, *USGS Water Supply Paper 2275:* 49–60.

Melillo, J. M., A. D. McGuire, D. W. Kicklighter, B. Moore III, C. J. Vorosmarty, and A. L. Schloss. 1993. Global climate change and terrestrial net primary production, *Nature* 363: 234–240.

Meybeck, M. 1979. Concentrations des eaux fluviales en èlèments majeurs et apports en solution aux ocèans, *Rev. Gèol. Dyn. Gèogr. Phys.* 21(3): 215–246.

Meybeck, M. 1980. Pathways of major elements from land to ocean through rivers. In *Proceedings of the Review and Workshop on River Inputs to Ocean-Systems,* ed. J.-M. Martin, J. D. Burton, and D. Eisma, pp. 18–30. Rome: FAO.

Meybeck, M. 1982. Carbon, nitrogen and phosphorus transport by world rivers, *Am. J. Sci.* 282: 401–450.

Meybeck, M. 1983. Atmospheric inputs and river transport of dissolved substances. In *Dissolved Loads of Rivers and Surface Water Quantity/Quality Relationships,* Proceedings of the Hamburg Symposium, August 1983. IAHS Publ. 141.

Meybeck, M. 1984. Les fleuves et le cycle gèochimique des èlèments. Thèse de Doctorat (no. 8435), Ecole Normal Supèrieure, Laboratoire de Gèologie, Université Pierre et Marie Curie, Paris 6, France.

Meybeck, M. 1987. Global chemical weathering of surficial rocks estimated from river dissolved loads, *Am. J. Sci.* 287: 401–428.

Meybeck, M. 1988. How to establish and use world budgets of river material. In *Physical and Chemical Weathering in Geochemical Cycles,* ed. A. Lerman and M. Meybeck, pp. 247–272. Dordrecht, Netherlands: Kluwer.

Meybeck, M. 1993. C, N, and P and S in rivers: From sources to global inputs. In *Interactions of C, N, P and S in Biogeochemical Cycles and Global Change,* ed. R. Wollast, F. T. Mackenzie, and L. Chou, pp. 163–193. Berlin, Heidelberg: Springer-Verlag.

Miller, A., J. C. Thompson, R. E. Peterson, and D. R. Haragan. 1983. *Elements of Meterology,* 4th ed. Columbus, Ohio: Chas. E. Merrill.

Miller, W. R., and J. I. Drever. 1977. Water chemistry of a stream following a storm, Absaroka Mountains, Wyoming, *Geol. Soc. Am. Bull.* 88: 286–290.

Milliman, J. D. 1980. Transfer of river-borne particulate material to the oceans. In *River Inputs to Ocean Systems,* SCOR/UNEP/UNESCO, Review and Workshop, ed. J. M. Martin, J. D. Burton, and D. Eisma, pp. 5–12. Rome: FAO.

Milliman, J. D., and R. H. Meade. 1983. World-wide delivery of river sediment to the oceans, *J. Geol* 91(1): 1–21.

Milliman, J. D., Y. S. Quin, M. E. Ren, and Y. Saito. 1987. Man's influence on the erosion and transport of sediment by Asian rivers: The Yellow River (Huanghe) example, *J. Geol.* 95: 751–762.

Milliman, J. D., and J. P. M. Syvitski. 1992. Geomorphic/tectonic control of sediment discharge to the ocean: The importance of small mountainous rivers, *J. Geol* 100: 525–544.

Möller, D. 1984. Estimation of the global man-made sulfur emission, *Atmos. Environ.* 18(1): 19–27.

National Research Council, 1972. *Accumulation of Nitrate.* Publication of Committee on Nitrate Accumulation. Washington, D.C.: National Academy of Sciences, National Research Council.

Nriagu, J. O. 1978. Production and uses of sulfur. In *Sulfur in the Environment,* Part 1, ed. J. O. Nriagu, pp. 1–21. New York: Wiley-Interscience.

Nriagu. J. O., and J. M. Pacyna. 1988. Quantitative assessment of worldwide contamination of air, water and soils by trace metals, *Nature* 333: 134–139.

Odén, S., and T. Ahl. 1978. The sulfur budget of Sweden. In *Effects of Acid Precipitation on Terrestrial Ecosystems,* ed. T. C. Hutchinson and M. Havas, pp. 111–122. New York: Plenum Press.

Pačes, T. 1985. Sources of acidification in Central Europe estimated from elemental budgets in small basins, *Nature* 315(6014): 31–36.

Pearce, F. 1991. A dammed fine mess, *New Scientist,* May 4: 36–39.

Peierls, B. L., N. F. Caraco, M. L. Pace, and J. J. Cole. 1991. Human influence on river nitrogen, *Nature* 350: 386–387.

Peters, N. 1984. Evaluation of environmental factors affecting yields of major dissolved ions in streams of the U.S., *USGS Water Supply Paper 2228.*

Pierrou, U. 1976. The global phosphorus cycle. In *Nitrogen, Phosphorus and Sulfur Global Cycles,* SCOPE Rep. 7, ed. B. H. Svenson and R. Söderlund, pp. 75–90. Stockholm: Ecol. Bull. 22.

Pinet, P., and M. Souriau. 1988. Continental erosion and large scale relief, *Tectonics* 7: 563–582.

Probst, J.-L., R. R. Nkounkou, G. Krempp, J. P. Bricquet, J. P. Thiebaux, and J. C. Olivry. 1992. Dissolved major elements exported by the Congo and the Ubangui rivers during the period 1987–1989, *J. Hydrol.* 135: 237–257.

Probst, J.-L., and Y. Tardy. 1989. Global runoff fluctuations during the last 80 years in relation to world temperature change, *Am. J. Sci.* 289: 267–285.

Reeder, S. W., B. Hitchon, and A. A. Levinson. 1972. Hydrogeochemistry of the surface waters of the Mackenzie River drainage basin, Canada: 1. Factors controlling inorganic compositions, *Geochim. Cosmochim. Acta* 26: 825–865.

Richey, J. E. 1983. The phosphorus cycle. In *The Major Biogeochemical Cycles and Their Interactions,* ed. B. Bolin and R. B. Cook, pp. 51–56. Chichester, U.K.: John Wiley.

Rosswall, T. 1983. The nitrogen cycle. In *The Major Biogeochemical Cycles and Their Interactions,* SCOPE 21, ed. B. Bolin and R. B. Cook, pp. 46–50. Chichester, U.K.: John Wiley.

Sarin, M. M., S. Krishnaswami, K. Dilli, B. L. K. Somayajulu, and W. S. Moore. 1989. Major ion chemistry of the Ganga-Bramaputra river systems, India, *Geochim. Cosmochim. Acta* 53: 997–1009.

Schindler, D. W. 1988. Effects of acid rain on freshwater ecosystems, *Science* 239: 149–157.

Schlesinger, W. H., and A. E. Hartley. 1992. A global budget for atmospheric NH_3, *Biogeochemistry* 15: 191–211.

Skinner, B. J. 1969. *Earth Resources.* Englewood Cliffs, N.J.: Prentice-Hall.

Smith, R. A., and R. B. Alexander. 1983. Evidence for acid-precipitation induced trends in stream chemistry at Hydrologic Bench-Mark Stations. *USGS Circ. 910.*

Smith, R. A., R. B. Alexander, and M. G. Wolman. 1987. Water-quality trends in the nation's rivers, *Science* 235: 1607–1615.

Söderlund, R., and B. H. Svensson, 1976. The global nitrogen cycle. In *Nitrogen, Phosphorus and Sulfur—Global Cycles,* ed. B. H. Svensson and R. Söderlund, pp. 23–73, SCOPE Report no. 7, Stockholm: *Eco. Bull.* 22.

Spitzy, A., and Leenheer, J. 1991. Dissolved organic carbon in rivers. In *Biogeochemistry of Major World Rivers,* SCOPE 22, ed. E. T. Degens, S. Kempe, and J. E. Richey, pp. 213–232. New York: John Wiley.

Stallard, R. F. 1980. Major element geochemistry of the Amazon River system. Ph.D. dissertation, MIT/Woods Hole Oceanographic Inst., WHOI-80-29.

Stallard, R. F. 1985. River chemistry, geology, geomorphology and soils in the Amazon and Orinoco Basins. In *The Chemistry of Weathering,* ed. J. I. Drever, pp. 293–316. Boston: D. Reidel.

Stallard, R. F., and J. M. Edmond. 1981. Geochemistry of the Amazon 1: Precipitation chemistry and the marine contribution to the dissolved load, *J. Geophys. Res.* 86(C10): 9844–9858.

Stallard, R. F., and J. M. Edmond. 1983. Geochemistry of the Amazon 2: The influence of the geology and weathering environment on the dissolved load, *J. Geophys. Res.* 88: 9671–9688.

Stallard, R. F., and J. M. Edmond. 1987. Geochemistry of the Amazon 3. Weathering chemistry and limits to dissolved inputs, *J. Geophys. Res.* 92(C8): 8293–8302.

Stumm, W. 1972. The acceleration of the hydrogeochemical cycling of phosphorus. In *The Changing Chemistry of the Oceans,* Nobel Symp. 20, ed. D. Dyrssen and D. Jagner, pp. 329–346. Stockholm: Almqvist and Wiksell.

Sugawara, K. 1967. Migration of elements through phases of the hydrosphere and atmosphere. In *Chemistry of the Earth's Crust,* vol. 2, ed. A. P. Vinogradov, pp. 501–510. Israel Program for Scientific Translation Ltd., Jerusalem. Reprinted in *Geochemistry of Water,* ed. Y. Kitano, pp. 227–237. New York: Halsted Press, 1975.

Telang, S. A., R. Pocklington, A. S. Naidu, E. A. Romankevitch, I. Gitelson, and M. L. Gladyshev. 1991. Carbon and mineral transport in major North American, Russian Arctic, and Siberian rilvers: The St. Lawrence, the Mackenzie, the Yukon, the Arctic Alaskan rivers, the Arctic Basin rivers in the Soviet Union and the Yenisei. In *Biogeochemistry of Major World Rivers,* SCOPE, ed. E. T. Degens, S. Kempe, and J. E. Richey, pp. 77–104. Chichester, U.K.: John Wiley.

Trimble, S. W. 1975. Denudation studies: Can we assume stream steady state? *Science* 188: 1207–1208.

Trimble, S. W. 1977. The fallacy of stream equilibrium in contemporary denudation studies, *Am. J. Sci.* 277: 876–887.

Trimble, S. W. 1983. A sediment budget for Coon Creek basin in the Driftless Area, Wisconsin, *Am. J. Sci.* 283: 454–474.

Van Der Weijden, C. H., and J. J. Middelburg. 1989. Hydrogeochemistry of the river Rhine: Long term and seasonal variability, elemental budgets, base levels and pollution, *Water Res.* 23: 1247–1266.

Velbel, M. A. 1993. Temperature dependence of silicate weathering in nature: How strong a negative feedback on long-term accumulation of atmospheric CO_2 and global greenhouse warming? *Geology* 21: 1059–1062.

Walling, D. E., and B. W. Webb. 1983. Patterns of sediment yield. In *Background to Paleohydrology,* ed. K. J. Gregory, pp. 69–100. Chichester, U.K.: John Wiley.

White, A. F., and A. E. Blum. 1995. Effects of climate on chemical weathering in watersheds, *Geochim. Cosmochim. Acta* 59: 1729–1747.

Wold, C. N., and W. W. Hay. 1990. Estimating ancient sediment fluxes, *Am. J. Sci.* 290: 1069–1089.

Wollast, R. 1993. Interactions of carbon and nitrogen cycles in the coastal zone. In *Interactions of C, N, P and S Biogeochemical Cycles and Global Change,* ed. R. Wollast, F. T. Mackenzie, and L. Chou, pp. 195–210. Berlin, Heidelberg: Springer-Verlag.

Yuretich, R. F., D. A. Crerar, D. J. J. Kinsman, J. L. Means, and M. P. Borcsik. 1981. Hydrogeochemistry of the New Jersey Coastal Plain, 1: Major element cycles in precipitation and river water, *Chem. Geol.* 33: 1–21.

Zehnder, A. J. B., and S. H. Zinder. 1980. The sulfur cycle. In *The Handbook of Environmental Chemistry,* ed. O. Hutzinger, vol. 1, pt. A, pp. 105–145. New York: Springer-Verlag.

Zhang, J. W., W. Huang, M. G. Lin, and A. Zhon. 1990. Drainage basin weathering and major element transport of two large Chinese rivers (Huanghe and Changjiang), *J. Geophys. Res.* 95: 13277–13288.

Zobrist, J., and W. Stumm. 1980. Chemical dynamics of the Rhine catchment area in Switzerland: Extrapolation to the "pristine" Rhine river input into the ocean. In *River Inputs to Ocean Systems,* SCOR/UNEP/UNESCO Review and Workshop ed. J. M. Martin, J. D. Burton, and D. Eisma, pp. 52–63. Rome: FAO.

Zverev, V. P., and V. Z. Rubeikin. 1973. The role of an atmospheric precipitation in circulation of chemical elements between atmosphere, hydrosphere and lithosphere, *Hydrogeochemistry* 1: 613–620.

CHAPTER 6

LAKES

INTRODUCTION

Although lakes constitute only about 0.01% of the total water at the earth's surface, they have received proportionately greater attention because of their importance to humans. Lakes are used as a source of drinking water, as a receptacle for sewage and agricultural runoff, for recreation, and for industrial purposes. Because of their generally small size, they can be severely altered by these activities; thus, any discussion of lake water chemistry must include the effects of humans. Also, lakes have received so much attention that a whole field devoted to their study, *limnology,* exists, and a large variety of phenomena unique to lakes have been discovered. Our goal in this chapter is to discuss some fundamental aspects of lakes and to show how lake water composition is affected by physical, biological, geological, and anthropomorphic processes. In the process, we hope again to demonstrate the necessity of using a multidisciplinary approach to the study of natural waters. For a more extensive discussion of limnology, the reader should refer to books devoted to the subject, such as those of Hutchinson (1957), Lerman (1978), Wetzel (1983), Stumm (1985), and Lerman et al. (1995).

PHYSICAL PROCESSES IN LAKES

Water Balance

Lakes can be considered in some respects as "little oceans." Like oceans they receive inflow from rivers, exhibit vertical stratification, undergo biological cycling and sedimentation, lose water through evaporation, and so forth. However, most lakes differ from oceans in one important aspect: They have outlets. Water is lost from the oceans only by evaporation (see Chapter 8), but in freshwater lakes the water also leaves via a surface or subsurface outlet. In this way a lake (with a surface outlet) is a sort of "fat" and "slow" portion of a river or, in other words, a portion of a drainage system where water is retained for considerably larger periods than in normal river channels. In some lakes, there is no outlet because of high aridity and interior drainage, and in this case the ocean simile is better. Waters entering such lakes leave only by evaporation and, as in the ocean, high salinities can result. Nevertheless, most lakes consist of fresh water, and this water is fresh because of the presence of outlets.

Table 6.1 summarizes the various inputs and outputs of water to and from lakes. The relative importance of each process depends on the lake. For example, as mentioned above, lakes in arid regions often have no outlet, and water is lost only by evaporation. Inputs can also vary considerably. Lake Victoria in Africa is a shallow but areally extensive lake with minor stream input and is located in a region of high rainfall. As a result, roughly three fourths of the water input to the lake is provided by rainfall on the lake surface (Hutchinson 1957). This is unusually high; For most larger lakes the major input is rivers and streams, with rainfall accounting for only a small percentage of the total. By contrast, many very small lakes (*spring-fed lakes*) are totally supplied by underground springs, and some, known as *karst lakes,* also lose water by underground seepage. Karst lakes develop in karst terrains, or regions of extensive underground dissolution of limestone (see Chapter 4), and because of the permeable nature of the underlying partly dissolved limestone bedrock, water is readily able to seep into and out of the lake floor. Most lakes, however, because of the presence of relatively impermeable clay sediment on their bottoms, do not undergo major underground seepage.

The volume of water in a lake, and consequently the level of its surface, depends on how the inputs and outputs of water balance. If there is a good balance, the water level fluctuates very little. However, because of year-to-year changes in rainfall and other climatic factors, some fluctuation is expected. The degree to which fluctuation affects the lake level depends on the size of the lake and the rate of water addition and removal. For lakes that receive most of their water from streamflow, this can be represented in terms of the time it would take to replace the water in the lake—that is, the time necessary, at the present rate of inflow, to fill the lake to its present volume.

Mathematically the *replacement,* or filling, time for water is defined as

$$\tau_w = \frac{V}{F_i} \tag{6.1}$$

TABLE 6.1 Processes of Water Input and Output in Lakes

Inputs	Outputs
1. Rainfall (and snowfall) on lake surface	1. Evaporation
2. Stream and river flow	2. Outflow at surface via natural outlets such as a stream, waterfall, etc. (usually only one outlet)
3. Spring discharge along lake margin	3. Outflow at surface via man-made conduits, irrigation channels, dams, etc.
4. Groundwater seepage through lake floor	4. Seepage out through lake floor
5. Artificial conduits	

Source: After Hutchinson 1957.

where τ_w = replacement time for water, V = volume of water in the lake, and F_i = rate of streamflow addition.

Low values of τ_w mean that short-term fluctuations in input are rapidly witnessed by lake level changes, whereas high τ_w values mean that the lake level (volume) is relatively impervious to rapid input changes. Not surprisingly, low τ_w values are characteristic of small lakes and ponds and high τ_w values are characteristic of large lakes. Large lakes are able to adjust their level to long-term averages of inputs and outputs. Values of τ_w for most lakes lie between 1 and 100 years.

If a lake maintains a constant volume because inputs and outputs are equal, it is said to be in a *steady state* with respect to its water content. In this case the replacement time can also be viewed as a *residence time* of water in the lake. Residence time is the average time a water molecule spends in the lake before being removed through the outlet. The concept of residence time is used extensively in the literature on both limnology and oceanography, but it is sometimes forgotten that it makes sense only when there is a steady state. Otherwise, more general concepts, such as that of replacement time, are preferable.

Thermal Regimes and Lake Classification

The chemistry of freshwater lakes depends to a large extent on circulation driven by temperature changes. The density of water in lakes is primarily a function of temperature. As temperature changes, density changes, and if less dense water becomes overlain by more dense water, convection, or lake water overturn, takes place. The effect of temperature on water density is shown in Figure 6.1. Note that an unusual situation occurs between 0 and 4°C. With heating,

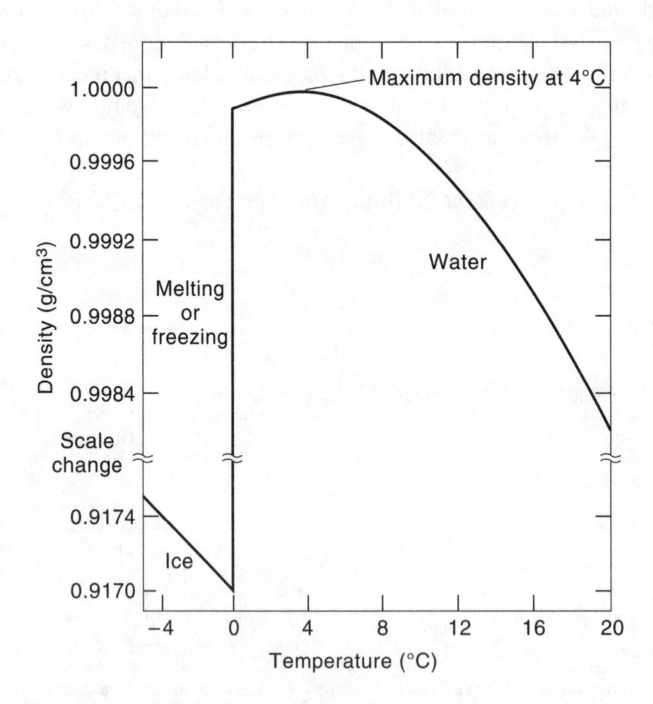

Figure 6.1. Density of water (and ice) as a function of temperature. Note maximum density of water at 4°C. (Data from Pauling 1953 and Hutchinson 1957.)

instead of decreasing in density like most liquids, dilute (fresh) water actually increases in density over this temperature range. At 4°C the water is at a maximum density, and above this temperature the density decreases with temperature like a normal liquid. Further unusual behavior is shown by water ice, which is less dense than cold liquid water and therefore floats on it. This unusual temperature-density behavior of water and ice provides the fundamental basis for classifying freshwater lakes and for describing their overall circulation behavior.

The reason for the strange density behavior of H_2O lies in its structure, which is based on hydrogen bonding (Pauling 1953). Ice, due to complete hydrogen bonding, has an open structure which gives it a low density. When ice melts, some of the hydrogen bonds are broken or bent, and the resulting non-hydrogen-bonded water molecules crowd more closely together. In this way liquid water at low temperatures is denser than ice. Normally, as a liquid is heated, its thermal energy is increased and the molecules bounce around more, occupying more space. In the case of water between 0 and 4°C, this effect is overshadowed by the breaking or bending of hydrogen bonds, and as liquid water is heated from 0 to 4°C its volume contracts as the hydrogen bonds are broken. At 4°C (strictly speaking, 3.94°C), water attains its maximum density; above 4°C it begins to behave like a normal liquid, continuously decreasing in density up to its boiling point at 100°C.

In order to provide a better understanding of the effects of temperature on lake circulation and classification, we shall discuss the effect of seasonal temperature changes on profiles of temperature versus depth for a typical medium-sized lake of temperate regions (for further details, consult Hutchinson 1957). This is shown in Figures 6.2 and 6.3. Let us start the discussion with late summer (Figure 6.2).

During late summer, air temperature is at a maximum and so is the surface water temperature. The depth of heating of the water depends on wind stirring and extends down to where wind

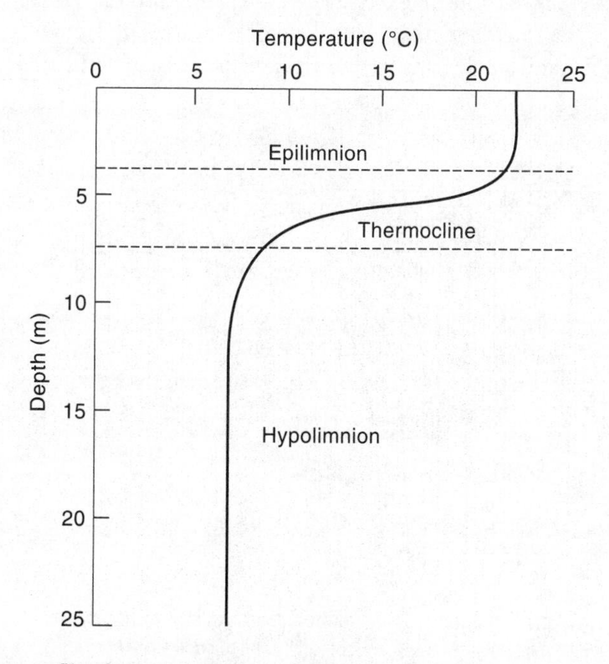

Figure 6.2. Temperature profile of a typical temperate freshwater lake in summer.

effects die out. The shallow, wind-stirred portion of the lake is called the *epilimnion.* Below the epilimnion is a region where temperature falls rather rapidly with depth. This is the *thermocline* or *mesolimnion.* Below the thermocline is the deep, cold portion of the lake known as the *hypolimnion.* Because the temperature of all the lake water is above 4°C, the coldest water is most dense and is located at the bottom. The situation depicted in Figure 6.2 represents stable density stratification, which means that there is little vertical circulation of water and the hypolimnion is effectively isolated from the atmosphere. This situation is characteristic of most of the oceans, but in temperate freshwater lakes it occurs only during the summer.

Let us examine (Figure 6.3) what happens as the air temperature drops during the autumn. As the air cools, the epilimnetic water also cools until its temperature matches that of the hypolimnion. In this case the thermocline disappears and there is a constant temperature from top to bottom. Further cooling of the surface water causes the density distribution to become unstable, with heavier water on top and, as a result, vertical convection arises. The lake mixes or *overturns* from top to bottom, and this overturn is aided by winds, which are generally strong in the autumn. As fall cooling continues, the constant top-to-bottom temperature drops with time (Figure 6.3) until 4°C is reached, at which point new behavior takes over.

Further cooling of surface water below 4°C causes it to become less dense than the deeper water. Stable density stratification develops and the top-to-bottom overturn ceases. Eventually the surface water reaches 0°C and freezes. During the winter, the cold surface water plus ice cap prevent any wind stirring of deeper water and we have a situation known as *winter stratification* (as opposed to summer stratification), where the deep water again becomes isolated from the atmosphere. As spring approaches, the ice melts and the surface water begins to warm up. When the surface water temperature reaches that of the deep water, we again have thermal instability and, as a result, there is a spring overturn. (The temperature of the bottom water can be somewhat less than 4°C due to excessive wind stirring prior to winter stratification and to conductive cooling during the winter.) The water then warms up from top to bottom during further spring warming until stable density stratification ensues and vertical mixing is inhibited. Further

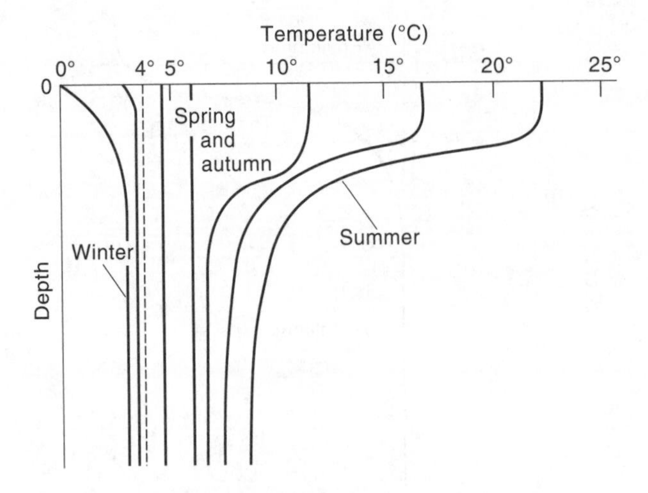

Figure 6.3. Schematic evolution of temperature-versus-depth profiles over the year for a typical dimictic (mixing twice a year) lake of temperate climates. The dashed line represents maximum density at 4°C. (See also Hutchinson 1957 and Wetzel 1983.)

heating on into the summer increases this stable summer-type stratification until our original situation of late summer is reattained and the seasonal cycle is completed.

The situation depicted above and in Figure 6.3, where there are two overturns per year, is typical of temperate lakes, and where this occurs the lakes are referred to as being *dimictic* (mixing twice a year). Other situations can arise in different climates. In warmer (e.g., Mediterranean-type) climates, the mean monthly air temperature does not drop below 4°C in the winter and there is no winter stratification or freezing. As a result, the water overturns throughout the winter until stratification sets in during the spring. The stratification is then maintained through the summer into the fall. Temperature-versus-depth profiles resemble only those to the right of the 4°C dashed line of Figure 6.3, and under these conditions warm, *monomictic* lakes arise, which mix only once a year. Typical examples are those of the Italian Lake District. Another kind of monomictic lake, the cold monomictic, occurs rarely in alpine or high-latitude cold climates. Here the water temperature never exceeds 4°C and, as a result, there is continuous mixing during the summer and ice cover plus winter stratification throughout the rest of the year. Temperature-versus-depth profiles resemble only those to the left of the 4°C dashed line of Figure 6.3.

In tropical regions the air temperature changes little over the year and, as a result, well-defined, cyclic temperature-versus-depth regions in lakes are not present. Overturn in this case depends on a variety of unpredictable factors, and stratification is variable from place to place. Such lakes are referred to as being *oligomictic.*

All density stratification in lakes is not brought about by seasonal temperature changes. In certain lakes that are fed by saline springs or rivers, a situation known as *meromixis* sets in. *Meromictic* lakes never undergo overturn due to the presence of dense, saline water at depth, and these lakes stand in fundamental contrast to all of the lakes discussed above, which overturn at least once a year and are therefore termed *holomictic.*

A summary of the types of lakes described above is shown in terms of the classification presented in Table 6.2. Added to these are two types of lakes that make up special cases because of their unusual morphology. One type is very shallow lakes. Some lakes are so shallow that they are always stirred by the wind and never develop stratification or a hypolimnion.

An opposite situation is provided by very deep lakes. Here, because of their great depth, wind stirring does not extend far enough downward, and little or no heating or cooling of the hypolimnion occurs. In other words, the deep water acts as a sort of thermal buffer which remains

TABLE 6.2 Classification of Freshwater Lakes

I. Holomictic (mixing between epilimnion and hypolimnion)
   A. Dimictic (mixes twice a year)
   B. Monomictic (mixes once a year)
      1. Warm monomictic
      2. Cold monomictic
   C. Oligomictic (mixes irregularly)
   D. Shallow lakes (continuous mixing)
   E. Very deep lakes (mixing only in upper portion of hypolimnion for most deep lakes)
II. Meromictic (no mixing between epilimnion and hypolimnion)

Source: See Hutchinson 1957 for details.

permanently isolated from the atmosphere. An example of this kind of lake is Lake Tanganyika, a rift lake in East Africa (and the second largest lake in the world). Lake Tanganyika has a maximum depth of 1410 m, has exhibited permanent thermal stratification with a constant deep-water temperature of 23.4°C for at least the last 50 years, and contains no dissolved oxygen below 150 m (see discussion below). There is cooling and vertical mixing of water into the upper oxic part of the thermocline during the trade-wind season, but there is no evidence for deeper mixing (Edmond et al. 1993). The lowest temperature reached by the surface waters is 23.9°C and the highest is 27 to 28°C

Some very deep temperate lakes do mix to the bottom but not annually or semiannually. Lake Baikal in eastern Siberia is the world's deepest lake (1623 m), but despite its great depth, it has a mean deep-water residence time of only about 8 years (Weiss et al. 1991). Mixing results because in very deep freshwater lakes the temperature of maximum density *decreases* with depth because of the large increase in pressure (see Figure 6.4). When the surface water is both colder than the deep water and below 4°C and the deep water just below the interface between the two water masses is colder than the temperature of maximum density for its depth, then stable stratification results (Figure 6.4a). But as the water becomes stirred downward by the wind, a point is reached (around 250 m depth in Lake Baikal) where the interface crosses the curve of maximum density versus depth. At this point the overlying colder water becomes unstable, and further deepening of the interface results in mixing all the way to the bottom (Figure 6.4b).

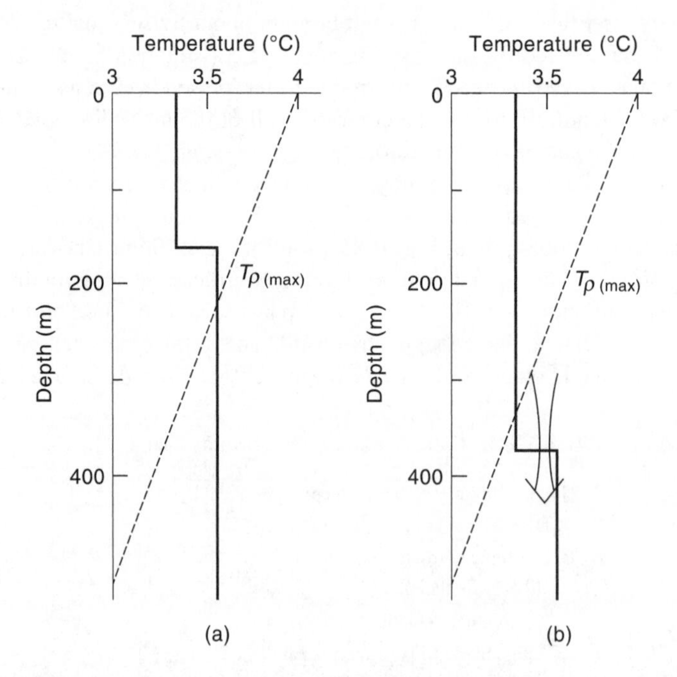

Figure 6.4. Mixing in very deep freshwater lakes. Dashed line shows how the temperature of maximum density ($T_{\rho max}$) increases with depth. (a) Stable stratification, where the interface between the surface and deep-water masses is sufficiently shallow that temperatures of both water masses are less than the temperature of maximum density. (b) Unstable stratification, where wind stirring has forced the surface layer to sufficiently great depths that it has become denser than the deeper layer and convection to the bottom results. (After Weiss et al. 1991, Fig. 2a & c.)

Another example of a deep lake that undergoes irregular mixing to the bottom is Lake Tahoe in the California-Nevada Sierra mountains. Lake Tahoe has a maximum depth of 505 m, and in the coldest winters the whole water columnn approaches 4°C. The lake does not mix to the bottom every year, but only when there are severe winter storms, which occur every several years (Goldman 1988).

Throughout this discussion of lakes we have mentioned that, during stratification, deep water is isolated from the atmosphere. This is especially important because isolation from the atmosphere, accompanied by biological oxygen consumption in the deep water, results in a lowering of dissolved O_2. If the isolation is prolonged, all O_2 may be consumed and, as a result, the deep water becomes anoxic, resulting in dramatic changes in water quality. These changes are important and will be discussed later in the present chapter. However, it is important to remember that it is the physical process of stratification that initiates these chemical changes.

LAKE MODELS

Like water, other substances are added to and removed from lakes. Rivers and groundwater carry dissolved constituents into a lake, where they may undergo chemical reactions, and this is followed by removal via water flow through an outlet. One way of quantitatively treating rates of addition and removal is by means of *box modelling*. In box modelling one assumes that a portion of a lake or the whole lake is so well stirred that it is homogeneous in composition and can be treated as a uniform "box." Rates of addition (or removal) to each box are slow enough, relative to mixing within the boxes, that high concentrations of added substances do not build up around each source. [This situation is not always obeyed; see Imboden and Lerman (1978).] The concentration of a given substance in a given "box" is controlled by the relative magnitude of inputs and outputs. If inputs balance outputs, there is a steady state and concentrations do not change with time. This is analogous to (but not necessarily connected with) the situation of steady-state water content discussed earlier.

The simplest kind of box model is that of a single box representing a whole lake. In this case we have input of a dissolved substance from streams (groundwater and rainwater inputs are neglected), output by a surface outlet, precipitation and removal to bottom sediments, and addition via dissolution, or bacterial regeneration, of suspended and sedimented solids (see Figure 6.5). Rates of these processes can be represented as

F_i = rate of water inflow from streams (volume per unit time)

F_o = rate of water outflow through outlet

M = total mass of dissolved substance in the lake

R_p = rate of removal via precipitation and sedimentation (mass per unit time)

R_d = rate of addition via dissolution of solids during sedimentation or while solids rest on the bottom (mass per unit time)

C_i = concentration of dissolved substance in stream water (mass per unit volume)

C = concentration in lake water

t = time

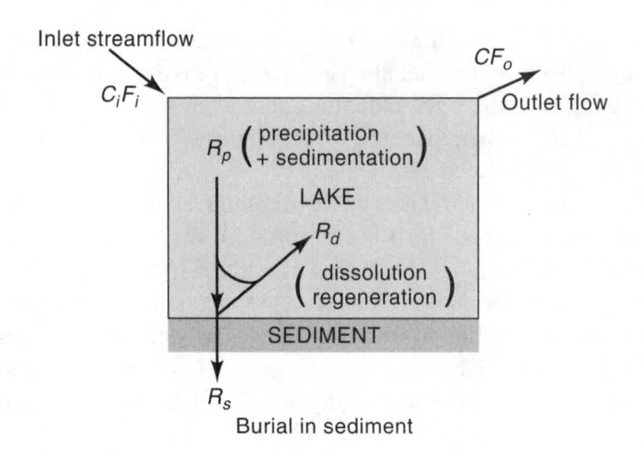

Figure 6.5. One-box model for lakes. (For explanation of symbols, see text.)

The rate of change of mass with time in the lake, $\Delta M/\Delta t$, is

$$\Delta M/\Delta t = C_iF_i - CF_o + R_d - R_p \tag{6.2}$$

If there is a steady state with respect to the dissolved substance, $\Delta M/\Delta t = 0$, then

$$C_iF_i - CF_o + R_d - R_p = 0 \tag{6.3}$$

Finally, if dissolution represents redissolution of the same material that was previously precipitated and sedimented toward the bottom, then

$$R_s = R_p - R_d \tag{6.4}$$

where R_s = rate of burial in sediments (mass per unit time).

Using Eq. (6.3), one can calculate, for example, from a knowledge of measured water flow and sedimentation rates, the maximum allowable input concentration (C_i) of a pollutant, P, to a lake, if the lake concentration of the pollutant (C) is not to exceed a certain level. Suppose the streamwater inflow rate (F_i) to the lake is equal to 100 m³/s (appropriate for a small river) and rainfall (minus evaporation) directly on the lake (averaged over the year) is 50 m³/s. Then the outflow rate (F_o) should be 150 m³/s in order to maintain constant lake volume. Let the sediment burial rate (sedimentation minus dissolution) (R_s) be 250 mg P/s and assume that the lake concentration (C) may not exceed 5 μg P/l (which equals 5 mg P/m³ and is sufficiently low that it should not promote eutrophication; see discussion below). Then, upon substituting in Eqs. (6.3) and (6.4),

$$\begin{aligned} C_i &= \frac{CF_o + R_s}{F_i} \\ &= \frac{(5 \text{ mg P/m}^3) \times (150 \text{ m}^3\text{/sec}) + (250 \text{ mg P/sec})}{100 \text{ m}^3\text{/sec}} \end{aligned}$$

or

$$C_i = 10\ \mu g\ P/l$$

Two useful concepts, already applied to water itself, are those of replacement time and residence time. Replacement time is the hypothetical time necessary to replace the mass of a dissolved substance, via the present rate of stream addition, if all of the substance were suddenly removed. It gives a measure of the sensitivity of lake concentration C to changes in input concentration, C_i, or water inflow, F_i. (However, it should not be confused with the actual time necessary to change the concentration in a lake by changing the input, which is a more complicated situation and which normally requires the use of calculus). The replacement time of a dissolved substance is defined as

$$\tau_r = \frac{\text{mass in lake}}{\text{rate of stream input to lake}} = \frac{M}{C_i F_i} \quad (6.5)$$

Recalling from Eq. (6.1) that the replacement time for water is

$$\tau_w = \frac{V}{F_i}$$

(V = volume of lake) and that $M = CV$, then Eq. (6.5) can be rewritten as

$$\tau_r = \left(\frac{C}{C_i}\right) \tau_w \quad (6.6)$$

If the lake is at a steady state with respect to the dissolved substance of interest (and water), then τ_r (and τ_w) can be viewed as residence times as well as replacement times. In other words, for steady state the value τ_r represents the average time spent by a dissolved species in the lake prior to removal either via sedimentation or through the outlet.

An additional concept, that of relative residence time, (Stumm and Morgan 1981) is very useful. Relative residence time is the residence time of a given dissolved substance relative to that of water,

$$\tau_{rel} = \frac{\tau_r}{\tau_w} \quad (6.7)$$

or, from Eq. (6.6),

$$\tau_{rel} = \frac{C}{C_i} \quad (6.8)$$

Relative residence time is an indication of the type of behavior to be expected for a given substance. A relative residence time of one indicates that the substance does not react chemically in the lake ($R_d = 0$; $R_p = 0$), and it simply accompanies water as it passes through the lake. In this case the substance acts as a tracer of water motion (dissolved Cl^- is a good example). If τ_{rel} is less than one, the substance tends to undergo removal via sedimentation in the lake ($R_p > 0$), indicating its chemical reactivity. (An example is dissolved Al.) If τ_{rel} is greater than one, the substance tends to be trapped in the lake while the water that brought it in is removed. This can take place if the substance is cycled within the lake, that is, it is precipitated and sedimented to the bottom ($R_p > 0$), then redissolved ($R_d > 0$), then reprecipitated, and so on. This is characteristic of elements involved in biological processes, for example, P, N, Si, and Ca, and such biological cycling within the lake can result in a relative residence time of each of these elements appreciably greater than 1.

Simple one-box models, although applicable to all lakes to express their *average* properties, are most accurate as representations of shallow lakes that do not undergo stratification. For the more usual case of stratified lakes, a two-box model is more appropriate (e.g., Imboden and Lerman 1978; Stumm and Morgan 1981). One box is used to represent the epilimnion and the other box the hypolimnion. This is shown in Figure 6.6. In the two-box model we have *fluxes* between the reservoirs (boxes) as well as inputs and outputs for the whole lake. Figure 6.6 represents the situation expected for a biological element such as phosphorus or nitrogen. There is input of dissolved material by streams to the epilimnion and output via an outlet. There is exchange of water containing the dissolved substance of interest between hypolimnion and epilimnion, which is represented by up and down (short) arrows. (Actual exchange occurs sporadically during seasonal overturn, but for modelling this is averaged over a year.) Finally, there are chemical reactions; in this case these include removal of the substances from the epilimnion via precipitation and transfer downward by sedimentation, injection of a portion to solution in the hypolimnion, and burial of the remainder.

Mathematical representation of the rates in a two-box lake model is similar to that presented above for a one-box lake. Besides the parameters defined for the one-box model we also have the following:

F_U = rate of water transfer from hypolimnion to epilimnion

F_D = rate of water transfer from epilimnion to hypolimnion

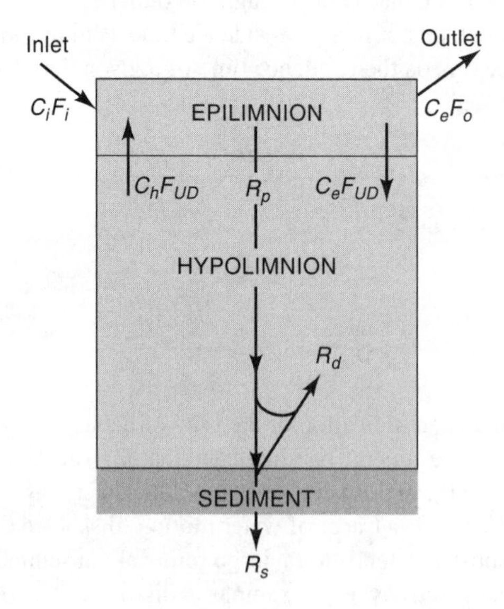

Figure 6.6. Two-box model for lakes (see text).

R_p = rate of removal, by precipitation and sedimentation, from epilimnion

R_d = rate of addition, via dissolution, to hypolimnion

M_e = mass dissolved in epilimnion

C_e = concentration in epilimnion (mass per unit volume)

M_h = mass dissolved in hypolimnion

C_h = concentration in hypolimnion (mass per unit volume)

If we have steady state with respect to water for both boxes (constant volumes of the epilimnion and hypolimnion, V_e and V_h, with time), then

$$F_U = F_D$$

and we can refer to either rate as F_{UD} (see Figure 6.6).

Using the above definition, and referring to Figure 6.6, we obtain the rates of change of mass in solution for each box $\Delta M_e/\Delta t$ and $\Delta M_h/\Delta t$, by summing inputs and outputs:

$$\frac{\Delta M_e}{\Delta t} = C_iF_i - C_eF_o + (C_h - C_e)F_{UD} - R_p \quad (6.9)$$

$$\frac{\Delta M_h}{\Delta t} = R_d - (C_h - C_e)F_{UD} \quad (6.10)$$

If in addition, we have steady state with respect to both the dissolved substance and its solid precipitated form, and dissolution represents redissolution of sedimenting solids, we have

$$C_iF_i - C_eF_o + (C_h - C_e)F_{UD} - R_p = 0 \quad (6.11)$$

$$R_d - (C_h - C_e)F_{UD} = 0 \quad (6.12)$$

$$R_s = R_p - R_d \quad (6.13)$$

From these equations we can thus gain information on rates of processes from measurements of other rates and concentrations and, in this way, the equations are most useful.

BIOLOGICAL PROCESSES IN LAKES AS THEY AFFECT WATER COMPOSITION

Photosynthesis, Respiration, and Biological Cycling

One of the main reasons why hypolimnia and epilimnia differ in chemical composition is that each is greatly, but differently, affected by biological processes. The starting point is photosynthesis.

In photosynthesis, CO_2 and H_2O are converted to organic matter by green plants, which use the energy of sunlight for this process. Organic compounds synthesized by the plants contain not only carbon, hydrogen, and oxygen, but also other essential nutrient elements, chief of which are nitrogen and phosphorus (see Table 4.1). The overall process of photosynthesis in lakes (and the ocean) can be represented by the following reaction (e.g., Stumm and Morgan 1981):

$$106CO_2 + 16NO_3^- + HPO_4^{2-} + 122H_2O + 18H^+ + \text{trace elements and energy}$$
$$\rightarrow C_{106}H_{263}O_{110}N_{16}P_1 + 138O_2 \qquad (6.14)$$

(Here the stoichiometric elemental composition of average marine plankton is used simply for the purposes of illustration.) Photosynthesis thus involves the uptake of dissolved CO_2, phosphate, and nitrate, and the production of O_2. (Nitrogen and phosphorus can be also taken up from other dissolved forms of these elements).

Hand in hand with photosynthesis is the process of aerobic respiration. Respiration occurs in all organisms, including plants, animals, and bacteria, and, in the presence of O_2, can be considered as the above reaction written backward. In other words, aerobic respiration is essentially the reverse of photosynthesis and involves the breakdown of organic matter, whereas photosynthesis involves its formation. Although in natural waters the rates of photosynthesis and respiration are closely matched, there is almost always a slight excess of photosynthesis. In lakes this excess in surface waters manifests itself as a downward flux of dead organic matter. Photosynthesis can occur only in water that is sufficiently shallow so that light, necessary for the process, can penetrate it. At greater depths light is absent. In sufficiently deep (or murky) lakes, respiration dominates over photosynthesis at depth, and this helps to destroy much of the dead organic matter sedimenting from above. In other words, in sufficiently deep lakes the processes of photosynthesis and respiration are in part separated from one another, with net photosynthesis at the surface and net respiration at depth. If the lake is also stratified, this separation can lead to dramatic changes in water composition.

Consider a stratified lake with photosynthesis confined to the epilimnion (reference here to the two-box-model diagram, Figure 6.6, is helpful). The nutrients phosphorus and nitrogen are removed from the epilimnetic water to form organic matter; some of this organic matter is eaten and respired by zooplankton, fish, and other organisms; some of it is decomposed by bacteria; and the remainder falls into the hypolimnion. Here it is further decomposed, especially by bacteria living in the water and on the lake bottom, and phosphorus and nitrogen are liberated to solution. Some remaining organic matter that escapes decomposition is buried in bottom sediments. The P and N liberated to solution in the hypolimnion is not reused by plants there because the water is too deep for photosynthesis. However, the P and N are eventually, but slowly on the average, returned to the epilimnion by occasional overturn. By these processes, concentrations of P and N in the hypolimnion, where there is net respiration, build up to much higher concentrations than exist in the epilimnion, where there is net photosynthesis, and continued biological cycling maintains these concentration differences. Only during overturn is there homogenization of concentrations.

Besides P and N, other elements undergo a differentiation of concentration due to biological cycling. Dissolved O_2 neither builds up nor is depleted in the epilimnion because it readily exchanges with the atmosphere. By contrast, in the hypolimnion there is no contact with the atmosphere and any O_2 lost by respiration is not immediately replaced. An example of O_2

depletion (and P enrichment) in the hypolimnion is shown in Figure 6.7. As a result, the concentration of O_2 in the hypolimnion is lowered and this lowering, depending on the rate of addition of dead organic matter, can be all the way to zero. Once all dissolved oxygen is removed, new anoxic, bacterially mediated chemical reactions can take place. This includes the reduction of collodial ferric and Mn^{4+} oxides to Fe^{2+} and Mn^{2+} in solution, the reduction of SO_4^{2-} to H_2S, and the production of methane, all of which accompany the anoxic decomposition of organic matter. Some examples of chemical changes brought about by anoxia are shown in Table 6.3. [For a further discussion of anoxic bacterial processes, see Chapter 8 and consult Claypool and Kaplan (1974); Fenchel and Blackburn (1979); and Berner (1980).] Also, the production of anoxia can have catastrophic effects on higher organisms dwelling in the deep water or on the bottom, as will be discussed later under eutrophication.

Accompanying changes brought about by organic matter itself are changes involving mineral matter synthesized by photosynthetic organisms. Opaline silica (not quartz, but a different, amorphous form of SiO_2 found in skeletal material) is secreted by microscopic floating organisms, chiefly diatoms. Calcium carbonate is secreted by some algae. Together these substances fall into deep water, where they ordinarily undergo dissolution. This dissolution, along with secretion in the epilimnion, leads to concentration differences in dissolved H_4SiO_4 and Ca^{2+} and HCO_3^-, between the epilimnion and hypolimnion. In addition, the nonbiogenic precipitation of $CaCO_3$ can also occur in the epilimnion. Dissolution of $CaCO_3$ in the hypolimnion is due primarily to a higher acidity there, arising from production of excess carbonic acid from respiratory CO_2. The reactions are

$$C_{organic} + O_2 \rightarrow CO_2$$

$$CO_2 + H_2O \rightarrow H_2CO_3$$

$$H_2CO_3 + CaCO_3 \rightarrow Ca^{2+} + 2HCO_3^-$$

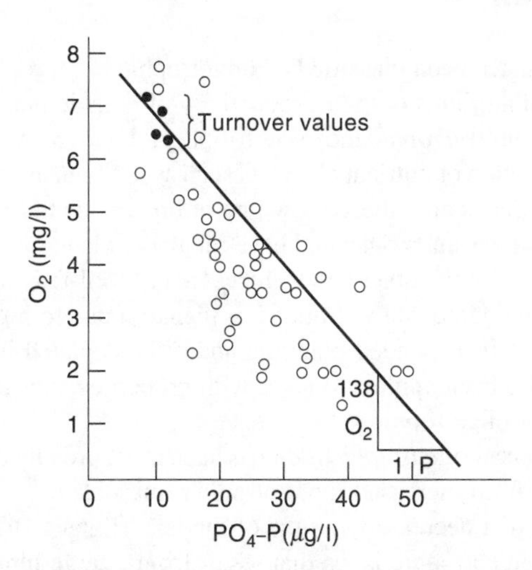

Figure 6.7. The correlation between measured concentrations of dissolved O_2 and PO_4-P in deep lake water (Lake Gersau, Switzerland). The line is for a ratio of 138 O_2:1 P, which is that expected for control by respiration [see Eq. (6.14)]. (Molar and weight ratios of O_2:P are the same because 1 mol of O_2 has nearly the same weight as 1 mol of P.) (From Stumm and Baccini, 1978, after H. Ambühl, 1975, *Swiss Journal of Hydrology* 37: 35–52, published by Birkhhuser Verlag AG, Basel, Switzerland.)

TABLE 6.3 Some Chemical Changes Brought About by Anoxic Condition in Natural Waters

Bacterial nitrate reduction (denitrification)

$$5CH_2O + 4NO_3^- \rightarrow 2N_2 + 4HCO_3^- + CO_2 + 3H_2O$$

Bacterial sulfate reduction

$$2CH_2O + SO_4^{--} \rightarrow H_2S + 2HCO_3^-$$

Bacterial methane formation

$$2CH_2O \rightarrow CO_2 + CH_4$$

Iron reduction

$$CH_2O + 7CO_2 + 4Fe(OH)_s \rightarrow 4Fe^{++} + 8HCO_3^- + 3H_2O$$

Manganese reduction

$$CH_2O + 3CO_2 + H_2O + 2MnO_2 \rightarrow 2Mn^{++} + 4HCO_3^-$$

Ferrous sulfide precipitation

$$Fe^{++} + H_2S \rightarrow FeS + 2H^+$$

Manganese and ferrous carbonate precipitation

$$Mn^{++} + 2HCO_3^- \rightarrow MnCO_3 + CO_2 + H_2O$$
$$Fe^{++} + 2HCO_3^- \rightarrow FeCO_3 + CO_2 + H_2O$$

Ferrous phosphate precipitation

$$8H_2O + 3Fe^{++} + 2PO_4^{---} \rightarrow Fe_3(PO_4)_2 \cdot 8H_2O$$

Note: CH_2O represents decomposing organic matter. (See also Table 8.4).

In the case of opaline silica, dissolution occurs because the water is undersaturated with silica throughout the lake. (Formation of opaline silica in shallow water occurs only because of the input of photosynthetic solar energy.)

Eutrophication

Historically, lakes have been classified as oligotrophic or eutrophic based on either their concentrations of plant nutrients or their productivity of organic matter (Hutchinson 1973; Vallentyne 1974; Rodhe 1969). *Trophic* means nutrition, and *oligotrophic* lakes are poorly fed, that is, have a low concentration of nutrient elements such as nitrogen and phosphorus. The lack of nutrients results in few plants and, thus, a low rate of organic matter production by photosynthesis. Oligotrophic lakes are usually deep and have relatively plankton-free clear water, which is well oxygenated at depth. On the other hand, *eutrophic* or "well-fed" lakes have high concentrations of plant nutrients and large concentrations of plankton due to high organic productivity. Their waters are murky with suspended plankton and often depleted in oxygen at depth. (The term *mesotrophic* has also been applied to lakes with properties intermediate between those that are termed eutrophic or oligotrophic.)

Recently, the process of *eutrophication* has been more broadly defined as high biological productivity resulting from increased input of either nutrients or of organic matter, with the ultimate development of a decreased volume of the lake (Likens 1972). Therefore, this definition also includes certain eutrophic lakes that result from a large input of organic matter from the surrounding area as opposed to internally produced organic matter. Natural eutrophication occurs as lakes gradually fill in with organic-rich sediments over a long period of time and eventually become swamps and then disappear. However, humans have greatly accelerated the process by artificially enriching lakes with too many nutrients and/or with excess organic matter; this has been called *cultural eutrophication* (Hasler 1947). The characteristics of eutrophic and oligotrophic lakes are summarized in Table 6.4. Mesotrophic lakes are between eutrophic and oligotrophic in characteristics.

The process of natural eutrophication (see Figure 6.8) can be described in terms of a typical lake in northern North America that was formed after the glacial retreat 10,000–15,000 years ago (Hutchinson 1973; Vallentyne 1974). The original oligotrophic lake is clear, with a clean bottom, and has a small population of phytoplankton (minute aquatic plants), zooplankton (small aquatic animals), and perhaps fish. When the organisms die, their remains, along with organic debris from the area surrounding the lake, settle to the bottom into the hypolimnion. Initially there is considerable erosion of glacial till into the lake, which rapidly buries the sedimenting organic matter and prevents return of its nutrients to the lake water. Later, after this first phase, a rather steady supply of nutrients from soil and organic remains is carried into the lake, supporting a rather constant biological productivity over the year. For a typical dimictic lake, when the lake overturns in the spring and fall, nutrients are carried to the surface, where they stimulate phytoplankton growth. The lake water becomes stratified in summer and dead plankton and organic debris accumulate and decompose on the bottom, consuming oxygen from the hypolimnion in the process and releasing nutrients. Gradually, a layer of organic-rich sediment builds up (10–15 m since glacial time; Vallentyne 1974). If the lake was originally deep, the hypolimnion has a large volume and can supply adequate oxygen for organic decomposition. In this case the lake tends to remain oligotrophic.

If, however, the lake was originally shallow, as it fills up the hypolimnion becomes too small to contain sufficient oxygen to counteract organic matter decomposition and, as a result, the

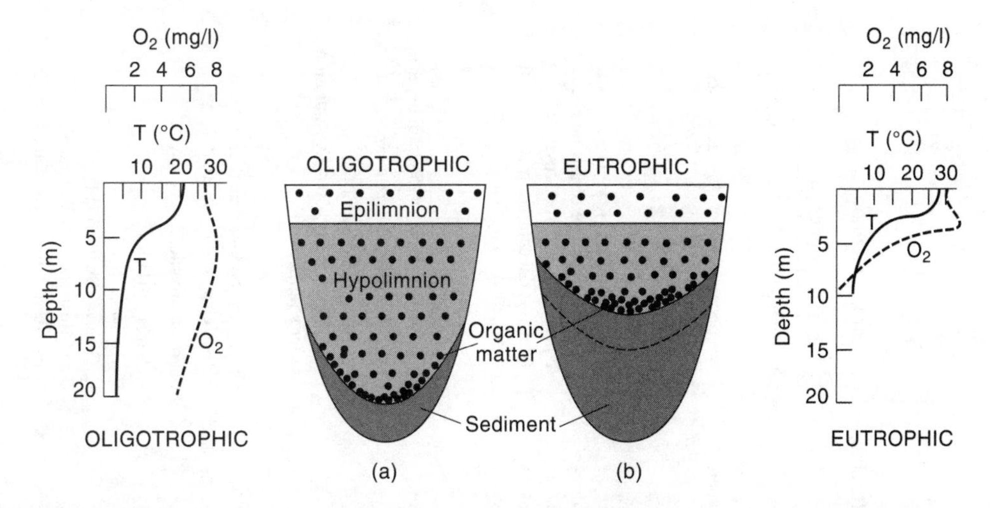

Figure 6.8. Comparison (in late summer) between an oligotrophic lake (a) and a naturally eutrophic lake (b) with moderate nutrient supply, showing organic matter failing through the epilimnion and hypolimnion, and organic matter and sediment accumulation at depth. (a) The oligotrophic lake has a large oxygenated hypolimnion and little drop in oxygen with depth (see graph). (b) The eutrophic lake is either a sediment-filled late stage of a deep lake (solid-line lake bottom) or a shallow contemporary lake (broken-line lake bottom). Breakdown of organic matter in the small hypolimnion has depleted the oxygen supply (see graph). (After G. E. Hutchinson. "Eutrophication," *Am. Sci.* 61: 270. © 1973 by American Scientist, reprinted by permission of the publisher.)

TABLE 6.4 Oligotrophic versus Eutrophic Lakes

Criteria	Oligotrophic	→	(Mesotrophic)	→	Eutrophic
1 Concentration of plant nutrients (N, P)	"Poorly fed"; low concentration of plant nutrients; $P<10 \mu g/l$[a]				"Well fed"; high concentration of plant nutrients; $P>20 \mu g/l$[a]
2. Organic matter enrichment	Low concentration of organic matter				High concentration of organic matter, either authochtonous (produced in lake) or allochthonous (transported from environment)
3. Biological productivity (organic matter productivity) in g org. $C/m^2/yr$; generally phytoplankton productivity via photosynthesis	Low productivity—few plants (phytoplankton) because of low nutrient concs. Primary productivity <150 g $C/m^2/yr$; concentration chlorophyll A (phytoplankton) <3 μg/l[a]				High productivity—large concentration of plants, particularly plankton (green and blue algae) and/or macrophytes (rooted plants); primary productivity >250 g $C/m^2/yr$; concentration chlorophyll A (phytoplankton) >6 μg/l[a]
4. Depth (volume) of lake	Deep, large volume (>15–25 m depth)				Shallow, small volume (≤10–15 m depth)
5 Water	Clear, blue; Secchi depth >5m[a]				Murky, dark turbid; Secchi depth <3m[a]
6. O_2 in hypolimnion in summer	Well-oxygenated hypolimnion (see Figure 6.6); >50% O_2 saturation after stratification[a]				Oxygen depleted in hypolimnion by high biological oxygen demand to decompose organic matter (see Figure 6.6); <10% O_2 saturation (~ <1mg/l O_2) after stratification[a]
7. Bottom fauna	Diversified bottom fauna, deep-water fish; char, whitefish, trout				Bottom fauna tolerant of low oxygen conditions, no deep-water fish; carp
8. Bottom sediment	Sandy, inorganic, low N				Organic rich muck, high in nitrogen
9. Examples	L. Superior (U.S.) L. Huron (U.S.) L. Geneva (Switzerland) Great Bear Lake (N.W. Canada) Great Slave Lake (N.W. Canada)				Western L. Erie (U.S.), L. Lugano (Switzerland-Italy)

[a] Rounded values from Chapra and Dobson 1981, primarily for offshore Great Lakes.
Sources: Chapra and Dobson 1981, where indicated; for other sources , see text.

bottom becomes depleted in oxygen in the summer. In addition, the organic-rich sediments are continually anoxic. The anoxic conditions enhance the solubilization of phosphate, thereby accelerating eutrophication. The extent to which solubilized nutrients can be resupplied to surface waters in such a lake is increased because of the greater bottom area relative to lake volume and better water circulation (the hypolimnion disappears in very shallow lakes).

A shallow lake is likely to become moderately eutrophic, and it may remain this way for thousands of years in the typical case of undisturbed natural eutrophication (see Figure 6.8; Hutchinson 1973). The biological productivity in such a lake is perhaps 10 times that in an oligotrophic lake but not nearly as excessive as that produced by cultural eutrophication. Ultimately, a very shallow, naturally eutrophic lake will fill up with enough sediments so that rooted plants can grow on the bottom, and the depletion of bottom oxygen will be so great that most fish cannot live there. The lake finally becomes a marsh or bog. Humans have often speeded up this "natural" moderate eutrophication by deforestation and cultivation of the land, which produces greater flow of water into the lake, carrying more sediment, nutrients, and organic matter. This accelerates both sedimentation and biological productivity in the lake.

Cultural eutrophication of a lake is vastly speeded-up eutrophication due to the accelerated input to the lake of nutrients and organic matter from sewage, agriculture, and industries. This results in greatly increased biological productivity. A naturally eutrophic lake might produce

Figure 6.9. Eutrophication in late summer in a small, shallow, North Haven, Connecticut, pond. The photographs show clear water in early spring, and murky, bright yellow-green, opaque water in late summer. (Tennis balls are for scale.)

75–250 g C/m²/yr, while a culturally eutrophic lake can support as much as 700 g C/m²/yr (Rodhe 1969). Mats of blue-green algae are typical of eutrophic lakes. Although many naturally eutrophic lakes support fish, in a culturally eutrophied lake the summer oxygen supply is often inadequate for most fish and bottom fauna, except those that are very tolerant of oxygen deprivation. The time scale of cultural eutrophication is greatly shortened and, in this way, the lake "dies." However, the process is reversible if the nutrient supply is reduced, whereas natural eutrophication is slow and irreversible since it involves filling of the lake basin with sediments. An example of a lake with cultural eutrophication in summer can be seen in Figure 6.9.

Attempts have been made to quantify the symptoms of cultural eutrophication in terms of surface water quality. The range of values shown in Table 6.4 are rounded numbers given by Chapra and Dobson (1981) for the offshore waters of the Great Lakes. The values used by other workers for the Great Lakes (Vollenweider et al. 1974) and other lakes (Wetzel 1983) are somewhat different, so rigid boundaries do not exist for all lakes.

Chapra and Dobson (1981) point out that the degree of depletion of dissolved oxygen in the hypolimnion during (summer) stratification should also be considered when characterizing the trophic state of a lake. Because 10% oxygen saturation (~1 mg/l dissolved O_2) is a threshhold for various biological and chemical processes, a lake is defined by them as being eutrophic when its hypolimnetic O_2 falls below this value during summer stratification. (However, cold-water fish require more than 50% O_2 saturation; Beeton 1965). Oligotrophy begins at or above 50% O_2 saturation (~6 mg/l dissolved O_2). Dissolved O_2 tends to be a problem in lakes with a thin hypolimnion, that is, shallow lakes which become stratified and have limited oxygen reserves. An example is shallow central Lake Erie (18 m), whose hypolimnion has a severe oxygen-depletion problem during summer stratification (eutrophic on the above O_2 scale) but whose surface water exhibits other characteristics (productivity, etc.) that place this portion of the lake in the mesotrophic category. Thus, the same area of a lake can be placed in different classifications depending on which criteria for trophic state are used.

Limiting Nutrients

Organic compounds produced by algal photosynthesis in lakes can be represented as consisting of carbon, hydrogen, oxygen, nitrogen, and phosphorus (see preceding section). Of the nutrient elements needed for photosynthesis, hydrogen and oxygen are readily available. Also carbon is generally available from atmospheric carbon dioxide. The major elements that are not always available are nitrogen and phosphorus. Relatively small amounts of N and P can produce relatively large amounts of organic matter. For instance, to synthesize 100 g (dry weight) of algae, only 7 g of N and 1 g of P [see Eq. (6.14)] are required. The total amount of organic matter produced will be determined by the nutrient element that is least available. This is the so-called *limiting nutrient.* A nutrient is limiting when there is an increase in organic production upon its addition, whereas there is no response to the addition of a nonlimiting nutrient.

Vallentyne (1974) has calculated the ratio of *demand* (amount required) by freshwater plants (algae, diatoms, rooted plants) to *supply* from average river water for various nutrients in a world average lake (in late winter before the spring algal growth bloom). This is shown in Table 6.5 for late winter. Those elements whose ratio of demand to supply is greater than 1500 for late winter include P, N, C, and Si. In midsummer, when the demand by organisms is greatest, the ratio of demand to supply for these elements is up to 100 times greater for P and N and 20% higher for C. Since the ratio of demand to supply for P is the greatest (80,000), this explains

TABLE 6.5 Concentrations of Essential Elements for Plant Growth in Living Tissues of Freshwater Plants (Demand), in Mean World River Water (Supply), and Plant/Water (Demand/Supply) Ratio of Concentrations

Element	Symbol	Demanded by Plants (%)	Supplied by Water (%)	Demand/Supply (Plant/Water) Ratio (approx.)
Oxygen	O	80.5	89	1
Hydrogen	H	9.7	11	1
Carbon[a]	C	6.5	0.0012	5,000
Silicon	Si	1.3	0.00065	2,000
Nitrogen[a]	N	0.7	0.000023	30,000
Calcium	Ca	0.4	0.0015	<1,000
Potassium	K	0.3	0.00023	1,300
Phosphorus[a]	P	0.08	0.000001	80,000
Magnesium	Mg	0.07	0.0004	<1,000
Sulfur	S	0.06	0.0004	<1,000
Chlorine	Cl	0.06	0.0008	<1,000
Soldium	Na	0.04	0.0006	<1,000
Iron	Fe	o.02	0.00007	<1,000
Boron	B	0.001	0.00001	<1,000
Mangenese	Mn	0.0007	0.0000015	<1,000
Zinc	Zn	0.0003	0.000001	<1,000
Copper	Cu	0.0001	0.000001	<1,000
Molybdenum	Mo	0.00005	0.0000003	<1,000
Cobalt	Co	0.000002	0.000000005	<1,000

[a] Concentrations in water for inorganic forms only.

Source: J. R. Vallentyne, *The Algal Bowl: Lakes and Man.* Copyright © 1974. Environment Canada. Reprinted by permission of the publisher.

why phosphorus is most often the limiting nutrient in lakes, followed by nitrogen (30,000). This is also why the addition of these nutrients can lead to greatly expanded biological productivity and eutrophication.

Schindler (1974, 1977) has studied the effects of adding various nutrients (phosphorus, nitrogen, and carbon) to experimental lakes in a remote area of Canada. He found that the standing crop of phytoplankton is proportional to the concentration of total phosphorus in most lakes and that phosphorus is the limiting nutrient even when the nutrient ratios of lake inputs (streams, rainfall, etc.) might be expected to favor nitrogen or carbon limitation. This is because biological mechanisms exist in lakes which can correct for carbon deficiencies and, in some cases, nitrogen deficiencies. Carbon dioxide from the atmosphere makes up for deficiencies in other carbon inputs to the lake. Nitrogen deficiencies (low N/P ratios) can be made up by development in the lake of blue-green algae which are capable of fixing N_2 from the atmosphere. In fact, 20%–40% of the total nitrogen input in the experimental lakes is from nitrogen fixation.

The importance of blue-green algae varies. In a study of a large number of lakes, Smith (1983) found that blue-green algae are abundant only when the ratio of total N/total P is less than 2.9.

Blue-green algae also make objectionable floating mats, but in lakes with adequate nitrogen they are supplanted by green algae, which do not form such mats.

Since phosphorus does not occur as a gas in the atmosphere, a lake has no way of compensating for phosphorus deficiencies, and phosphorus thus becomes the limiting nutrient. Schindler (1977) suggests that only when sudden or very large increases of phosphorus input occur during cultural eutrophication do lakes show temporary carbon or nitrogen limitation, which will ultimately be corrected by biological and environmental mechanisms. This results in phytoplankton growth being proportional to phosphorus concentration. In the water of the lakes he studied, the molar ratios of C/P (450) and N/P (68) were more than four times that required for the formation of average (marine) plankton [C/P = 106 and N/P = 16—the so-called Redfield ratio—see Eq. (6.14)], showing that carbon and nitrogen were present to excess. Also, plots of N versus P for many other lakes indicate that P is exhausted before N.

Hecky et al. (1993) measured the C:N:P ratios for particulate matter from small and large freshwater lakes from the arctic to the tropics and concluded that ratios of C:N and N:P were higher and more variable than the marine ratio (which tends to be quite uniform). The median molar N:P ratio was 19:1, with ratios > 22 assumed to be indicative of severe P deficiency. Although freshwater lakes most commonly are P deficient, exceptions can be found. Tropical lake particulate matter is closer to the Redfield ratio, and these lakes can have either N or P limitation or neither N nor P limitation (in the latter case, some other factor must limit organic production).

Sources of Phosphorus in Lakes

Since phosphorus is a limiting nutrient in most lakes, it is instructive to delineate its sources. The primary sources of phosphorus (and nitrogen) in lakes are direct rainfall and snowfall on the lake and runoff from the surrounding drainage area. In oligotropic lakes, most phosphorus in runoff comes from rock weathering and soils. However, in areas influenced by humans there are additional sources of phosphorus, including agricultural runoff (containing phosphorus from fertilizers and animal wastes) and sewage (containing phosphorus from human wastes, detergents, and industrial wastes), which are discharged directly into the lake or its inlet tributaries. (For a more detailed discussion of phosphorus on land, sources of phosphorus pollution, and the phosphorus cycle, see Chapter 5).

Atmospheric precipitation may be a very important source of phosphorus (and nitrogen) for oligotrophic lakes, particularly those in areas of granitic terrain with low contributions of nutrients from weathering and those lakes whose area is large compared to the drainage area (for example, Lake Superior). Likens et al. (1974) found that 50% of the phosphorus in some oligotrophic lakes comes from precipitation. As anthropogenic influences increase, runoff becomes more important, and for eutrophic lakes the average precipitation contribution is only 7% of the phosphorus and 12% of the nitrogen. Table 6.6 summarizes nutrient sources for various lakes.

Chapra (1977) has modelled changes in the sources of phosphorus over the last 150 years for tbe U.S.–Canada Great Lakes. Those lakes most influenced by human activities, Lakes Michigan, Erie, and Ontario, are depicted in Figure 6.10. There are two phases of increased phosphorus input. The first began around 1850, with much greater land runoff of phosphorus resulting from the conversion of forests to agricultural land. The second phase, which began about 1945, is a result of population growth, accompanied by the use of sewers, which has introduced human wastes and phosphate detergents directly into the lakes. Lake Erie has the most phosphorus loading and is eutrophic in its western basin, and mesotrophic in its central and eastern basins (Chapra

TABLE 6.6 Phosphorus and Nitrogen Sources for Selected Lakes as a Percent of Total Annual Input

Lake	Phosphorus				Nitrogen			
	Precipitation	Urban Runoff and Waste	Rural Runoff and Waste	Total Runoff	Precipitation	Urban Runoff and Waste	Rural Runoff and Waste	Total Runoff
Disturbed lakes								
Lake Erie	4			96	18			82
Lake Ontario	4			96	28			72
Lake Ontario[a]	10			90				
European lakes[b]	1	70	29	99	3	37	60	97
Lake Mendota	6	35	59	94	17	11	66	77
Lake Canadagua	2	46	52	98	3	6	91	97
Eutrophic lakes[c]	7			93	12			88
Undisturbed lakes								
Lake Superior	46			54	47			53
Lake Huron	27			73	62			38
Oligotrophic lakes[c]	50			50	56			44

[a] From Robertson and Jenkins 1978.

[b] From Vollenweider 1968; see also Stumm 1972.

[c] Summary of 18 lakes.

Source: Likens et al. 1974, except as noted.

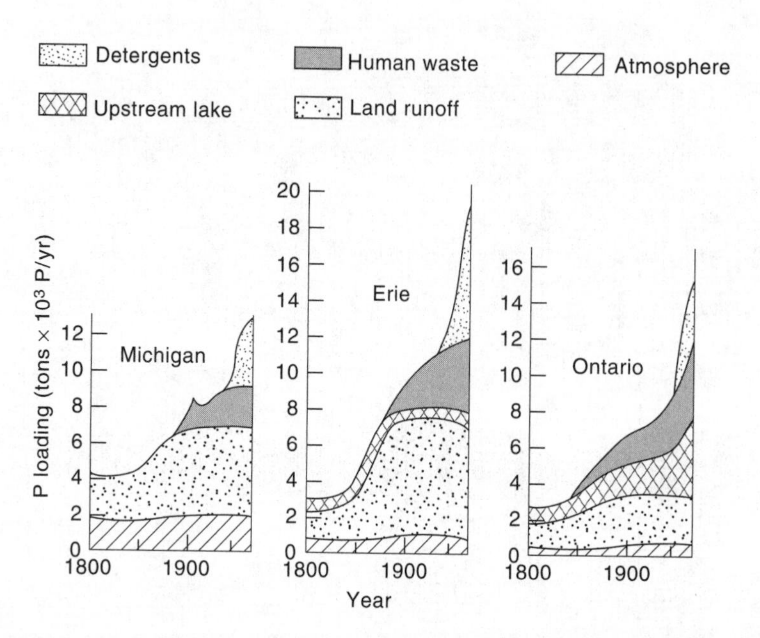

Figure 6.10. Historical loading of total phosphorus (in 10^3 tons/yr) from 1800 to 1970 for three Great Lakes, Michigan, Ontario, and Erie, based on model calculation. [From Stumm and Morgan, 1981, after S. C. Chapra, "Total Phosphorus Model for the Great Lakes," *Journal of Div. of Environmental Engineering* 103(EE2): 153. Copyright © 1977. American Society Civil Engineering, reprinted by permission of the publisher.]

and Dobson 1981). Since it is upstream from Lake Ontario, it greatly increases the phosphorus load in the latter, which is mesotrophic offshore (Chapra and Dobson 1981).

Vollenweider (1968) plotted annual phosphorus inputs to lakes versus mean depth (Figure 6.11) as a means of predicting which lakes should become eutrophic. He also included predictions of how much the phosphorus load of Lake Erie and Lake Ontario could be reduced between 1968 and 1986 with phosphate input control, and the growth of the phosphorus load without phosphate control. His diagrams and predictions called attention in the United States and Canada to the importance of human activities as a control on lake water composition and the dominant (but reversible) role played by phosphorus pollution.

Since Vollenweider made his predictions there have been considerable changes in the composition of the Great Lakes. The peak phosphorus loading in the Great Lakes Basin occurred in 1972 (see Figure 6.11). In 1972, the United States and Canada established a phosphorus control program for municipal sewage water entering the Great Lakes (Lee et al. 1978); the phosphorus concentration in sewage was reduced from 6 mg P/l in the late 1960s to 2 mg P/l from 1975 to 1980 and 1 mg P/l after 1980 (Chapra 1980). In addition, in 1973 the P content in detergents in the Great Lakes drainage basin was greatly reduced.

The results of the Great Lakes phosphorus control program could be seen by 1980 in Lake Ontario, which has a 7-year water residence time and deep, well-oxygenated bottom waters (avoiding anaerobic P feedback from bottom sediments) (Chapra 1980). Phosphorus concentrations in Lake Ontario dropped significantly, from about 25 μg P/l (eutrophic) in 1973 to about 16 μg P/l (upper-level mesotrophic) in 1980 (Kwiatkowski 1982). This 35% drop in P was

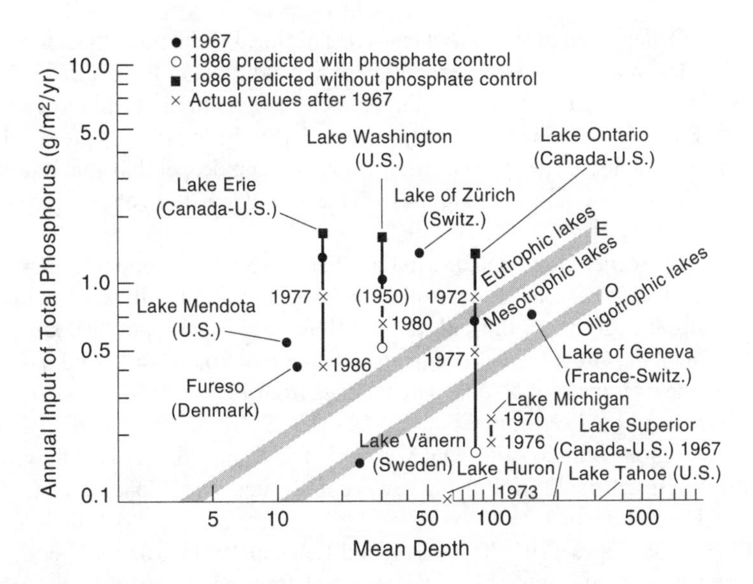

Figure 6.11. Annual phosphorus input versus depth for some North American and European lakes. Oligotrophic lakes lie below line O, eutrophic lakes above line E, with mesotrophic (intermediate) lakes in between. (Diagram, 1967 data, and 1986 predictions from Vollenweider 1968; diagram modified by Vallentyne 1974. Great Lakes data after 1967: Lake Huron in 1973 from Vollenweider et al. 1974; Lake Michigan in 1970 from Chapra 1977, and in 1976 from Eisenreich et al. 1977; Lake Ontario in 1972 from Chapra 1980, and in 1977 from Fraser 1980; Lake Erie in 1977 from Vallentyne and Thomas 1978 (see also Charlton 1980) and in 1986 from Laws 1993. Lake Washington data from Edmonson and Lehman 1981.)

accompanied by a 15% reduction in open-lake algal biomass concentrations. Lake Ontario could be maintained at upper-level mesotrophy (15–20 μg P/l) until 2000 by the phosphorus controls in effect in 1980. But in order to restore oligotrophy in Lake Ontario (i.e., P concentrations < 10 μg P/l), it would be necessary to both reduce diffuse land P runoff (a more difficult task) and reduce the P concentrations in Lake Erie, which supplies 40% of Lake Ontario's phosphorus (Chapra 1980).

Lake Erie, the shallowest of the Great Lakes, is mesotrophic as a whole (10–20 mg/m^3 P), and there have been considerable improvements in lake quality. Actual P loading in 1986 was 0.39 g/m^2 (Laws 1993), which was what Vollenweider predicted for 1986 with P controls (see Figure 6.11). The P concentration in 1986 was 13 mg/m^3, about 60% of its 1970 value (Laws 1993). Cyanobacteria (blue-green algae) were replaced by diatoms and green algae.

Lake Erie is often considered as three separate basins. Western Lake Erie (8 m deep) was still eutrophic (30–35 μg P/l) in 1977–1978 despite a 50% drop in P concentration from 1970 and a dramatic reduction in green and blue-green algae. This was due largely to a reduction in P loading from the Detroit River (Nicholls et al. 1980). Because of its shallowness, western Lake Erie does not have thermal stratification and thus has no problems with O_2 depletion. As we noted earlier, the central and largest basin of Lake Erie has mesotrophic surface waters, but in the recent past severe oxygen-depletion problems occurred in the bottom waters during summer stratification (often leading to anoxia) (Chapra and Dobson 1981). In 1985, late summer anoxia caused P release from the sediments into the water, but, because of continuing efforts to curb inputs of

P, by 1988 the O_2 depletion in the central basin had declined to the point that anoxia did not occur (Laws 1993). The maintenance of year-round oxic conditions in the central Lake Erie bottom waters is still not assured because of the thinness of the hypolimnion and its consequently small O_2 reserves (Barica 1982; Charlton 1980). Anoxia has to be avoided to prevent P regeneration from bottom sediments. Eastern Lake Erie, which is much deeper, has much less of a problem with O_2 depletion, and, as a result, exhibits both mesotrophic bottom and surface waters (Chapra and Dobson 1981).

Seventy-five percent of the P loading to Lake Erie is now from nonpoint sources such as agricultural runoff, which have to be controlled by erosion control and better animal waste management, a considerably more difficult project (Laws 1993). N input has been less controlled than P input in all the Great Lakes (and tends to come less from sewage and more from runoff). For example, the river input of N to the Great Lakes from 1974 to 1981 *increased* by 36%, while the P input declined by 7% (Smith et al. 1987).

Lake Washington, near Seattle, which is shown in Figure 6.11, also showed considerable improvement in water quality following sewage diversion. The P loading of Lake Washington was reduced from a high of 1.7 $g/m^2/yr$ in 1963 to 0.64 $g/m^2/yr$ in 1980, following sewage diversion in 1963–1968 (Laws 1993; Edmonson and Lehman 1981). The number of cyanobacteria dropped, and the lake concentration of P dropped from 65 mg/m^3 (clearly eutrophic) to 17 mg/m^3, which would qualify it as being mesotrophic by containing less than 20 mg/m^3 (see Table 6.4) but would not qualify if the criteria of Figure 6.11 are used. Since sewage contributed only 35% of the N, with most of the N coming from streams, N did not decline as much as P after sewage diversion. Because of this the lake, which had been N-limited during its eutrophic stage, shifted back to its prior P-limited state after sewage diversion.

Lake Tahoe (also shown in Figure 6.11) is a very deep, ultraoligotrophic lake in the Sierra Nevada Mountains, California-Nevada, which is showing the early signs of cultural eutrophication. In a 30-year period from 1959 to 1988, the population of the basin increased by 10 times, which correlated with a doubling of primary productivity in the lake and a significant increase in the NO_3-N content of the lake (Goldman 1988). The N comes from fertilizer and dry deposition, and there is less opportunity for terrestrial plant uptake of N with more rapid runoff. The P content of the lake did not increase because sewage was exported from the basin. As a result, the lake, which was initially N and Fe limited, has an increased ratio of N:P and is becoming more P sensitive.

POLLUTIVE CHANGES IN MAJOR LAKES: POTENTIAL LOADING

The predicted input of pollutional substances to a lake can be expressed in terms of potential loading (Stumm and Morgan 1981). Potential loading is equated to human energy consumption per unit volume of lake water (for example, watts per cubic meter as in Table 6.7). The expression for potential loading is

$$\text{Potential loading} = \frac{\text{drainage area}}{\text{lake area}} \times \frac{1}{\text{lake depth}} \times \frac{\text{inhabitants}}{\text{drainage area}} \times \frac{\text{energy consumption}}{\text{inhabitants}}$$

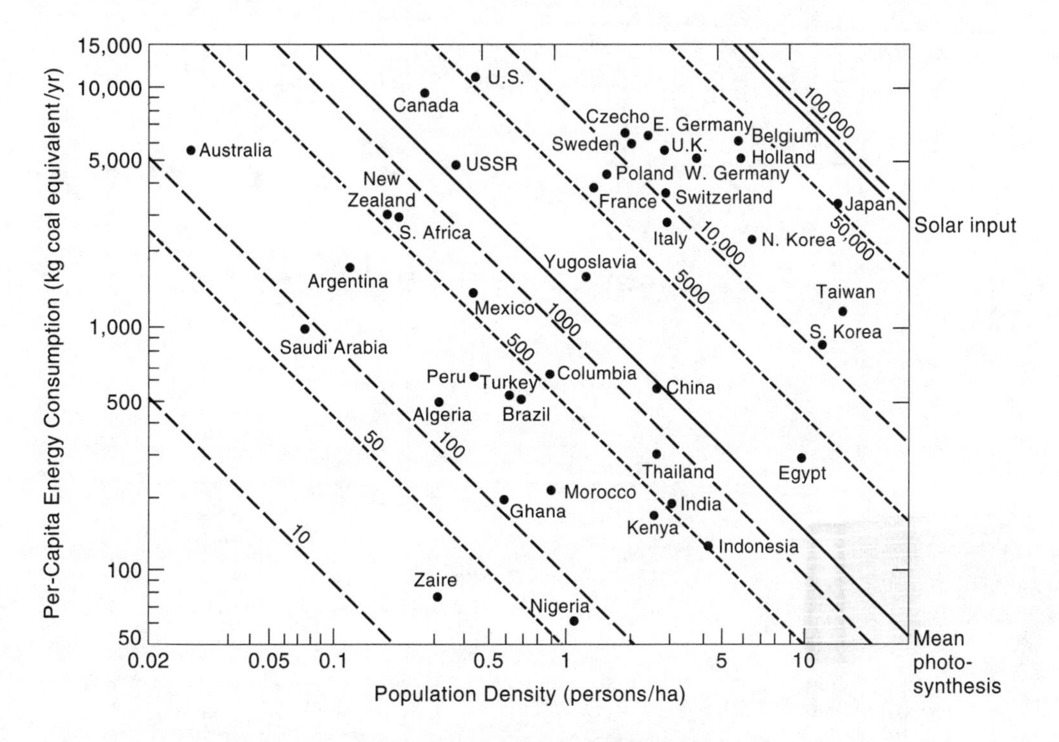

Figure 6.12. Relationship between per-capita energy consumption (kilograms coal equivalent per year) and population density (people per hectare; 1 ha = 10^4 m^2). One kilogram coal equivalent per year per hectare equals about 1.1×10^{-4} W/m^2 (Stumm and Morgan 1981). Diagonal lines represent constant values of energy consumption per unit area. [From Stumm and Baccini, 1978, after Y. H. Li, 1976, "Population Growth and Environmental Problems in Taiwan (Formosa): A Case Study," *Environmental Conservation* 3: 176. Reprinted by permission of the publisher.]

This expression assumes that waste production or "pollution" will be roughly proportional to energy consumption and considers the fact that the energy consumption per capita is much greater in heavily industrialized countries such as the United States than it is in underdeveloped countries (see Figure 6.12). For any one country where the energy consumption per capita is roughly constant, the population density per lake volume will be the major criterion for potential pollutional loading. This concept of potential loading is useful for various pollutional inputs (Cl, SO_4, etc.) in addition to those nutrients which cause eutrophication (P, N). Table 6.7 gives loading parameters for various lakes. The first six lakes (including Lake Erie) are or have been eutrophic (Stumm and Morgan 1981).

The potential pollutional loading for the Great Lakes can be compared to actual changes in their chemistry. For this purpose the non-nutrient ions chloride, sulfate, calcium, and sodium, plus potassium (Figure 6.13), which are sensitive to pollution in lakes as they are in rivers (see Chapter 5), can be used. These ions are especially useful because there is an abundance of concentration data on them covering the past 140 years. Lake Superior, which is the least affected by human activities, shows little change in chemistry with time, and the concentrations of Lake Superior water resemble incident precipitation (Beeton 1969). Lake Michigan shows considerable increases in Cl^- and particularly SO_4^{2-} from 1880 to 1967. These undoubtedly would have

TABLE 6.7 Loading Parameters for Some Lakes

Lake	Country	Drainage Area/ Lake Area	Mean Depth (m)	Inhabitants per m^2 of Drainage Area	Energy Consumption (10^3 W/inhab.)[c]	Potential Loading[a] = Energy Consump./ Lake Vol. (W/m^3)[c]
Greifensee[b]	Switzerland	15	19	441	5.2	1.81
Washington[b]	United States	~15	18	~50	11.4	0.48
Constance[b]	Switzerland-Germany-Austria	19	90	114	5.0	0.12
Lugano[b]	Switzerland-Italy	11	130	264	4.9	0.11
Erie[b]	United States	1.34	19	293	10.0	0.21
Biwa[b]	Japan	4.5	41	~150	4.4	0.07
Winnipeg	Canada	35	13	~3	8.6	0.07
Ontario[b]	United States-Canada	3.2	85	108	10.0	0.041
Michigan	United States-Canada	2.3	84	42.6	10.0	0.012
Huron	United States-Canada	2.0	59	16.9	10.0	0.0057
Titicaca	South America	14	~100	~40	0.2	0.001
Victoria	Africa	3	40	~70	0.4	0.002
Baikal	USSR	17	730	~5	0.8	0.0005
Tanganjika	Africa	4	572	~50	0.3	0.0001
Superior	United States-Canada	1.5	145	~5	10.0	0.0005

[a] Potential loading $= \frac{\text{drainage area}}{\text{lake area}} \times \frac{1}{\text{lake depth}} \times \frac{\text{inhabitants}}{10^6\ \text{m}^2\ \text{drainage area}} \times \frac{\text{energy consumption}}{\text{inhabitants}} = \frac{\text{energy consumption}}{\text{lake volume}}$

[b] Eutrophic lakes, presently or in past.

[c] W = watts.

Source: Modified from W. Stumm and J. J. Morgan, *Aquatic Chemistry*, 2nd ed., p. 693.

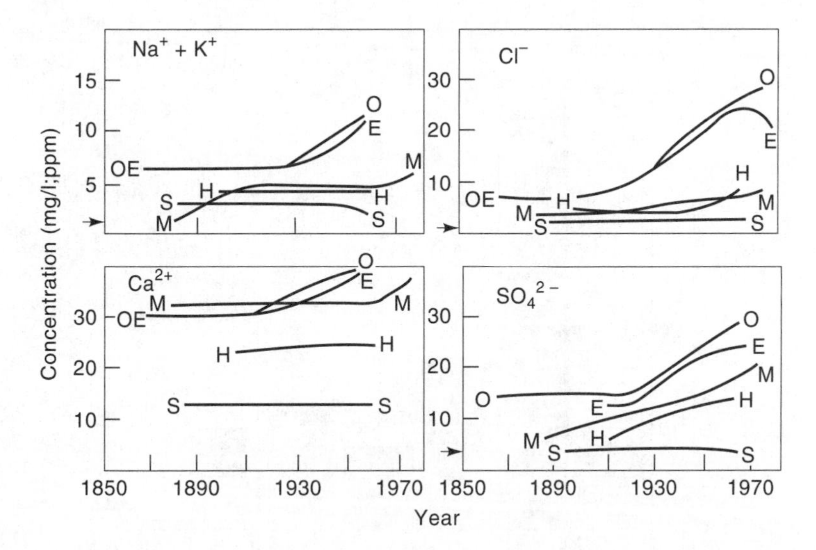

Figure 6.13. Changes in the chemical composition of the Great Lakes from 1850 to 1967. S is Lake Superior, M is Lake Michigan, H is Lake Huron, E is Lake Erie, and O is Lake Ontario. Arrows represent concentration of ions in Lake Superior precipitation (Beeton 1969). [Data from Beeton (1969). Diagram modified from Vallentyne (1974). Later data: Lake Michigan in 1976 from Bartone and Schelske (1982); Lake Ontario Cl in 1978 from Fraser (1980); northeastern Lake Erie Cl in 1970 and 1978 from Heathcote et al. (1981).]

been greater except for the fact that during this period sewage from the city of Chicago was diverted into the Chicago River, away from the lake. Lake Huron, which is fed by both Lake Superior and Lake Michigan, tends to have concentrations of all ions intermediate between them. However, increased Cl^- and SO_4^{2-} concentrations in Lake Huron are due partly to effects from its own drainage area in addition to input from Lake Michigan. Lake Erie, as noted previously, is heavily polluted, and this is reflected not only in large increases in sulfate and chloride but in other ions as well. Lake Ontario, as mentioned previously for phosphorus, shows smaller increases in all ion concentrations than Lake Erie, which feeds it, but the changes mirror Lake Erie changes. Robertson and Jenkins (1978) attribute 94% of the Cl^- input to Lake Ontario as coming from Lake Erie. Thus, cleaning up Lake Erie would clean up Lake Ontario as well.

In summary, it can be seen that, in addition to influences of the surrounding drainage area, shown by potential loading, the chemistry of the Great Lakes (particularly Lakes Huron and Ontario) is complicated by pollution from upstream lakes.

ACID LAKES

During the past several decades, many lakes have been found to exhibit large decreases in pH, and this has proven to be a serious environmental problem. The acidification of dilute freshwater lakes and streams in large areas of southern Scandinavia, southeastern Canada, and the northeastern United States is linked to the increasingly acid precipitation (pH 4.0–4.6) received in these areas (Wright and Gjessing 1976; National Research Council 1986; Schindler 1988; Wright 1988; Brakke et al 1988; for further discussion of acid rain and its environmental consequences consult Chapter 3).

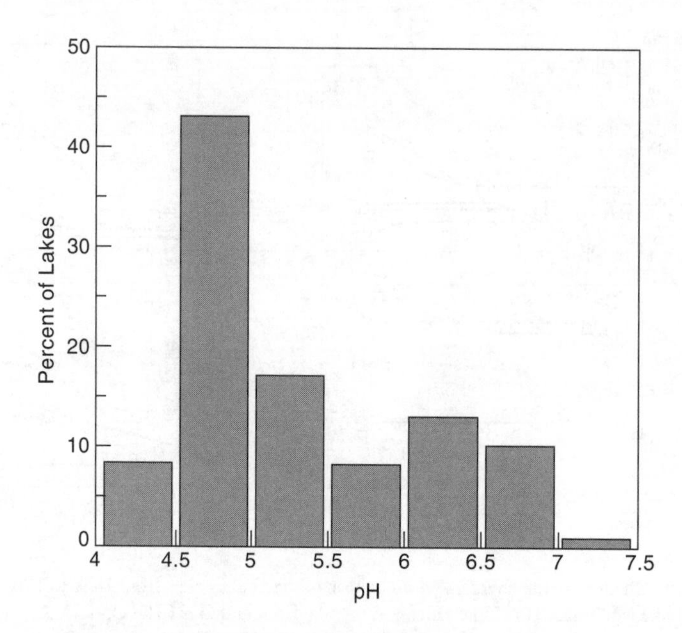

Figure 6.14. Frequency distribution of pH in 216 lakes in the Adirondack Mountains, New York, in 1975. (After Wright and Gjessing 1976; based on data of Schofield 1976a.)

The lakes of the Adirondack Mountains of New York State have been cited as outstanding examples of lake acidification due to the input of acid from acid rain. Figure 6.14 shows that a large proportion of the lakes in the Adirondacks were highly acid (pH < 5.5) in 1975, whereas earlier results for the 1930s show many fewer lakes with such low pH values (Wright and Gjessingd 1976). Although the earlier pH values have been challenged as to their accuracy, one can still demonstrate well-documented recent drops in pH for two individual Adirondack Lakes, Big Moose Lake and Upper Wallface Lake (Figure 6.15a).

Further verification of recent changes in pH of Adirondack lakes can be seen from the study of diatoms (siliceous algae), which are very sensitive to changes in pH, and which are preserved in the lake sediments. Diatom evidence shows that slow, long-term *natural* acidification of these lakes appears to have taken place, with a decrease in pH from 6.0 to 5.0 occurring over hundreds to thousands of years. However, diatom-inferred further decreases in pH of 0.5–1.0 pH units during the past 50 years have occurred in several lakes (Charles and Norton 1986; see Figure 6.15b). Diatom-inferred pH data from 10 lakes in the Adirondack Mountains, showed that 6 of them which had current pH values of 5.2 or lower had become more acid in the period from 1930 to 1970, presumably due to acid deposition. Most of these lakes also had data on measured pH (see Figure 6.15a), and fish populations (see below) which tend to confirm the diatom data (National Research Council 1986). (The four lakes with current pH values above 5.2 showed no pH trend with time.)

The most convincing evidence for the acidification of lakes, even where no long-term records of pH exist, has been their declining fish populations (many species of fish do not survive in acid lakes). For example, 90% of the Adirondack lakes that had a pH less than 5 in 1975 had no fish (Schofield 1976a). In several Adirondack lakes a change in pH from 5.2 to 4.8 was accompanied

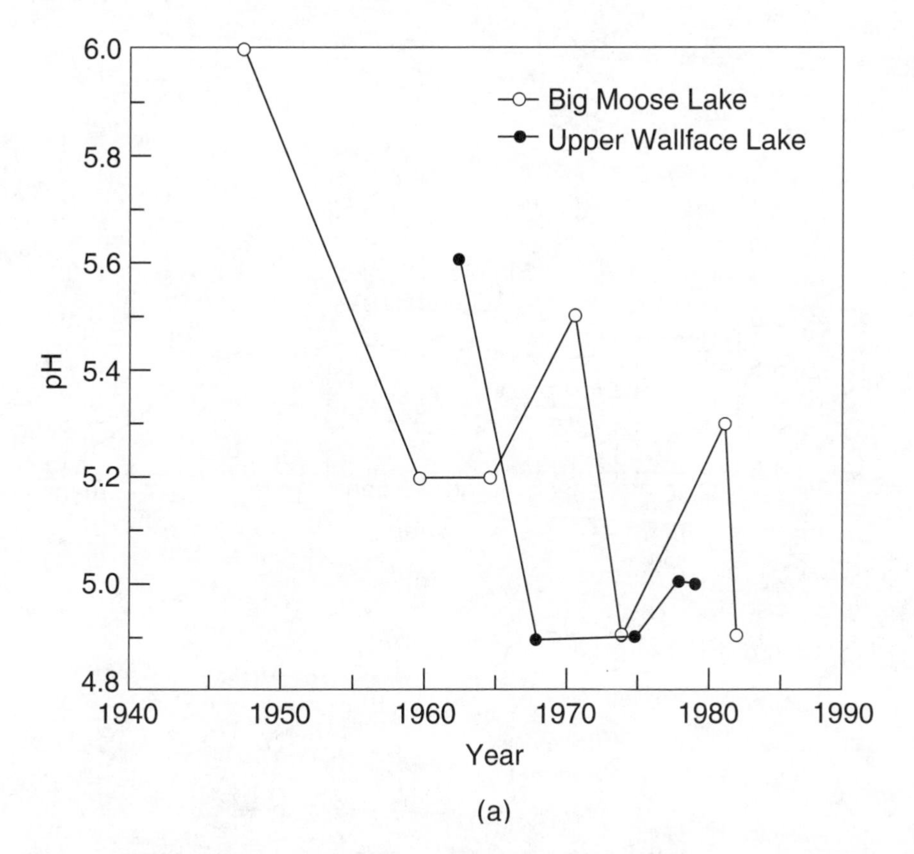

Figure 6.15. Changes with time of pH in lakes of the Adirondack Mountains, New York State. (a) Rapid decreases in (measured) pH in Big Moose Lake and Upper Wallface Lake. (Data from NRC 1986.)

by a large drop in fish population (National Research Council 1986). Apparently, fish are very sensitive to pH in this range and, in addition, acidification is acompanied by both an increase in dissolved aluminum, which is toxic to fish, and also a decrease in dissolved organic carbon (Davis et al. 1985).

A group of remote lakes about 65 km from the metal smelters in Sudbury, Ontario, experienced very rapid acidification (change in pH from 6.3 to 4.9) in the 1960s (Beamish et al. 1975), and this was accompanied by a simultaneous loss of fish in these lakes. (This change of lake pH occurred after the building of a higher smokestack at Sudbury, which spread SO_2 over a larger downwind area.) However, since the SO_2 emissions at Sudbury have been reduced by two thirds, there have been rapid rises in alkalinity and pH in these lakes (Schindler 1988).

Rapid declines in lake pH such as those shown in Figure 6.15 are not necessarily correlated exactly in time with rapidly increasing sulfur dioxide emissions in the eastern United States, although sulfate deposition is considered to be most important in lake acidification. Particularly in the case of the diatom-inferred pH changes in Big Moose Lake (Figure 6.15b), major increases in SO_2 emissions had occurred in 1920 and 1940, well before changes in the lake pH, which did correspond to another rise in emissions (National Research Council 1986). Schindler (1988) suggests a number of reasons for the lake damage occurring later: (1) Large smokestacks

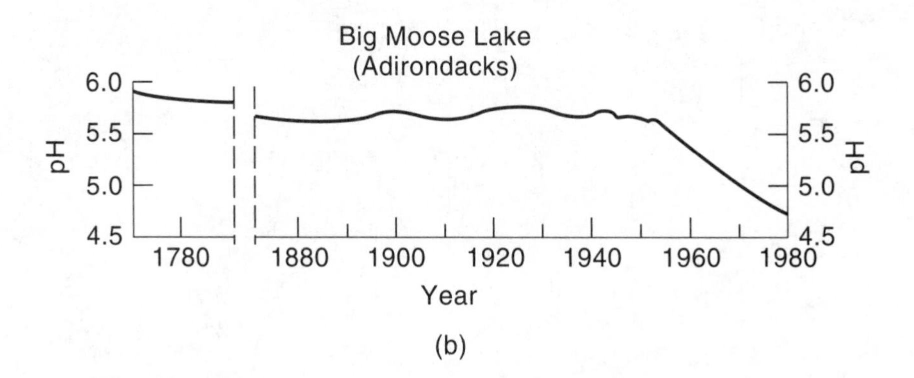

Figure 6.15. *Continued.* (b) Diatom-inferred changes in pH of Big Moose Lake from 1770 to 1980 showing a rapid drop in pH around 1950. (Adapted From National Research Council 1986.)

were built for power plants in the mid-1950s, transforming a local problem into a regional one; (2) alkaline ash material has been removed from emissions to control particulates; (3) emissions of NO_x, which contribute to acid deposition, have also increased; (4) it took several years to deplete the acid-neutralizing capacity of the lakes.

In addition to chronically acidified lakes, there are also episodic pH drops in poorly buffered, but usually nonacidic, lakes caused by acid snowmelt runoff in the spring. This comes about because acid precipitation accumulates in the snow over the winter, leading to large concentrations of sulfuric acid at the time of spring runoff due to snow melt. One Adirondack lake dropped from pH 7 to pH 5.9 over a period of a week or two and temporarily had greatly increased Al concentrations, resulting in fish mortality (Schofield 1980).

Although lake acidification is a problem in the Adirondacks, there are other regions of the United States where lakes have not become more acidic over recent decades. A careful analysis of alkalinity and pH data from three large U.S. lake surveys was undertaken by Kramer et al. (1986). They found that some lakes had increases in pH and alkalinity and some decreases. Wisconsin lakes, on average, showed *increases* in pH and alkalinity (HCO_3^- concentration) since the 1930s (which may be due to human disturbances such as road building and construction; Schindler 1988). New Hampshire lakes show *no obvious change* in alkalinity, although they may have increased in pH. However, New York lakes have on average had *decreases* in pH and alkalinity. Depending on the assumptions made about the data, the drop in pH was from 0.74 to 0.12 pH units. In addition, a number of lakes in Ontario have shown decreases in alkalinity (Schindler 1988).

Acid lakes are also a common phenomenon in Scandinavia because of the input of acid rain from industrialized areas of central and western Europe to the south (see Chapter 3). For example, in southern Norway in 1986, 70% of the lakes had lost their bicarbonate buffering capacity (i.e., were acid) as a result of receiving acid precipitation with a pH < 4.7, and in some areas as low as 4.3 (Henriksen et al. 1988).

Acidification of lakes occurs in areas that are unusually sensitive to acid precipitation because of a characteristic bedrock geology and soil. They are underlain by weathering-resistant igneous and metamorphic rocks or noncalcareous sandstones and have thin, patchy acid soils, neither of which are conducive to acid neutralization. Also, calcium carbonate, which has the buffering ability to provide carbonate and bicarbonate ions to solution and thus neutralize acid precipitation, is generally lacking. (The role of calcium carbonate in neutralizing acid deposition is dicussed in detail below.) By contrast, other areas that are also receiving acid precipitation but have calcareous sedimentary rocks (limestone and calcareous sandstone) contain lakes whose pH values are essentially unaffected by acid precipitation (Norton 1980).

Figure 6.16a shows areas in North America that would be sensitive to acid precipitation based on the concentration of dissolved bicarbonate ion in lakes as a reflection of bedrock geology. The criterion used is natural lake concentrations less than 0.5 mEq/l HCO_3^- (However, many scientists consider areas with < 0.2 mEq/l HCO_3^- to be acid sensitive; Schindler 1988). Areas with acid lakes in Scandinavia have similar geology. Pristine lakes in these sensitive noncalcareous areas are dilute because of slow chemical weathering and, as a result, they have low bicarbonate concentrations and the pH values, before acidification, are between 6 and 7. By contrast, lakes developed in areas with calcareous sedimentary rocks can be expected to have both higher bicarbonate concentrations and natural (preacidification) pH values of around 8.0. Thus, initial (natural) lake composition can be used as a guide to susceptibility to acidification. Consistently acid lakes (pH < 5), as opposed to those that have episodes of acidity, seem to develop in areas with susceptible bedrock geology that are also receiving precipitation more acid than about pH 4.6 (Henriksen 1979).

Figure 6.16b shows areas of northeastern North America where lake acidification has already occurred (Schindler 1988). Evidence for acidification is given in terms of the ratio of alkalinity (HCO_3^-) to the sum of the concentrations of Ca^{2+} and Mg^{2+}. The ratio decreases with the addition of acid rain as HCO_3^- is replaced by SO_4^{2-} (from sulfuric acid), and Ca^{2+} and Mg^{2+} are often increased. The average equivalent ratio of HCO_3^- to Ca^{2+} plus Mg^{2+} is 0.6–1.1 in surface waters of unaffected acid-sensitive areas. Average values of the ratio for sensitive lakes in areas receiving highly acid deposition are < 0.2. Note that the areas where lake acidification has been greatest in Figure 6.16b correspond well with those predicted to have been susceptible to acidification in Figure 6.16a.

The areas in the United States that have considerable numbers of acid lakes in 1988 are listed in Table 6.8. In the northeastern United States, in addition to the Adirondacks, the Poconos/Catskill mountains and southern New England (eastern Massachusetts) are particularly affected (see also Figure 6.16b). In the upper Midwest, the Upper Peninsula of Michigan has the largest percentage of acid lakes. Florida has the largest percentage in the country, but at least half of these are due predominantly to organic acidity.

Figure 6.16. Lake Acidification in North America. (a) Regions in North America containing lakes that would be sensitive to potential acidification by acid precipitation (shaded areas). These areas have igneous or metamorphic bedrock geology which results in dilute lakes with low HCO_3^- concentrations (<0.5 mEq HCO_3^-/l). Unshaded areas have calcareous or sedimentary bedrock geology. (From J. N. Galloway and E. B. Cowling, "The Effects of Precipitation on Aquatic and Terrestrial Ecosystems, A Proposed Precipitation Chemistry Network," *Journal of the Air Pollution Control Association* 28(3): 233. Copyright © 1978. J. of the Air Pollution Control Assoc. Reprinted by permission of the publisher.]

The effect of acid rain on lakes varies depending on what part of it falls directly on the lake and what part is runoff from the land. Soils generally are better able to neutralize acid precipitation than lake water. In Norway, runoff averages around 0.2–0.9 pH units higher than precipitation, and lakes are consequently 0.3–1.0 unit higher than precipitation (Wright and Gjessing 1976). Thus, even in susceptible areas receiving acid precipitation, lakes will vary in acidity depending on how much contact their input water has had with the soil and what the local characteristics of the soil and bedrock are. [This can be seen in Figure 6.14 in the distribution of pH of Adirondack lakes in 1975, where although 51% of the lakes had pH < 5.0, another 11% had pH > 6.5; Schofield (1976a).]

The local characteristics affecting the neutralization of rain acidity, or the lack thereof, include the geometry of the lake basin and the thickness and type of soil. A lake with a small area and volume, whose input comes from a large watershed, will receive runoff that has had considerable

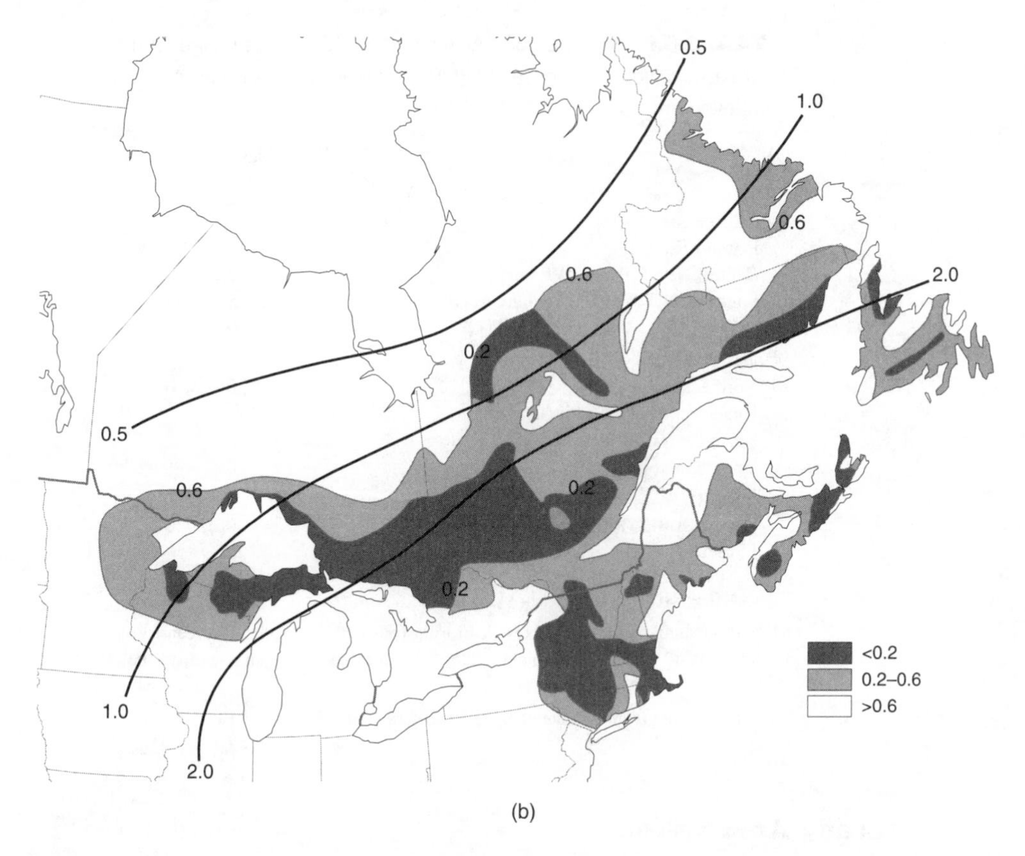

Figure 6.16. *Continued.* (b) Areas of northeastern North America, showing the ratio of alkalinity (HCO_3^-)(or acid-neutralizing capacity) to Ca^{2+} and Mg^{2+} of included lakes. Average ratios in pristine lakes are 0.6–1.1, while lakes which have been affected by acid deposition have average ratios less than 0.2, due to HCO_3 loss and Ca and Mg gain. Only lakes with <200 μEq/l Ca + Mg are included. Heavy lines numbered 0.5, 1.0, and 2.0 indicate sulfate deposition in g/m²/yr. [After Schindler(1988), *Science* 239, p. 150. Copyright 1988 by the AAAS.]

soil contact. Such a lake is less likely to become acid. In contrast, acid lakes are favored by steep slopes and exposed bedrock with little vegetation and soil development; in this way they receive precipitation virtually unaffected by the soil (Galloway and Cowling 1978). Also, lakes fed by headwater streams, which are more acid than larger streams, will tend to become more acid (Johnson 1979).

The soil cover in susceptible areas is frequently absent or thin, and the soil varies in grain size and composition. Carbonate minerals in rock outcrops, in soils, or in calcareous glacial till will prevent lake acidity by buffering acid precipitation. In areas of noncalcareous soils, the presence of unconsolidated sediment, which has a greater surface area for minerals to contact acid precipitation, aids acid neutralization. For example, Northern Ontario lakes in a quartz sandstone area receiving acid precipitation have a pH of 3.8–4.4 when on bedrock, while lakes with gravel- or sand-sized sediment have a pH > 5 (Kramer 1978). In addition to grain size, the presence of clays and organic matter in soils also helps to provide acid neutralizing capacity. [See Gorham and McFee (1978) for further details on neutralization.]

TABLE 6.8 Percentage of Acidic Lakes, Defined as Lakes That Have Acid-Neutralizing Capacity ANC[a] ≤ 0, in Various Eastern and Midwestern U.S. Areas in 1988.[a]

Area	Percentage Acid Lakes
Northeast	
Adirondack Mts.	11
Poconos/Catskills	5
Central New England (Vt, N.H.)	2
Southern New England (Conn., Mass.)	5
Maine	0.5
Upper Midwest	
Northeast Minnesota	0
Upper Peninsula, Michigan	10
North central Wisconsin	3
Upper Great Lakes area	0
Southeast	
Southeast Blue Ridge	0
Florida[b]	22

[a] $ANC = [HCO_3^-] - [H^+]$.

[b] About half of the acid lakes in Florida (particularly in the Okefeenokee Swamp and in the Florida Panhandle) have dominantly organic acidity (see text).

Source: Data from Brakke et al. 1988; Ellers et al. 1988a, and 1988b.

Naturally Acid Lakes

As pointed out by Krug and Frink (1983) and by others, in the eastern North American and northern European areas that are receiving acid rain, many soils and lakes may be naturally acid due to the presence of organic acids, particularly humic acid. Forest regrowth occurring in these areas tends to produce acid soils with thick organic layers. Thus, how can one decide whether or not any given lake owes its acidity to natural processes as opposed to pollution? If the organic acids from soils persist without oxidation as groundwaters enter lakes, then the criteria discussed in Chapter 5 that indicate the presence of organic acidity in rivers should also be true for lake acidity that is due partially or totally to (natural) organic acids. If acidity is organic in origin, there should be a high concentration of dissolved organic carbon relative to dissolved inorganic matter. Also, the sum of the major inorganic cations should be greater than the sum of the major inorganic anions, because organic anions supply part of the negative charge.

Most organic acids become oxidized microbially to carbonic acid and CO_2 in the subsoil and after leaving the soil, and as a result they are not found in lake water to any extent. Even in this case it is still possible to test the argument of Krug and Frink (Drever 1988). If low pH of a soil is due to high concentrations of organic acids, upon oxidation by microbes high concentrations of H_2CO_3 and dissolved CO_2 result (see Chapter 4). In this case the pH of the soil water can change appreciably upon entering a lake or stream. Water with high dissolved CO_2 will degas and lose the CO_2 to the atmosphere upon contacting air (which is low in CO_2). This causes the pH to rise via the reaction

$$HCO_3^- + H^+ \rightarrow CO_2\uparrow + H_2O$$

Thus, if soil acidity is due to organic acids (and carbonic acid), then acid soil water may become neutralized by CO_2 loss upon entering a lake. In this case acid soils do not lead to acid lakes. By contrast, if the low pH of the soil is due to the presence of strong acids, such as H_2SO_4 and HNO_3 from acid rain, then there is little or no HCO_3^- available to sequester H^+ during this degassing reaction. (Lack of HCO_3^- comes about because H^+ from the strong acids has already converted it in the soil to H_2CO_3 and CO_2.) In this way, if soil acidity is due to acid rain, the pH does not go up as CO_2 is degassed to the atmosphere, and acid lakes do result from acid soils.

In a study of Norwegian lakes, Brakke et al. (1987) found that both organic acids and sulfuric acid contributed to the acidity but that in those lakes with a pH < 5.3, sulfuric acid was superimposed on organic acidity. Organic acidity made lakes more sensitive to atmospheric sulfate deposition, which has increased with time. In addition, changes in acidity of lakes due to organic acids tend to be very slow (hundreds or thousands of years to change one pH unit). It has been suggested that earlier deforestation decreased organic acidity, which is now increasing due to reforestation. However, deforestation at forest test sites has been found actually to bring about *higher* acidity than forest regrowth, because of increased nitrification and oxidation of reduced sulfur compounds accompanying the deforestation (Schindler 1988).

Florida has a large number of acid lakes (22% of all Florida lakes), and about half of these have anions from organic acids as a dominant anion. The organic acid lakes occur in the Okefenokee Swamp (where DOC averages 36 mg/l), and there also are many in the Florida Panhandle, where although the DOC averages only around 3 mg/l, the base cation concentrations are also proportionately lower (Ellers et al. 1988b).

The *neutral salt effect,* the input of NaCl from atmospheric deposition of sea salt, followed by soil exchange of Na^+ for H^+ in acid soils and release of H^+ to surface waters, has been suggested as a cause of lake acidity in coastal areas. However, the salt effect is only a short-term process, and there seems to be little evidence for it in long-term acidification of northeastern lakes, most of which have had Na:Cl ratios similar to sea salt (Sullivan et al. 1988).

Chemical Composition of Acid Lakes

Acid lakes have a distinctive chemical composition. In general, they have hydrogen-calcium-magnesium sulfate waters as opposed to unaffected lakes, which have calcium-magnesium bicarbonate waters (Wright and Gjessing 1976; Wright 1988). Table 6.9 gives the chemical composition of two sets of lakes, both located in similar areas, one in Norway and the other in Ontario, Canada. The acid lakes (defined as lakes with pH < 5.2) in each case receive very acidic precipitation (pH < 4.5), while the unaffected lakes receive less acidic precipitation (pH > 4.8). The sulfate concentration in the acid lakes is three to five times that in unaffected lakes. All the bicarbonate (HCO_3^-) has been lost in the acidic Ontario lakes. While the acidic Norwegian lakes have a slightly lower HCO_3^- concentration than the "nonacidic" Norwegian lakes, both concentrations are low and the "nonacidic" lakes probably already have been affected by acid precipitation.

The increased Ca^{2+} and Mg^{2+} concentrations in acid lakes are often accompanied by increased dissolved Al and heavy metal concentrations (Wright and Gjessing 1976). The cation increase reflects the release of these ions from rocks and soils in neutralizing acid precipitation. [How-

TABLE 6.9 Mean Chemical Composition of Acid Lakes in Areas Receiving Highly Acidic Precipitation and of Otherwise Similar Nonacidic Lakes in Areas Not Receiving Highly Acidic Precipitation

Area	Number of Lakes	Lake pH	Rain pH	Concentration (mg/l)						
				Na^+	K^+	Ca^{++}	Mg^{++}	HCO_3^-	SO_4^{--}	NO_3
Norway:										
Southern	26	4.76	<4.5	9	4	50	25	11	92	4
West Central	23	5.2	>4.8	9	3	16	7	13	30	5
Ontario, Canada:										
La Cloche Mtns., S.E. Ontario	4	4.7	<4.5	9	10	150	65	0	290	—
Experimental Lakes Area, N.W. Ontario	40	5.6–6.7	>4.8	4	10	80	65	60	55	<1.5

Note: Data corrected for sea salt on the basis of Cl^-

Source: R. F. Wright and E. T. Gjessing, "Changes in the Chemical Composition of Lakes," *Ambio,* 5(5–6); 220–221. Copyright © 1976 by the Royal Swedish Academy of Sciences, reprinted by permission of the publisher.

ever, acid-sensitive lakes tend to have low Ca^{2+} and Mg^{2+} concentrations prior to acidification, corresponding to their low HCO_3^- concentrations (Brakke et al. 1988).] Sulfate concentrations are higher in acid lakes because of the nature of acid precipitation, which is predominantly sulfuric acid (H_2SO_4) (see the section on acid rain in Chapter 3). Because it is the balance between sulfuric acid and basic cations that is important, even if sulfuric acid input decreases, the pH of a lake can continue to drop if cation input also decreases (Dillon et al. 1987). This is because, in the absence of H_2SO_4, cations are normally supplied along with bicarbonate ion, which is (weakly) basic, so that fewer cations means less bicarbonate and, thus, lower pH.

Bicarbonate loss in acid lakes represents a loss of buffering in the lake. Buffering in lake water is the ability of the water to neutralize input of acid (or base). If a lake is buffered, its pH is not changed greatly by the addition of moderate quantities of acid (or base). Most lakes are buffered by carbonate species (bicarbonate in particular), and for most lakes the most effective bicarbonate buffering occurs in the pH range 6.0–8.5 (Hem 1970). Bicarbonate (HCO_3^-) buffering of acid comes about via the reaction of H^+ with HCO_3^- to produce neutral carbonic acid (H_2CO_3):

$$HCO_3^- + H^+ \leftrightarrows H_2CO_3$$

Thus, the buffering ability of a lake depends on its concentration of HCO_3^-.

Figure 6.17 shows the relative molar concentrations of H_2CO_3 and HCO_3^- at different pH values in a fairly dilute solution of total carbonate ($H_2CO_3 + HCO_3^- + CO_3^{2-}$) = 10^{-3} mol/l, typical of fresh water. The concentrations of HCO_3^- and H_2CO_3 are equal at pH 6.3, and at pH values greater than that there is little change in the concentration of HCO_3^- for a change in pH (i.e., there is good buffering). At pH values less than 6, the concentration of HCO_3^- drops very rapidly with a decrease in pH. There is no buffering at all below the point where the concentration of H^+ equals the concentration of HCO_3^-, which for the situation of Figure 6.17 is at pH 4.65. (For a more dilute solution with total concentration of 10^{-4} mol/l, represented by the dashed lines

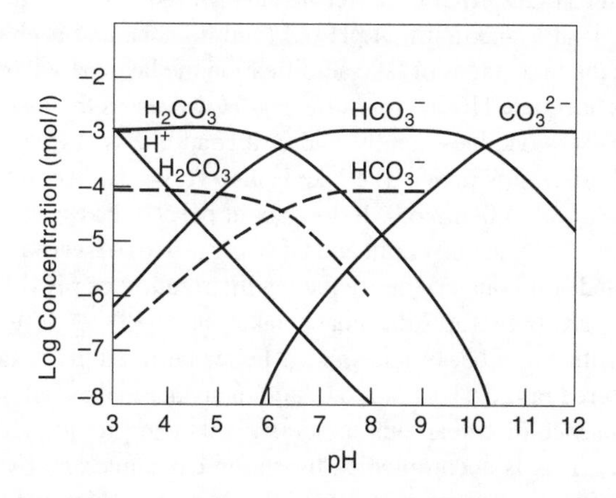

Figure 6.17. Concentration of carbonate species (H_2CO_3, HCO_3^- and CO_3^{2-}) in fresh water at 25°C. (Total carbonate concentration = $[H_2CO_3] + [HCO_3^-] + [CO_3^{2-}] = 10^{-3}$ mol/l). Concentration of H^+ is also shown. Dashed lines show how the curves shift for total carbonate concentration = 10^{-4} mol/l. (Redrawn from W. Stumm and J. J. Morgan, *Aquatic Chemistry,* 2nd ed., p. 176. Copyright © 1981. John Wiley & Sons, Inc. Reprinted by permission of John Wiley & Sons, Inc.)

in Figure 6.17 and found in some lakes that are susceptible to acidification, the buffer capacity would be lost at about pH 5.15; for rainwater, which is in equilibrium with atmospheric CO_2 and is even more dilute, this point is at pH 5.65.)

Kramer (1978) shows why the buffering ability of lakes in a calcareous area is much greater than that in a noncalcareous area. A lake in a calcareous regime has carbonate minerals (such as calcite, $CaCO_3$) that can resupply HCO_3^- previously lost by reaction with H^+. A lake in a calcareous area in equilibrium with atmospheric CO_2 would have a pH of 8.4 and a molar concentration of HCO_3^- of about 10^{-3} (Garrels and Christ 1965). This pH and HCO_3^- concentration favor a high buffering capacity (see Figure 6.17). In addition, dissolution of $CaCO_3$,

$$H^+ + CaCO_3 \rightarrow Ca^{2+} + HCO_3^- \qquad (6.15)$$

replaces HCO_3^- previously lost from solution by reaction with acid rain. In this way $CaCO_3$ greatly raises the buffering capacity of the lake.

The reaction of H^+ with $CaCO_3$ [Eq. (6.15)] also releases Ca^{2+} ions; if the carbonate mineral is dolomite, $CaMg(CO_3)_2$, it also releases Mg^{2+} ions. This is part of the reason for the increased Ca^{2+} and Mg^{2+} concentration in acidic lakes in comparison to similar nonacidic lakes.

By contrast to calcareous lakes, dilute lakes in noncalcareous regions have a much lower natural pH of 6 to 7, and much lower HCO_3^- concentrations (about 10^{-4} mol/l). If the only input were acid rain of pH 4 ($H^+ = 10^{-4}$ M), the HCO_3^- would be used up upon addition of a volume of rain equal to that in the lakes. (Small amounts of HCO_3^- are resupplied from the weathering of silicate minerals by carbonic and organic acids—see Chapter 4—but this is usually insufficient to neutralize the acid rain). Thus, lakes in noncalcareous area, which are poorly buffered and have little ability to neutralize incoming H^+ ions, tend to become acid. Other terrain and soil factors discussed previously therefore become very important in such areas.

Lakes in susceptible noncalcareous areas with an initial pH of 6 to 7 tend to show a characteristic pattern of acidification in response to acid precipitation (Wright and Gjessing 1976). This is determined by the H_2CO_3–HCO_3^- buffering curve (Figure 6.17). The concentration of HCO_3^- relative to H_2CO_3 is at a maximum at pH 6.3 (and higher), and it changes very little with pH changes. Thus, in the first stages of lake acidification the lake pH will change slowly toward pH 6 because there is adequate HCO_3^- for buffering. However, as the lake pH drops below 6.0, the concentration of HCO_3^- declines rapidly, and further additions of H^+ result in a much more rapid drop in pH (i.e., buffering is poor). The lake is also very sensitive to temporary changes in pH in the pH range between 5.0 and 6.0. Below about pH 5.0, lakes are unbuffered due to loss of HCO_3^- and are chronically acid. Wright and Gjessing (1976) observed that, because of local variations in terrain and soil contact, the frequency distribution of pH of lakes receiving acid precipitation (see Figure 6.14 for Adirondack lakes in 1975) tends to reflect the bicarbonate buffering curve, with a number of lakes in the better-buffered pH range above pH 6, few lakes in the poorly buffered range of pH 5.5–6.0, and many acid lakes below pH 5.0.

Even in the absence of calcareous minerals, soils can provide some buffering or neutralization of H^+ ions. This is accomplished by cation exchange and by chemical weathering of silicates (Norton 1980; Galloway et al. 1981). Acid neutralization via cation exchange involves uptake of H^+ to replace cations associated with clays and organic (humic) substances. It is more rapid than weathering because exchangeable cations are held less strongly by the host phases. However, silicate weathering is the only process that will ultimately neutralize acidity, because weathering is needed to replace the base soil cations used in cation exchange (Drever 1988).

In a recently glaciated Adirondack Mountain, New York, watershed, the present silicate weathering rate is adequate to replace cations lost from the soil exchange pool, so the watershed is not in danger of acidification from acidic atmospheric deposition because of readily weatherable bedrock minerals (plagioclase feldspar and hornblende) (see below). Using Sr isotopes, and Sr as a proxy for base cations (Ca, Mg, K, Na), it was found that of the Sr lost from the watershed in streamwater, 70% came ultimately from bedrock mineral weathering and 30% from soil-cation-exchange Sr derived from other sources (Miller et al. 1993). (The soil-cation-exchange Sr not coming from bedrock weathering was derived from atmospheric input and glacial till weathering, which had a distinctly different Sr isotopic composition from the bedrock.)

Normal silicate weathering (Chapter 4) involves the reaction between naturally produced H^+ and silicate minerals with the release of cations and silica to solution, but not aluminum. Highly acidic weathering, e.g., from acid deposition, by contrast, involves the release of dissolved Al via the the dissolution of aluminous phases (see Chapter 4). For example:

$$Al_2Si_2O_5(OH)_{4\ \text{kaolinite}} + 6H^+ \rightarrow 2Al^{3+} + 2H_4SiO_4 + H_2O$$

$$4H_2O + NaAlSi_3O_{8\ \text{plagioclase}} + 4H^+ \rightarrow Na^+ + Al^{3+} + 3H_4SiO_4$$

$$Al(OH)_{3\ \text{gibbsite}} + 3H^+ \rightarrow Al^{3+} + 3H_2O$$

Most of these reactions release Al^{3+} [or hydroxylated species such as $Al(OH)^{2+}$) in solution in exchange for the H^+ ions they neutralize. Since aluminum solubility increases at low pH (below pH 5; Cronan and Schofield 1979), this aluminum can be transported in solution in soil water to lakes and, as a result, the aluminum concentration of acid lakes tends to be higher than in similar nonacid lakes. This is shown in Figure 6.18. The excessive Al in acid lakes probably contributes to fish mortality (Cronan and Schofield 1979; Schofield 1980).

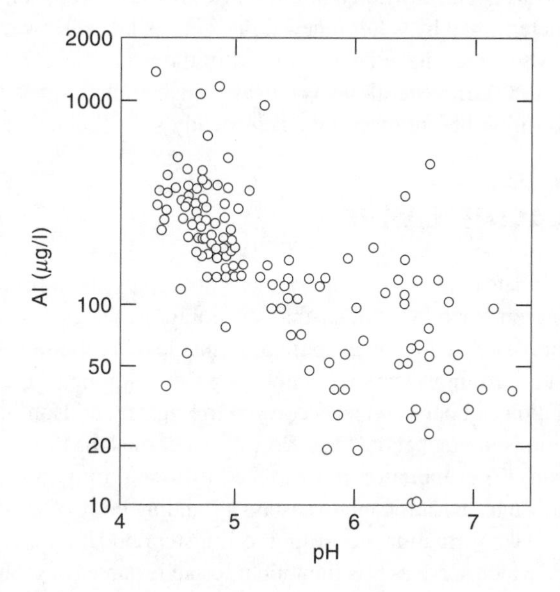

Figure 6.18. Aluminum versus pH for 217 high-altitude lakes in the Adirondack Mountains, New York. (From Galloway et al. 1981, generalized from Schofield 1976b.)

Schindler et al. (1985) artificially acidified a small experimental lake from pH 6.8 to 5.0 over an 8-year period to determine changes in lake biota in response to increased acidity. These changes occurred in direct response to increased H^+ ion, since all other ion concentrations were not greatly increased. Decline in the fish population was caused by reproductive failure (below pH 5.4) and loss of key species in the fish food chain. Algal mats appeared in shallow areas where fish normally spawned. Contrary to expectations, however, acidification did not cause decreases in primary productivity or rates of decomposition, nor did decreases in phosphorus concentration result.

Eutrophic lakes, which have anoxic bottom water during summer stratification, may be protected against becoming acid. In two of the Canadian experimental lakes that were artificially made eutrophic (see earlier discussion of nutrients), a model calculation shows that enough persistent alkalinity could be produced by bacterial processes in the hypolimnion to neutralize typical levels of acid deposition (Kelly et al. 1982). This comes about because acid deposition results in higher levels of sulfate and nitrate deposition in addition to H^+. In anoxic portions of lakes (hypolimnia and sediments), bacteria reduce SO_4^{2-} to H_2S and NO_3^- to N_2 because of the presence of high concentrations of oxidizable organic matter. Both of these bacterial processes result in the production of alkalinity (HCO_3^-) (for reactions, see Table 6.3), which is capable of neutralizing H^+ in a lake. However, the alkalinity produced under anoxic bottom-water conditions during summer stratification may be only temporary unless H_2S is converted to FeS_2 (pyrite) and permanently removed to sediments, and unless N_2 escapes. Otherwise, when aerobic conditions are restored upon lake overturn, H_2S is reoxidized to SO_4^{2-} and N_2 to NO_3^{2-}, and the HCO_3^- alkalinity is accordingly consumed. Nevertheless, Schindler et al. (1986) found that bacterial sulfate reduction and bacterial nitrate reduction did bring about acid neutralization (see also Cook et al. 1986).

Lakes and streams can lose acidity or recover their alkalinity after the acidity of precipitation is reduced. This has been shown in a number of Canadian lakes (Schindler 1988). The rate of lake recovery is determined by water renewal rates of the lakes. However, in cases where the base cations (Ca^{2+}, Mg^{2+}, etc.) have been depleted in the watershed, recovery may take much longer (Dillon et al. 1987). In general, lakes must be flushed three times with input waters of a new chemical composition before reaching a new steady state (Schindler 1988).

SALINE AND ALKALINE LAKES

In arid to semiarid climates, lakes often are saline. This comes about primarily because there is no outflow of water from such lakes other than evaporation. Waters containing dissolved salts flow in, but only pure water is lost by evaporation, thus leaving the salts behind to accumulate in the lake. Arid conditions, however, do not always produce saline lakes. Necessary conditions for saline lake formation and persistence, according to Eugster and Hardie (1978: 237–238), are as follows (our additions in brackets): (1) outflow of water must be restricted, as it is in a hydrologically closed basin; (2) evaporation must exceed inflow [during initial stages]; and (3) [for persistence] the inflow must be sufficient to sustain a standing body of water. An unusually favorable locale for saline lake formation, according to Eugster and Hardie, is in arid basins located near high mountains, which serve as precipitation traps and sources of groundwater. Some examples are the numerous saline lakes found in intermontane basins of the western United States, including the Great Salt Lake. The world's largest saline lake is the Caspian Sea of central Asia.

In general, the volume of water in saline lakes fluctuates considerably, both seasonally and from year to year depending on climatic conditions. Nevertheless, for highly saline lakes one can visualize a (quasi-) steady-state water balance of input by springs and streams and output by evaporation and a steady-state salt balance of input by springs and streams and output by the precipitation of saline minerals. In other words, dissolved salts cannot build up forever in saline lakes, and ultimately saturation is reached with respect to soluble minerals. One of the characteristic features of saline lakes is the unusualness of the minerals formed from them and found in their sediments. Some examples are shown in Table 6.10. (Besides the soluble phases listed in the table, some insoluble silicate minerals, e.g., sepiolite and smectite, form by the reaction of saline solutions with silicate detritus; see below). Pathways of evaporation necessary to form different minerals and to bring about different lake water compositions are discussed and summarized by Eugster and Hardie (1978), Drever and Smith (1978), and Eugster and Jones (1979).

Saline lakes are often also highly alkaline and exhibit a high pH. Whereas most freshwater lakes (excluding the special case of acid lakes) have pH values ranging from roughly pH 6 to 8 (Baas Becking et al. Kaplan, and Moore 1960), the pH of saline lakes can rise to values greater than 10. The cause for high pH has been studied by several workers (e.g., Garrels and Mackenzie 1967) and can be explained fundamentally in terms of the natural processes of weathering, evaporation, and CO_2 gas equilibration. The reasoning begins as follows: In areas that are underlain by acid igneous rocks, the weathering of feldspars and volcanic glass by carbonic acid (see Chapter 4) results in the production of groundwaters containing dissolved HCO_3^-, which is balanced by Na^+ and K^+ as well as by Ca^{2+} and Mg^{2+}. In other words, the concentration of HCO_3^- is more than twice the concentration of Ca^{2+} plus Mg^{2+}. This means that upon evaporation of the groundwater after passing into a lake, the lake cannot precipitate all HCO_3^- as Ca^{2+} or Mg^{2+} carbonates, and some HCO_3^- remains behind to be concentrated by evaporation.

Let us follow the course of composition change during the evaporation of a typical igneous-derived groundwater according to the scheme proposed by Garrels and Mackenzie (1967). This is shown in Figure 6.19. As the water undergoes initial evaporative concentration, it reaches saturation with calcium carbonate and, at about the same time, also with magnesium silicate

TABLE 6.10 Some Typical Minerals Formed from Saline Lakes

Mineral	Composition
Halite	$NaCl$
Gypsum	$CaSO_4 \cdot 2H_2O$
Calcite	$CaCO_3$
Dolomite	$CaMg(CO_3)_2$
Thenardite	Na_2SO_4
Mirabilite	$Na_2SO_4 \cdot 10H_2O$
Glauberite	$CaNa_2(SO_4)_2$
Trona	$Na_2CO_3 \cdot NaHCO_3 \cdot 2H_2O$
Nahcolite	$NaHCO_3$
Pirssonite	$CaNa_2(CO_3)_2 \cdot 2H_2O$
Gaylussite	$CaNa_2(CO_3)_2 \cdot 5H_2O$
Aphthitalite	$K_3Na(SO_4)_2$

Note: Many other minerals have been identified depending on lake composition (e.g., borates), but, for lack of space, are not listed here.

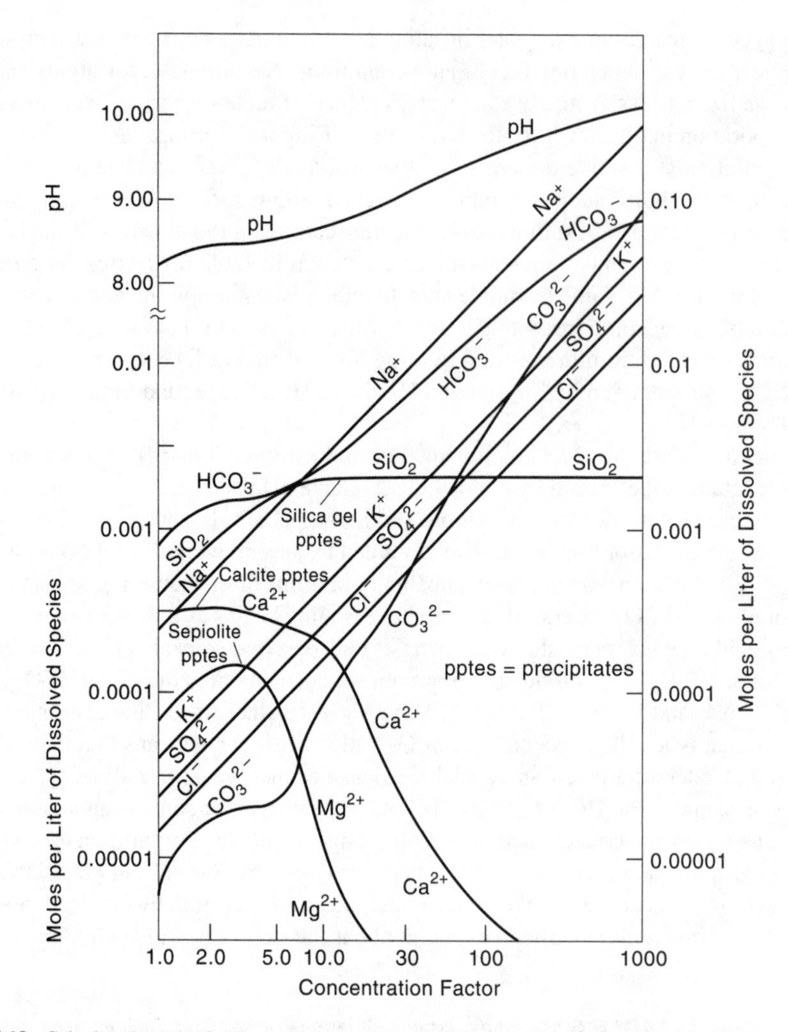

Figure 6.19. Calculated results for the evaporation of typical springwater, from a granitic terrain, which is in equilibrium with atmospheric CO_2. (After R. M. Garrels and F. T. Mackenzie, "Origin of the Chemical Compositions of Some Springs and Lakes," in *Equilibrium Concepts in Natural Water Systems,* Advanvaces in Chemistry Series 67, p. 239. Reprinted with permission from the American Chemical Society. Copyright © 1967 by the American Chemical Society.)

(represented in Figure 6.19 by the mineral sepiolite). Further evaporation results in $CaCO_3$ precipitation (see below) and in Mg-silicate precipitation according to the generalized reaction

$$2Mg^{2+} + 4HCO_3^- + 3H_4SiO_4 \rightarrow Mg_2Si_3O_8{\cdot}nH_2O + 4CO_2 + 8H_2O$$

Note that Mg-silicate precipitation (along with the removal of silica) results in the removal of two HCO_3^- ions for each Mg^{2+} ion. If the concentration of silica exceeds that of Mg^{2+}, and the concentration of HCO_3^- exceeds twice the concentration of Mg^{2+}, on a molar basis, then eventually evaporation should result in a loss of most all the Mg^{2+}. This is shown in Figure 6.19. During and after Mg-silicate precipitation, $CaCO_3$ precipitation results in the removal of Ca^{2+} and HCO_3^- according to the reaction:

$$2HCO_3^- + Ca^{2+} \rightarrow CaCO_3 + CO_2 + H_2O$$

Note that, analogous to Mg-silicate formation, HCO_3^- is removed in a ratio with Ca^{2+} of 2:1. If the remaining concentration of HCO_3^- exceeds twice that of Ca^{2+}, as would be the case for our typical igneous-derived water, then upon further evaporation all Ca^{2+} is removed and eventually an alkaline lake results.

After Mg-silicate and $CaCO_3$ formation, further evaporation causes the concentration of HCO_3^- to rise at about the same rate as other ions, such as Cl^-, which are not involved in precipitation reactions. While this is going on the pH is also rising (Figure 6.19). This is so because increase in HCO_3^- concentration causes the reaction

$$H^+ + HCO_3^- \rightarrow H_2O + CO_2\uparrow$$

to be driven to the right resulting in the loss of hydrogen ions. (Carbon dioxide does not build up in solution to back-react, but instead is readily lost to the atmosphere.) In this way an alkaline lake comes about. Because the rise in pH causes some HCO_3^- to be converted to CO_3^{2-} (see Figure 6.17), the concentration of CO_3^{2-} ions also rises and the rise is faster than that expected for simple evaporative concentration (see Figure 6.19). The reaction for the conversion of HCO_3^- to CO_3^{2-} is

$$2HCO_3^- \rightarrow CO_3^{2-} + CO_2\uparrow + H_2O$$

If evaporation continues to a very great extent, not shown in Figure 6.19, eventually saturation with alkali carbonates, for example $NaHCO_3$, may be reached and in this way rare soda lakes can arise.

From the above one can see that without weathering of Na-K silicates, especially volcanic glass (which weathers rapidly), the evolution of alkaline lakes would not follow the path shown in Figure 6.19. In the situation where HCO_3^- concentration does not exceed that necessary to precipitate all Ca^{2+} and Mg^{2+}, no HCO_3^- can build up on continued evaporation and no high pH results. This explains why many salt lakes (e.g., the Great Salt Lake) are not highly alkaline.

The evaporation path shown in Figure 6.19 was merely calculated by Garrels and Mackenzie (1967). However, this "thought experiment" has actually been duplicated using real water. Gac et al. (1978) have evaporated dilute water from the River Chari, an input river to Lake Chad in Africa, and found that it underwent concentration changes during evaporation that more or less agree with those shown in Figure 6.19. One modification that Gac et al. found was that the Garrels and Mackenzie predictions were better matched if detrital clays, collected with the river water, were left in the water during evaporation. Removal of the clays resulted in much delayed and subdued precipitation of Mg-silicate. Apparently the aluminous clays reacted with Mg^{2+} and H_4SiO_4 upon evaporation to form a Mg-aluminosilicate, such as smectite, which precipitates more easily than pure Mg-silicates (i.e., sepiolite).

REFERENCES

Ambühl, H. 1975. Versuch der Quantifizierung der Beeinflussing der Oekosystems durch chemische Faktoren: Stehen Gewäisser, *Schweiz. Z. Hydrol.* 37: 35–52.

Baas Becking, L. G. M., I. R. Kaplan, and D. Moore. 1960. Limits of the natural environment in terms of pH and oxidation-reduction potentials, *J. Geol.* 68: 243–284.

Barica, J. 1982. Lake Erie oxygen depletion controversy, *J. Great Lakes Res.* 8(4): 719–722.

Bartone, C. R., and C. L. Schelske. 1982. Lake-wide seasonal changes in limnological conditions in Lake Michigan in 1976, *J. Great Lakes Res.* 8(3): 413–427.

Beamish, R. J., W. L. Lockhart, J. C. Van Loon, and H. H. Harvey. 1975. Longterm acidification of a lake and resulting effects on fishes, *Ambio* 4(2): 98–104.

Beeton, A. M. 1969. Changes in the environment and biota of the Great Lakes. In *Eutrophication: Causes, Consequences and Correctives,* pp. 150–187. Natl. Acad. Sci./Natl. Res. Council Publ. 1700.

Berner, R. A. 1980. *Early Diagenesis: A Theoretical Approach.* Princeton, N.J.: Princeton University Press.

Brakke, D. F., A. Henriksen, and S. A. Norton. 1987. The relative importance of acidity sources for humic lakes in Norway, *Nature* 329: 432–434.

Brakke, D. F., D. H. Landers, and J. M. Ellers. 1988. Chemical and physical characteristics of lakes in the northeastern United States, *Environ. Sci. Technol.* 22: 155–163.

Chapra, S. C. 1977. Total phosphorus model for the Great Lakes, *J. Div. Environ. Eng.,* Am. Soc. Civ. Eng., 103(EE2): 147–161.

Chapra, S. C. 1980. Simulation of recent and projected total phosphorus trends in Lake Ontario, *J. Great Lakes Res.,* 6(2): 101–112.

Chapra, S. C., and H. F. H. Dobson. 1981. Quantification of the lake trophic typologies of Naumann (surface quality) and the Thienemann (oxygen) with special reference to the Great Lakes, *J. Great Lakes Res.* 7(2): 182–193.

Charles, D. F., and S. A. Norton. 1986. Paleolimnological evidence for trends in atmospheric deposition of acids and metals. In *Acid Deposition: Long-Term Trends,* ed. G. H. Gibson, pp. 231–299. Washington, D.C.: National Academy Press.

Charlton, M. N. 1980. Oxygen depletion in Lake Erie: Has there been any change? *Can. J. Fish. Aquatic Sci.* 37: 72–81.

Claypool, G., and I. R. Kaplan. 1974. The origin and distribution of methane in marine sediments. In *Natural Gases in Marine Sediments,* ed. I. R. Kaplan, pp. 99–139. New York: Plenum Press.

Cook, R. B., C. A. Kelly, D. W. Schindler, and M. A. Turner. 1986. Mechanisms of hydrogen ion neutralization in an experimentally acidified lake, *Limnol. Oceanogr.* 31: 134–148.

Cronan, C. S., and C. L. Schofield. 1979. Aluminum leaching in response to acid precipitation: Effects on high elevation watersheds in the northeast, *Science* 204: 304–306.

Davis, R., D. Anderson, and F. Berge. 1985. Loss of organic matter, a fundamental process in lake acidification: Paleolimnological evidence, *Nature* 316: 436–438.

Dillon, P. J., R. A. Reid, and E. de Grosbois. 1987. The rate of acidification of aquatic ecosystems in Ontario, Canada, *Nature* 329: 45–48.

Drever, J. I. 1988. *The Geochemistry of Natural Waters, 2nd ed.* Englewood Cliffs, N.J.: Prentice-Hall.

Drever, J. I., and C. L. Smith. 1978. Cyclic wetting and drying of the soil zone as an influence on the chemistry of ground water in arid terranes, *Am. J. Sci.* 278: 1448–1454.

Edmond, J. M., R. F. Stallard, H. Craig, V. Craig, R. F. Weiss, and G. W. Coulter. 1993. Nutrient chemistry of the water column of Lake Tanganyika, *Limnol. Oceanogr.* 38:725–738.

Edmonson, W. T., and J. T. Lehman. 1981. The effects of changes in the nutrient income on the condition of Lake Washington, *Limnol. Oceanogr.* 26: 1–29.

Eisenreich, S. J., P. J. Emmling, and A. M. Beeton. 1977. Atmospheric loading of phosphorus and other chemicals to Lake Michigan, *J. Great Lakes Res.* 3: 291–304.

Ellers, J. M., D. F. Brakke, and D. H. Landers. 1988a. Chemical and physical characteristics of lakes in upper Midwest United States, *Environ. Sci. Technol.* 22: 164–172.

Ellers, J. M., D. H. Landers, and D. F. Brakke. 1988b. Chemical and physical characteristics of lakes in the southeastern United States, *Environ. Sci. Technol.* 22: 172–177.

Eugster, H. P., and L. A. Hardie. 1978. Saline lakes. In *Lakes: Chemistry, Geology, Physics,* ed. A. Lerman, pp. 237–293. New York: Springer-Verlag.

Eugster, H. P., and B. F. Jones. 1979. Behavior of major solutes during closed-basin brine evolution, *Am. J. Sci.* 279: 609–631.

Fenchel, T., and T. H. Blackburn. 1979. *Bacteria and Mineral Cycling.* New York: Academic Press.

Fraser, A. S. 1980. Changes in Lake Ontario total phosphorus concentrations 1976–1978, *J. Great Lakes Res.* 6(1): 83–87.

Gac, J.-Y., D. Badaut, A. Al-Droubi, and Y. Tardy. 1978. Comportement du calcium, du magnèsium, et de la silice en solution. Prècipitation de calcite magnèsienne, de silice amorphe et de silicates magnèsiens au cours de l'èvaporation des eaux du Chari (Tchad), *Sci. Gèol. Bull. Strasbourg,* 31: 185–197.

Galloway, J. N., and E. B. Cowling. 1978. The effects of precipitation on aquatic and terrestrial ecosystems, a proposed precipitation chemistry network, *J. Air Pollut. Control Assoc.* 28(3): 229–235.

Galloway, J., S. A. Norton, D. W. Hanson, and J. S. Williams. 1981. Changing pH and metal levels in streams and lakes in the eastern U.S. caused by acid precipitation. In *Proc. EPA Conf. on Lake Restoration,* pp. 446–452.

Garrels, R. M., and C. L. Christ. 1965. *Solutions, Minerals and Equilibria.* New York: Harper.

Garrels, R. M., and F. T. Mackenzie. 1967. Origin of the chemical compositions of some springs and lakes. In *Equilibrium Concepts in Natural Water Systems,* pp. 222–242. Am. Chem. Soc. Adv. Chem. Ser. 67.

Goldman, C. R. 1988. Primary productivity, nutrients, and transparency during the early onset of eutrophication in ultra-oligotrophic Lake Tahoe, California-Nevada, *Limnol. Oceanogr.* 33: 1321–1333.

Gorham, E., and W. W. McFee. 1978. Effects of acid deposition upon outputs from terrestrial to aquatic ecosystems. In *Effects of Acid Precipitation on Terrestrial Ecosystems,* ed. T. C. Hutchinson and M. Havas, pp. 465–480. New York: Plenum Press.

Hasler, A. D. 1947. Eutrophication of lakes by domestic drainage, *Ecology* 28: 383–395.

Heathcote, I. W., R. R. Weiler, and J. W. Tanner. 1981. Lake Erie nearshore water chemistry at Nanticoke, Ontario, 1969–1978, *J. Great Lakes Res.* 7(2): 130–135.

Hecky, R. E., P. Campbell, and L. L. Hendzel. 1993. The stoichiometry of carbon, nitrogen and phosphorus in particulate matter of lakes and oceans, *Limnol. Oceanogr.* 38: 709–724.

Hem, J. D. 1970. Study and interpretation of the chemical characteristics of natural water. *USGS Water Supply Paper 1473.*

Henriksen, A. 1979. A simple approach for identifying and measuring acidification of freshwater, *Nature* 278: 542–545.

Henriksen, A., L. Tien, T. S. Traaen, I. S. Sevaldrud, and D. F. Brakke. 1988. Lake acidification in Norway—Present and predicted chemical status, *Ambio* 17: 259–266.

Hutchinson, G. E. 1957. *A Treatise on Limnology,* vol. 1. New York: John Wiley.

Hutchinson, G. E. 1973. Eutrophication, *Am. Sci.* 61: 269–279.

Imboden, D., and A. Lerman. 1978. Chemical models of lakes. In *Lakes: Chemistry, Geology, Physics,* ed. A. Lerman, pp. 341–356. New York: Springer-Verlag.

Johnson, N. M. 1979. Acid rain: Neutralization within the Hubbard Brook ecosystem and regional implications, *Science* 204: 497–499.

Kelly, C. A., J. W. M. Rudd, R. B. Cook, and D. W. Schindler. 1982. The potential importance of bacterial processes in regulating rate of lake acidification, *Limnol. Oceanogr.* 27(5): 868–882.

Kramer, J. R. 1978. Acid precipitation. In *Sulfur in the Environment, Part 1: The Atmospheric Cycle,* ed. J. R. Nriagu, pp. 325–369. New York: John Wiley.

Kramer, J. R., A. W. Andren, R. A. Smith, A. H. Johnson, R. B. Alexander, and G. Oehlert. 1986. Streams and Lakes. In *Acid Deposition: Long-term Trends,* ed. H. Gibson, pp. 231–299. Washington, D.C.: National Academy Press.

Krug, E. C., and C. R. Frink. 1983. Acid rain on acid soil: A new perspective, *Science* 221: 520–525.

Kwiatkowski, R. E. 1982. Trends in Lake Ontario surveillance parameters, 1974–1980, *J. Great Lakes Res.* 8(4): 648–659.

Laws, E. A. 1993. *Aquatic Pollution,* 2nd ed. New York: John Wiley.

Lee, G. F., W. Rast, and R. A. Jones. 1978. Eutrophication of water bodies: Insights for an age-old problem, *Environ. Sci. Technol.* 12(8): 900–908.

Lerman, A. (ed.). 1978. *Lakes: Chemistry, Geology, Physics.* New York: Springer-Verlag.

Lerman, A., J. Gat, and D. Imboden (eds.). 1995. *Physics and Chemistry of Lakes,* 2nd ed. New York: Springer-Verlag.

Li, Y. H. 1976. Population growth and environmental problems in Taiwan (Formosa): A case study, *Environ. Conserv.* 3: 171–177.

Likens, G. E. 1972. Eutrophication and aquatic ecosystems. In *Nutrients and Eutrophication,* ed. G. E. Likens, pp. 3–13. Am. Soc. Limnol. and Oceanogr. Spec. Symp., vol. 1.

Likens, G. E. J. S. Eaton, and J. N. Galloway. 1974. Precipitation as a source of nutrients for terrestrial and aquatic ecosystems. In *Precipitation Scavenging,* ed. R. G. Semonen and R. W. Beadle, pp. 552–570. ERDA Symp. Ser. 41.

Miller, E. K., J. D. Blum, and A. J. Friedland. 1993. Determination of soil-exchangeable-cation loss and weathering rates using Sr isotopes, *Nature* 362: 438–441.

Mohnen, V. A. 1988. The challenge of acid rain, *Sci. Am.* 259: 30–38.

National Research Council. 1986. *Acid Deposition: Long-Term Trends,* ed. G. H. Gibson, Washington, D.C.: National Academy Press.

Nicholls, K. H., D. W. Standen, and G. J. Hopkins. 1980. Recent changes in the near-shore phytoplankton of Lake Erie's Western Basin at Kingsville, Ontario, *J. Great Lakes Res.* 6(2): 146–153.

Norton, S. A. 1980. Geologic factors controlling the sensitivity of ecosystems to acidic precipitation. In *Atmospheric Sulfur Deposition: Environmental Impact and Health Effects,* pp. 521–531. Ann Arbor, Mich.: Ann Arbor Science.

Pauling, L. 1953. *General Chemistry,* 2nd ed. San Francisco: W. H. Freeman.

Robertson, A., and C. F. Jenkins. 1978. The joint Canadian-American study of Lake Ontario, *Ambio* 7(3): 106–112.

Rodhe, W. 1969. Crystallization of eutrophication concepts in northern Europe. In *Eutrophication: Causes, Consequences, and Corrective,* pp. 50–64. Natl. Acad. Sci./ Natl. Res. Council Publ. 1700.

Schindler, D. W. 1974. Eutrophication and recovery in experimental lakes: Implications for lake management, *Science* 184: 897–899.

Schindler, D. W. 1977. Evolution of phosphorus limitation in lakes, *Science* 195: 260–262.

Schindler, D. W. 1988. Effects of acid rain on freshwater ecosystems, *Science* 239: 149–157.

Schindler, D. W., K. H. Mills, D. F. Malley, D. L. Findlay, J. A. Shearer, I. J. Davies, M. A. Turner, G. A. Linsey, and D. R. Cruikshank. 1985. Long-term ecosystem stress: The effects of years of experimental acidification on a small lake, *Science* 228: 1395–1401.

Schindler, D. W., M. A. Turner, M. P. Stainton, and G. A. Linsey. 1986. Natural sources of acid neutralizing capacity in low alkalinity lakes of the Precambrian shield, *Science* 232: 844–847.

Schofield, C. L. 1976a. Acid precipitation: Effects on fish, *Ambio* 5(5–6): 228–230.

Schofield, C. L. 1976b. Dynamics and management of Adirondack fish populations. Final report, Proj. F-28-R, State of New York.

Schofield, C. L. 1980. Processes limiting fish populations in acidified lakes. In *Atmospheric Sulfur Deposition: Environmental Impact and Health Effects,* pp. 345–355. Ann Arbor, Mich.: Ann Arbor Science.

Smith, R. A., R. B. Alexander, and M. G. Wolman. 1987. Water-quality trends in the nation's rivers, *Science* 235: 1607–1615.

Smith, V. H. 1983. Low nitrogen to phosphorus ratios favor dominance by blue-green algae in lake phytoplankton, *Science* 221: 669–671.

Stumm, W. 1972. The acceleration of the hydrogeochemical cycling of phosphorus. In *The Changing Chemistry of the Oceans,* ed. D. Dryssen and D. Jagner, Nobel Symp. 20, pp. 329–346. Stockholm: Almqvist and Wicksell.

Stumm, W. (ed.). 1985. *Chemical Processes in Lakes.* New York: John Wiley.

Stumm, W., and P. Baccini. 1978. Man-made chemical perturbation of lakes. In *Lakes: Chemistry, Geology, Physics,* ed. A. Lerman, pp. 91–126. New York: Springer-Verlag.

Stumm, W., and J. J. Morgan. 1981. *Aquatic Chemistry,* 2nd ed. New York: John Wiley.

Sullivan, T. J., C. T. Driscoll, J. M. Ellers, and D. H. Landers. 1988. Evaluation of the role of sea salt inputs in the long-term acidfication of coastal New England lakes, *Environ. Sci. Technol.* 22: 185–190.

Vallentyne, J. R. 1974. *The Algal Bowl: Lakes and Man.* Ottawa: Environment Canada.

Vallentyne, J. R., and N. A. Thomas. 1978. Fifth year review of Canada-United States Great Lakes water quality agreement. Report of Task Group III, a technical group to review phosphorus loadings. Windsor, Ontario: International Joint Commission (IJC).

Vollenweider, R. A. 1968. Scientific fundamentals of the eutrophication of lakes and flowing waters with particular reference to nitrogen and phosphorus as factors in eutrophication. OECD Rep. DAS/CSI/68.27. Paris, France: OECD.

Vollenweider, R. A., M. Munawarily, and P. Stadelmann. 1974. A comparative review of phytoplankton and primary production in the Laurentian Great Lakes, *J. Fish Res. Bd. Can.* 31(5): 739–762.

Weiss, R. F., E. C. Carmack, and V. M. Koropalov. 1991. Deep-water renewal and biological production in Lake Baikal, *Nature* 349: 665–669.

Wetzel, R. G. 1983. *Limnology,* 2nd ed. Philadelphia: W. B. Saunders.

Wright, R. F. 1988. Acidification of lakes in the eastern United States and southern Norway: a comparison, *Environ. Sci. Technol.* 22: 178–182.

Wright, R. F., and E. T. Gjessing. 1976. Changes in the chemical composition of lakes, *Ambio* 5(5–6): 219–223.

CHAPTER 7

MARGINAL MARINE ENVIRONMENTS

Estuaries

INTRODUCTION

Marginal marine environments encompass all those bodies of seawater that have salinities decidedly different from that of the open ocean. Throughout most of the ocean, seawater is remarkably uniform in salinity (for further details, see Chapter 8). However, along many coastlines it undergoes mixing with river water and glacial meltwater to produce subsaline or brackish marginal marine bodies of water. Such bodies of water vary greatly in size, ranging from brackish ponds and small lagoons to such large water masses as Hudson's Bay and the Baltic and Black Seas. Of special interest are the class of drowned river mouths known as estuaries. Here river water meets seawater and the resulting processes that take place provide an important control on cycling of the elements. Thus, much attention will be paid in this chapter to the subject of estuarine chemistry.

Marginal marine environments may also be more saline than the open ocean. In regions of limited runoff and high evaporation, seawater can undergo extensive loss of H_2O with the consequent concentration of dissolved salts. Where mixing with the ocean is impeded, this gives rise to supersaline bodies of water and an antiestuarine circulation. A large-scale example (of moderate supersalinity) is the Mediterranean Sea. Although studied much less than brackish marginal marine environments, supersaline environments provide a modern-day analog for ancient evaporite basins where, in the geologic past, vast beds of salt and gypsum were formed by evaporation and mineral precipitation. Therefore, they will also receive some attention here.

Estuaries: Circulation and Classification

Estuaries are drowned river valleys filled with brackish (diluted) seawater. There is a large variation in estuarine circulation patterns, depending on the relative magnitude of the river flow and of oceanic tidal currents. [The following discussion is based on reviews by Bowden (1967),

Pickard and Emery (1982), and Pritchard and Carter (1973), which should be consulted for further details.] In the simplest case with minimal tidal mixing, the river tends to flow seaward as a lighter freshwater layer over the denser seawater. However, the tides, although they do not produce net water transport, mix some seawater upward into the fresh water and thus, a portion of the seawater is carried out of the estuary along with the river flow. In order to conserve water, since the estuary is neither filling nor emptying, there is an inward flow of seawater at depth to replace the saline water being lost along with fresh water at the surface. This is the typical estuarine circulation with fresh-to-brackish water flowing out at the top and saline water flowing in at the bottom. Salt must also be conserved in an estuary if it is to maintain a constant salinity. Thus, the amount of salt lost by mixing into the outward-flowing upper layer must be replaced by the inflow of saline ocean water in the deeper layer.

In addition to the river flow and tidal currents, estuaries are also affected by the earth's rotation (Coriolis force; see Chapter 1). The effect of the Coriolis force is greatest in wide estuaries, where it causes flow variations across the width of the estuary (i.e., lateral variations). Facing in the direction of flow, both the seaward freshwater flow at the surface and the landward saltwater flow at depth are stronger to the right in the Northern Hemisphere (and to the left in the Southern Hemisphere). This causes the interface (between seawater and fresh water) to slope downward to the right looking toward the sea (in the Northern Hemisphere), as exhibited, for example, by the Mississippi River estuary and Long Island Sound.

Open estuaries can be classified by their circulation pattern and the resulting distribution of salinity within the estuary (Stommel and Farmer 1952; see also Pritchard and Carter 1973; Pickard and Emery 1982.) This is shown in Table 7.1 and Figure 7.1. (The symbol ‰ refers to

TABLE 7.1 Classification of Estuaries

Type	Water Circulation	Physical Properties	Examples
A. Open estuaries			
1. Salt wedge	Salt wedge below river flow	River flow dominant	Mississippi River estuary
2. Highly stratified	Two-layer flow with entrainment (upward mixing)	River flow modified by tidal currents	Fjords—with deep sill below upper layer; deep estuaries
3. Slightly stratified	Two-layer flow with vertical mixing (upward and downward)	River flow and tidal mixing	Thames River estuary; Long Island Sound, Chesapeake Bay, shallow estuaries
4. Well mixed	Vertically homogeneous (a) with lateral variations (b) laterally homogeneous	Tidal currents dominant (shallow, with Coriolis; deep, no Coriolis)	Severn estuary
B. Silled estuaries (with shallow sill)	Surface layer flow with entrainment (upward mixing); restricted saltwater influx at depth which is rare if still is shallow (stagnant anoxic water may result)	River flow modified by tidal currents	Fjords with shallow sill; Black Sea

Source: Modified from Stommel and Farmer 1952; Bowden 1967; Pickard and Emery 1982.

parts per thousand or grams per kilogram.) These types grade into one another, and the type of estuarine circulation can vary considerably throughout the year and with varying river discharge. In *salt wedge estuaries*, river flow, which is very large, dominates and there is very little tidal mixing. The saline ocean water enters the estuary as a wedge beneath the lighter fresh river water flowing out at the surface. A small amount of this saline water is mixed upward by entrainment into the fresh outflowing upper layer, which becomes somewhat more saline seaward, but there is no mixing of fresh water downward so the salt wedge retains its original oceanic salinity. Thus, there is a sharp salinity change at depth between layers (Figure 7.1a). In order to compensate for the small amount of salt water being transported outward by the surface layer, there is weak flow of seawater inward at depth. The Mississippi River is an example of this type of estuary.

In a *highly stratified (entrainment) estuary* (Figure 7.1 b), the river flow still dominates but the tidal currents cause more mixing of saline water upward (entrainment) into the seaward-flowing surface layer. The surface layer becomes progressively more saline and its volume increases seaward as a result, but because there is very little downward mixing of fresh water, the bottom layer is still nearly of oceanic salinity. Thus, a strong salinity gradient still exists between the surface and bottom layers. The volume of water flowing seaward in the surface layer is often 10 to 30 times the river flow itself because of the entrainment of saline water from the bottom layer (Pickard and Emery 1982). Deep, narrow estuaries have this type of water circulation (including some fjords with deep sills).

Slightly stratified estuaries grade into highly stratified estuaries. In slightly stratified estuaries, river flow and tidal mixing are both important. Considerable vertical mixing occurs with fresh river water being mixed downward as well as seawater being mixed upward. Although there is still a seaward-flowing surface layer and a landward-flowing bottom layer, the salinity gradient between the surface layers and the bottom is not very sharp (see Figure 7.1c) and the surface layer is only a little less saline than the bottom layer. The James River in Chesapeake Bay, Long Island Sound, and the Thames River estuary are of this type. Salinity increases from the head to the mouth of the estuary in both the surface and deep layers.

Well-mixed estuaries are dominated by tidal currents, which overwhelm the river flow. These are shallow estuaries, and the Severn River estuary in England is of this type. The result is an estuary that is well mixed vertically, with no salinity gradient from the surface layer to deeper layers. There is a gradual increase in salinity horizontally from the head to the mouth of the estuary. In addition, although a complicated current system may develop, there is net flow seaward at all depths (see Figure 7.1d). The river water must still be discharged seaward, and there is also net downstream transport of salt (which is balanced by turbulent diffusion of salt upstream from the sea) (Pritchard and Carter 1973). In wide estuaries of this type, the Coriolis force may produce a strong seaward current on the right side of the estuary and a landward current on the left side (facing downstream in the Northern Hemisphere). In narrow estuaries, however, the lateral current variation is absent because the Coriolis effect is not as strong.

Silled or restricted estuaries are partially closed to the ocean by a shallow near-surface sill and cannot develop a full estuarine circulation (Figure 7.2). The sill or shallow area is generally near the seaward end, and it restricts flow into the deeper portions of the estuary. Fresher water flows seaward in the surface layer, but if the sill is sufficiently shallow, inward return flow of saline ocean water is blocked at depth, and occurs only near the surface in the lower part of an entrainment-type surface layer. There is a very strong halocline, with the bottom water being much more saline and dense than the surface water (silled estuaries are an extreme case of highly stratified estuaries). The strong density stratification prevents vertical mixing between surface and deep water. As a result, analogous to stratified lakes, oxygen depletion in bottom waters

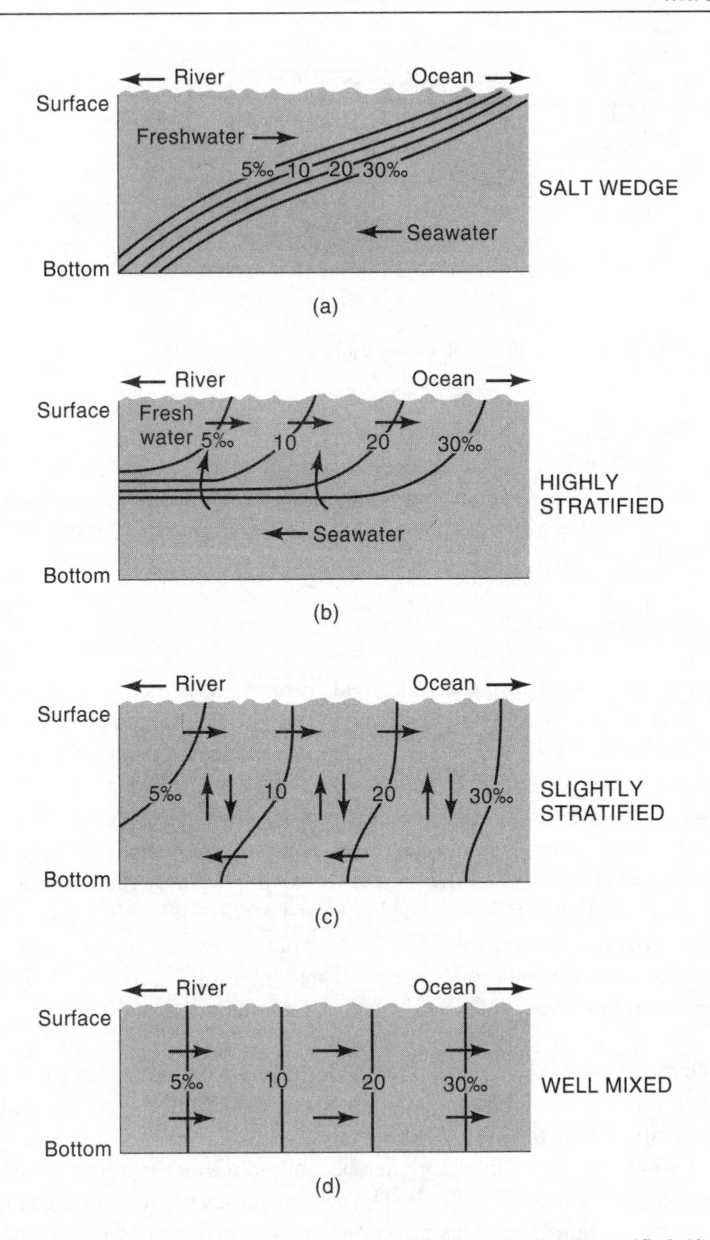

Figure 7.1. Types of estuaries: (a) salt wedge; (b) highly stratified; (c) slightly stratified; (d) well mixed. In all cases, generalized salinity contours (in ‰) are drawn for an idealized longitudinal cross section down the estuary. Arrows represent net water flow, i.e., tidally average. (Adapted from G. L. Pickard and W. J. Emery. *Descriptive Physical Oceanography.* 4th ed., p. 220. Copyright © 1982 by G. L. Pickard, reprinted by permission of the author.)

often takes place. Only occasionally does outside ocean water flow over the sill into the bottom of the basin.

Estuaries with a shallow sill, restricted circulation, and anoxic bottom waters (at least periodically) are typified by fjords. (Much larger-scale examples are provided by the Black Sea, and

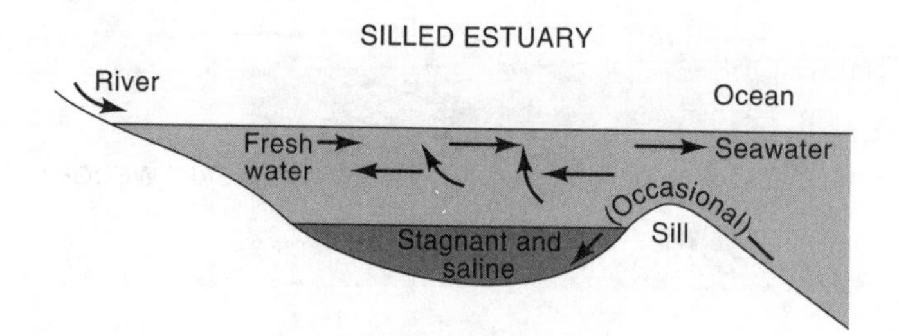

Figure 7.2. Generalized representation of a silled estuary.

parts of the Baltic Sea.) Fjords are deep, long basins with a U-shaped cross section and a sill between the estuary and the ocean (Pritchard and Carter 1973). They are characteristic of the coastal regions of Norway, the Canadian west coast, and Chile, and were formed by the scouring activity of preexisting glaciers. Those fjords that develop anoxic bottom waters have a shallow sill and low river runoff (Pickard and Emery 1982).

The Black Sea

The Black Sea is a gigantic example of a silled "estuary." It is fed by a number of major rivers and is connected to the Mediterranean Sea through the Bosporus and Dardanelles passages, which have shallow sills (40–100 m). Very little saline water from the Mediterranean ever flows over the Bosporus sill to replenish the deeper layers of the Black Sea. As a result, there is a strong density stratification and a sharp halocline between outflowing, less saline (18‰) surface waters, which contain a large component of river water, and inflowing, more saline (22‰) deep water. Because of this strong salinity stratification, little mixing occurs across the halocline and a large mass of anoxic water, rich in H_2S, is found between a depth of 80–130 m and the bottom at about 2000 m. This gives rise to the *chemocline* or boundary between surface oxygenated water and the deep anoxic water. Because of its large size and the limited input of Mediterranean seawater, the deep water has a very long residence time, on the order of 500–1000 years (Falkner et al. 1991).

The depth of the Black Sea chemocline below the water surface has been shown to vary considerably with time, by 30–80 m on the scale of months (Kempe et al. 1990), by 40–50 m for an extended period during the the last 300 or more years (Lyons et al. 1993), and more than this over the past 9000 years since the Black Sea became saline as a result of the postglacial rise in sea level (Sinninghe Damste et al. 1993). The natural causes suggested for changes in the depth of the chemocline include incursions of very saline Mediterranean water into the deeper anoxic layers, interannual and decadal changes in the fresh riverine input to the surface layer, changes in coastal currents due to storms, internal waves in the chemocline, and changes in both plankton productivity and species distribution due to changes in salinity (Murray et al. 1989; Kempe et al. 1990; Sinninghe Damste et al. 1993).

Because the riverine freshwater input into the surface layers has been reduced by 15% since the 1950s due to Russian dams and irrigation, and because there has been greater nutrient input from agriculture and industry (which stimulates planktonic production), several authors (Murray et al. 1989; Falkner et al. 1991) have suggested anthropogenic causes for a recent rise in the chemocline. However, larger changes occur naturally in the Black Sea, so it is not necessary to invoke anthropogenic causes for every change in chemocline depth. Because the Black Sea chemocline is so sensitive to changes, both natural and human induced, it is difficult to distinguish between natural and anthropogenic effects without a great deal of additional information.

Estuarine Chemistry: Conservative versus Nonconservative Mixing

Estuaries and similar marginal marine waters are the principal places where the two major types of earth surface water meet: fresh land-derived water (predominantly river water) and saline ocean water. As we have seen, estuaries vary considerably in how these waters mix, depending largely on the relative influence of river input and tidal mixing combined with basin geometry. Salinity changes are variable both between different estuaries and within any one estuary. In addition, there are temporal variations in salinity due to changes in the amount of river runoff and in the tides. River runoff varies both seasonally and annually, depending on the amount of rainfall and the incidence of floods. Because of time variations, chemical measurements in estuaries have to be made over the whole year and over several different years to be representative (Aston 1978).

Besides mixing of fresh and saline water in estuaries, there are internal processes within the estuary itself that can change the chemical composition of the water. Exchange of both dissolved and particulate matter occurs between the sediments on the bottom of the estuary and the overlying water. In addition, considerable biological activity occurs in the estuarine water, in the surrounding marsh tidal areas, and in the bottom sediments. Nutrients (C, N, P, Si) are cycled biologically within the estuary and, as a result, dissolved and particulate organic matter is both produced and consumed. Humans cause changes in estuaries both in the amount and type (e.g., sewage sludge) of suspended sediment and of dissolved material reaching estuaries through rivers and land runoff from surrounding urban and rural areas (see Chapter 5). Nutrients are particularly affected by pollution, and estuaries, because they retain water for appreciably long periods, can become eutrophic in a manner similar to lakes (see Chapter 6). There is also concern about trapping of anthropogenic trace metals in estuaries.

The time that a river-borne dissolved constituent or pollutant spends in an estuary obviously affects how available it is for sediment exchange or biological processes. A measure of how long it would take to remove a pollutant that does not undergo sediment exchange or biological cycling is given by the flushing time. The *flushing time* is defined as the length of time required to replace the existing volume of fresh water in the whole estuary, or some part of the estuary, at the river discharge rate (Aston 1978). Thus, the flushing time τ is analogous to the replacement time for water in lakes, as given in Chapter 6, and is equal to the total volume of fresh water in the estuary (V_f) divided by the rate of river discharge into the estuary (R):

$$\tau = \frac{V_f}{R}$$

A representative average flushing time for a vertically well-mixed estuary is of the order of days (1–10 days). This is less than the residence time of most lakes, which are measured in years, but longer than that for many rivers.

It is desirable to know from the point of view of geochemical cycling whether element fluxes calculated from river water concentrations really represent the flux reaching the ocean after passing through estuaries, or whether the river flux of an element is reduced and/or added to in passage through the estuary. In the idealized model for mixing of river water and seawater in an estuary (Boyle et al. 1974; Liss 1976; Officer 1979; Loder and Reichard 1981; Kaul and Froelich 1984), measured concentrations of the dissolved river water constituent that is being studied are plotted against corresponding measured values of a dissolved estuarine constituent which is assumed to behave *conservatively* (i.e., to show no loss or gain during mixing). Measurements are generally made on a series of samples collected along the length of the estuary from the river mouth to the ocean. The conservative constituent is generally either total salinity or chloride concentration. In the case of a constituent of interest that is also conservative, its measured concentration plotted against increasing chloride concentration (as a measure of mixing with oceanic water) should lie on a straight line between the concentration of the element in river water (C_R) and its concentration in the oceanic water (C_S) with which it is mixing. This is the theoretical dilution line for a conservative constituent (Liss 1976), which is illustrated in Figure 7.3.

Some elements behave *nonconservatively* in estuaries; that is, they are removed from or added to solution during mixing. If the dissolved component is being added as the salinity increases, then a concave-down curve above the dilution line will result. Conversely, if the constituent is being removed during mixing, a concave-up curve below the dilution line will result. This is also illustrated in Figure 7.3. To generalize, then, a constituent that is conservative in estuaries will show a straight-line plot versus salinity (or chloride concentration), and the plot of a nonconservative component will show curvature.

These simple mixing models assume that there are only two well-defined end-member concentrations (river and oceanic). However, if there is a third end member such as another tributary to the estuary, then behavior of a dissolved constituent may appear erroneously nonconservative (Boyle et al. 1974). In addition, the concentrations of the oceanic and riverine end members often are either not well known or variable. The oceanic end member in coastal or shelf water may have an intermediate salinity, less than that of the open ocean, and by erroneously assuming ocean salinity (35‰), one can introduce curvature into an otherwise straight-line plot (Boyle et al. 1974). Also, it is often difficult to determine the exact concentration of the component of interest in the oceanic end member because of problems in locating and sampling the oceanic end of the salt gradient. Furthermore, river concentrations may vary with time: regularly in some estuaries, such as in the Florida Ochlocknee estuary, where nutrients show a yearly sinusoidal concentration curve (Kaul and Froelich 1984); or irregularly, such as is represented by pollutants and silica in the Tamar estuary of England (Morris et al. 1981). This riverine variation can introduce curvature into mixing plots, particularly in estuaries that are not rapidly flushed (Officer and Lynch 1981; Loder and Reichard 1981). As these problems demonstrate, one should apply caution when using mixing plots of the types shown in Figure 7.3 to elucidate estuarine chemical processes.

ESTUARINE CHEMICAL PROCESSES

The dissolved constituents of estuarine water can be divided into two groups (Liss 1976):

1. Those which are more abundant in seawater than in river water (e.g., Ca, Mg, Na, K, Cl, and SO_4)
2. Those which are more abundant in river water than in seawater (e.g., Fe, Al, P, N, Si, dissolved organic matter)

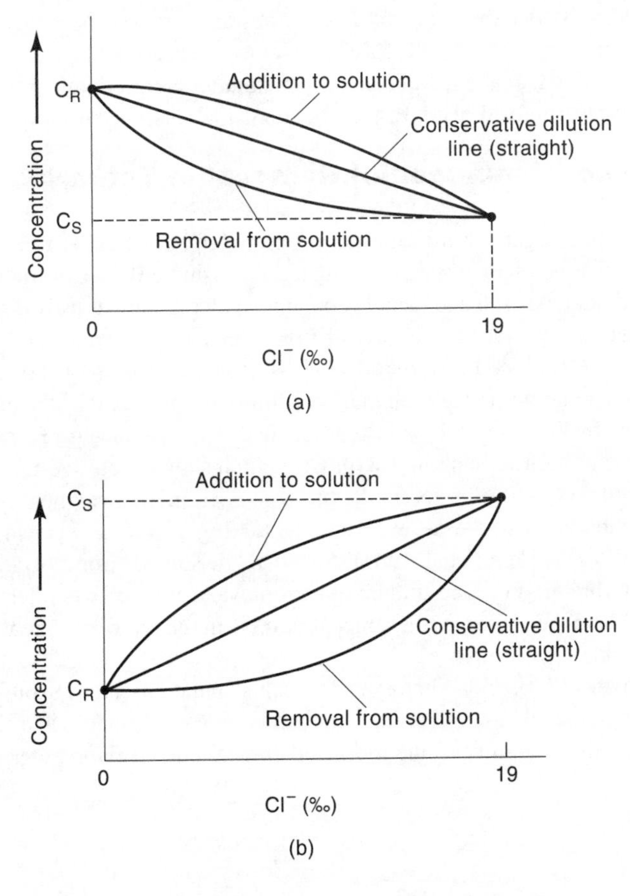

Figure 7.3. Idealized plots for estuarine water of the concentration of dissolved components versus chloride (which serves as conservative measure of the degree of mixing between fresh water and seawater). C_R = concentration in river water; C_S = concentration in seawater. (a) Component whose concentration in fresh water is greater than it is in seawater (for example, P, N, Si). (b) Component whose concentration in fresh water is less than it is in seawater (for example, Ca, Mg, K). (Modified from P. S. Liss, "Conservative and Non-Conservative Behavior of Dissolved Constitutents During Estuarine Mixing," in *Estuarine Chemistry,* ed. J. D. Burton and P. S. Liss, p. 95. ©1976 by Academic Press, reprinted by permission of the publisher.)

Since seawater has a much greater salinity than river water, most of the major dissolved elements have a greater concentration in seawater than in river water. However, metals such as Fe, Al, Mn (and trace metals such as Zn, Cu, Co, etc.) as well as nutrients such as P, N, Si, and dissolved organic matter (DOM) generally have a greater concentration in river water than in ocean water. Here we shall mention only briefly the behavior of the elements that are more abundant in seawater because they are discussed in detail in Chapter 8. In this chapter we shall focus instead on those elements that have greater concentrations in river water than in seawater.

Dissolved constituents that have a greater concentration in river water than in seawater must be removed either in estuaries, where the original mixing of seawater and river water occurs, or later, in the oceans (Mackenzie and Garrels 1966). Thus, there is a reason to suspect that removal of elements such as Fe and Al might occur in estuaries.

The elements that are removed in estuaries are removed by either inorganic (nonbiogenic) processes or by biogenic processes. The elements Fe, Al, and Mn are involved mainly in inorganic removal, whereas Si, N, P, and organic matter are predominantly biogenic. Under certain circumstances, removal of Si, P, and organic C may also occur inorganically.

Inorganic (Nonbiogenic) Removal in Estuaries

Discussion of inorganic removal is here confined to Fe, Al, Si, P, and organic matter. Use of the term "dissolved" refers arbitrarily to material passing a 0.45-μm filter and in fact may consist of fine colloidal material and complex organic matter as well as truly dissolved inorganic species. The inorganic removal of "dissolved" iron in estuaries is well documented (see summary in Boyle et al. 1977; Liss 1976; Aston 1978; Burton 1988). Removal occurs rapidly upon mixing of river water and ocean water in the low salinity (0–5‰) part of the estuary, and most removal is complete by the time 15‰ salinity is reached. The evidence for Fe removal comes from plots of Fe concentration versus salinity (or Cl), which show a concave-up curve below the theoretical mixing line (see discussion in the previous section). An example is shown in Figure 7.4. In addition, laboratory studies of experimental mixing of river and ocean water (Sholkovitz 1976; Boyle et al. 1977; Crerar et al. 1981) show that "dissolved" iron flocculates or precipitates from solution during mixing. The amount of Fe removal estimated by either of these methods is large (50–95%), and the higher the Fe concentration is in the rivers, the greater is the total Fe removal (Boyle et al. 1977).

Sholkovitz (1976), in laboratory experiments on the mixing of Scottish river water with ocean water, found flocculation of Fe, Al, Mn, P, and organic substances in a similar low salinity range. He felt that the solubility of the inorganic constituents in river water and their flocculation in

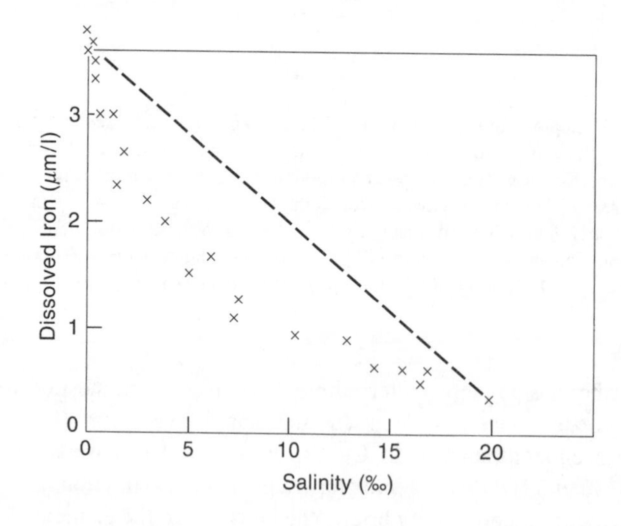

Figure 7.4. Total dissolved iron (μm/l = micromoles per liter) versus salinity in the Merrimack Estuary., Massachusetts. Data points in order of increasing salinity fall on a concave-up (iron removal) curve. Dashed line indicates theoretical conservative mixing between fresh and ocean end members. (Modified from E. A. Boyle, R. Collier, A. T. Dengler, J. M. Edmond, A. C. Ng, and R. F. Stallard, "On the Chemical Mass Balance in Estuaries," *Geochimica et Cosmochimica Acta,* 38, p. 1722 Copyright © 1974 by Pergamon Press, reprinted by permission of the publisher.)

seawater was due to their association with organic matter. The mechanism suggested (Boyle et al. 1977) is the flocculation of mixed iron oxide-organic matter colloids, which are stabilized in river water by the organic matter, but which undergo neutralization by seawater cations of their negative colloid charges. Boyd et al. felt that river-borne "dissolved" Fe in the east coast estuaries of the United States is almost entirely colloidal (Fe oxide particles coated with an organic film) and not truly dissolved. In the laboratory experiments, a large percentage (>50%) of the total Fe, Mn, and P was precipitated, as was a considerable part of the Al (10–70%), but only a small part (3–11%) of the dissolved organic matter. The latter was explained as being due to the removal of only the high-molecular-weight (humic acid) part of the DOM (Sholkovitz et al. 1978). It is worth noting that this removal of DOM is too small to be apparent on DOC (dissolved organic carbon)-versus-salinity plots.

When highly acid rivers reach the ocean in estuaries, the consequent rise in pH can bring about precipitation of some of the dissolved species. Crerar et al. (1981) found that estuarine removal of Fe, Al, and dissolved organic matter from the organic-rich, acid Pine Barrens rivers (pH 4–5) of New Jersey all occurs by a common physical-chemical mechanism: acid neutralization combined with flocculation. They found that more than 50% of the Fe and Al in these acid rivers was truly dissolved (predominantly as dissolved inorganic Fe with a small amount of soluble Fe-organic complexes), and the rest was colloidal (mixed Fe oxyhydroxide and organic colloids). Upon increase of pH during mixing with seawater, the dissolved inorganic Fe and Al become supersaturated and precipitate as Fe and Al oxyhydroxide floccules along with the preexisting Fe and Al colloids and high-molecular-weight humics. Although a large percentage of the Fe and Al is removed, only about 10% of the DOM is removed [similar to Sholkovitz's (1976) estimate].

In highly polluted estuaries, such as the Belgian-Dutch Scheldt Estuary (Wollast 1983), where anoxic bottom waters exist, reactive Fe^{3+} hydroxides carried by the river are reduced to dissolved Fe^{2+} ions. Phosphate, previously adsorbed by the Fe^{3+} hydroxides, is released to solution. When the estuarine waters become more oxygenated farther downstream, the ferrous (Fe^{2+}) iron is reprecipitated as ferric (Fe^{3+}) hydroxide, which removes dissolved phosphate by readsorption. Similar effects occur at the sediment–water interface, where sediments are exposed to alternately anoxic and oxygenated conditions (Krom and Berner 1981; Klump and Martens 1981). This is especially true of deltaic estuarine regions, such as that of the Amazon River, where sediments undergo extensive resuspension due to strong tidal currents (Fox et al. 1986).

Hydes and Liss (1977) found that 30% of "dissolved" riverine Al was removed in the Conway estuary (U.K.) during the early stages of mixing (i.e., < 8‰ salinity). They have suggested that the mechanism for removal of Al is the flocculation of very fine clay particles containing adsorbed Al which are suspended in fresh water but which are irreversibly coagulated upon entering the estuary. Dion (1983) found that, in the Connecticut River and Amazon River, Al seems to be associated with humic acids, probably as organic complexes. This would suggest that Al removal by flocculation of humics (Sholkovitz 1976) is the likely removal mechanism. However, as pointed out by Dion, when humic material (including complexed Al) is adsorbed on clay particles, the two mechanisms become the same.

An alternative model for the flocculation of Al has been suggested by Mackin and Aller (1984). In this case, Al is displaced from river-borne organic complexes and adsorption sites by the increase in cation concentrations and rise in pH in estuaries. The displaced Al reacts in solution with H_4SiO_4 and cations to form *authigenic* (newly formed) aluminosilicates, which precipitate out. In addition, bottom sediments can act as either a sink or a source for Al. Al can diffuse into the sediments and react with Si in solution to form authigenic aluminosilicates. However, if bottom sediments are resuspended in silica-depleted waters, the clays dissolve and dissolved Al is released back to solution.

Because of the documentation of extensive "dissolved" Fe and Al, and lesser DOM removal in estuaries, the river flux of these elements cannot necessarily be considered as the ocean input. However, the question arises as to whether the estuarine removal is permanent and whether material is remobilized and carried out of estuaries (Boyle et al. 1977; Bewers and Yeats 1980). Although river Fe flocculates extensively in the Pine Barrens estuaries, it is not found concentrated in the bottom sediments, leading to the asssumption that most flocculated Fe is carried out onto the shelf by currents (Coonley et al. 1971; Crerar et al. 1981). Thus, basically, the further removal of Fe from estuaries is controlled by the removal of particulates. In addition, ferric iron that is precipitated onto the bottom sediments or originally deposited as Fe coatings on clay minerals can be remobilized as ferrous iron by reactions in reducing sediments and then released either by wave, current, and biological stirring of bottom sediments or by diffusion into bottom water. Thus, there is potentially an Fe source in estuarine bottom sediments.

Although silica removal in estuaries is predominatly affected by biological processes, under certain circumstances it can be nonbiological. (Inorganic removal is roughly 10–20% of total silica removal where it occurs, and in all cases less than 30%.) Liss (1976) noted that nonbiological silica removal can occur in some estuaries, but only during the early stages of fresh water–seawater mixing within the low-salinity (0–5‰) part of the upper estuary. Also, the presence of suspended matter as well as a high dissolved silica (roughly >14 mg/l SiO_2) in the river runoff seems to be required. Liss suggested that some sort of buffering mechanism is involved which requires high silica concentrations, which are present only before river water has been mixed appreciably with seawater. The buffering mechanism may involve adsorption on colloidal ferric and aluminum hydroxides formed during early river and ocean water mixing (Faxi 1980), which could help explain the importance of suspended matter.

Estuarine studies of the behavior of Na, K, Mg, Ca, and SO_4 which are much more abundant in seawater than in river water, have shown that they are essentially conservative upon mixing of seawater with river water (see reviews by Liss 1976 or Aston 1978). This does not mean that reactions do not occur, but only that, because of their large concentrations in the seawater end member, small changes in concentration during estuarine mixing are very difficult to detect via standard mixing models. (Documentation of reactions involving the major ions, upon the addition of river-borne clays to seawater, are given in Chapter 8 based on laboratory experiments and studies of estuarine and marine sediments.)

Biogenic Nutrient Removal in Estuaries: Nitrogen and Phosphorus

Nutrients, specifically N and P (Si will be discussed later as a special case), are generally more abundant in river water than in oceanic water, partly because of their removal by organisms from oceanic water and partly because of the input to rivers from pollutants and land weathering. As a result, rivers serve as nutrient sources for estuaries and here biogenic removal is often found. Minute phytoplankton (unicellular floating organisms), particularly diatoms and algae, are the organisms most often responsible for the removal of dissolved nutrients from estuarine surface water. The amount of phytoplankton activity is measured in terms of the net primary production in grams of carbon fixed (in organic matter) per square meter of water surface area per year (g $C/m^2/yr$). The less productive parts of the open ocean, where there is no upwelling, tend to have a primary productivity of around 50 g $C/m^2/yr$ or less (see Figure 8.4). The primary productivity of estuarine and coastal waters is much higher; a typical value is 230 g $C/m^2/yr$ (Wollast 1993). The amount of phytoplankton productivity will generally be higher in polluted areas, which have a greater nutrient supply.

In order for phytoplankton "blooms" or large concentrations of the organisms to occur in estuaries (which occur seasonally in temperate climates), the surface water must be fairly clear so that the light necessary for photosynthesis can pass through. Thus, large amounts of suspended sediment tend to inhibit organic growth as in the Yangtze and Amazon estuaries, and diatom blooms tend to occur in areas where the suspended load has been deposited and the water has cleared (Milliman and Boyle 1975). Because of sediment resuspension, turbulent water is also less favorable for phytoplankton growth than relatively placid water, where the sediment can settle out.

Another factor that influences the quantity of phytoplankton (i.e., primary productivity) in an estuary is the flushing time (see definition above) for fresh water. A longer flushing time favors greater phytoplankton growth and therefore greater biogenic nutrient removal. The flushing time for fresh, nutrient-rich water is longer in a well-mixed estuary than in a stratified estuary, where the river water rapidly flows out in the surface layer. Flushing time also affects the amount of recycling of nutrients that occurs. If the estuary is flushed rapidly, there is not enough time for appreciable nutrient regeneration because planktonic debris is carried out of the estuary. Also, in a highly stratified estuary, nutrients that are remineralized and returned to the bottom water will accumulate there. In this way the bottom water acts as a so-called nutrient trap (Redfield et al. 1963).

In addition to dissolved nutrients carried in freshwater river runoff into estuaries, there are a number of other *external* nutrient sources for phytoplankton in the surface waters of estuaries and coastal waters. These sources, which vary from estuary to estuary, include (1) fixation of atmospheric N_2 as plant N [Wollast (1983) for the North Sea], (2) atmospheric deposition primarily of nitrogen in rain (particularly in heavily populated areas), (3) regeneration (transfer to solution) of nutrients from particulate organic and inorganic material carried by rivers [e.g., Edmond et al. (1981) and Berner and Rao (1994) for the Amazon] and (4) coastal upwelling or lateral advection of deeper nutrient-enriched oceanic water, which then enters the estuary [e.g., van Bennekom et al. (1978) for the Zaire]. *Internal* estuarine nutrient sources include (1) regeneration of nutrients from the breakdown of internally produced biogenic debris as it passes down through the water column, and (2) benthic biogenic nutrient regeneration from estuarine bottom sediments (e.g., Nixon 1981).

The importance in estuaries and coastal waters of nutrient sources other than rivers on a worldwide basis can be calculated from flux data. An example of the budget for N in the coastal zone (including estuaries) is shown in Figure 7.5. For coastal regions worldwide, less than 3% of N (Wollast 1993) is provided *directly* to phytoplankton by dissolved N carried in rivers. A similarly low proportion is also likely for P. Most of the N and P is supplied by internal recycling within the estuary, and much of this recycled N and P was originally supplied by rivers. Thus, as an *ultimate* source of nutrients, rivers are much more important than what appears in Figure 7.5.

Permanent removal of dissolved nutrients from estuaries can occur by (1) sedimentation and burial of biogenic debris in bottom sediments, (2) denitrification with loss of N_2 and N_2O to the atmosphere, (3) inorganic adsorption of phosphate on particles plus burial in sediments (see previous section), and (4) passage out to sea, the dominant mechanism. Some of the biogenic debris formed in estuaries also may be carried seaward, where it falls into deeper water and undergoes regeneration. Since this also involves the liberation of carbon to subsurface waters, it has been

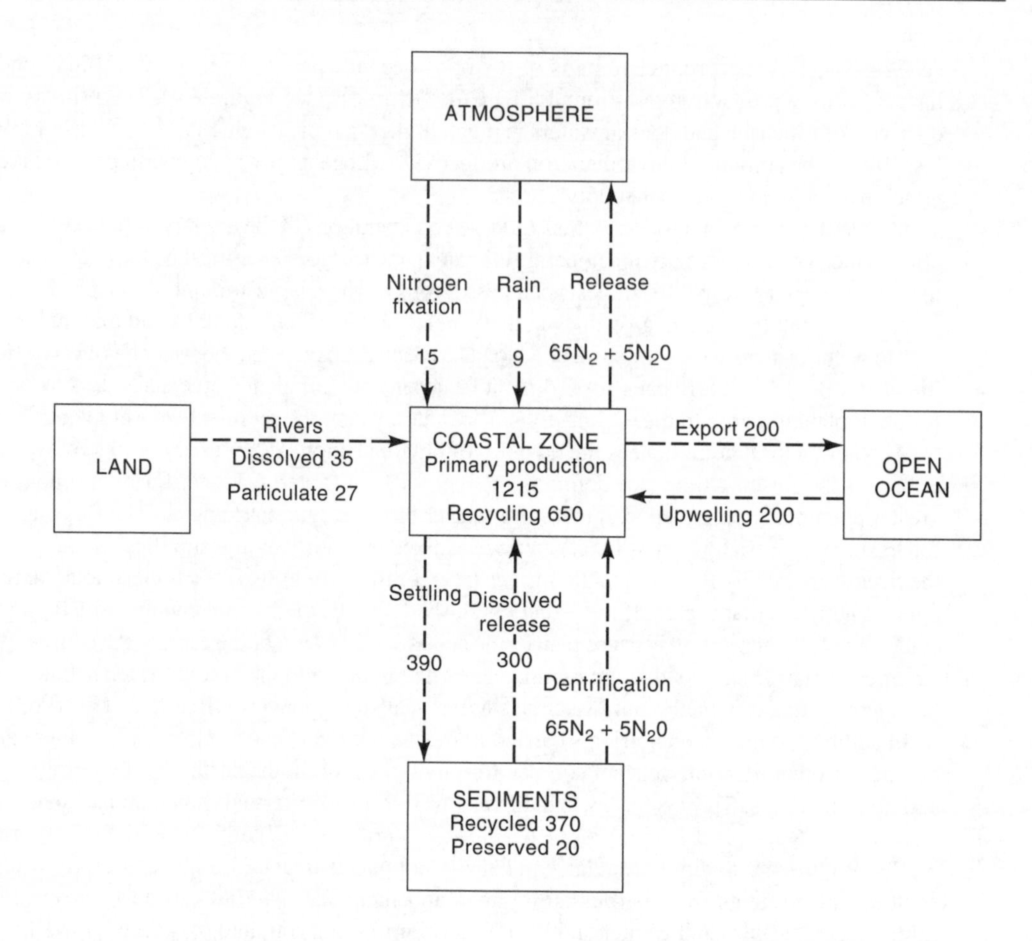

Figure 7.5. Tentative global budget of nitrogen in the coastal zone. Fluxes are given in 10^{12} g N/yr. (Adapted from Wollast 1993.)

mentioned as a possible way of sequestering C that would otherwise be converted to atmospheric CO_2 (Walsh 1991; Wollast 1993).

Nitrogen, Phosphorus, and Limiting Nutrients

In forming organic matter, the nutrients nitrogen and phosphorus are used by phytoplankton, in definite ratios to carbon. The average composition for marine plankton, given in terms of the classic Redfield ratio, $C_{106}N_{16}P_1$ (see Chapter 8), is an idealized ratio, and the actual marine phytoplankton nutrient utilization ratios can vary from 5N:1P to 16N:1P (Ryther and Dunstan 1971), depending on the availability of nutrients in the water and the kind of phytoplankton growth. Ryther and Dunstan (1971) estimate that in North American coastal waters, the average phytoplankton N:P utilization ratio is about 10:1. Apparently, phytoplankton will use more phosphorus relative to nitrogen if more phosphorus is available—the so-called *luxury P consumption* (Redfield et al. 1963).

By analogy with lakes, P might also be expected to be limiting in estuaries. (See Chapter 6 for a discussion of limiting nutrients.) However, nitrogen, not phosphorus, is generally found to be the limiting nutrient in temperate estuaries and coastal waters (Ryther and Dunstan 1971; Howarth 1988; Howarth et al. 1994). This situation has three principal causes. First, the ratio of nitrogen to phosphorus (N:P) in a number of rivers, especially polluted rivers, is lower than that of estuarine plankton, so there is excess P left over in the estuary upon consumption of all N (see Figure 5.10). Second, due to denitrification (the bacterial reduction of dissolved NO_3^- to N_2 and N_2O), nutrient nitrogen can be selectively lost from the estuary (Rowe et al. 1975; McElroy et al. 1978; Seitinger et al. 1980, 1984). Note that in typical coastal waters (Figure 7.5), denitrification represents a loss of 5% of the primary production of N, about equal to all outside input fluxes (rivers, rain, and fixation) except upwelling. Third, upon deposition in sediments, N is regenerated much more slowly than P, such that the N:P ratio of the regenerated nutrients is lower than that used by plankton (e.g., Krom and Berner 1981). (Krom and Berner also found a considerable flux of dissolved PO_4 out of sediments due to the reduction of iron minerals by H_2S in anoxic portions of the sediment.) Finally, temperate estuarine and coastal seas have low rates of N fixation due to light limitation (Howarth 1988).

Because the inputs of N and P and biogeochemical processes in coastal waters vary so much from place to place, there are many exceptions to any generalizations about which nutrient is limiting in any given situation. For example, nitrogen is not limiting in the South Bight of the North Sea, where P is the limiting nutrient in the spring; see van Bennekom et al. 1975. Also, many oligotrophic tropical estuaries and coastal seas tend to be P limited (Howarth 1988; Howarth et al. 1994) due to P adsorption by carbonate sediments and high N fixation by vegetation.

Pollution is a probable contributing factor to nitrogen deficiency, and to nitrogen becoming a limiting nutrient, in many estuaries. Polluted river water along the U.S. east coast formerly was generally enriched in phosphorus over nitrogen, with an average ratio of 5N:1P (Ryther and Dunstan 1971). This ratio is decidedly less than that taken up by plankton in the same area, 10N:1P, causing nitrogen to be the limiting nutrient. However, the relative input of N and P to east Atlantic coastal rivers changed from 1974 to 1981, with nitrate increasing by about 30% while phosphorus generally remained constant or even decreased (Smith et al. 1987.). Thus, the nature of which nutrient is limiting may be changing in this area.

Additional pollutional input comes from the direct input of N to coastal waters from atmospheric deposition (Paerl 1985; Fisher at al. 1988; Paerl et al. 1990). Duce (1991) has estimated the global atmospheric input as about 10–25% of the total N input. Since nitrate is limiting, increased N due to pollution can cause increased primary production, but increasingly higher N:P input could cause P to become limiting.

To obtain the global average N:P ratio added to the oceans from rivers, one can use our river input estimates (Chapter 5) of total N (49–62 Tg/yr) and total P (22 Tg/yr). This results in an atomic ratio of 5–6N:1P. This ratio, however, is misleading because much of the N and most of the P are in solids that are not available to organisms. A better measure of nutrient input is the ratio of reactive N to reactive P, where "reactive" means dissolved N and P plus that N and P added to solution by release from suspended solids carried by rivers. The flux for reactive N is 28–42 Tg N/yr and for reactive P it is 5 Tg P/yr, from which we obtain the ratio 13–19N:1P, a value that is rather near the Redfield ratio of 16:1. Most of this river input goes into estuaries and coastal waters, where the nutrients are taken up by photosynthesis. However, efficient nutrient regeneration results in most of this material being released to solution and carried farther seaward (see Figure 7.5).

A few studies of N and P cycling have been done in unpolluted estuaries of different types which should be more representative of the natural situation. In the well-mixed Ochlockonee estuary in Florida, Kaul and Froehlich (1984) found that N limitation and the low measured sediment regeneration ratio of $3NO_3$-N:$1PO_4$-P is due to both nitrate losses and phosphate gains. Some 20% of the river NO_3^-N flux to the estuary does not reach the ocean because all nitrate removed by diatoms in the estuary is not regenerated to solution, whereas phosphate removed by diatoms is completely regenerated. In fact, the dissolved phosphate flux from the bottom is actually greater than that from diatom removal due to bottom regeneration of river-borne particulate P. Nitrogen limitation has also been reported from the estuaries of the Amazon (Edmond et al. 1981) and Zaire (Van Bennekom et al. 1978).

Biogenic Silica Removal in Estuaries

Some planktonic organisms—for example, diatoms—remove silica from solution as well as P and N. Silica removal associated with diatom blooms has been observed in the Amazon River estuary (Milliman and Boyle 1975) and also in a number of other estuaries (Ocklochonee in Florida, Scheldt in Netherlands, San Francisco Bay, Connecticut River, etc.). Milliman and Boyle (1975) and Knappet al. (1981) found that some 20–35% of the Amazon River silica load is removed by diatoms. However, siliceous diatom tests do not seem to be accumulating to any degree in the Amazon estuarine or shelf sediments, most probably because of redissolution (Knapp 1981). As a result, estuarine biogenic silica removal amounts to only about 10% or less of the total Amazon silica load supplied by the river. The estuarine circulation of the Amazon is complex, and part of the silica redissolved in bottom water may be transported by longshore currents or toward the river mouth before it returns to the surface water by upwelling or vertical mixing.

Dion (1983) and Kaul and Froelich (1984) have also observed evidence for extensive redissolution of sedimented diatoms in estuaries of the Connecticut River and Ochlockonee River (Florida), respectively. DeMaster (1981) has estimated that, on a global basis, 20% of riverine silica is removed (both biogenically and inorganically) in estuaries. However, because of evidence for redissolution of silica in a number of estuaries, this may be a maximum value; the actual net Si removal may be less than 20% of the river flux. An exception is eutrophic estuaries, where removal of as much as 50–60% of riverine Si input may occur (Billen et al. 1991).

Eutrophication and Organic Matter Pollution of Estuaries

Humans introduce a variety of dissolved pollutants into estuaries, for example, hydrocarbons, heavy metals, and bacteria. Here discussion is confined to estuarine pollution arising from (1) organic matter enrichment and (2) nutrient enrichment. The human addition of nutrients (which brings about the artificial enhancement of planktonic production) and organic wastes can be referred to together as *cultural eutrophication* (Likens 1972), and we shall adopt this definition here. (For a discussion of definitions and further details on eutrophication, as applied to lakes, consult Chapter 6.)

Direct organic matter enrichment results from the addition to the estuary of large quantities of dissolved and particulate organic carbon and organic nitrogen, mainly from sewage. As we have discussed in previous chapters, respiration attending bacterial decomposition of this organic matter consumes dissolved O_2, so that one result of organic matter enrichment in estuaries is the depletion of dissolved oxygen, particularly if there is not rapid resupply of O_2 by the estuarine circulation combined with air–water exchange. Oxygen depletion is greatest in bottom waters, leading in some extreme cases to totally anoxic conditions at depth. An example of the latter is the Scheldt estuary in Belgium-Holland (Wollast 1983), where anoxic conditions extend over a length of over 30 km along the bottom. Some U.S. estuaries that exhibit oxygen depletion due to organic matter enrichment from sewage include the Delaware estuary, the Houston ship channel, the Hudson River estuary, and New York Harbor (O'Connor, et al. 1975; Simpson et al. 1975). Under unusual circumstances, a high degree of oxygen depletion can occur even in unpolluted estuaries which receive large quantities of natural organic matter—for example, in the Zaire estuary (Van Bennekom et al. 1978).

In water-quality papers, the amount of decomposable organic matter in water is often referred to in terms of its *biochemical oxygen demand* (BOD), which is a measure of the amount of dissolved oxygen consumed during decomposition of the organic matter in the water by microorganisms. Sewage treatment plants can reduce the biochemical oxygen demand of waste waters. For example, with secondary sewage treatment, two thirds of the oxygen demand of sewage waters entering New York Harbor is removed.

Nutrient enrichment in estuaries, primarily increases in dissolved inorganic nitrogen and phosphorus, leads to excessive phytoplankton (or algal) growth, which in turn leads to increased O_2 depletion at depth. Nutrient enrichment usually involves the dissolved inorganic nitrogen forms NH_4^+, NO_3^-, and NO_2^- along with dissolved phosphate (and occasionally Si). Sewage, even treated sewage, is a particularly rich source of ammonia and phosphate, because the removal of dissolved organic matter during secondary waste treatment still leaves the P and N in solution (Simpson et al. 1975).

The response of phytoplankton to increased pollution-derived nutrients (N, P, and Si) is more complex than bacterial response to increased organic matter and varies from estuary to estuary depending on the estuarine circulation and original nutrient supply and balance. (See earlier discussion on estuarine nutrients.) In many eastern U.S. estuaries and coastal waters, where nitrogen tends to be limiting and the waters have lower N:P ratios than are normally required by phytoplankton, nitrogen from pollution is often quickly consumed by phytoplankton, leaving excess PO_4 in the water (Ryther and Dunstan 1971; Nixon 1981). For this reason, excess PO_4 in coastal waters off the eastern United States tends to be a tracer of nutrient and organic pollution. In this case, reductions in pollutive P (e.g., in detergents) will not greatly decrease algal growth, but reductions in pollution-derived N will (Ryther and Dunstan 1971; Howarth et al. 1994). A feedback mechanism to remove N exists so that in polluted waters, higher productivity, accompanied by greater organic matter deposition to sediments, tends to favor denitrification, with N loss to the atmosphere limiting the regeneration of N to solution. This occurs until sediments become anoxic and denitrification is limited (Wollast 1993); thus, denitrification cannot completely compensate for excessive nitrogen loading (Seitzinger and Nixon 1985).

Other factors besides the concentrations of nitrogen and phosphorus are also important in determining the extent of algal growth. For example, in New York Harbor there is a low algal population, despite large pollutive concentrations of nitrogen and phosphorus, because there are high suspended matter concentrations which limit light and also because of rapid nutrient flushing times by the estuarine circulation. Phosphate, for example, has a residence time of only a week or less (Simpson et al. 1975).

In Europe, large-scale algal blooms occur in the South Bight of the North Sea along the Dutch coast, where some 50% of the water from the polluted Rhine, Meuse, and Scheldt rivers is transported. Van Bennekom et al. (1975) found that for the summer diatom blooms, silica is the limiting nutrient; whereas during the short spring algal blooms, phosphorus is limiting. Some estuaries (such as Chesapeake Bay) shift seasonally between N and P limitation (Howarth et al. 1994). This contrasts with the nitrogen-limited estuaries discussed earlier and points out both the difficulties in making generalizations about pollution-derived nutrient overloading in estuaries and the need for more studies of individual estuaries.

SUSPENDED SEDIMENT DEPOSITION IN MARGINAL MARINE ENVIRONMENTS

Most river-borne suspended sediment is deposited in deltas, estuaries, and other coastal marine environments (Gibbs 1981; Eisma 1988; Milliman 1991). Thus, from the point of view of geochemical cycling, it is important to study how this deposition comes about. River-borne suspended sediment consists of both inorganic particles (clays, iron oxide aggregates, etc.) and organic particles. We have already discussed the cycling of Fe, Al, Si, N, and P in estuaries and will confine the discussion here to the removal of the bulk of fine-grained river-borne inorganic and organic particulate material or, in other words, suspended sediment. (A number of heavy trace metals, such as Cu, Zn, and Pb, tend to be associated with fine suspended sediments and thus will be removed with them; see Turekian et al. 1980).

There are a number of sources of coastal and estuarine suspended sediment besides river-borne material (Bokuniewicz and Gordon 1980; Postma 1980; Eisma 1988). Coastal erosion is a sediment source, as is offshore oceanic suspended sediment carried to the coasts (Meade 1972; Milliman 1991). In addition, organisms produce suspended organic particulate material. However, we shall mainly discuss the removal of river-borne suspended sediment, since this is most interesting from the point of view of overall geochemical cycling.

Large-scale deposition of fine river-borne suspended sediment occurs in many deltas and estuaries and on the continental shelves. Several factors influence the transport and deposition of such suspended sediment: (1) physicochemical processes that cause flocculation and aggregation of river-borne suspended particles on the transition from fresh water to seawater; (2) estuarine circulation and other hydrologic processes such as variations in current velocity due to alternating tides and changes in cross-sectional area; and (3) agglomeration by organisms as fecal pellets. The relative importance of these processes varies depending on conditions and has been the subject of some debate (Dyer 1972, 1986; Meade 1972; Kranck 1973; Krone 1978; Burton 1976; Aston 1978; Bokuniewicz and Gordon 1980; Postma 1980; Eisma 1988; Burton 1988).

Flocculation (aggregation) of river-borne suspended particles involves the formation of large particle aggregates that have a greater settling rate than the original suspended particles. Flocculation and aggregation are tied up with two important factors: cohesion of the particles and collision of the particles (Kranck 1973, Kranck 1984; Krone 1978; Eisma 1986).

Clay minerals (kaolinite, illite, and smectite) that are carried in colloidal suspension in river water have a net negative charge on their faces due to a variety of causes. [This discussion is largely from Van Olphen (1977), to which the reader is referred for further details.] Each clay particle attracts a layer (Gouy layer) of positively charged cations around it to balance this negative charge. This results in excessive cation concentration around each clay particle relative to the bulk fresh water between particles. Thus, colliding clay particles repel one another and they do not aggregate. There is also an attractive force between the clay particles (van der Waals force), but it is less strong in fresh water than the repulsive force due to the excess cation concentration around each particle. Since saline water has a greater concentration of dissolved charged ions (ionic strength) than fresh water, when river-borne clay particles encounter greater salinity, the concentration of ions in the water between particles increases. This brings about a thinning, or collapse, of the Gouy layer so that the charged particles may come closer together before any repulsion occurs. Thus, in saline water, the attractive or cohesive (van der Waals) forces become stronger than the repulsive forces due to positive charge, and cohesion between clay particles can occur. Since clay particles are always cohesive at a salinity greater than 1–3‰, increases in salinity beyond the initial mixing of fresh and saline water are not important in bringing about cohesion (Krone 1978).

The stability of clay suspensions against flocculation, when subjected to saline water, is dependent on not only the salinity of the water but also on the type of clay involved. In experiments with brackish estuarine water, Edzwald (1974) found that, at the same salinity, kaolinite was less stable toward flocculation than illite. In the Pamlico River estuary (North Carolina), Edzwald et al. attributed the distribution of kaolinite and illite in the bottom sediments to their relative colloidal stability. Kaolinite, being less stable, flocculated more rapidly than illite and was found to be more concentrated than illite in the sediments near the river mouth. Krone (1978) quotes work that shows that the flocculation sequence with increasing salinity is kaolinite before illite before smectite, with all flocculating at low salinities (<3‰).

Gibbs (1977) found a sequence of clay minerals deposited off the Amazon River mouth that was reminiscent of what would be expected for differential flocculation. Smectite increased with distance offshore, while illite decreased greatly and kaolinite decreased less so. However, Gibbs attributed the laterally changing concentrations of clay minerals not to clay flocculation but to physical sorting of sediments by size. Differential clay flocculation was felt to be less important because possible natural organic and metal oxyhydroxide coatings on clay particles could modify the surface properties of the clays. Organic coatings (particularly of humic acid) probably produce changes in surface chemistry (Burton 1976), although organics have a net negative charge in fresh water similar to clay minerals. Kranck and Milligan (1980) found that organic matter combined with mineral particles causes flocculation and greatly increases the settling rate. (See earlier, under inorganic removal, for a discussion of the estuarine removal of iron and organics.)

In order for clay particles to flocculate fully, they must also collide with other clay particles. Collisions between suspended particles can occur due to several mechanisms (Krone 1978): (1) Brownian motion due to thermally induced water molecule motion, (2) differential settling velocities of different suspended particles, and (3) velocity gradients that cause particles of different speeds to collide with one another. With very high suspended sediment concentrations, the first two mechanisms for particle collisions, Brownian motion and differential settling, are important and cause the formation of floccules or aggregates of cohesive clay particles. Particles settle out in restricted quiet areas or at times in the tidal cycle when there is little motion. Since these aggregates are weak, they tend to be broken up by high velocities (Krone 1978).

Fluid mud, a dense fluid layer with very high concentrations of suspended clay floccules (2000–20,000 mg/l; Dyer 1972), often forms on the bottom from collisions due to Brownian motion and differential settling. Fluid mud forms when there is a very large sediment supply (and low population of benthic animals), such as in the estuaries of the Amazon, Chao Phraya, Thames, and Severn rivers (Kineke and Sternberg 1992; Eisma 1988; Wells 1983; Dyer 1972; Meade 1972). This fluid mud is ephemeral and can be disturbed by strong currents, as in the Amazon estuary (Kineke and Sternberg 1992; Kuehl et al. 1992).

The velocity gradient is very high in the mixing zone of estuaries where fresh water first meets saline water under tidal conditions. This can cause collisions and the rapid formation of strong aggregates when the suspended sediment concentration is lower than that needed for floccule growth by Brownian motion and differential settling. These strong aggregates, formed in areas of large velocity gradients, are resistant to being broken down as long as they do not become too large (Krone 1978). When the suspended sediment concentration is very low (<300 mg/l; Dyer 1972), collisions become too infrequent for appreciable flocculation via any of the above-mentioned processes, and sedimentation occurs instead by means of additional processes such as biogenic removal.

The importance of estuarine circulation processes in the nearshore transport of suspended sediment has been emphasized by Meade (1972) and Postma (1980). In moderately stratified estuaries (Figure 7.1), where there is both vertical mixing and a net landward flow at the bottom of the estuary, sediment while settling out is carried landward and temporarily trapped in the estuary. As a result, in many moderately stratified estuaries there is a zone of maximum suspended sediment concentration (turbidity) bounded by lower concentrations both landward and seaward (see Figure 7.6). This turbidity maximum occurs near the farthest (landward) extent of the bottom landward-flowing saline water, where it meets seaward-flowing fresh water (Figure 7.7). The turbidity maximum is often accompanied by a zone of maximum sediment accumulation, which is also a place of high velocity gradients. Since these high velocity gradients tend to promote more collisions between suspended clay particles, the high sediment deposition may be a result of both estuarine circulation and velocity gradient-induced flocculation and aggregation. A turbidity maximum can also develop in an estuary with dominant tidal flow and tidal assymmetry, where the flood tide in carries more sediment than the ebb flow out (Postma 1980).

In salt wedge estuaries, where river flow is dominant over tidal forces, little mixing occurs between the fresh, seaward-flowing river water in the upper layer and the inward-flowing salt wedge below. Meade (1972) points out that when Mississippi River water, carrying a large suspended sediment load, meets the salt wedge, most of the suspended sediment is carried over the salt wedge by the river water. Because of the steep density gradient and turbulence at the

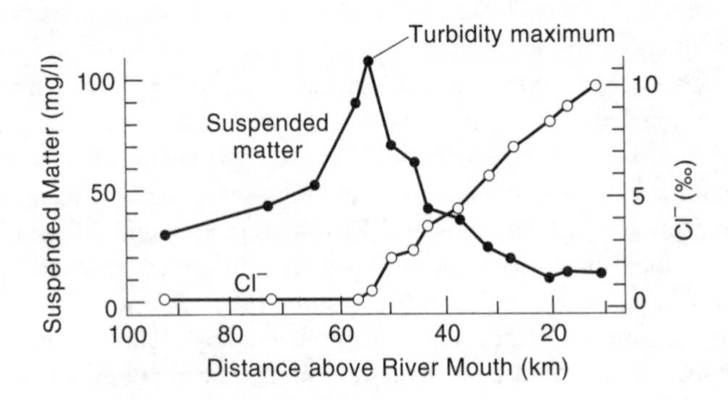

Figure 7.6. Concentration of suspended matter versus chlorinity in York River, Virginia, showing a *turbidity maximum* at the landward limit of sea-salt penetration into the estuary, e.g., where the chloride concentration drops to nearly zero. [After R. H. Meade, 1972, "Transport and Deposition of Sediment in Estuaries," *Geol. Soc. Am. Mem.* 133: 100; and B. N. Nelson, 1960, "Recent Sediment Studies in 1960," *Va. Polytech. Inst. J.* 7(4): 1–4.]

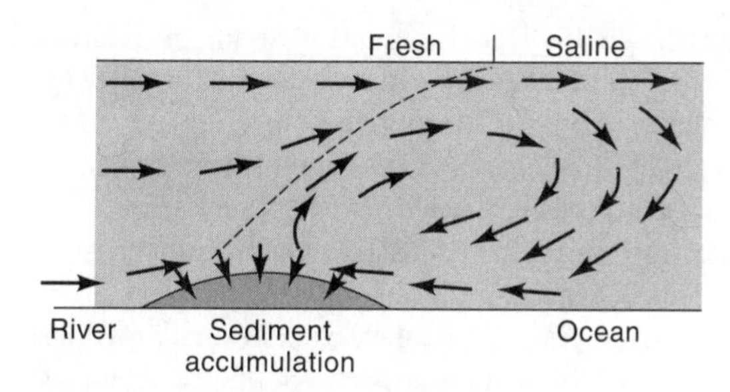

Figure 7.7. Suspended matter transport (arrows) and sediment accumulation near landward limit of seawater mixing in an estuary. Dashed line is fresh–saline water boundary. (After Meade 1972.)

fresh–saline boundary, the river suspended sediment does not settle out and is carried out to sea to be deposited in the submarine Mississippi Delta and on the Gulf of Mexico shelf. This is also true of the Zaire River (Eisma et al. 1978; van Bennekom et al. 1978), where most of the large particulate organic load is carried through the estuary in the surface layer and deposited at the head of a submarine canyon offshore. In some salt wedge estuaries (Zaire, Rhone), part of the suspended sediment may be dropped at the head of the estuary and form a seaward-moving turbid layer just above the bottom (Eisma 1988).

The ability of moving water to transport sediment is dependent on its velocity. As river water flows into an estuary and the cross section for flow widens, the water velocity decreases, its sediment transport ability diminishes, and sediment deposition is likely to occur. The reverse occurs when the flow cross section narrows. These changes affect both river-borne suspended sediment carried into estuaries and sediment carried by tidal flow in and out of an estuary (Dyer 1972).

Extreme hydrologic conditions are often very important in near-shore sedimentation (Dyer 1972). In river floods, there is much greater sediment transport than usual; the sediment load transported in several days may be greater than the total load for an average year. In addition, during river floods, the river flow dominates any estuarine circulation so that, in some estuaries such as the Mississippi River, the salt wedge moves out of the estuary altogether and there is no bottom flow inward (Meade 1972). This means that the coarser suspended sediment is carried out of the estuary, and that sediment previously deposited in the estuary is eroded, flushed out by the dominant river flow (Dyer 1972), and deposited in submarine deltas and on the continental shelf.

Another major process that results in sediment accumulation in coastal regions is biogenic agglomeration. A number of bottom-living (benthic) organisms feed on suspended particulate matter and agglomerate fine-grained suspended matter into fecal pellets that are larger and denser and thus have a higher settling velocity than the original suspended particles (Rhoads 1974). Fecal pellets deposited on the bottom subsequently may be broken down by deposit feeders and either resuspended or agglutinated to the bottom by invertebrates (or plants). Also, organic matter coatings resulting from biogenic reworking increase the tendency of particles to stick together by acting as a type of glue (Eisma 1988). The benthic organisms involved in fecal pellet formation include various bivalves (oysters, clams, mussels, scallops), copepods, tunicates, and barnacles (Rhoads 1974). Bottom feeders are capable of reworking and pelletizing the bottom sediment at a rate greater than the sediment deposition rate, as in Buzzard's Bay (Massachusetts) and Long Island Sound, so that the top centimeter of the bottom sediment may consist entirely of a layer of fecal pellets (Rhoads 1974). In Long Island Sound the biological processing rate is so fast that sediment thickness is not related to estuarine circulation (Bokuniewicz and Gordon 1980).

Over the long run, most sediment brought by rivers passes out of estuaries due to their small size and small storage capacity, and this is evidenced by the paucity of estuarine sediments in the geological record. However, at present, relatively deep, unfilled estuaries are rather common as a result of the drowning of many coastlines by the rapid postglacial rise of sea level. As a result there are numerous instances of estuaries that trap a large proportion of the sediments brought to them by rivers. Examples include the Scheldt estuary (Wollast and Peters 1978), the Gironde estuary (Allen et al. 1976), and Long Island Sound (Bokuniewicz and Gordon 1980).

Most suspended sediment from rivers is presently being deposited on the continental margins (Burton 1988; Eisma 1988; Milliman 1991), which are unusually wide by having been drowned after the last glaciaton. (As a result, most organic carbon is also buried there—see Berner 1982.) Milliman (1991) estimates that for the 10 largest sediment-transporting rivers, only 25–30% of the total riverine sediment reaches deep water off the shelf. The only large rivers to discharge appreciable sediment to deep water are the Mississippi and the Ganges-Bramaputra. Much of the sediment carried onto broad coastal shelves, along passive margins (such as bordering the Atlantic Ocean) is trapped in estuaries and on the shelves, while much of the sediment that is carried onto narrow shelves along active margins (Pacific and northwest Indian Ocean) or which reaches submarine canyons (Zaire River) reaches deeper water. Shallow water deltas form about river mouths as a result of high sediment loads, low tidal range, and limited wave and current action. Only during low stands of sea level in glacial periods is terrigenous sediment discharged directly to the deep sea, probably as turbidity currents resulting from short meltwater bursts.

Human changes that affect the transport of river-borne suspended sediment through estuaries include dredging or dumping of sediment, construction of piers or barriers, and diversion of freshwater flow from the river by dams or reservoirs. Added to this is increased river sediment transport because of deforestation and agriculture (Dyer 1972; Meade 1972; Simpson et al. 1975; Milliman 1991). Dredging of river channels to aid navigation is common and often alters estuarine circulations by allowing the bottom landward-flowing salt wedge, in stratified estuaries, to penetrate further into the estuary. Since sediment deposition often occurs where the saline water meets the fresh river flow, this moves the maximum sediment deposition zone further inland (Dyer 1972; Meade 1972). Conversely, dumping of sediment (as in the Hudson River estuary) interferes with the inward flow of the bottom layer and results in localized natural sediment deposition (Simpson et al. 1975). The construction of piers and barriers results in the reduction of current velocity and consequent sediment deposition. Less fresh river water flow and less downriver sediment transport accompanying the building of dams (for example, the Aswan Dam on the Nile; Milliman 1991), besides causing increased salinity, less nutrient input, and reduced marine productivity in coastal areas, can result in lowered sediment input, less sediment accumulation, greater coastal erosion, and flooding of coastal areas.

ANTIESTUARIES AND EVAPORITE FORMATION

The term *antiestuary* refers to a supersaline coastal body of water that has a circulation that is the reverse of the typical estuarine circulation; that is, seawater flows into the estuary in the surface layer instead of in the bottom layer (see Figure 7.8). Such a reverse flow develops in a coastal embayment or basin because of certain conditions: (1) the basin has a high *net* rate of evaporation (evaporation – inflow from rain and rivers) and (2) circulation with the open ocean is restricted by the presence of a sill, sand bar, reef, etc., at the entrance to the basin. Net evaporation requires a warm, arid climate. The net water loss is made up for by surface seawater flow into the basin from the open ocean in order to maintain constant water volume in the basin (note the landward-sloping water surface in Figure 7.8). The evaporation of seawater in the surface layer of the basin leaves the excess salts behind and, since the barrier prevents mixing with the open ocean, the surface water becomes denser and tends to settle toward the bottom of the basin. If the basin is to maintain a constant salinity, there must be a reverse flow at depth of denser, more saline water out over the sill to remove the excess salts. In this situation, the combination of evaporative water loss and deep return flow of water to the ocean is equal to surface inflow of seawater. The Mediterranean Sea is the best-known present-day example of this type of reverse flow. The salinity of the Mediterranean Sea is maintained at a somewhat greater value (37–38‰; Pickard and Emery 1982) than that of average ocean water (35‰).

If a restricted evaporitive basin becomes completely isolated from the resupply of seawater, water is lost via evaporation and the salinity increases. If the salinity becomes great enough, precipitation of salts takes place. The sequence of salts that precipitates from the evaporation of seawater was determined originally by Usiglio in 1849. When normal seawater has been evaporated to 19% of its original volume, gypsum ($CaSO_4$-$2H_2O$) precipitates and halite (NaCl) precipitates at around 10% of the original volume. Further evaporation will also produce a whole series of Mg and K salts known as bitterns. The relative thicknesses of gypsum:halite:Mg + K salts produced by complete seawater evaporation are approximately 1:20:5 (Hsü 1972). (Some $CaCO_3$ may form via evaporation, but it is practically negligible compared to that removed from seawater by other means such as skeletons of organisms; and much of this skeletal debris can be converted to dolomite $CaMg(CO_3)_2$ by the Mg-rich brines.) The precipitation of gypsum and halite provides an important mechanism for the removal of Na, Cl, and SO_4 from seawater (see Chapter 8).

A major problem with the complete evaporation of seawater in a closed basin is that in order to deposit a 1-m thickness of $CaSO_4$, some 1700 m of seawater must be evaporated (Hsü 1972). In addition, for every meter of $CaSO_4$ deposited, one would expect 20 times as much halite and 5 times as much bitterns. Such ratios are rarely found in ancient evaporite deposits. Instead, one often encounters essentially monomineralic beds of $CaSO_4$ or of halite. In order to provide for greater evaporite deposition without the need to evaporate such great columns of water, and for the formation of thick monomineralic deposits, the barred-basin theory of evaporate deposition with its various modifications has been proposed. (For a discussion of the historical development of theories of evaporite deposition, see Hsü 1972.)

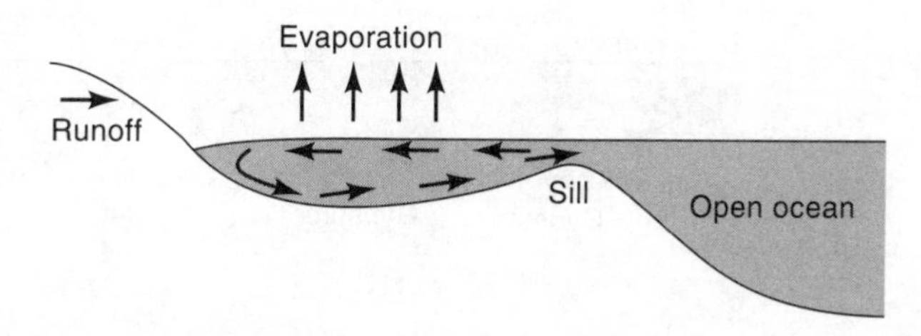

Figure 7.8. Antiestuarine circulation. Arrows indicate direction of water flow.

The theoretical barred basin has reverse circulation of the type described for an antiestuary. Evaporite deposition can occur, even though the water body is not completely cut off from the ocean, because the salinity of the basin water builds up to saturation with saline minerals faster than the water can be replaced by fresh seawater. Water balance is maintained by seawater influx, a very high evaporation rate, and subsurface return to the ocean of very saline brines (which are missing that part of the salts that has precipitated). In this way, a constant concentration of salts necessary for continual deposition of $CaSO_4$, for example, is maintained. As a result, large thicknesses of a single evaporite mineral can be deposited over a period of time without having to evaporite an extremely large thickness of seawater. Variations on this model also permit the deposition of different evaporite mineral facies in different parts of the basin where the salinity is different (see Figure 7.9). For example, as the entering seawater becomes more and more saline, the mineral sequence $CaCO_3$ (calcite) or $CaMg(CO_3)_2$, (dolomite), $CaSO_4 \cdot 2H_2O$ (gypsum), NaCl (halite) develops along the length of the basin with increasing distance from the mouth.

Thick, areally extensive evaporite deposits were formed at various times in the geologic past, for example, the late Miocene evaporites under the Mediterranean basin (Hsü 1972), the well-known Permian Zechstein in Europe (Schmalz 1970), and the Silurian Salina in the Michigan Basin (Briggs 1958). However, most present-day marine evaporite formation is severely limited both in areal extent and in the total removal of major ions from seawater.

There are very few notable present-day depositional environments where evaporite minerals are forming naturally from marine waters. However, the Bocana de Virrilà on the arid Peruvian coast provides one example of a small modern restricted basin. It is a narrow, winding body of water with very restricted circulation (although not with an actual bar), which is precipitating evaporites in lateral zones along its length (Morris and Dickey 1975; Brantley et al. 1984). Seawater enters from the ocean, evaporates, and becomes more saline as it moves inland. Calcite deposition occurs first at the mouth, followed by gypsum precipitation (at salinities greater than 160‰) and then halite as surface water becomes progressively more saline toward the head. The inflow of marine water at least partly balances the water loss by evaporation. However, since the circulation pattern has not been studied, it is not known whether bottom reverse flow of more saline estuarine water also occurs. No Mg and K salts seem to be precipitating, which means that these seawater components must somehow be removed.

The best-known environment of modern evaporite formation is the coastal *sabhka* or *salina*, which is particularly well developed along the Trucial Coast of the Persian Gulf (e.g., Kinsman

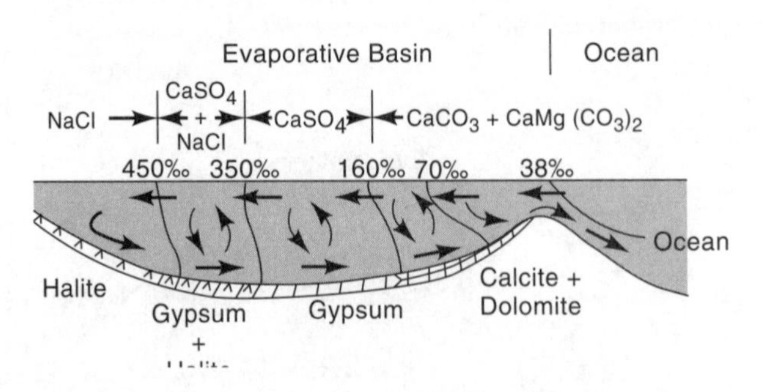

Figure 7.9. Restricted evaporite basin showing zoned deposition of evaporite minerals. Arrows indicate water flow. Contours are for approximate salinities (in ‰). (Modified from Scruton 1953.)

1969; Butler 1969; Shearman 1966, 1978). Here the main evaporite minerals forming are anhydrite ($CaSO_4$) and gypsum ($CaSO_4 \cdot 2H_2O$), and deposition occurs on vast mud flats above the reach of normal tides. Persian Gulf sabhkas occur in areas of very low relief in a very arid climate. The water table is never more than 1–2 m below the surface and consists primarily of seawater-derived brines near the coast, grading into continentally derived brines inland. Seawater occasionally floods the sabhkas during storms, enters the sediments, and then evaporates, resulting in the formation of gypsum and anhydrite within the sediments. Dolomite and additional gypsum is formed by reactions between $CaCO_3$ in the sediment and the residual brines. However, because of redissolution, halite does not accumulate.

REFERENCES

Allen, G. P., G. Sauzay, and J. H. Castaing. 1976. Transport and deposition of suspended sediment in the Gironde estuary, France. In *Estuarine processes,* ed. M. Wiley, Vol. 2:63–81. New York: Academic Press.

Aston, S. R. 1978. Estuarine chemistry. In *Chemical Oceanography,* 2nd ed., ed. J. P. Riley and R. Chester, pp. 361–440. New York: Academic Press.

Berner, R. A. 1982. Burial of organic carbon and pyrite sulfur in the modern ocean: Its geochemical and environmental significance, *Am. J. Sci.* 282: 451–473.

Berner, R. A., and Rao, J.-L. 1994. Phosphorus in sediments of the Amazon River and Estuary: Implications for the global flux of P to the sea, *Geochim. Cosmochim. Acta* 58:2333–2339.

Bewers, J. M., and P. A. Yeats. 1980. Behavior of trace metals during estuarine mixing. In *Proceedings of the Review and Workshop on River Inputs to Ocean Systems,* ed. J.-M. Martin, J. D. Burton, and D. Eisma, pp. 103–115. Rome: FAO.

Billen, G., C. Lancelot, and M. Meybeck. 1991. Chemical exchange at the air–coastal sea interface. In *Ocean Margin Processes in Global Change,* ed. R. F. C. Mantoura, J.-M. Martin, and R. Wollast, pp. 19–44. Chichester, U. K.: John Wiley.

Bokuniewicz, H. J., and R. B. Gordon. 1980. Sediment transport and deposition in Long Island Sound. In *Estuarine Physics and Chemistry: Studies in Long Island Sound,* ed. B. Saltzman, Adv. in Geophysics, vol. 22, pp. 69–106. New York: Academic Press.

Bowden, K. F. 1967. Circulation and diffusion. In *Estuaries,* ed. G. H. Lauff, AAAS Publ. 83, p. 1536. Washington, D.C.: American Association for the Advancement of Science.

Boyle, E. A., R. Collier, A. T. Dengler, J. M. Edmond, A. C. Ng, and R. F. Stallard. 1974. On the chemical mass-balance in estuaries, *Geochim. Cosmochim. Acta* 38: 1719–1728.

Boyle, E. A., J. M. Edmond, and E. R. Sholkovitz. 1977. The mechanism of iron removal in estuaries, *Geochim. Cosmochim. Acta* 41: 1313–1324.

Brantley, S. L., N. E. Moller, D. A. Crerar, and T. H. Weare. 1984. Geochemistry of a modern marine evaporite: Bocana de Virrila, Peru, *J. Sed. Petrol.* 54(2): 0447–0462.

Briggs, L. I. 1958. Evaporite facies, *J. Sed. Petrol.* 28: 46–56.

Burton, J. D. 1976. Basic properties and processes in estuarine chemistry. In *Estuarine chemistry,* ed. J. D. Burton and P. S. Liss, pp. 1–35. London: Academic Press.

Burton, J. D. 1988. Riverborne materials and the continent–ocean interface. In *Physical and Chemical Weathering in Geochemical Cycles,* ed. A. Lerman and M. Meybeck, pp. 299–321. Dordrecht, The Netherlands: Kluwer.

Butler, G. P. 1969. Modern evaporite deposition and geochemistry of coexisting brines, the sabkha, Trucial Coast, Arabian Gulf, *J. Sed. Petrol.* 39: 70–89.

Coonley, L. S., Jr., E. B. Baker, and H. D. Holland. 1971. Iron in the Mullica River and in Great Bay, New Jersey, *Chem. Geol.* 7: 51–63.

Crerar, D. A., J. L. Means, R. F. Yuretich, M. P. Borcsik, J. L. Amster, D. W. Hastings, G. W. Knox, K. E. Lyon, and R. F. Quiett. 1981. Hydrogeochemistry of the New Jersey coastal plain, 2. Transport and deposition of iron, aluminum, dissolved organic matter, and selected trace elements in stream, ground, and estuary water, *Chem. Geol.* 33: 23–44.

DeMaster, D. J. 1981. The supply and accumulation of silica in the marine environment, *Geochim. Cosmochim. Acta* 45: 1715–1732.

Dion, E. P. 1983. Trace elements and radionuclides in the Connecticut River and Amazon River Estuary. Ph.D. dissertation, Yale University, New Haven, Conn.

Duce, R. A. 1991. Chemical exchange at the air–coastal sea interface. In *Ocean Margin Processes in Global Change,* ed. R. F. C. Mantoura, J.-M. Martin, and R. Wollast, pp. 91–109. Chichester, U.K.: John Wiley.

Dyer, K. R. 1972. Sedimentation in estuaries. In *Estuarine Environments,* ed. R. S. K. Barnes and J. Green, pp. 10–32. London: Applied Science.

Dyer, K. R. 1986. *Coastal and Estuarine Sediment Dynamics,* Chichester, U.K.: John Wiley.

Edmond, J. M., E. A. Boyle, B. Grant, and R. F. Stallard. 1981. The chemical mass balance in the Amazon plume, I: The nutrients. *Deep-Sea Res.* 28A(11): 1339–1374.

Edzwald, J. K., T. B. Upchurch, and C. R. O'Melia. 1974. Coagulation in estuaries, *Environ. Sci. Technol.* 8(1): 58–63.

Eisma, D. 1986. Flocculation and de-flocculation of suspended matter in estuaries, *Neth. J. Sea Res.* 20(2/3): 183–199.

Eisma, D. 1988. Riverborne materials and the continent-ocean interface. In *Physical and Chemical Weathering in Geochemical Cycles,* ed. A. Lerman and M. Meybeck, pp. 273–298. Dordrecht, The Netherlands: Kluwer.

Eisma, D., J. Kalf, and S. J. Van Der Gaast. 1978. Suspended matter in the Zaire estuary and adjacent Atlantic Ocean, *Neth. J. Sea Res.* 12: 382–406.

Falkner, K. K., D. J. O'Neill, J. F. Todd, W. S. Moore, and J. M. Edmond. 1991. Depletion of barium and radium-226 in Black Sea surface waters over the past thirty years, *Nature* 350: 491–494.

Faxi, Li. 1980. An analysis of the mechanisms of removal of reactive silicate in the estuarine zone. In *Proceedings of the Review and Workshop on River Inputs to Ocean Systems,* ed. J. M. Martin, J. D. Burton, and D. Eisma, pp. 200–210. Rome: FAO.

Fisher, D., J. Ceraso, and M. Oppenheimer. 1988. *Polluted Coastal Waters: The Role of Acid Rain.* New York: Environmental Defense Fund.

Fox, L. E., Sager, S. L., and Wofsy, S. C. 1986. The chemical control of soluble phosphorus in the Amazon estuary, *Geochim. Cosmochim. Acta* 50: 783–794.

Gibbs, R. J. 1977. Clay mineral segregation in the marine environment, *J. Sed. Petrol.* 47(1): 237–243.

Gibbs, R. J. 1981. Sites of river-derived sedimentation in the ocean, *Geology* 9: 77–80.

Howarth, R. W. 1988. Nutrient limitation of net primary production in marine ecosystems. *Ann. Rev. Ecol. Sys.* 19: 89–110.

Howarth, R. W., H. Jensen, R. Marino, and H. Postma. 1994. Transport and processing of phosphorus in estuaries and oceans. In *Phosphorus Cycling in Terrestrial and Aquatic Ecosystems,* SCOPE, ed. H. Tiessen, in press.

Hsü, K. T. 1972. Origin of saline giants: A critical review after the discovery of the Mediterranean evaporite, *Earth-Sci. Rev.* 8: 371–396.

Hydes, D. J., and P. S. Liss. 1977. The behavior of dissolved aluminum in estuarine and coastal waters, *Est. Coast. Mar. Sci.* 5: 755–769.

Kaul, L. W., and P. N. Froelich, Jr. 1984. Modeling estuarine nutrient geochemistry in a simple system, *Geochim. Cosmochim. Acta* 48: 1417–1433.

Kempe, S. S., G. Liebezett, A.-R. Diercks, and V. Asper. 1990. Water balance in the Black Sea, *Nature* 346: 419.

Kineke, G.C., and R. W. Sternberg. 1992. Fluid muds on the Amazon continental shelf, *EOS* (abstr.) 73: 268.

Kinsman, D. J. J. 1969. Modes of formation, sedimentary associations, and diagnostic features of shallow-water and supratidal evaporites, *Bull. Am. Assoc. Petrol. Geol.* 53: 830–840.

Klump, J. Val, and C. S. Martens. 1981. Biogeochemical cycling in an organic-rich coastal marine basin, 2: Nutrient sediment-water exchange processes, *Geochim. Cosmochim. Acta* 45: 101–121.

Knapp, G. B., D. J. DeMaster, and C. A. Nitrouer. 1981. Processes affecting the uptake and accumulation of silica on the Amazon continental shelf, *Geol. Soc. Am. Abstr.* 13: 488–489.

Kranck, K. 1973. Flocculation of suspended sediment in the sea, *Nature* 246: 348–350.

Kranck, K. 1984. The role of flocculation in the filtering of particulate matter in estuaries. *In The Estuary as Filter,* ed. V. S. Kennedy, pp. 159–175. New York: Academic Press.

Kranck, K., and Milligan, 1980. Macroflocs: Production of marine snow in the laboratory, *Mar. Ecol. Progr. Ser.* 3: 19–24.

Krom, M. D., and R. A. Berner. 1981. The diagenesis of phosphorus in a nearshore marine sediment, *Geochim. Cosmochim. Acta* 45: 207–216.

Krone, R. B. 1978. Aggregation of suspended particles in estuaries. In *Estuarine Transport Processes,* ed. B. Kjerfve, pp. 177–190. Columbia, S.C.: University of South Carolina Press.

Kuehl, S. A., T. D. Paccioni, J. M. Rine, and C. A. Nittrouer. 1992. Seabed dynamics of the inner Amazon continental shelf: Temporal and spatial variability of surface mixed layer, *EOS* (abst.) 73: 268.

Likens, G. E. 1972. Eutrophication and aquatic ecosystems. In *Nutrients and Euthrophication,* ed. G. E. Likens, pp. 3–13. Am. Soc. Limnol. Oceanogr. Spec. Symp. 1: 3–13.

Liss, P. S. 1976. Conservative and non-conservative behavior of dissolved constituents during estuarine mixing. In *Estuarine Chemistry,* ed. J. D. Burton and P. S. Liss, pp. 93–130. New York: Academic Press.

Loder, T. C., and R. P. Reichard. 1981. The dynamics of conservative mixing in estuaries, *Estuaries* 4(l): 64–69.

Lyons, T. W., R. A. Berner, and R. F. Anderson. 1993. Evidence for large pre-industrial pertubations of the Black Sea chemocline, *Nature* 365: 538–540.

Mackenzie, F. T., and R. M. Garrels. 1966. Chemical mass balance between rivers and oceans, *Am. J. Sci.* 264: 507–525.

Mackin, J. E., and R. C. Aller. 1984. Processes affecting the behavior of dissolved Al in estuarine waters, *Mar. Chem.* 14: 213–232.

McElroy, M. B., J. W. Elkins, S. C. Wofsy, C. E. Kolb, A. P. Durdin, and W. A. Kaplan. 1978. Production and release of N_2O from Potomac Estuary, *Limnol. Oceanogr.* 23(6): 1168–1182.

Meade, R. H. 1972. Transport and deposition of sediment in estuaries, *Geol. Soc. Am. Mem.* 133: 91–120.

Meybeck, M. 1982. Carbon, nitrogen and phosphorus transport by world rivers, *Am. J. Sci.* 282: 401–450.

Milliman, J. D. 1991. Flux and fate of fluvial sediment and water in coastal seas. In *Ocean Margin Processes in Global Change,* ed. R. F. C. Mantoura, J.-M. Martin, and R. Wollast, pp. 69–89. Chichester, M.K.: John Wiley.

Milliman, J. D., and E. Boyle. 1975. Biological uptake of dissolved silica in the Amazon River Estuary, *Science* 189: 995–997.

Morris, A., A. J. Bale, and R. J. M. Howland. 1981. Nutrient distributions in an estuary: Evidence of chemical precipitation of dissolved silicate and phosphate, *Est. Coast. Shelf Sci.* 12: 205–217.

Morris, R. C., and P. A. Dickey. 1975. Modern evaporite deposition in Peru, *Am. Assoc. Petrol. Geol. Bull.* 41: 2467–2474.

Murray, J. W., H. W. Jannasch, S. Honjo, R. F. Anderson, W. S. Reeburgh, Z. Trop, G. E. Friederich, L. A. Codispoti, and E. Izdar. 1989. Unexpected changes in the oxic/anoxic interface in the Black Sea, *Nature* 338: 411–413.

Nelson, B. N. 1960. Recent sediment studies in 1960, *Va. Polytech. Inst. J.* 7 (4) 1–4.

Nixon, S. W. 1981. Remineralization and nutrient cycling in coastal marine ecosystems. In *Estuaries and Nutrients,* ed. B. J. Neilson and L. E. Cronin, pp. 111–138. Clifton, N.J.: Humana Press.

O'Conner, D. J., R. V. Thomann, and D. M. Di Toro. 1975. Water-quality analyses of estuarine systems. In *Estuaries, Geophysics and the Environment,* Natl. Res. Council Geophysics of Estuaries Panel, Natl. Acad. Sci, pp. 71–83. Washington, D.C.: National Academy of Sciences.

Officer, C. B. 1979. Discussion of the behavior of nonconservative constituents in estuaries, *Est. Coast. Mar. Sci.* 9: 91–94.

Officer, C. B., and D. R. Lynch. 1981. Dynamics of mixing in estuaries, *Est. Coast. Shelf Sci.* 12: 525–533.

Paerl, H. W. 1985. Enhancement of marine primary production by nitrogen enriched acid rain, *Nature* 315: 747–749.

Paerl, H. W., J. Rudek, and M. A. Malin. 1990. Stimulation of phytoplankton production in coastal waters by natural rainfall inputs: Nutritional and trophic implications, *Mar. Biol.* 107: 247–254.

Pickard, G. L., and W. J. Emery. 1982. *Descriptive Physical Oceanography,* 4th ed. New York: Pergamon Press.

Postma, H. 1980. Sediment transport and sedimentation. In *Chemistry and Biogeochemistry of Estuaries,* ed. E. Olausson and I. Cato, pp. 153–186. New York: John Wiley.

Pritchard, D. W., and H. H. Carter. 1973. Estuarine circulation patterns. In *The Estuarine Environment: Estuaries and Estuarine Sedimentation,* J. R. Schubel, Convener, pp. iv, 1–7. AGI Short Course. Washington, D.C.: AGI.

Redfield, A. C., B. H. Ketchum, and R. A. Richards. 1963. The influence of organisms on the composition of sea-water. In *The Sea,* ed. M. N. Hill, vol. 2, 26–77. New York: Wiley-Interscience.

Rhoads, D. C. 1974. Organism–sediment relations on the muddy sea floor, *Oceanogr. Mar. Biol. Ann. Rev.* (ed. H. Barnes) 1–2: 263–300.

Rowe, G. T., C. M. Clifford, K. L. Smith, Jr., and P. L. Hamilton. 1975. Benthic nutrient regeneration and its coupling to primary production in coastal waters, *Nature* 255: 215–217.

Ryther, J. H., and W. M. Dunstan. 1971. Nitrogen, phosphorus and eutrophication in the coastal marine environment, *Science* 171: 1008–1013.

Schmalz, R. F. 1970. Environment of marine evaporite deposition, *Miner. Ind.* 35(8): 1–7.

Scruton, P. C. 1953. Deposition of evaporites, *Bull. Am. Assoc. Petrol. Geol.* 37: 2498–2512.

Seitzinger, S. P., and S. W. Nixon. 1985. Eutrophication and the rate of denitrification and N_2O production in coastal marine sediments, *Limnol. Oceanogr.* 30(6): 1332–1339.

Seitzinger, S. P., S. W. Nixon, and M. E. Q. Pilson. 1984. Dentrification and nitrous oxide production in a coastal marine ecosystem, *Limnol. Oceanogr.* 29: 73–83.

Seitzinger, S. P., S. Nixon, M. E. Q. Pilson, and S. Burke. 1980. Denitrification and N_2O production in nearshore marine sediments, *Geochim. Cosmochim. Acta* 44: 1853–1860.

Shearman, D. J. 1966. Origin of marine evaporites by diagenesis, *Inst. Mining Met. Trans.* 375: 207–215.

Shearman, D. J. 1978. Evaporites of coastal sabhkas. In *Marine Evaporites,* ed. W. F. Dean and B. C. Schreiber, pp. 6–20. SEPM Short Course no. 4. Oklahoma City, Okla.: SEPM.

Sholkovitz, E. R. 1976. Flocculation of dissolved organic and inorganic matter during the mixing of river and seawater, *Geochim. Cosmochim. Acta* 40: 831–845.

Sholkovitz, E. R., E. A. Boyle, and N. B. Price. 1978. The removal of dissolved humic acids and iron during estuarine mixing, *Earth Planet. Sci. Lett.* 40: 130–136.

Simpson, H. J., S. C. Williams, C. R. Olsen, and D. R. Hammond. 1975. Nutrient and particulate matter budgets in urban estuaries. In *Estuaries, Geophysics and the Environment,* Natl. Res. Council Geophysics of Estuaries Panel, Natl. Acad. Sci., pp. 94–103. Washington, D.C.: National Academy of Sciences.

Sinninghe Damste, J. S., S. G. Wakeham, M. E. L. Kohnen, J. M. Hayes, and J. W. de Leeuw. 1993. A 6000-year sedimentary record of chemocline excursions in the Black Sea, *Nature* 362: 827–829.

Smith, R. A., R. B. Alexander and M. G. Wolman. 1987. Water-quality trends in the nation's rivers, *Science* 235: 1607–1615.

Stommel, H., and H. G. Farmer. 1952. On the nature of estuarine circulation. *Woods Hole Tech. Rep.* 52–63, part 3, chap. 7.

Turekian, K. K., J. K. Cochran, L. K. Benninger, and R. C. Aller. 1980. The sources and sinks of nuclides in Long Island Sound. In *Estuarine Physics and Chemistry: Studies in Long Island Sound,* ed. B. Saltzman, Advances in Geophysics, vol. 22, pp. 129–164. New York: Academic Press.

Van Bennekom, A. J., G. W. Berger, W. Helder, and R. T. P. De Vries. 1978. Nutrient distribution in the Zaire estuary and river plume, *Neth. J. Sea Res.* 12: 296–323.

Van Bennekom, A. J., W. C. Gieskes, and S. B. Tijesen. 1975. Eutrophication of Dutch coastal waters, *Proc. R. Soc. London B* 189: 359–374.

Van Olphen, H. 1977. *An Introduction to Clay Colloid Chemistry,* 2nd ed. New York: John Wiley.

Walsh, J. J. 1991. Importance of the continental margins in the marine biogeochemical cycling of carbon, and nitrogen, *Nature* 350: 53–55.

Wells, J. T. 1983. Dynamics of coastal fluid muds in low-, moderate-, and high-tide environments, *Can. J. Fish Aquat. Sci.* 40(suppl. 1): 130–142.

Wollast, R. 1983. Interactions in estuaries and coastal waters. In *The Major Biogeochemical Cycles and Their Interactions,* ed. B. Bolin and R. B. Cook, pp. 385–407. Chichester, U.K.: John Wiley.

Wollast, R. 1993. Interactions of carbon and nitrogen cycles in the coastal zone. In *Interactions of C, N, P and S Biogeochemical Cycles and Global Change,* ed. R. Wollast, F. T. Mackenzie, and L. Chou, pp. 195–210. Berlin-Heidelberg: Springer-Verlag.

Wollast, R., and J. J. Peters. 1978. Biogeochemical properties of an estuarine system: The river Scheldt. In *Biogeochemistry of Estuarine Sediments,* ed. E. D. Goldberg, pp. 279–293. Paris: UNESCO.

CHAPTER 8

THE OCEANS

INTRODUCTION

In this chapter we shall discuss the largest portion of the world's water, that contained in the oceans. The principal and defining characteristic of the open oceans separating them from the marginal marine environments discussed in the previous chapter is their relatively uniform chemical composition. Seawater, as compared to all other natural waters, is amazingly constant in composition. Variations in total dissolved solids are small and, for well over 95% of the oceans, the salinity (total dissolved solids) ranges no more than ±7% from its mean value of 35 parts per thousand (Sverdrup et al. 1942). In addition, the ratios of major ions to one another are even more constant. This uniformity of composition allows one to discuss many aspects of seawater chemistry in terms of average properties and still have the discussion pertain to most of the world's oceans. This is what is done here. We shall first present chemical data on the composition of seawater and then discuss how physical, biological, and geological factors both disturb and maintain this composition.

To acquaint the reader with seafloor morphology and thereby provide a frame of reference during our discussion of seawater chemistry, a schematic diagram is presented in Figure 8.1, showing the principal physiographic divisions of the ocean floor.

CHEMICAL COMPOSITION OF SEAWATER

The major dissolved constituents of seawater are almost the same ones encountered in continental waters as discussed in earlier chapters: Na^+, Ca^{2+}, Mg^{2+}, K^+, Cl^-, SO_4^{2-}, and HCO_3^-. Concentrations of major components in average seawater of 35‰ salinity, where ‰ represents parts per thousand or grams per kilogram, are given in Table 8.1. Although the concentrations shown can range by roughly ±10% due to changes in total salinity, the ratios of one ion to another vary by

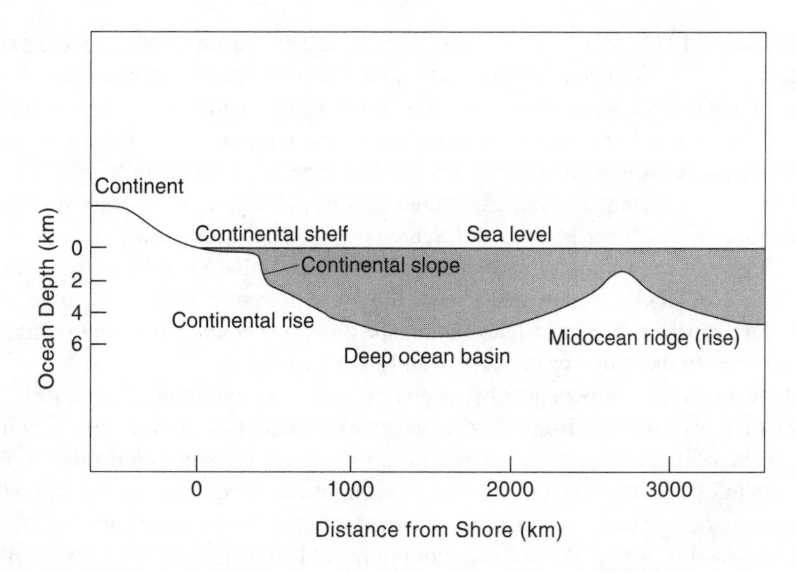

Figure 8.1. Generalized schematic cross section of the oceans showing major physiographic features. Note large differences in vertical and horizontal scales.

less than 1% (Wilson 1975). (The concentration of bicarbonate varies somewhat more, 5% to 10%, due to carbonate reactions; see later in this chapter.) This constancy of seawater composition was first well documented by W. Dittmar (1884) as a part of the pioneering and world-famous Challenger Expedition, and has since been thoroughly corroborated. Because of such constancy, the total salinity of seawater has historically been determined by measuring the concentration of the most abundant species, Cl^-, and multiplying this value by appropriate constants representing the ratios of the other major ions to chloride. (Salinity today, however, is normally determined by measuring electrical conductivity; see, e.g., Wilson 1975 for details.)

The major components listed in Table 8.1 are commonly assumed to be present in the ionic forms shown. However, according to a prominent theory of seawater composition (e.g., Millero

TABLE 8.1 Major Dissolved Components of Seawater for a Salinity of 35‰

	Concentration		Percent
Ion	g/kg	mM[a]	Free Ion
Cl^-	19.354	558	100
Na^+	10.77	479	98
Mg^{++}	1.290	54.3	89
SO_4^{--}	2.712	28.9	39
Ca^{++}	0.412	10.5	99
K^+	0.399	10.4	98
HCO_3^- [b]	0.12	2.0	80

[a] mM = millimoles per liter at 25°C.

[b] For pH = 8.1, $P = 1$ atm, $T = 25$°C.

Sources: Wilson 1975; Skirrow 1975; Millero and Schreiber 1982.

and Schreiber 1982), these "ions" can be viewed as being made up of a number of different individual species whose concentrations add up to the total concentration given for each element. The individual species consist of the actual ion shown (free ion) plus ion pairs formed with ions of opposite charge. For example, for SO_4^{2-} the concentration shown represents the sum of concentrations of the following actual species: SO_4^{2-} $NaSO_4^-$, $CaSO_4^0$ and $MgSO_4^0$ (the superscript 0 represents an uncharged ion pair). Some idea of the degree of ion pairing for each element, taken from the results of Millero and Schreiber (1982), is also shown in Table 8.1. In addition to the high degree of ion pairing exhibited by sulfate, Millero and Schreiber show that 15% of CO_3^{2-}, 29% of HPO_4^{2-}, and only 0.15% of PO_4^{3-} are present as free, or unpaired, ions. This high degree of ion pairing helps partly to explain the different chemical behavior of highly paired ions as compared to those that are present mainly as free ions.

Many dissolved components of seawater of lesser concentration, in contrast to the major elements, do show variation, from place to place, in concentration ratios to chloride; in other words, they are nonconservative as compared to the conservative major elements. Some minor constituents (>1 μM) are shown along with concentration ranges in Table 8.2. Note that the extent of variability changes from element to element. Many other elements are dissolved in seawater in trace quantities (<1 μM), but they are not listed here because discussion of them is outside the scope of this book. The interested reader is referred to the review paper of Bruland (1983).

TABLE 8.2 Minor Dissolved Components of Seawater (Excluding Trace Components <1μM), Showing Ranges in Concentration

Component	Concentration Range	
	μg/kg or ppb	μM[a]
Br^-	66,000–68,000[b]	840–880
H_3BO_3	24,000–27,000[b]	400–440
Sr^{++}	7,700–8,100[b]	88–92
F^-	1,000–1,600[b]	50–85
CO_3^{--}	3,000–18,000	50–300
O_2	320–9,600	10–300
N_2	9,500–19,000	300–600
CO_2	440–3,520	10–80
Ar	360–680	9–17
H_4SiO_4-Si	<30–5,000	<0.5–180
NO_3^-	<60–2,400	1–40
NO_2^-	<4–170	<0.1–4
NH_4^+	<2–40	<0.1–2
Orthophosphate[c]	<10–280	<0.1–3
Organic carbon	300–2,000	—
Organic nitrogen	15–200	—
Li^+	180–200	26–27
Rb^+	115–123[b]	1.3–1.4

[a] μM = micromoles per liter.

[b] For a salinity of 35‰.

[c] Includes PO_4^{3-}, HPO_4^{--} and $H_2PO_4^-$; concentrations expressed as μg P/kg.

Sources: Wilson 1975; Kester 1975; Spencer 1975; Brewer 1975; Skirrow 1975; Williams 1975.

The reason that some elements are conservative and others are nonconservative is twofold. First, the major elements are all conservative because there is so much of them in seawater. Fluctuations resulting from river inputs or reactions in seawater are undetectably small due to the relatively large masses of these elements and the fact that potential variations are dissipated by mixing of the oceans accompanying the general circulation of seawater. In other words, the major elements have very long replacement times in seawater relative to the time scale of oceanic homogenization via mixing of 1000–2000 years (Broecker and Peng 1982). (Replacement time is the time, at the present rate of addition by rivers, necessary to build up to average concentration levels; see Chapter 6.) This allows the major elements to be throughly mixed throughout the oceans. By contrast, some trace elements (such as Fe) have such short replacement (residence) times relative to oceanic mixing that they can exhibit spatially varying concentrations. A list of replacement times for oceanic constituents of interest in the present chapter is shown in Table 8.3.

The other principal reason for nonconservative behavior is that biological processes acting within the oceans tend to deplete certain nutrient elements in surface waters by biological uptake and to return these elements to solution at depth due to death, settling out, decomposition, and dissolution. Some of the chief elements involved are nitrogen, phosphorus, and silicon. Due to the rapidity of these biological processes compared to the rate of vertical mixing of seawater, strong vertical concentration gradients of the nutrient elements result. They are accompanied by corresponding gradients in dissolved O_2 and CO_3^{2-}, and together they provide striking testimony to the efficacy of biological activity as a major process for the alteration of seawater composition. Much more on this important subject will be presented later in the present chapter.

Pressure and temperature also exert some effect on the composition of seawater. Because the oceans are deep compared to other water bodies, they are subjected to much higher pressures. The mean depth of the oceans (excluding adjacent seas) is 4100 m (Sverdrup et al. 1942), which,

TABLE 8.3 Replacement Time with Respect to River Addition, τ_r, for Some Major and Minor Dissolved Species in Seawater

	Concentration (μM)		τ_r [a]
Component	River Water	Seawater	(1,000 yr)
Cl	230	558,000	87,000
Na^+	315	479,000	55,000
Mg^{++}	150	54,300	13,000
SO_4^{--}	120	28,900	8,700
Ca^{++}	367	10,500	1,000
K^+	36	10,400	10,000
HCO_3^-	870	2,000	83
H_4SiO_4	170	100	21
NO_3^-	10	20	72
Orthophosphate	1.8[b]	2	40

[a] $\tau_r = ([SW]/[RW])\ \tau_w$, where τ_w = replacement (residence) time of H_2O = 36,000 yr; RW = river water; SW= seawater, and [ ] = concentration in μmoles per liter = μM.

[b] Includes input from solubilization of solids.

Sources: Based on Tables 8.1 and 8.2 and data of Meybeck (1979, 1982) for world average river water.

at the bottom, corresponds to a pressure of about 400 atm. High pressure exerts an influence on the composition of seawater, mainly by bringing about the dissolution of biogenic calcium carbonate falling to the bottom. Temperatures in the ocean, in general, decrease with depth, and this too exerts an influence on seawater composition by restricting vertical mixing due to thermally induced density stratification (see Chapter 2).

The pH of seawater is controlled, on an oceanic time scale (thousands of years), by the dissolved bicarbonate-carbonate buffer system. (Variations in the ratio $[HCO_3^-]/[CO_3^{2-}]$, however, are not large, and as a result, the pH of the open oceans ranges only between approximately 7.8 and 8.4 (see Skirrow 1975). The oxidation potential (Eh, pe) of seawater cannot be measured accurately (e.g., Stumm and Morgan 1981), but there is no doubt that it is controlled by dissolved O_2. The open ocean practically everywhere contains enough dissolved oxygen to enable O_2 control of its redox state.

MODELLING SEAWATER COMPOSITION

Sillèn's Equilibrium Model

Several approaches to explaining the chemical composition of seawater have been used by previous workers. One approach, adopted by Sillèn (1967) and others, assumes that the ocean represents a simple chemical equilibrium between seawater, the atmosphere, and solids deposited on the ocean floor. In other words, the oceans plus atmosphere plus sediments are thought of as a giant closed reaction vessel (Figure 8.2), where the aqueous solution portion has attained equilibrium with the included minerals via dissolution and precipitation, and with the overlying air space via gas exchange. The resulting solution composition is calculated by combining a large number of equilibrium expressions (e.g., the solubility product of $CaCO_3$) with an equation expressing the charge balance between cations and anions, and by specifying a given chloride content. Sillèn was hindered by a lack of sufficiently accurate data for the various equilibrium constants, but he presupposed that once proper data were obtained, the resulting calculated major

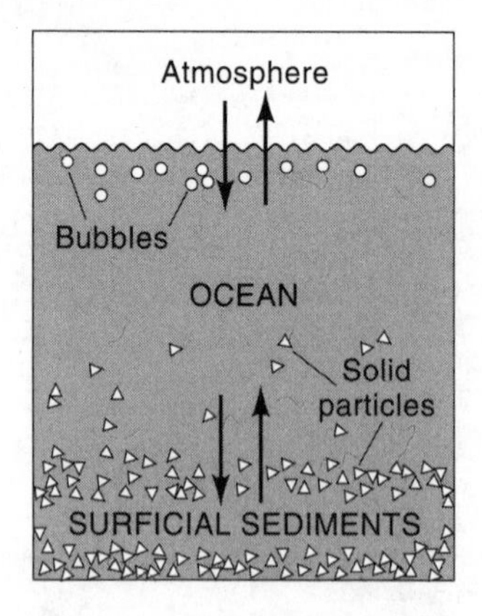

Figure 8.2. Sillèn equilibrium model for the ocean. Large arrows represent exchange between gases, solids, and seawater. Exchange and mixing of the oceans are assumed to be sufficiently rapid that chemical equilibrium between all phases is attained.

ion concentrations of seawater would approximate those actually found and would represent equilibrium with the atmosphere and the following minerals: quartz, calcite, kaolinite, illite (muscovite), smectite, chlorite, and either dolomite or phillipsite. (For compositions of these minerals, see Tables 4.3 and 4.4; phillipsite is a hydrous Ca, Na, K aluminosilicate found occasionally in deep-sea sediments.)

Sillèn's model is readily amenable to testing. If minerals, especially silicate minerals, react readily with seawater to maintain an equilibrium composition, then one would expect to find evidence of this reaction. For example, if a given detrital mineral carried into the oceans by rivers were not in equilibrium with seawater, it should dissolve and its components reprecipitate to form an equilibrium mineral. Thus, the equilibrium mineral should contain some evidence that it was formed from seawater. In addition, nonequilibrium minerals should not be present.

Most tests of these predictions have proven negative. The isotopic compositions of most clay minerals are not those expected for formation from seawater, but instead record the isotopic composition of the original rocks and weathering solutions from which they were formed on the continents. Also, nonequilibrium silicates, such as potassium feldspar, opaline silica, and gibbsite, are often found in surficial marine sediments. A further problem is that the biological activity of marine organisms is ignored. Because of the impact of solar energy resulting in photosynthesis, organisms can produce compounds that are not in thermodynmic equilibrium with seawater.

Even though most tests of the Sillèn model have been negative, this has not diminished the usefulness of the model as an idealized concept. Given sufficient time, unstable minerals should react with seawater to form stable minerals. Thus, the Sillèn model points to the importance of silicate reactions on seawater composition, even though total chemical equilibrium at any one time may not be attained.

Oceanic Box Models

Since an equilibrium model does not work well to explain oceanic composition, recourse must be made to other models. The most widely used approach (e.g., Broecker and Peng 1982; Garrels et al. 1975) is that of box modelling, which has already been applied to lakes in Chapter 6. In box modelling the oceans are divided into several regions ("boxes") of uniform composition, and rates of change of concentration within each box are computed as the difference between input and output rates. Except for human perturbations, such as fossil fuel burning, the input and output rates are generally assumed to balance one another so that there is no change with time of concentrations within each box. This is known as steady-state box modelling.

In box modelling of the major, or conservative, elements, the entire ocean is treated as a single, well-mixed, and therefore compositionally homogeneous box. This is illustrated in Figure 8.3. The difference between inputs, mainly from rivers, and outputs, mainly by sedimentation to the bottom and reaction with volcanics, results in changes in oceanic composition. If there is a steady state, then inputs and outputs balance one another and replacement times, as shown in Table 8.3, can be viewed as mean residence times. (For a detailed discussion of residence time, see Chapter 6.)

For nonconservative elements, especially those that are heavily involved in biological processes, the oceans are divided into a number of boxes, each representing a region of more or less homogeneous concentration. The simplest approach is that of a two-box ocean divided into surface seawater and deep seawater, with the dividing line being the permanent thermocline or narrow zone of rapid vertical temperature change. This is similar to the model that is normally

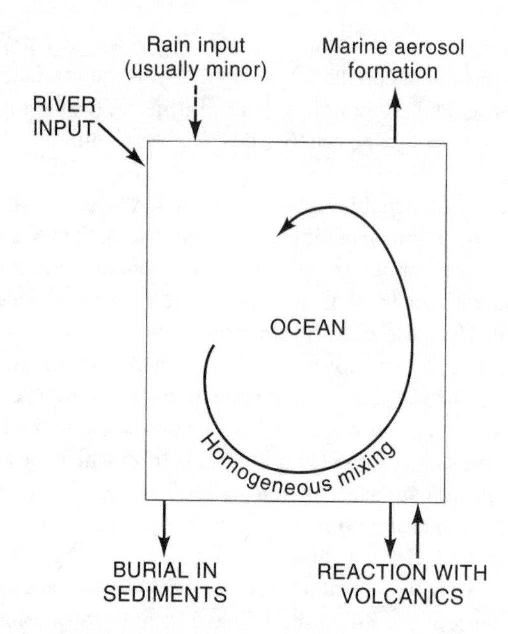

Figure 8.3. Simple box model appropriate for conservative elements in seawater. Note that, compared with the Sillèn model of Figure 8.2, fluxes enter and leave the box, and the atmosphere and sediments are considered to be outside the box. Also note that, in contrast to lakes, there is no outlet, so dissolved materials carried in by rivers can be removed only by sea–air transfer (marine aerosol formation), burial in sediments, or reaction with volcanics.

applied to lakes. Because of density stratification, the thermocline acts as a barrier to rapid exchange between deep and surface seawater and therefore behaves as a good box boundary. Also, vertical concentration-versus-depth profiles of nonconservative biological elements in general parallel the temperature profile, so that distinctly different concentrations exist in deep and surface waters (e.g., see Figure 8.5). This makes such elements particularly amenable to two-box modelling.

A closer look at "deep water" shows that, based on temperature and salinity, it can be subdivided into a number of separate, internally homogeneous water masses (Antarctic Bottom Water, North Atlantic Deep Water, etc.; see Chapter 1). Thus, a more realistic box model of the oceans requires a number of boxes, one for each water mass. However, the use of many boxes leads to difficulty in that the rate of water flow and material exchange between water masses is not known accurately. Box modelling generally strikes a compromise between multibox sophistication and a desire to describe the processes under consideration in as simple a manner as possible.

Studies of ancient sedimentary rocks suggest that the chemical composition of the oceans has not changed drastically over the past 200 million years (e.g., Holland 1978). Therefore, as a first approximation, in constructing simple box models one may assume that a steady state has been established and that long-term inputs and outputs are in balance. This helps to explain the prevalent use of steady-state models both in chemical oceanography (e.g., Broecker and Peng 1982) and in geochemical cycling (Garrels and Mackenzie 1971; Holland 1978). However, the effects of worldwide pollution by humans constitute a sudden change in the rate of addition of certain substances to the ocean (e.g., CO_2, SO_4^{2-}, pesticides), and modelling of consequent concentration changes of these components requires a non-steady-state approach. A good example of non-steady-state modelling is that for the rate of excess CO_2 addition to the ocean which has resulted from the burning of fossil fuels (see Sarmiento 1993). Because of anthropogenic perturbations, more such modelling will be needed in the future.

Continuum Models

The ultimate and most accurate models for both conservative and nonconservative elements in the oceans are the so-called continuum models. In effect, the continuum models divide the ocean into an infinite number of boxes and treat concentration changes as continuous variations from place to place. Such models may be one-dimensional, whereby only depth variations in concentration are treated (see, e.g., Craig 1969), two-dimensional (e.g, lateral variations only), or three dimensional (see, e.g., Fiadeiro and Craig 1978). Concentration changes at a given point are described in terms of a differential equation which expresses the effects of turbulent diffusion, advective transport, and chemical (biochemical, radioactive decay, etc.) processes on concentration variation. Mixing and advection rates can be obtained by such modelling by applying a differential equation to temperature and conservative species (for example, Cl^-) which undergo no chemical reaction and whose concentration variations are due solely to salinity variations. These rates can then be used in solving additional differential equations for the accompanying nonconservative elements. Although they are important, these models are mathematically complex and beyond the scope of the present book.

ENERGY SOURCES FOR CHEMICAL REACTIONS

Chemical reactions take place in the oceans, like everywhere else, as a result of chemical disequilibrium. Creation of this disequilibrium is brought about by the input of energy from two sources: the sun, and the interior of the earth. Solar energy, by means of photosynthesis, enables the formation by marine plants of organic matter and hard parts (e.g., opaline silica) which are out of equilibrium with seawater. In addition, marine animals consume photosynthetically produced organic matter and also produce chemically unstable skeletal hard parts. As a consequence of both plant and animal activity, surface waters become depleted in certain elements, and deep waters, upon death and settling out of organic remains, become enriched in the same elements in an attempt to reattain chemical equilibrium via dissolution. Photosynthetic activity is concentrated in surface waters (top ~200 m) because of the lack of penetration of sunlight to greater depths. (Synthesis of organic matter in the dark by nonphotosynthetic marine organisms living at depth near volcanic centers on midocean ridge crests also occurs, but it is extremely rare.)

The other major source of energy for chemical disequilibrium in the oceans is heat contained within the earth. This heat manifests itself, among other ways, as submarine volcanism. Basaltic minerals and volcanic glass, orginally formed from hot magmas poor in H_2O, are suddenly thrust into contact with seawater, in which they are chemically unstable. As a result, silicate–seawater reactions occur. Reactions may occur at high temperatures or at low, seafloor temperatures. Temperature gradients near oceanic ridge crests result in the convective circulation of seawater through the ridges and the extensive heating of this water at depth. Basalt–seawater reactions are rapidly accelerated by increased temperature and, consequently, the seawater that eventually exits from the ridges is extensively altered in chemical composition.

Solar energy and that contained within the earth also have important *indirect* effects on oceanic composition. Solar energy brings about the evaporation of seawater, the transport of water vapor to the continents, and the formation of rainwater, which then falls on the continents to form soil water, groundwater, lake water, and river water. Meanwhile, the interior energy of the earth, manifested by tectonic uplift and volcanism, causes rocks formed at depth to be uplifted into the zone of weathering on the continents. Here the rocks are unstable and react with soil waters and groundwaters via the mediation of (photosynthetically driven) green plants, to bring about the formation of dissolved elements and new minerals. The minerals are (crudely) in equilibrium with continental waters but not necessarily in equilibrium with seawater, which has a distinctly different composition. Consequently, terrestrially formed weathering products (dissolved and suspended solids) are carried by rivers into the oceans, where they undergo a variety of chemical reactions with seawater. In addition, any chemically unstable igneous and metamorphic minerals that managed to escape chemical weathering are also added to the oceans and may also undergo chemical reactions. In these ways, processes that take place on the continents in response to solar and earth interior energy inputs can also affect the composition of seawater.

MAJOR PROCESSES OF SEAWATER MODIFICATION

Processes affecting the chemistry of the oceans, besides addition of elements by rivers, can be classified into six categories. The first three, alluded to in the previous section, are the most important and will be discussed in detail here. They are (1) principal biological processes (secretion of hard parts and the production and decomposition of organic matter), (2) volcanic–seawater reaction, and (3) reaction of seawater with solid materials transported from the continents. Two additional processes, sea-to-air transfer of cyclic salts and the precipitation of evaporite minerals, have already been discussed in previous chapters, on rainwater (Chapter 3) and on marginal marine environments (Chapter 7), respectively, and will not be discussed further here. The sixth category constitutes special processes that generally affect only one or two elements, and includes pore-water burial of chloride and sodium, and a number of processes unique to the nitrogen and phosphorus cycles (denitrification, nitrification, N_2 fixation, addition of NO_3^- in rain, adsorption of phosphate on ferric oxides, and authigenic apatite formation). These special processes are also not discussed in this section but instead are treated under the chemical budgets for each appropriate element. The goal of this section is to deepen the reader's understanding of the three major processes that are mentioned time and again when discussing chemical budgets of the elements.

Biological Processes

Chemical reactions in the ocean which are intimately intertwined with life processes constitute major controls on the concentrations of the following seawater constituents: Ca^{2+}, HCO_3^-, SO_4^{2-}, H_4SiO_4, CO_2, O_2, NO_3^-, and orthophosphate. [For recent reviews of oceanic biogeochemistry and its role in global cycles, consult Berger et al. (1989) and Wollast et al. (1993).] Biological activity also strongly affects many trace elements, for example, copper and nickel (Boyle et al. 1977; Sclater et al. 1976). Three principal processes can be recognized: (1) the synthesis of soft tissues or organic matter, (2) the bacterial decomposition of organic matter upon death, and (3) the secretion of skeletal hard parts.

Essentially all organic matter is ultimately formed in surface waters by the process of photosynthesis. (Very rarely, organic matter is also synthesized near deep volcanic vents.) This process requires the presence of light and therefore takes place only at water depths where sunlight can penetrate, that is, in the top few hundred meters of the oceans. The organisms involved, known as *phytoplankton,* are minute floating marine plants that all contain chlorophyll, which is necessary for photosynthesis.

Redfield (1958) has shown that the average elemental composition of phytoplankton from the open ocean in terms of the major components carbon, nitrogen, and phosphorus can be represented by the molar ratio C:N:P = 106:16:1. Therefore, one can represent marine photosynthesis by the reaction

$$106CO_2 + 16NO_3^- + HPO_4^{2-} + 122H_2O + 18H^+ \xrightarrow{\text{light}} C_{106}H_{263}O_{110}N_{16}P + 138O_2$$

This reaction shows that photosynthesis involves not only the removal of CO_2 from solution and the production of O_2, but also the uptake of nutrients such as nitrate and orthophosphate. Other nutrient elements, such as trace metals, are also taken up, but their exact role in marine photosynthesis is not well established (an exception is Fe, which is discussed below). Because of the omnipresence of CO_2, H_2O, and light in most surface waters, the limiting factors in how much photosynthesis can occur generally are nitrate and (ortho-)phosphate, both of which occur at low concentrations. (Strictly speaking, other forms of nitrogen and phosphorus are also used for photosynthesis, e.g., NH_4^+, NO_2^-, and dissolved organic phosphorus, but they are normally present in seawater at concentrations even lower than that of NO_3^- and orthophosphate.)

Because of highly varying nitrate and phosphate concentrations in surface waters, the rate of photosynthesis, known as planktonic productivity, can vary considerably. For example, productivity in central gyres of the ocean, such as the Sargasso Sea, may be as little as one tenth as fast as that occurring in coastal upwelling regions such as in the Pacific off Peru. This is illustrated in Figure 8.4. High marine productivity is brought about in the open ocean by mixing processes that bring deep, nutrient-rich water to the surface (and, in nearshore water, by river flow to the sea; see Chapter 7). Since phosphate and nitrate are highly enriched in subsurface waters (see Figures 8.5 and 8.6), any process that brings these waters up into the zone of light penetration will aid in photosynthesis. Two major processes, discussed in Chapter 1, are coastal upwelling and high-latitude mixing attending the formation of deep water.

Recent work (e.g., see Martin et al. 1990) has shown that in the vast oceanic region around Antarctica the limiting nutrient is neither nitrogen nor phosphorus, but rather iron. Apparently, extensive resupply of N and P to surface waters results in the exhaustion of other key nutrients, in this case Fe. Because of this Fe limitation, it has been suggested (Martin et al. 1990) that additional Fe be added to Antarctic waters to enable plankton to take up some of the extra atmospheric CO_2 that has resulted from the burning of fossil fuels. However, this proposal, as a practical plan of action, has proven to be controversial.

Phytoplankton are eaten by zooplankton, which are in turn eaten by fish, and so on up the food chain. All during this activity, respiration is taking place both by these higher organisms as well as by bacteria living on their dead remains. Respiration (used here in the strict sense of O_2 respiration) is the reverse of photosynthesis, and can be viewed essentially as the photosynthetic reaction given above written backwards. In other words, oxygen is consumed and CO_2, NO_3^-, and orthophosphate are liberated to solution; the rates of photosynthesis and respiration are so well adjusted that they almost balance one another in surface waters, but not quite. A small

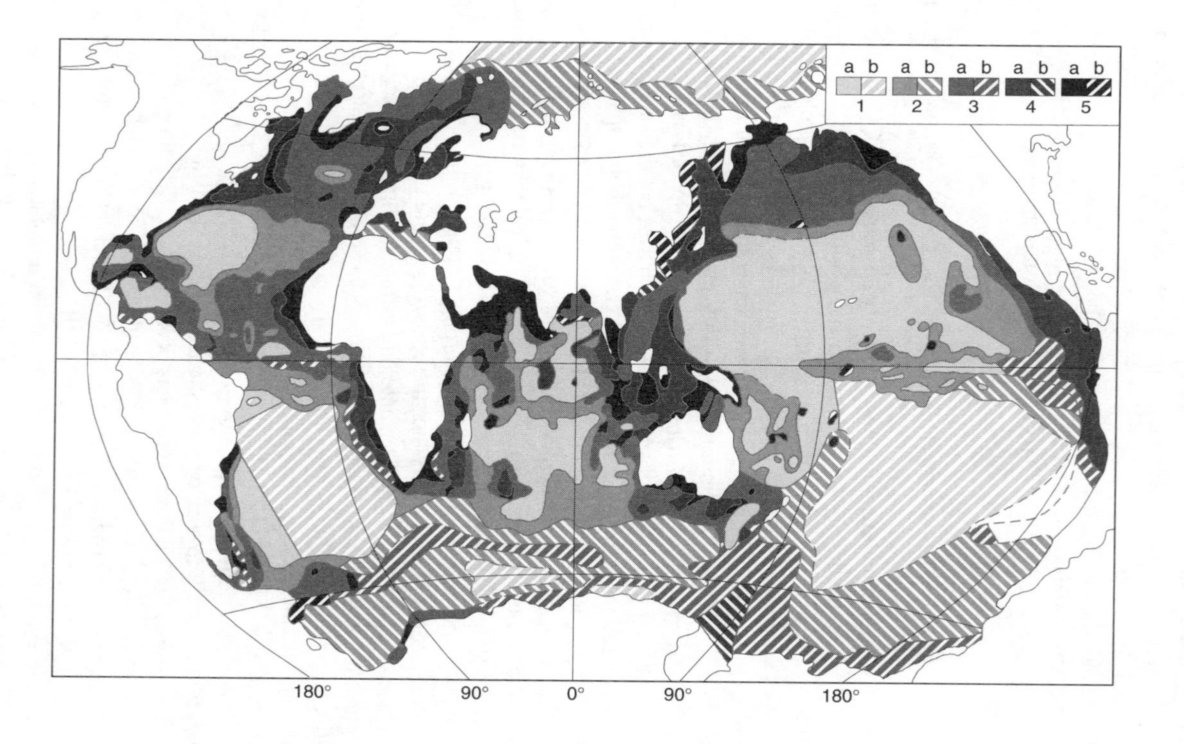

Figure 8.4. Rates of organic matter production (in mg C/m^2/day) for the oceans: (1) less than 100; (2) 100–150; (3) 150–250; (4) 250–500; (5) more than 500. a = data from direct ^{14}C measurements; b = data from phytoplankton biomass, hydrogen, or oxygen saturation. (After O. J. Koblentz-Mishke, V. V. Volkovinsky, and J. G. Kabanova, "Plankton Primary Production of the World Ocean." In *Scientific Exploration of the South Pacific,* ed. W. S. Wooster, p. 185. Copyright © 1970 by the National Academy of Science, reprinted by permission of the publisher.)

amount of dead organic matter sinks into deeper waters, and this represents a net gain of photosynthesis over respiration. This organic matter is further oxidatively respired by bacteria in deep waters to produce CO_2, NO_3^-, and HPO_4^{2-}, and to consume O_2, but in the absence of photosynthesis. As a result, there is a net input of nutrients, and this helps to explain why higher concentrations of NO_3^-, HPO_4^{2-}, and CO_2 and lower concentrations of O_2 are found at depth. Some typical depth distributions of these species are shown in Figures 8.5 and 8.6. (Note in Figure 8.5 that higher phosphate concentrations are succeeded at even greater depths by somewhat lower concentrations, leading to the concept of the nutrient maximum and oxygen minimum. This concentration reversal results from the input of high-O_2, low-nutrient surface waters to great depths accompanying the general deep-water thermohaline circulation; see Chapter 1.)

In the open ocean at subsurface depths, the ratio of dissolved NO_3^- to dissolved phosphate is almost the same as that found in average plankton (Redfield 1958) demonstrating stoichiometric decomposition of the plankton. Also, at many open ocean locations, during photosynthesis both nutrients are removed in approximately the same proportions as their concentrations such that both become exhausted simultaneously; that is, both nutrients are limiting. (Actually, there is in general a very slight deficiency of nitrate relative to phosphate, leading to N-limitation in most offshore waters; Fanning 1989.) Simultaneous N and P limitation contrasts with the situation of most lakes, where phosphate is limiting (Chapter 6), and many marginal marine areas,

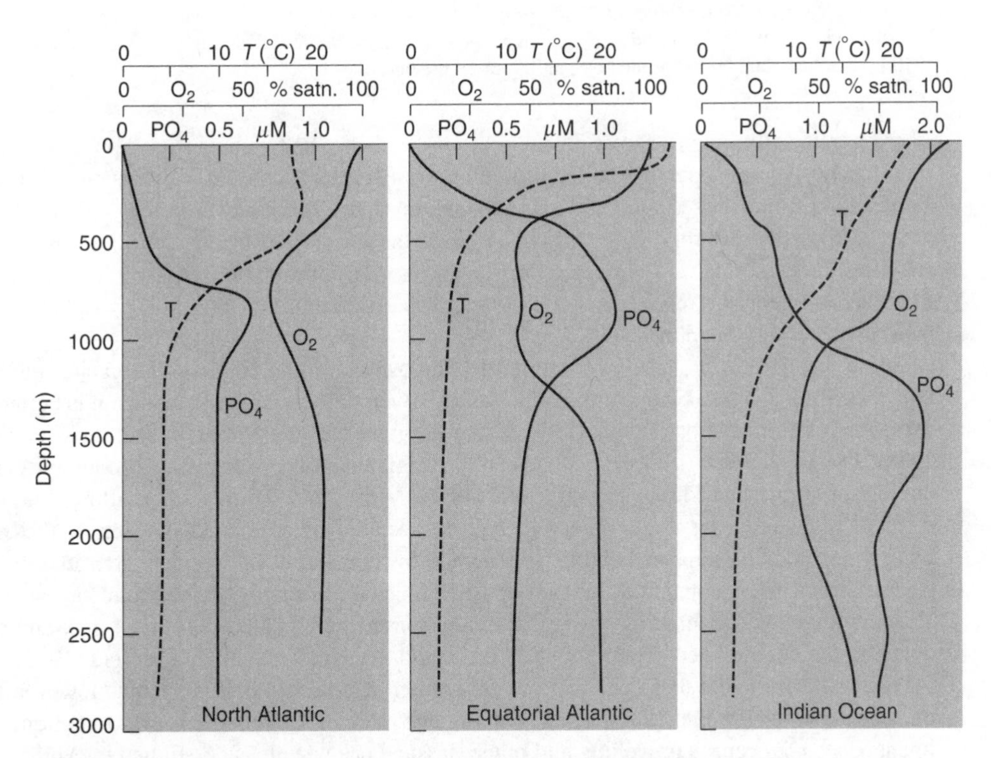

Figure 8.5. Depth profiles of dissolved oxygen (O_2) and phosphate (PO_4) and temperature (T) at stations in the north and equatorial Atlantic Ocean and in the Indian Ocean. Note the close anticorrelation between O_2 and PO_4^{-3}. (Adapted from J. P. Riley and R. Chester, *Introduction to Marine Chemistry,* p. 173. Copyright © 1971 by Academic Press, reprinted by permission of the publisher.)

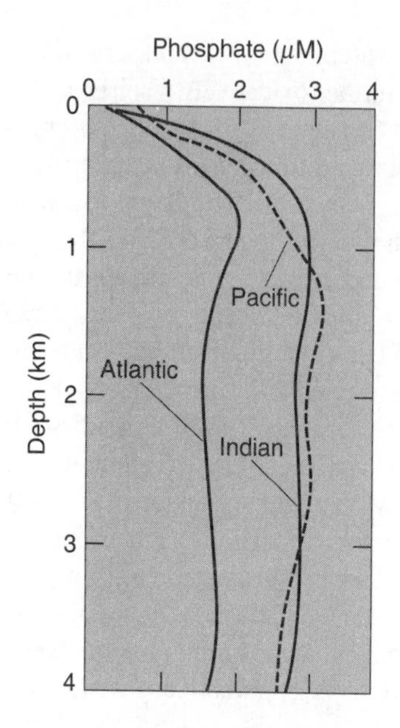

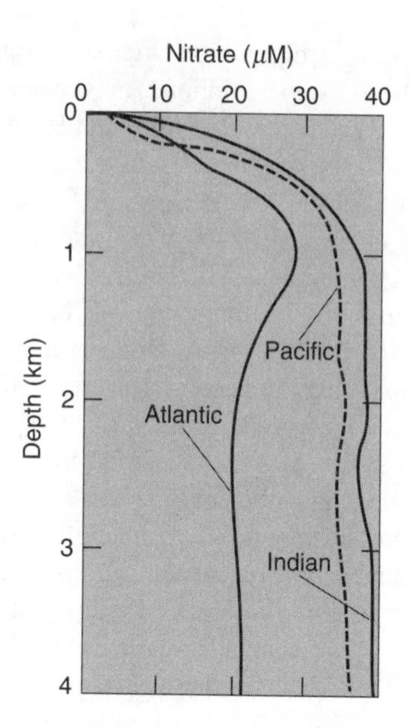

Figure 8.6. Average depth profiles of dissolved phosphate and nitrate in the major world oceans. (After H. V. Sverdrup, M. W. Johnson, and R. H. Fleming, *The Oceans,* pp. 241–242. Copyright © 1942, renewed 1970 by Prentice-Hall, Inc., reprinted by permission of the publisher.)

where nitrate is limiting (Chapter 7). It is not obvious why the two nutrients should go to zero simultaneously in the open ocean, but it might represent the evolutionary adjustment of the elemental composition of seawater and marine plankton to one another over geological time (Redfield 1958). Presently this adjustment may be in the process of being distrurbed by human activities. In the northern Atlantic and Pacific Oceans, the addition of excess nitrate in acid rain from the continents has resulted in a the situation where the limiting nutrient is phosphate rather than both N and P (Fanning 1989).

Some organic matter survives bacterial decomposition during sinking and reaches the bottom, where additional oxic decomposition occurs. Eventually, a very small portion is buried by subsequent sedimentation—about 0.3% of that originally formed via photosynthesis (Holland 1978). The amount that is buried depends on the original amount falling (i.e., productivity) and the rate of burial via sedimentation of total solids. More rapid sedimentation allows burial of organic compounds that would otherwise be destroyed during prolonged exposure to the overlying water, and this helps to explain the generally lower contents of organic matter in deep-sea pelagic sediments, as compared to those deposited much more rapidly on the continental shelves. The relation between amounts of organic matter buried, sedimentation (burial) rate, and productivity has been studied extensively by Müller and Suess (1979) and by Canfield (1993), and the reader is referred to these publications for further details. It should be noted, however, that one of us (Berner 1982) has shown that most organic matter loss from the oceans to bottom sediments does not occur in upwelling and other classical oceanic areas, as studied by Müller and Suess, but rather in deltaic and other areas of very high sedimentation rate near river mouths, where a simple relation between organic content, sedimentation rate, and productivity is not found.

The decomposition of organic matter by bacteria using dissolved O_2 continues after burial in sediments, but only for a short time. In most sediments there is sufficient organic carbon burial that oxygen present in the interstitial waters is completely used up by this process within tens of centimeters of the sediment–water interface. In other words, most sediments are anoxic. Under anoxic conditions, further organic matter decomposition is accomplished by bacteria which use oxygen bound in a variety of other compounds. They are dissolved nitrate, manganese oxides, iron oxides, dissolved sulfate, and organic matter itself. The oxygen-containing substances become reduced, while the organic matter is oxidized to CO_2. These compounds are generally attacked successively, until each is completely consumed in the order (see Table 8.4): nitrate reduction (denitrification), manganese reduction, iron reduction, sulfate reduction, and fermentation (methane formation). [For a discussion of anoxic bacterial processes and their succession in sediments, consult Berner (1980) and Canfield (1993).] Of these processes, the one of major interest to the present chapter is bacterial sulfate reduction.

In sediments that contain appreciable concentrations of organic matter, including most nearshore and continental margin sediments, the dominant process of organic matter decomposition is bacterial sulfate reduction (Goldhaber and Kaplan 1974; Canfield 1993). The bacteria that accomplish this are strict anaerobes (they are killed by even traces of dissolved O_2), and the overall reaction can be represented as

$$2CH_2O + SO_4^{2-} \rightarrow H_2S + 2HCO_3^-$$

where CH_2O is a generalized representation of organic matter. Evidence for this reaction in sediments is provided by decreasing concentrations of interstitial dissolved sulfate, with depth. Most hydrogen sulfide produced by sulfate reduction migrates out of the sediment and is subsequently oxidized, by O_2 in seawater, back to sulfate while the remainder reacts with the detrital iron minerals in the sediment to form a series of iron sulfides which are ultimately transformed to, pyrite, FeS_2. (A minor amount of H_2S also reacts with organic matter.) The overall process of sedimentary pyrite formation is summarized in Figure 8.7. This process, once pyrite is permanently buried, constitutes a major mechanism for the removal of sulfate from seawater.

TABLE 8.4 Major Processes of Organic Matter Decomposition in Marine Sediments; Reactions Succeed One Another in the Order Written as Each Oxidant is Completely Consumed

Oxygenation (oxic)

$$CH_2O + O_2 \rightarrow CO_2 + H_2O$$

Nitrate reduction (mainly anoxic)

$$5CH_2O + 4NO_3^- \rightarrow 2N_2 + CO_2 + 4HCO_3^- + 3H_2O$$

Manganese oxide reduction (mainly anoxic)

$$CH_2O + 2MnO_2 + 3CO_2 + H_2O \rightarrow 2Mn^{++} + 4HCO_3^-$$

Ferric oxide (hydroxide) reduction (anoxic)

$$CH_2O + 4Fe(OH)_s + 7CO_2 \rightarrow 4Fe^{++} + 8HCO_3^- + 3H_2O$$

Sulfate reduction (anoxic

$$2CH_2O + SO_4^{--} \rightarrow H_2S + 2HCO_3^-$$

Methane formation (anoxic)

$$2CH_2O \rightarrow CH_4 + CO_2$$

Note: Organic matter schematically represented as CH_2O.

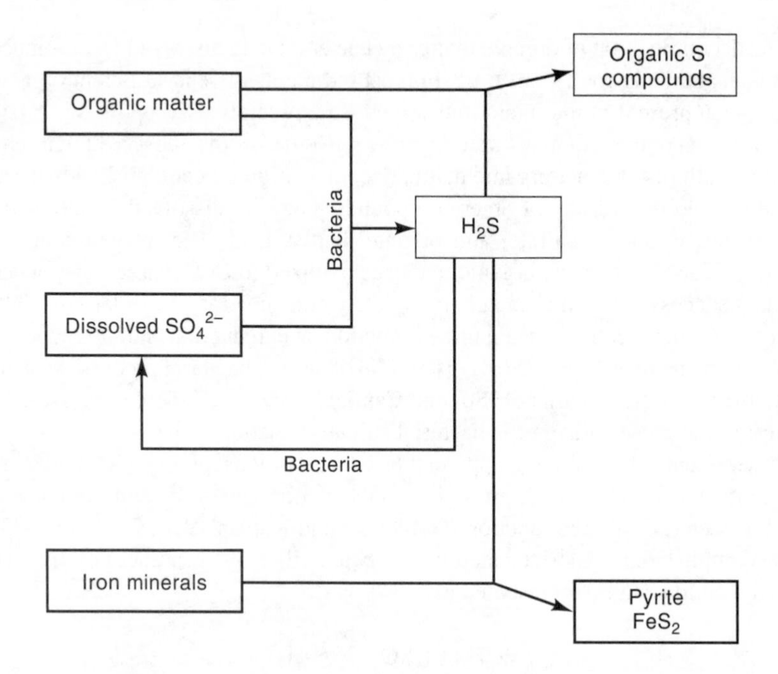

Figure 8.7. Simplified version of the process of sedimentary pyrite formation. [Intermediate steps involving elemental sulfur and iron monosulfides omitted; see also Berner (1984).]

As shown in Figure 8.7, the amount of sulfur removed from the oceans as pyrite depends on the availability of organic matter, sulfate, and iron minerals. Organic matter is most important in that, without it, there can be no sulfate reduction. In addition, the reactivity or metabolizability of the organic matter dictates how fast sulfate is reduced, and therefore how fast pyrite can form. Abundant sulfate is present in seawater at the very shallow depths where most pyrite forms, but upon burial the sulfate is ultimately removed by bacterial reduction. Iron minerals are necessary for pyrite formation but, in terrigenous sediments (those derived from weathering on the continents), the iron minerals are generally present to excess, so the limiting factor here is the availablity of organic matter for sulfate reduction. This is shown in Figure 8.8 by a positive correlation between organic carbon and pyrite sulfur.

The form of iron most readily reactive toward H_2S is present as terrigenous fine-grained iron oxyhydroxides which enter the sediment as colloidal coatings on detrital clays, feldspars, and so forth (Canfield 1989). Lack of terrigenous iron in organic-rich $CaCO_3$ sediments helps to explain the relative absence of pyrite in these sediments, which otherwise are high in H_2S. In summary, maximum pyrite formation is favored by the burial of high concentrations of fine-grained iron minerals and readily metabolized organic matter in sediments where rapid sulfate reduction occurs at shallow enough depths to enable replenishment of sulfate from the overlying water. In this way, an appreciable flux of sulfur into the sediment and its removal as pyrite is maintained. [For a further discussion of sedimentary pyrite formation consult Goldhaber and Kaplan (1974) and Berner (1970), (1984).]

So far we have dwelt upon biological activity only as it affects the synthesis and decomposition of organic matter. Another major biological process, important from both a chemical and a

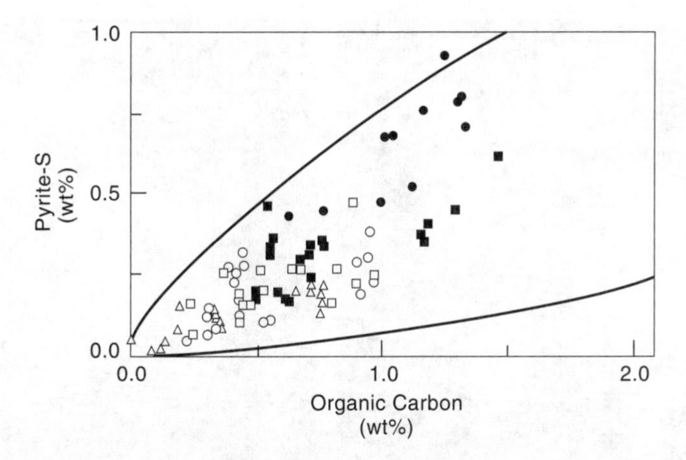

Figure 8.8. Relationship between pyrite-S and organic carbon in sediments (Mississippi River delta shelf slope (●), Texas-Louisiana continental shelf slope (■), western (○), southwestern (Δ), and southern (□) Gulf of Mexico sediments. [After S. Lin and J. W. Morse 1991, *American Journal of Science* 291, p. 74. Reprinted by permission of American Journal of Science.]

geological viewpoint, is the secretion of skeletal hard parts. Although a wide variety of minerals and mineraloids are known to be secreted by organisms (see, e.g., Lowenstam 1981), only quantitatively important ones will be discussed here. They are calcite ($CaCO_3$), aragonite ($CaCO_3$), magnesian calcite (defined here as calcite with more than 10 mol% $MgCO_3$ in solid solution), and opaline silica (SiO_2). A summary of the types of organisms that secrete each substance is listed in Table 8.5, and photographs of some common organisms are shown in Figure 8.9.

Besides mineralogy, an important distinction between organisms that secrete hard parts is that between benthos (bottom dwellers) and plankton (small, microscopic floating organisms). All

TABLE 8.5 Quantitatively Important Plants and Animals That Secrete Calcite, Aragonite, Mg-calcite, and Opaline Silica

Mineral	Plants	Animals
Calcite	Coccolithophorids[a]	Foraminifera[a]
		Molluscs
		Bryozoans
Aragonite	Green algae	Molluscs
		Corals
		Pteropods[a]
		Byozoans
Mg-calcite	Coralline (red) algae	Benthic foraminifera
		Echinoderms
		Serpulids (tubes)
Opaline silica	Diatoms[a]	Radiolaria[a]
		Sponges

[a] Planktonic organisms.

Sources: For further information on skeletal mineralogy consult Lowenstam 1981.

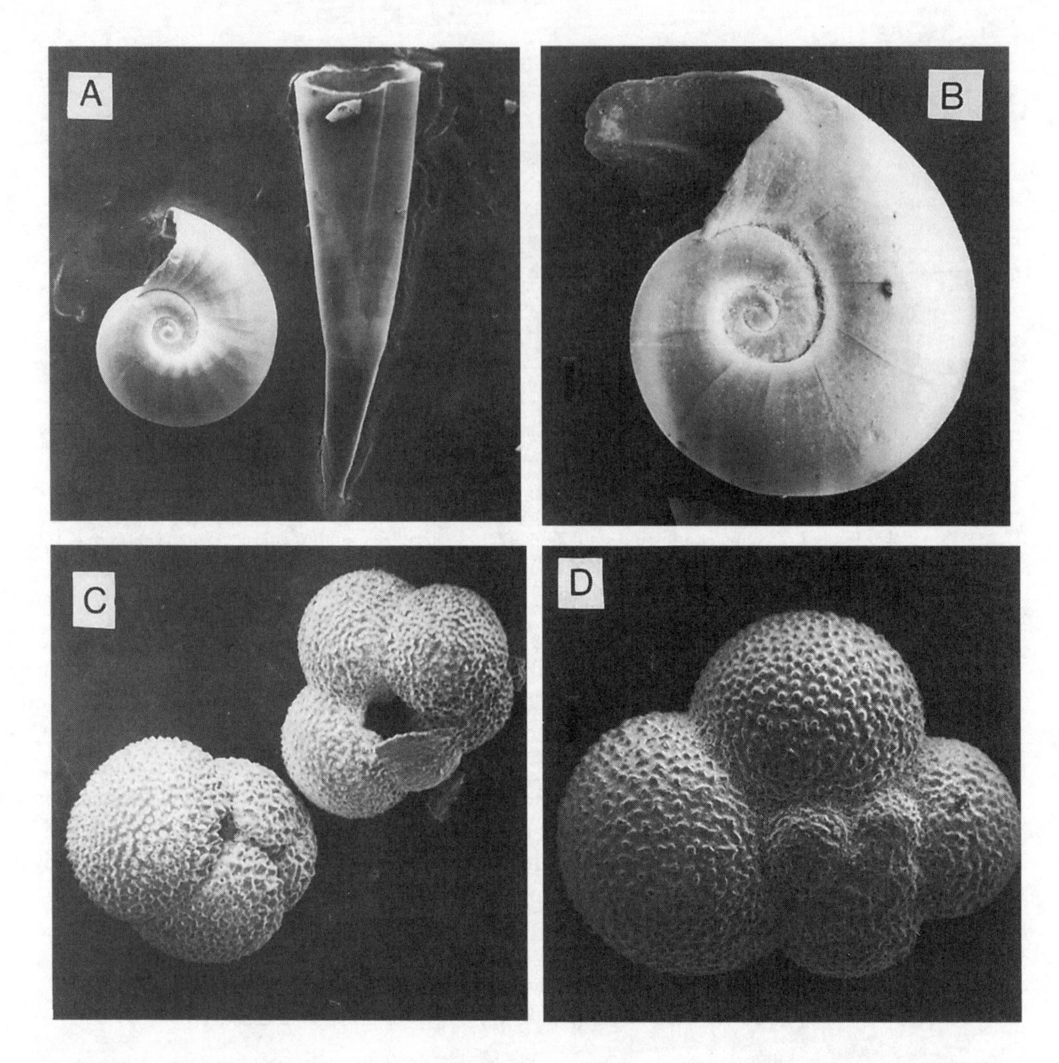

Figure 8.9. Photomicrographs of some planktonic $CaCO_3$-secreting organisms: (a) pteropod shells (aragonite) × 20; (b) pteropod shell (aragonite) × 10; (c) foram tests (calcite) × 70; (d) foram test (calcite) × 100.

important carbonate- and silica-secreting organisms (i.e., both benthos and plankton) live in surface waters (<200 m) where photosynthesis can occur and where abundant photosynthetically produced food is present. As a result, most benthic skeletal debris accumulates in sediments overlain by shallow waters, such as banks, atolls, and continental shelves, whereas plankton debris, because plankton can live over any depth of water, can also fall to the deep-sea floor. In this way deep-sea sediments differ from shallow-water sediments in containing far higher proportions of planktonic skeletal debris, and this results in distinct mineralogical differences between the two sediment types. Deep-sea carbonate sediments contain mainly coccolith and planktonic foraminiferal calcite (pteropod aragonite is mainly dissolved away before burial), while shallow-water sediments are dominated by coral-algal-mollusc aragonite and magnesian calcite.

(Strictly speaking, magnesian calcite with more than 10 mol% Mg and calcite are the same mineral, but because of their different composition and different dissolution behavior, they are distinguished from one another.)

The fate of calcium carbonate is also different in the two environments. In the deep sea, much of the coccolith and foram calcite and almost all of the pteropod aragonite is dissolved prior to burial, whereas in shallow water (bank tops, etc.), little or no dissolution of $CaCO_3$ occurs. This is because the oceans are supersaturated with respect to calcite and aragonite in shallow water and undersaturated at depth. [For further discussion of $CaCO_3$ chemistry in the oceans, consult Morse and Mackenzie (1991) and the section below on the oceanic budget for calcium.]

Opaline silica secreted by diatoms and radiolaria also undergoes dissolution upon death, but at all water depths. Opaline silica is distinctly more soluble than quartz and is undersaturated everywhere in the oceans. It is only because of the input of photosynthetic energy that plankton are able to remove opaline silica from surrounding seawater even though the water is undersaturated with respect to amorphous silica. As a result of the activity of diatoms and radiolaria, silica removal occurs in shallow water and, upon death and sinking, much of this silica is returned to solution at depth by dissolution, an overall process resembling the behavior of both organic matter and calcium carbonate.

The removal of biogenic elements in shallow water and their return in deep water can be described quantitatively in terms of box modelling. This has been done by Broecker (1971) (see also Broecker and Peng 1982) using a two-box model, illustrated here by Figure 8.10. Broecker divides the Atlantic and Pacific oceans each into two compartments, one representing shallow

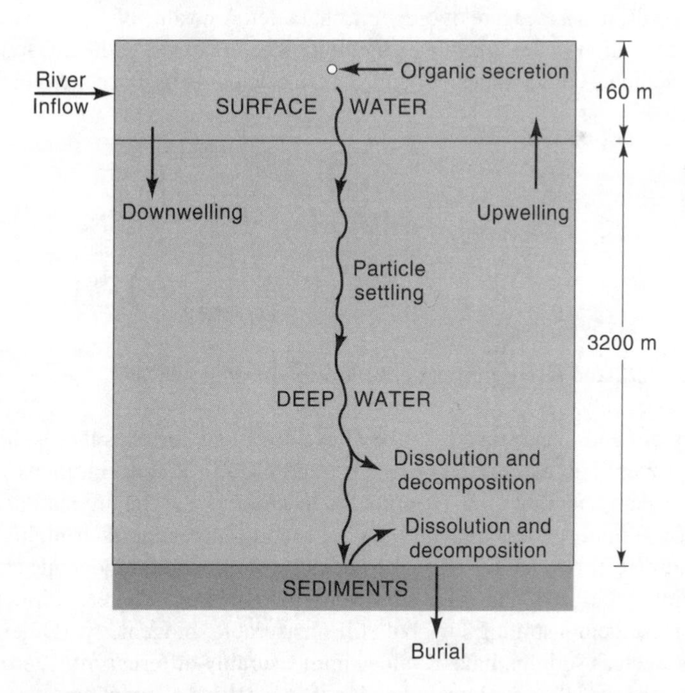

Figure 8.10. Two-compartment box model for the oceans as used by Broecker (1971). Arrows represent fluxes to and from each "box."

water, where photosynthesis and skeletal secretion occur, and the other deep water, which is dominated by organic matter decomposition and mineral dissolution. The deep reservoir is assumed to be 20 times the volume of the shallow one (see Figure 8.10). Water is exchanged between deep and shallow reservoirs by upwelling and downwelling accompanying normal oceanic circulation, and the average residence time of water at depth, which is also the time for a complete overturn of the oceans, is 1600 years. Combining this value with the volume of the deep reservoir and the rate of water flow by rivers to the sea, Broecker obtains a ratio of upwelling to river-water inflows to the shallow reservoir of 20:1. Finally, organic matter and hard parts fall, by sinking of particles, from the shallow to the deep reservoir to complete the cycle.

By assuming a steady state for both shallow and deep reservoirs, Broecker was able to calculate the values for two useful parameters, denoted as *f* and *g*, from concentrations of each element in deep seawater, shallow seawater, and average river water. The parameters are defined as:

$$f = \frac{\text{particle flux into sediments}}{\text{particle flux from shallow to deep water}}$$

$$g = \frac{\text{particle flux from shallow to deep water}}{(\text{river input flux}) + (\text{upwelling input flux})}$$

The parameter *f* represents the fraction of a biogenic element falling into deep water that survives decomposition and dissolution to become buried in bottom sediments. In other words, it is a removal indicator. The parameter *g* represents the fraction of an element delivered to surface water by rivers and upwelling that is removed by biological secretion plus particle fallout. In other words, it is a measure of biogenic character. A *g* value of 1 means that all of the element carried into shallow water is removed by biological processes, with none left over to be removed by downwelling. Actual values of *f* and *g* were calculated by Broecker via the equations

$$f = \frac{1}{20\left(\frac{[D]-[S]}{[R]}\right)+1}$$

$$g = 1 - \left(\frac{(20[S]/[R]}{20\,[D]/[R]+1}\right)$$

where *[S]*, *[D]*, and *[R]* represent concentration in shallow water, deep water, and average river water, respectively.

Values of *f* and *g*, calculated via the above equations for phosphorus, nitrogen, silicon, calcium, and inorganic carbon (CO_2, HCO_3^-, and CO_3^{2-}) at concentrations found in the Pacific Ocean are shown in Table 8.6. (Results for the Atlantic Ocean are similar.) As expected, high values of *g* for phosphorus and nitrogen are found, illustrating their highly biogenic character. High values are also found for silica. Lesser values are found for inorganic carbon, demonstrating its moderately biogenic character relative to downwelling, whereas a low value of $g = 0.01$ is found for Ca, demonstrating low, but still measurable, biogenicity. (Other nonbiogenic major elements, such as sodium, have *g* values immeasurably different from zero.)

Values of *f* are all rather low, indicating that for these elements most of the biogenic material falling into deep water is returned to solution. Calcium and inorganic carbon do show a

TABLE 8.6 Average Concentration in Shallow and Deep Ocean Reservoirs (Using Pacific Ocean Data) and Corresponding *f* and *g* Values for Some Biogenic Elements According to the Broecker (1971) Two-Box Steady-State Model

	Concentration (μM)				
Element	Shallow Water [S]	Deep Water [D]	River Water [R]	*f*	*g*
P	0.2	2.5	0.7	0.015	0.92
N	3	35	20	0.03	0.92
Si	2	180	170	0.05	0.99
C_{inorg}	2,050	2,480	870	0.09	0.19
Ca	10,000	10,090	367	0.17	0.01

Sources: Concentration values from Broecker 1971; Tables 8.1 and 8.2; Meybeck 1979, 1982. See also Broecker and Peng 1982.

somewhat greater survival potential, however, and this may be the result of $CaCO_3$ being closer to equilibrium with deep seawater than the substances (organic matter, opaline silica) containing the other elements. At any rate, the Broecker model (which can be applied to many additional elements) provides a useful way of characterizing the behavior of biogenic elements in the oceans.

Volcanic–Seawater Reaction

Volcanic activity in the oceans is extensive, and its products consist of submarine basaltic lava flows as well as widespread ash falls. The most abundant products of this volcanism include volcanic glass, pyroxenes, calcium plagioclase, and olivine, all of which are chemically unstable in seawater. As a result of their instability, and their widespread distribution, these substances react with seawater, altering its composition and producing new minerals in a variety of high-temperature and low-temperature environments. Along with biological processes (and river input), volcanic–seawater reactions constitute the two best-documented mechanisms by which the composition of the modern oceans is created and maintained.

Volcanic–seawater reactions occur primarily as a result of seafloor spreading. [For a further discussion of seafloor spreading and associated plate tectonics, consult, e.g., Kennett (1982).] New sea floor, consisting mainly of basalt, is created by igneous activity at midocean rises and ridges and then is conveyed away from the rises by lateral spreading. Because of heating from below, a convective circulation of seawater occurs along the axes and flanks of the rises. Cool water descends, becomes heated, rises, and then returns to the ocean. If sediment overlying the basalt is not too thick, and the depth in the basalt is not too great, the amount of circulating seawater can be appreciable. Because of chemical reactions between seawater and basalt, this circulation leads to major changes in the composition of both the basalt and the circulating seawater.

Quantification of the effects of the reaction between basalt and seawater is very difficult because there is no general agreement as to the amount and location of the hydrothermal circulation. Location is important, because high-temperature reactions, which occur near the axes of the rises, result, for many elements, in different changes in seawater composition from those reactions taking place at the lower temperatures encountered on the rise flanks (e.g., Thompson 1983). From geophysical modelling, Sleep et al. (1983) have estimated that nine tenths of the total heat flow due to hydrothermal convection on the rises occur along the flanks, with only one tenth along the axes; this is in strong disagreement with the calculation of Edmond et al. (1979), who attribute essentially all heat flow, and therefore all chemical reaction, to the higher-temperature axial region.

The high-temperature (200–400°C) reactions between basalt and seawater at rise axes have been well documented, from the viewpoint of both basalt composition and seawater composition. Evidence for seawater compositional alteration is derived from studies of heated water obtained from geothermal drill holes on Iceland (Holland 1978), from natural submarine hot springs in the Atlantic and Pacific oceans (for a summary, consult Von Damm 1990), and from laboratory studies of high-temperature basalt–seawater reaction (for a summary, consult Mottl 1983). The average of results for hot water emerging from submarine vents at 11°N, 13°N, and 21°N on the East Pacific Rise, the Galapagos spreading center, and the Juan de Fuca Ridge of the eastern Pacific and from one location on the Mid-Atlantic ridge are given in Table 8.7.

Results, in terms of compositional differences from ordinary seawater, agree reasonably well with those for the Icelandic waters and with results of laboratory hydrothermal experiments. All indicate that hot basalt–seawater reaction at depth involves the complete removal of Mg^{2+} and SO_4^{2-} from seawater and the addition of Ca^{2+}, H_4SiO_4, and K^+. Based on experimental studies (Bischoff and Dickson 1975; Mottl et al. 1979), initially sulfate removal occurs almost entirely via $CaSO_4$ precipitation (later, at higher temperatures, some of the sulfate may be reduced to H_2S), and the source of calcium for this is the basalt. Therefore, the total amount of calcium released to seawater is equivalent to the increase in seawater Ca^{2+} concentration plus the drop in seawater SO_4^{2-} concentration due to $CaSO_4$ precipitation, in other words, ΔCa^{2+} minus ΔSO_4^{2-} (see Table 8.7). Note that this total calcium release is equivalent, on a molar basis, to Mg^{2+} uptake. This result, as we shall see, will prove to be very useful.

Seawater also reacts with basalt at lower temperatures as the basalt spreads away from the rise axes. In this case there is much less data on water chemistry, due both to a paucity of clearly documented springs in rise flanks and to the slowness of laboratory reactions at low temperatures. Evidence has been obtained mostly from the study of altered basalts obtained by deep drilling. A summary of studies of low-temperature basalt alteration is given by Thompson (1983).

TABLE 8.7 Concentration Change of Some Major Seawater Constituents upon Reacting with Basalt at High Temperatures

Constituent	Concentration (mM)		
	Seawater	Hot Springs[a]	Δ (mM)[b]
Mg^{++}	54	0	−54
Ca^{++}	10	36	26
K^+	10	26	16
SO_4^{--}	28	0	−28
H_4SiO_4	~0	20	~20
ΔCa^{++} minus ΔSO_4^{--}	—	—	54[c]

Note: mM = millimoles per liter.

[a] Averages for 24 vents at six widely separated oceanic locations taken from the compendium by Van Damm (1990).

[b] Δ = concentration difference between average hot spring water and seawater.

[c] ΔCa^{++} minus ΔSO_4^{--} = total Ca^{++} released to solution.

As far as the elements of Table 8.7 are concerned, Thompson concludes that, overall, off-axis reactions result in uptake of K^+ and release of Ca^{2+}, Mg^{2+}, and H_4SiO_4 to seawater solution (data on sulfate are not given). Thus, for K^+ and Mg^{2+} there appears to be a reversal of the sense of reaction found for high-temperature basalt–seawater reaction. This is also true for a number of other elements, such as Li^+, Rb^+, and Ba^{2+}, which are released to solution at high temperatures and taken up by basalt at low temperatures.

Other studies are in general agreement, qualitatively, with the results summarized by Thompson, but not those for Mg^{2+}. First, the low-temperature (50°C) laboratory experiments of Crovisier et al. (1983) provide evidence for the initial uptake of Mg^{2+} from seawater, in the form of microscopic layers of Mg-containing minerals on the surfaces of altered basaltic glass. Second, several workers (e.g., Perry et al. 1976; Gieskes and Lawrence 1981) have shown that during long-term burial to hundreds of meters, sediments containing volcanic ash (fine fragments of basalt; see below) show consistent decreases in dissolved Mg^{2+} (and K^+) with depth and increases in dissolved Ca^{2+}, which are essentially equivalent with and which mirror Mg^{2+} decreases. Accompanying this is the formation of smectite and the alteration of the ^{18}O content of the interstitial water, indicating reactions with silicate minerals. An example of these changes is shown in Figure 8.11. These diagenetic depth changes are best explained in terms of the reaction of volcanic material with interstitial seawater to form smectite, and the reactions may occur either within the sediment or below it. (Some of the interstitial water compositional changes can be

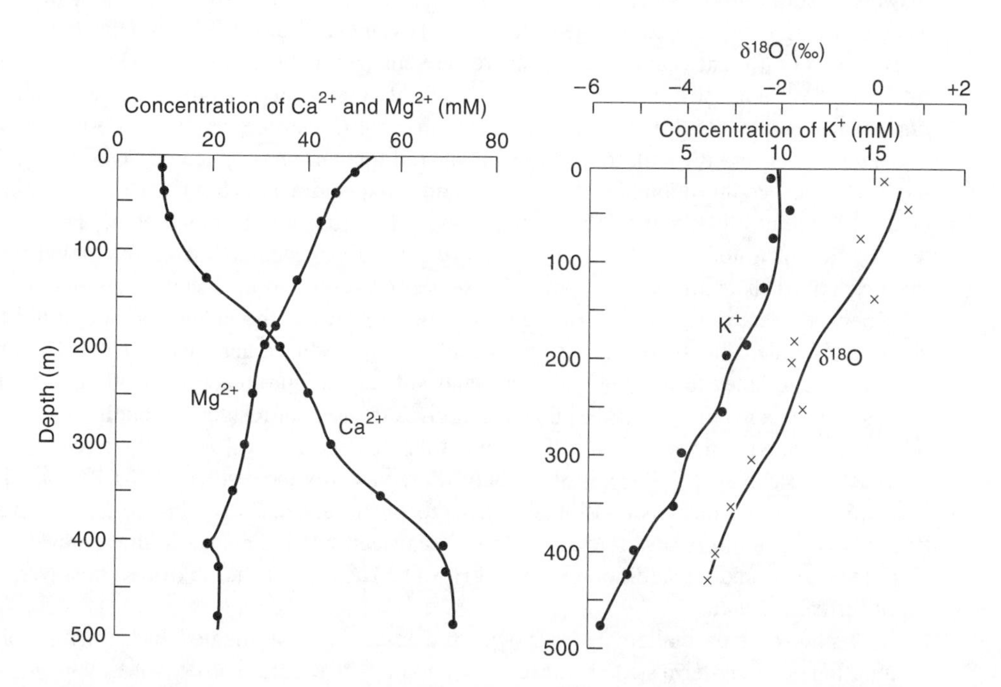

Figure 8.11. Depth distribution of concentrations of dissolved Ca^{2+}, Mg^{2+}, and K^+ in interstitial water and the oxygen isotopic composition of the water (in terms of $\delta^{18}O$ relative to Standard Mean Ocean Water) for sediments from DSDP site 336. (Modified after J. M. Gieskes and J. R. Lawrence, "Alteration of Volcanic Matter in Deep Sea Sediments: Evidence from the Chemical Composition of Interstitial Waters from Deep Sea Drilling Cores," *Geochimica et Cosmochimica Acta* 45: 1694. Copyright © 1981 by Pergamon Press, reprinted by permission of the publisher.)

explained in terms of low-temperature alteration of basalts underlying the sediments.) In either case, low-temperature volcanic–seawater reaction again involves the uptake of dissolved Mg^{2+}, a corresponding release to solution of Ca^{2+}, and a lesser uptake of K^+. Third, the experiments of Seyfried and Bischoff (1979) indicate uptake of Mg^{2+} by reaction with basalt at 70°C. Finally, the data of Staudigel and Hart (1983) indicate that failure to quantify the precipitation of Mg^{2+} to form smectite along veins in altered basalt may lead to erroneous conclusions regarding Mg^{2+} release. Based on these points, we feel that it is more likely that Mg^{2+} is taken up from solution rather than released during low-temperature basalt–seawater reactions.

Another problem with low-temperature volcanic–seawater reaction is that, as alluded to above, it does not always involve basalt layers associated with seafloor spreading. Volcanic ash, consisting of fine fragments of basaltic (and other volcanic rock) constituents, is formed by explosive subaerial volcanism on oceanic islands and is subsequently carried by winds and water currents and sedimented to the sea floor. The ash sitting on the bottom reacts with seawater, with the formation of smectitic clay minerals as a major consequence. This process is especially common in the South Pacific Ocean (Peterson and Griffin 1964), but its quantitative importance, on a worldwide basis, has not been estimated. At the rate by which sediments are buried in the deep (eupelagic) portions of the South Pacific, the process can be shown to be quantitatively unimportant, but for areas of both high explosive volcanic activity and high sedimentation rate, such as the East Indies, further quantitative estimates are needed. At least, the chemical changes in seawater resulting from volcanic ash–seawater reaction are known from studies of interstitial water concentration in buried sediments (e.g., Gieskes and Lawrence 1981), as pointed out above.

Because of differences in uptake-versus-release behavior between high- and low-temperature (including volcanic ash) reaction, as well as lack of agreement as to the relative quantitative importance of axial (high-temperature) versus off-axial (lower-temperature) subsurface water circulation, it is very difficult to quantify the overall total effect of volcanic–seawater reaction on the chemical composition of seawater. An attempt at quantification has been made by Edmond et al. (1979), but it is based solely on high-temperature reaction. Edmond et al. also assumed that the flux of water through the mid-ocean ridges can be calculated from the global flux of mantle-derived ^3He into the ocean from hydrothermal vents and the assumption of a constant ratio between heat flow and ^3He flux. From subsequent work, the latter assumption has been shown to be untenable. Nevertheless, their results are reproduced here in Table 8.8. Thompson (1983) has attempted to include both high- and low-temperature reactions, and one set of his results is also shown in Table 8.8. [Thompson gives two sets of results, depending on whether all circulation, e.g., the approach of Edmond et al. (1979), or only 10% of the circulation, the approach of Sleep et al. (1983), is axial; here we report only the results for his 10% axial calculation.] Although there is considerable disagreement between the Edmond and Thompson results, one can still say that, overall, volcanic–seawater reaction results in the removal of Mg from the oceans and the addition of Ca and H_4SiO_4. The sign of the flux of K, however, must await further research.

In this book we have attempted our own crude quantitative estimates, based mainly on the assumption of a long-term steady-state composition for seawater; in other words, volcanic–seawater reaction is used to balance budgets. Details of the methods used to calculate fluxes are summarized later, in the sections devoted to each of the elements; for the sake of comparison, however, results are also shown here in Table 8.8. Also, because of the possible importance of both biogenic and volcanic influences on the oceanic chemistry of dissolved sulfate and silica, additional discussion of these two species is presented here.

TABLE 8.8 Removal or Addition Fluxes of Some Major Seawater Constituents as a Result of Basalt-Seawater Reaction Near Midocean Ridges

Constituent	Flux (Tg/yr) Edmond et al. (1979)	Thompson (1983)	Present Study
Mg^{++}	−187	−60	−119
Ca^{++}	140	73	191
K^{+}	51	−27	53
SO_4^{--} as S	−120	—	—[a]
H_4SiO_4 as Si	90	82	56

Note: Removal values shown as negative numbers. Tg = 10^{12}g.
[a] Less than 10% of Edmond et al. (1979) value

For sulfate there are serious problems with the removal value given by Edmond et al. (1979). Appreciable concentrations of $CaSO_4$ in altered basalt, as necessitated by the results of experimental studies, simply are not found when examining altered basalt samples obtained by drilling or dredging. Probably this $CaSO_4$ is redissolved later by other circulating waters (Mottl et al. 1979). In addition, the sulfur budget of the oceans can be balanced (on a geologic time scale) by considering sulfur burial only in sediments; inclusion of such a large additional removal by reaction with basalt results in an unbalanced budget and unlikely changes in the sulfate concentration of seawater over time (see section in this chapter on the oceanic sulfur budget).

There are additional problems with the sulfur removal value of Edmond et al. (1979). Over geologic time, good agreement is found when one compares measured values of the carbon isotopic composition of the oceans (as recorded by limestones) with values calculated by a theoretical model (Garrels and Lerman 1984) that considers removal of sulfate from the oceans only as sedimentary pyrite and $CaSO_4$, and not by reaction with basalt. Put another way, if the Edmond et al. basaltic sulfate removal flux occurs mainly by reduction to sulfide, then gross imbalances in atmospheric O_2 occur over geologic time. Finally, mass balance calculations based on the isotopic composition of metal sulfides formed from H_2S exiting from submarine hydrothermal vents (summarized by Von Damm 1990), indicates that sulfate reduction, via reaction with basalt, cannot be nearly as important as envisaged by Edmond et al. (1979).

Dissolved silica is also a problem. Thompson (1983) suggests a large release of H_4SiO_4 to solution during low-temperature reaction, but the data of Staudigel and Hart (1983) suggest that most of this silica is reprecipitated to form veins of smectite within the basalt. Furthermore, based on Icelandic hydrothermal waters, flux values less than one third that given for silica addition by submarine springs (Table 8.7) were obtained by Holland (1978). For these reasons we have calculated a separate value for the hydrothermal silica flux (56 Tg/yr), which is based on our estimated hydrothermal Mg flux (Table 8.8) and the observed inverse proportionality between Mg and silica concentrations at the vent sites (Von Damm 1990).

Regardless of whose quantitative estimates are adopted, there is no doubt that the oceanic geochemical cycles of Mg, Ca, K, and many minor elements are appreciably affected by volcanic activity [see Von Damm (1990) for a discussion of Li^+, Rb^+, Ba^{2+}, etc.]. In addition, the common occurrence of sodium-enriched basalts known as spilites, which are believed to form by the reaction of seawater at high temperatures with basalt, point to the possibility that Na^+ removal at high temperatures is also involved. Compositional data from basalt–seawater experiments and from studies of altered basalts, and hydrothermal solutions, however, are inconclusive as to the quantitative effects on seawater sodium, and more research on this topic is needed. [There is slight evidence for Na removal at high temperature for most of the hydrothermal vent sites (Von Damm 1990).]

Interaction with Detrital Solids

Detrital materials carried to the oceans by rivers consist to a large extent of silicate minerals, especially clay minerals, which are not in equilibrium with seawater. Therefore, upon their entering the oceans, chemical reactions take place. Such reactions may involve the entire silicate mineral, or only its surface. In the former case we have formation of a new, generally more cation-rich mineral from the old detrital one, and the process, because it resembles weathering, is referred to as *reverse weathering* (Mackenzie and Garrels 1966). In the latter case, because of slowness of bulk reaction, only chemical changes involving species on the mineral surface take place, and this process is referred to as *adsorption-desorption* or, if ions are involved, *ion exchange.* Together, reverse weathering, adsorption-desorption, and ion exchange comprise all major reactions between river-borne silicate detritus and seawater.

The concept of reverse weathering was developed by Mackenzie and Garrels (1966) to provide a ready mechanism for the removal of several species, added to the oceans by rivers, for which at that time there were no well-documented removal processes. This includes Na^+, K^+, Mg^{2+}, HCO_3^-, and H_4SiO_4. The reasoning goes as follows: Aluminosilicate weathering products, which are depleted in cations and silica relative to the primary silicates from which they originally formed, upon entering seawater take up cations and silica and in the process convert HCO_3^- to CO_2. The overall reaction, written, for example, in terms of Na^+, can be expressed in a very generalized manner as

$$Na^+ + HCO_3^- + H_4SiO_4 + \text{Al-silicate} \rightarrow \text{NaAl-silicate} + CO_2 + H_2O$$

This reaction is very similar to those written for weathering reactions (see Chapter 4), except for going in the reverse direction, and this is how the term reverse weathering arose. However, one major difference from true reverse weathering is that the cation-enriched aluminosilicate formed is not the same as the primary silicates involved in weathering. In other words, feldspars, pyroxenes, and so forth, are not formed. (They are unstable in seawater.) Besides helping to remove cations and silica added to the oceans by rivers, reverse weathering is especially useful as a mechanism for the conversion of HCO_3^-, by the reaction with cation-depleted or "acid" aluminosilicates, to CO_2 so as to balance the removal of CO_2 by the opposite process, which occurs during weathering.

As a major process affecting the bulk mineralogy of marine sediments, the concept of reverse weathering early on did not hold up well under testing. Russell (1970), by closely examining changes in the chemical composition of highly degraded and cation-depleted clays carried by a Mexican river, found that, upon addition to and deposition in seawater, there was no measurable net uptake of cations by the clay. The only change in clay composition observed was the exchange of river-borne cations for new seawater cations on the clay particle surfaces. This observation agrees with those cited earlier in this chapter, when discussing the Sillèn equilibrium model, that little evidence, whether mineralogical, compositional, or isotopic, exists for the large-scale formation of new clay minerals from old detrital clay minerals in the marine environment. Furthermore, the results of DeMaster (1981) indicate that the riverine flux of silica can be removed from seawater almost entirely as opaline silica skeletal remains, without the necessity of invoking large amounts of reverse weathering, as can the Mg^{2+} flux be removed by volcanic–seawater reaction.

More recent work, however, has suggested that reverse weathering may be more important than is commonly believed. Possible support for reverse weathering has been found by Sayles (1979, 1981). Using a very precise chemical technique, he reports that small but measurable chemical concentration gradients exist in the concentration of dissolved Mg^{2+}, K^+, Na^+, Ca^{2+}, and HCO_3^- across the sediment–water interface of many different pelagic (deep-sea) sediments of the Atlantic Ocean. Some of his results are shown in Table 8.9. Note that the changes in Mg^{2+}, K^+, and Ca^{2+} are all in the same direction as those for volcanogenic sediments discussed earlier. In other words, there is evidence for Mg^{2+} and K^+ uptake and Ca^{2+} release from the sediment. Most of the Ca^{2+} concentration gradient is due to $CaCO_3$ dissolution within the sediment, as witnessed by corresponding increases in HCO_3^-, but some excess Ca^{2+} remains which cannot be explained by carbonate dissolution. Since the Mg^{2+}, K^+, and Ca^{2+} changes are all in the same direction as predicted for low-temperature volcanic ash–seawater reaction, it is tempting to ascribe these subtle changes at the seawater–sediment interface to this process as well. However, most of the sediments are not obviously volcanogenic, and it is possible that the observed concentration gradients in K^+ and Mg^{2+} are, instead, brought about by reverse weathering.

Adding to the results of Sayles are the studies of Mackenzie et al. (1981) and Rude and Aller (1989). Mackenzie et al. found geochemical and mineralogical evidence for the transformation of amorphous aluminosilicates and ferric oxyhydroxides to Fe-rich and Mg-containing smectite (nontronite) in sediments of Kaneohe Bay, Oahu, Hawaii. On a larger scale, Rude and Aller have observed the uptake of Mg by ferric hydroxide coatings on sand grains in Amazon Delta sediments, and they speculate that this represents the formation of new authigenic Fe-Mg silicates. They calculate that a large proportion of the Mg carried by the Amazon River could be removed in the form of such silicates. In addition, Rude and Aller (1994) have observed the removal of K^+ and F^- from interstital solution within the Amazon sediments to form what they believe are additional authigenic silicate minerals.

TABLE 8.9 Change in Concentration in Interstitial Water for Various Ions Versus Depth in a Sediment from the Brazil Basin, South Atlantic Ocean (Station CH 115–DD)

Sediment	Concentration Change (Pore Water – Overlying Seawater) (mM)						
Depth (cm)	pH	ΔNa^+	ΔMg^{++}	ΔCa^{++}	ΔK^+	ΔHCO_3^-	ΔSO_4^{--}
0	7.4	0.00	0.00	0.00	0.00	0.00	0.00
5	7.5	0.07	–0.04	0.17	0.05	0.19	0.05
15	7.3	0.06	–0.35	0.45	–0.11	0.25	0.04
30	7.5	0.46	–0.42	0.50	–0.08	0.34	0.06
60	7.5	0.45	–0.58	0.76	–0.11	0.68	0.06
100	7.2	0.56	–0.78	0.97	–0.16	0.82	–0.01
195	7.4	0.95	–1.09	1.18	–0.26	1.12	–0.13

Note: Negative Δ values refer to uptake by the sediment (loss from pore water). mM = millimoles per liter.

Source: Adapted from F. L. Sayles 1979.

An important factor, mentioned originally by Garrels and Mackenzie and reemphasized by Rude and Aller, is that for reverse weathering to serve as an important sink for K, Mg, etc., only a very small proportion of the total material deposited in areas of heavy sedimentation, such as major river deltas, need represent new minerals formed by this process. If the results of Rude and Aller are found to be applicable to deltaic sediments in general (globally, most sediment is deposited in and near deltas), then the process of reverse weathering may eventually rival or surpass basalt–seawater reaction in importance as the principal process for the removal of Mg and K from the oceans. Nevertheless, until the quantitative significance of reverse weathering is conclusively demonstrated on a global basis, this process will be ignored in this book as a major process of removal of cations and silica from seawater.

There is no doubt that the surfaces of clay minerals carried into the marine environment by rivers undergo reaction with seawater. From the viewpoint of the present chapter, the reaction of greatest interest is that of cation exchange. Clay minerals rapidly (hours to days) undergo exchanges of cations on their external surfaces and, in some clays such as smectites, in interlayer positions as well, when transferred from a water of one composition to another of a different composition. In this case of transfer from river water to seawater, the degree of exchange is large because of the large difference in concentrations between the two waters.

An example of cation exchange upon transfer of a common clay mineral, smectite, from river water to seawater, taken from the work of Sayles and Mangelsdorf (1977), is shown in Table 8.10. (Later work of Sayles and Mangelsdorf using natural, untreated river clay from the Amazon gave similar results.) Note that the principal change is the uptake of Na^+ and release of Ca^{2+} from the clay, with a lesser exchange of K^+ for H^+. Holland (1978) has stated that the cationic exchange capacity of average river suspended sediment is approximately 18 mEq/100 g. Using this value, the value of 20,000 Tg for the mass of suspended sediment carried to the sea by rivers each year (see Chapter 5), and assuming that the proportional changes for each element found by Sayles and Mangelsdorf for smectite can be extrapolated to all river-borne sediment, we have calculated present-day removal rates of Ca^{2+}, Mg^{2+}, Na^+, and K^+ from seawater as a result of

TABLE 8.10 Cation Exchanges on Smectite upon Transfer from River Water to Seawater and Oceanic Removal Rates

	Surface Concentration[a] (mEq/100 g dry wt.)			Oceanic Removal (−) or Addition (+)
Ion	Equilibrated with River Water	Equilibrated with Seawater	Change on Clay	Rate[b] (Tg/yr)
Ca^{++}	57.6	15.7	−41.9	+37
Mg^{++}	18.4	18.1	−0.3	+0.2
Na^+	2.3	44.3	+42.0	−42
K^+	0.6	2.7	+2.1	−4
H^+	2.0	0	−2.0	—
All ions (CEC)[c]	80.9	80.8	0	—

[a] Values are based on smectite reacting for 7 days with average seawater after being previously equilibrated for 7 days with synthetic average river water (mEq = milliequivalents of +1 charge).

[b] Rates are based on a river flux of 20,000 Tg of sediment per year and an average cation-exchange capacity of 18mEq/100 g dry wt. (Tg = teragram = 10^{12} g).

[c] CEC = cation exchange capacity.

Source: Sayles and Mangelsdorf 1977.

cation exchange. These values are also shown in Table 8.10, and, for sodium, the rate represents an appreciable fraction (16%) of the total input rate of sodium to the ocean.

The values of Table 8.10 are based on the rate of suspended sediment input by present-day rivers. As we have shown in Chapter 5, this rate is unusually high due to increased erosion resulting from human agricultural activity. It may also be high because of an abundance of easily erodible debris produced by the last Pleistocene glaciation. In fact, a number as low as 6000 Tg of solids per year, roughly one third of the present rate, has been used by Garrels and Perry (1974) for the pre-Pleistocene long-term input rate. Our own calculation in Chapter 5 indicate a pre-human value of about one half that at present, which we will use for constructing long-term budgets (millions of years). In this case the numbers listed in Table 8.10 must be divided by 2 to obtain the proper values for the effects of cation exchange.

In addition to simple cation exchange, a slightly different exchange process occurs upon the transfer of certain clays from fresh water to the marine environments. This is potassium fixation. Whereas simple cation exchange involves the replacement of one hydrated cation on the clay surfaces or in interlayer positions by another from solution, in potassium fixation the addition of potassium to the clay involves dehydration of the added potassium ion, replacement of hydrated cations by potassium only in interlayer positions, and a consequent loss of exchangeability. This process involves mainly micas which have previously lost some K^+ during weathering (degraded mica) and which take up K^+ readily upon addition to seawater. It is an irreversible process as compared to simple cation exchange, which is reversible (The K^+ taken up from seawater is not readily lost upon reexposure of the mica to low-K fresh water.)

Quantitative evaluation of the importance of K fixation as a control of seawater composition is provided by the work of Hoffman (1979). It was found that micalike clay (illite) from the Mississippi River, when deposited in seawater on the Mississippi Delta, took up K^+ in measurable quantities. Extrapolating these results to total river-borne suspended sediment (20,000 Tg/yr), we obtain a worldwide K^+ uptake rate of about 6 Tg/yr. This represents an appreciable but minor portion (10%) of the K^+ input by rivers.

CHEMICAL BUDGETS FOR INDIVIDUAL ELEMENTS

Summary of Processes

Through the remainder of this chapter we shall attempt to estimate quantitatively, using the simple box model presented earlier (Figure 8.3), the various inputs and outputs for each of the major components (and some minor components) of seawater. Two time scales will be used: one for the present (past tens of years), and one for geologic time (past tens of millions of years). Where sufficient independent data are available, we shall check to see if there is a steady-state balance between inputs and outputs on the million-year time scale. Otherwise we shall assume a steady state in order to quantify long-term outputs from inputs. In the case of biological elements, we shall also discuss how areally varying concentration differences are attained and maintained. As an aid to the reader, a summary of the major processes affecting each element is presented in Table 8.11.

During the upcoming discussion of major seawater species, the amounts added by rivers and amounts removed via aerosol (sea-spray) transfer across the air–seawater interface and not returned by marine rainfall (i.e., that transferred via the atmosphere from the oceans to the

TABLE 8.11 Major Processes Affecting the Concentration of Specific Components of Seawater Numbered in Order of Approximate Decreasing Importance

Component	Input Processes	Output Processes
Chloride (Cl^-)	1. River-water addition (including pollution)	1. Evaporative NaCl deposition (in past) 2. Net sea–air transfer 3. Pore-water burial
Sodium (Na^+)	1. River-water addition (including pollution)	1. Evaporative NaCl deposition (in past) 2. Net sea–air transfer 3. Cation exchange 4. Basalt–seawater reaction 5. Pore-water burial
Sulfate (SO_4^{--})	1. River-water addition (including pollution) 2. Polluted rain and dry deposition	1. Evaporative $CaSO_4$ deposition (in past) 2. Biogenic pyrite formation 3. Net sea–air transfer
Magnesium (Mg^{++})	1. River-water addition	1. Volcanic–seawater reaction 2. Biogenic Mg-calcite deposition 3. Net sea–air transfer
Potassium (K^+)	1. River-water addition 2. Volcanic-seawater reaction (high temp.)	1. Low-temperature volcanic–seawater reaction or slow K^+ fixation or reverse weathering 2. Fixation on clays near river mouths 3. Net sea–air transfer
Calcium (Ca^{++})	1. River-water addition 2. Volcanic–seawater reaction 3. Cation exchange	1. Biogenic $CaCO_3$ deposition 2. Evaporitic $CaSO_4$ deposition (in past)
Bicarbonate (HCO_3^-)	1. River-water addition 2. Biogenic pyrite formation	1. $CaCO_3$ deposition
Silica (H_4SiO_4)	1. River-water addition 2. Basalt–seawater exchange	1. Biogenic silica deposition
Phosphorus (HPO_4^{--}, PO_4^{-3}, $H_2PO_4^-$, organic P)	1. River-water addition (including pollution) 2. Rain and dry fallout	1. Burial of organic P 2. $CaCO_3$ deposition 3. Adsorption on volcanogenic ferric oxides 4. Phosphorite formation
Nitrogen (NO_3^-, NO_2^-, NO_4^-, organic N)	1. N_2 fixation 2. River-water addition (including pollution) 3. Rain and dry deposition	1. Denitrification 2. Burial of organic N

continents) must often be known. As an additional aid to the reader, a summary of river addition (from Chapter 5) and net sea-salt transfer is presented here (Table 8.12). Rates of sea–air transfer are derived from the rates given in Chapter 5 for the cyclic salt contributions to rivers.

Chloride

There is little doubt that the chloride budget in the present ocean is badly out of balance and that the concentration of chloride is increasing with time, albeit very slowly because of the long replacement time of Cl^- of 87 million years. This has come about in two ways. First, there are essentially no modern evaporite basins where appreciable NaCl (halite) precipitation is occurring and, thus, no corresponding removal of the large riverine influx of chloride derived from the dissolution of old salt beds. Second, about a third of the input of Cl^- is due to pollution (see Chapter 5), and the oceans have had far too little time to adjust to this extra addition. Combined, these two factors have caused a large chloride imbalance.

An idea of the present-day Cl^- imbalance is provided by a quantitative comparison between known inputs and outputs. This is shown in Table 8.13. The only important output processes are net sea-salt aerosol transfer and burial of interstitial water in fine-grained sediments. The sea–air transfer rate is given in Table 8.12. The value for the burial rate is based on the average Cl^- content of shales of 0.12% (Holland 1978) and the sedimentation rate (riverine delivery rate) of fine-grained solids of 20,000 Tg/yr for the present day and half that value for the geologic past. (Shales should represent the ultimate burial of salts contained in interstitial water because essentially all H_2O is squeezed out by compaction, while some salts are trapped by selective filtration.)

Evaporite basins where NaCl (halite) is deposited can form only where there is both an arid climate and restriction of water exchange with the open ocean so that high salinities can be

TABLE 8.12 Rates of Addition via Rivers of Major Elements to the ocean (as Dissolved Species) and Rates of Net Loss from the Ocean by Transfer of Sea Salt to the Continents via the Atmosphere

Species	Rate of Addition from Rivers[a] (Tg/yr)	Rate of Net Sea Salt Loss to Atmosphere (Tg/Yr)
Cl^-	308	40
Na^+	269	21
SO_4-S	143	4
Mg^{++}	137	3
K^+	52	1
Ca^{++}	550	0.5
HCO_3^-	1980	—
H_4SiO_4-Si	180	—

Note: Tg = 10^{12} g.

[a] Based on river water input of 37,400 km^3/yr: includes pollution

Sources: River-water data from Meybeck 1979; cyclic salt data from Chapter 5.

attained (see Chapter 7). This combination of conditions is fortuitous and, thus, evaporite basin formation is basically a stochastic process, dependent on both tectonic and climatological factors. It occurs sporadically, and the present happens to be a time when there are no major evaporite basins. Also, once evaporite basins are formed, precipitation of NaCl (and $CaSO_4$) is very rapid (e.g., King 1947), so that at times of major evaporite deposition the concentration of Cl^- in seawater would be expected to decrease suddenly. These factors, combined, would be expected to produce considerable fluctuation in the Cl^- content, and thus in the salinity of seawater. However, the salinity of the oceans over the past 600 million years cannot have deviated too greatly from the value found today, as evidenced by the presence of marine organisms in rocks deposited over this time which could not have survived unusually high or low salinities. Furthermore, even if all evaporite deposition ceased over a long time and the present existing mass of halite were all transferred via weathering to the ocean, the result, which is a maximum effect, would only be a doubling of the present Cl^- concentration.

Over the long term, the average chloride content and salinity of seawater probably represents a steady-state balance. But superimposed on this balance one would expect small fluctuations brought about by the formation, or lack of formation, of evaporite basins. Overall a chloride content-versus-time curve should look something like that depicted in Figure 8.12. Assuming steady state, a time-averaged balanced budget, using the present-day natural riverine input rate, is shown in Table 8.13. Sodium chloride deposition is used to balance the budget. Thus, averaged over geologic time, the major process for Cl^- removal from the oceans would be evaporitic NaCl deposition, with sea–air seasalt transfer a distant second.

Sodium

By contrast to chloride, sodium is a more complex element. It is involved in the same processes as chloride, as NaCl, but because it is a major constituent of silicate minerals, it is also involved in rock weathering, cation exchange, and possibly volcanic–seawater reaction and reverse weathering. (The role of sodium in the latter two processes is not well established.) It provides a most interesting link between the geochemical cycles of silicates and evaporites. The present-day

TABLE 8.13 The Oceanic Chloride Budget (Rates in Tg Cl^-/yr)

Present-Day Budget			
Inputs		**Outputs**	
Rivers (natural)	215	Net sea–air transfer	40
Rivers (pollution)	93	Pore-water burial	25
Total	308	Total	65
Long-Term (Balanced) Budget			
Inputs		**Outputs**	
Rivers	215	NaCl evaporative deposition	163
		Net sea–air transfer	40
		Pore-water burial	12
		Total	215

Note: Tg = 10^{12} g. Replacement time for Cl^- is 87 million years.

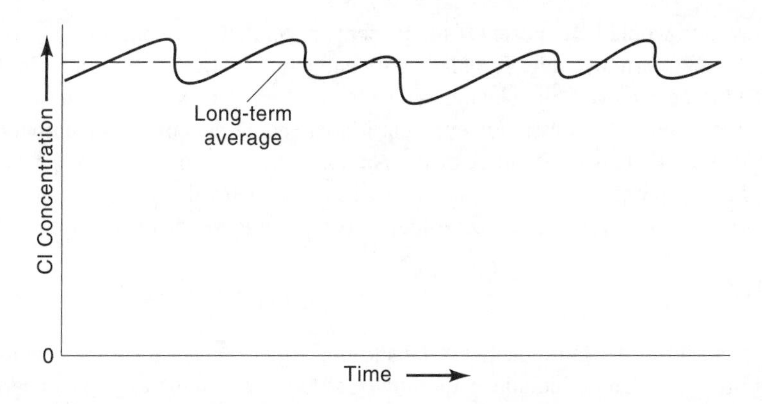

Figure 8.12. Schematic representation of change of chloride (Cl^-) concentration in seawater with geologic time. Sudden drops are due to the rapid precipitation of NaCl in evaporite basins.

oceanic budget of sodium is shown in Table 8.14, where only known, quantifiable removal mechanisms are listed. The cation-exchange value is taken from Table 8.10, and the sea–air transfer value from Table 8.12. Pore water burial is that accompanying Cl^- burial as given in Table 8.13. Because of a lack of evaporative NaCl formation and a pollutional input of NaCl by rivers, it can be seen that the present ocean is badly out of balance with respect to Na^+, as it is with Cl^-. Also, over geological time, the fluctuations predicted for Cl concentration must also be mirrored by Na^+ concentration.

The long-term budget of sodium is also shown in Table 8.14. Here sodium removal, as evaporitic NaCl, is that corresponding to the rate of chloride removal used to balance the chloride budget (Table 8.13). Also, the rate of removal by cation exchange is one half the value for the present

TABLE 8.14 The Oceanic Sodium Budget (Rates in Tg Na^+/yr)

Present-Day Budget			
Inputs		Outputs	
Rivers (natural)	193	Cation exchange	42
Rivers (pollution)	76	Net sea–air transfer	21
		Pore-water burial	16
Total	269	Total	79
Long-Term Budget			
Inputs		Outputs	
Rivers	193	NaCl deposition	106
		Net sea–air transfer	21
		Cation exchange	21
		Pore-water burial	8
		Basalt–seawater reaction	37
		Total	193

Note: Tg = 10^{12} g. Replacement time for Na^+ is 55 million years.

ocean, which would be expected if the present suspended sediment transport by rivers is about twice the long-term average. To balance the budget, a relatively minor amount (19%) of Na^+ is assumed to be removed by volcanic–seawater reaction. This volcanic removal must be considered a very tentative number, however, until independent data on the quantitative importance of this process are obtained. Studies of the Na content of hydrothermal vent waters (Von Damm 1990) have so far provided no consistent evidence of Na uptake from seawater, but the existence of Na-rich basalts (spilites) in the geological record suggests that some uptake does occur.

Sulfate

Sulfate is another versatile element like sodium. It is removed from the oceans by several different processes. These include evaporative $CaSO_4$ precipitation as gypsum and anhydrite, biogenic pyrite (FeS_2) formation, net sea–air transfer, and possibly high-temperature basalt–seawater reaction. Like chloride and sodium, a large proportion of sulfate entering the oceans is derived from worldwide pollution. Unlike sodium and chloride, this pollutive addition also occurs by way of the atmosphere as a result of fossil fuel burning.

The sulfur budget for the present-day ocean is shown in Table 8.15. The value for biogenic pyrite formation is that derived by one of us (Berner 1982) from the well-documented constancy of the organic carbon-to-pyrite sulfur ratio in anoxic marine sediments and the integrated rate of organic carbon removal from the modern ocean. Most, by far, of the sulfur removal as pyrite occurs in continental margin sediments. (For a further discussion of biogenic pyrite formation, consult the section on biogenic processes.) The values for rain (and dry deposition) addition and net sea–air transfer are taken from the budget presented in Chapter 3. It is assumed (for the purposes of checking this assumption) that sulfate removal during basalt–seawater reaction is negligible. As can be seen, there is a very bad imbalance in the modern ocean budget, which is due to excessive inputs of pollutive sulfur and a lack of removal as $CaSO_4$ in evaporites. (This is yet another consequence

TABLE 8.15 The Oceanic Sulfate Budget (Rates in Tg S/yr)

Present-Day Budget			
Inputs		**Outputs**	
Rivers (natural)	66	Biogenic pyrite formation	39
Rivers (pollution)	77	Net sea–air transfer	
Rain and dry deposition[a]	17	(sea salt 4; H_2S 4)	8
Total	160	Total	47
Long-Term (Balanced) Budget			
Inputs		**Outputs**	
Rivers	66	Biogenic pyrite formation	20
Rain (unpolluted)	4	Evaporitic $CaSO_4$ deposition	42
		Net sea–air transfer	8
Total	70	Total	70

Note: Tg = 10^{12} g. To convert to sulfate, simply multiply by 3. The replacement time for SO_4^{--} (from river addition only) is 8.7 million years.

[a] Equal to input to marine air via transport from the continents, 75% of which is due to pollution; see Table 3.12.

of the absence of large modern-day evaporite basins.) As will be shown, the imbalance is not likely to be caused also by the omission of basalt–seawater reaction as a removal process.

The long-term budget for sulfate is also shown in Table 8.15. Pollution is excluded, pyrite burial is reduced by half to account for half the rate of burial of total sediment, and the budget is balanced via removal of $CaSO_4$ in evaporite basins. The $CaSO_4$ removal rate, thus derived, is reasonable. The range of estimates by various workers for the ratio of $CaSO_4$ sulfur to pyrite sulfur in sedimentary rocks lies betwen 2:1 and 1:2 (R. M. Garrels, personal communication). If the weathering rate of $CaSO_4$ (per unit mass) is about twice the rate for pyrite (see Chapter 5), then the two minerals would be expected to contribute fluxes to rivers in a ratio ranging from 1:1 to 4:1. To balance these fluxes, they would be expected to undergo removal from seawater in the same ratio. Thus, the ratio of 2 gypsum sulfur to 1 pyrite sulfur, obtained by balancing the long-term budget only with evaporitic $CaSO_4$ deposition, is not out of line with expectations.

Obtaining a reasonable long-term balance with evaporitic $CaSO_4$ deposition means that basalt–seawater reaction is not needed as a removal mechanism. If the total basalt–water removal value obtained by Edmond et al. (1979) were used (120 Tg S/yr), the long-term budget would be badly out of balance. For this reason and others given earlier in the section on volcanic–seawater reaction, it is believed that removal of SO_4^{2-} by heated basalt is not a major mechanism for the removal of SO_4^{2-} from seawater.

Magnesium

In contrast to sodium and sulfate, magnesium removal from the oceans is greatly affected by volcanic–seawater reaction. In fact, it probably is the most important removal process. The only other significant processes of Mg removal are biogenic magnesian calcite secretion and net sea–air transfer, and both are quantitatively much lower.

As pointed out earlier (see Table 8.8), there is a lack of agreement as to the rate by which Mg is removed from the oceans via volcanic–seawater reaction. All one can say is that the process is probably important. Here we assume it to be the dominant process and derive a removal rate by subtracting all other rates of removal from the total input so as to bring about steady state (see Table 8.16). The rate of removal in biogenic magnesian calcite was obtained from the approximate rate of deposition of $CaCO_3$ in shallow-water sediments, 1000 ± 300 Tg/yr (Milliman 1974; Hay and Southam 1977) and an average Mg content of 1.5 wt% Mg in such sediments (Milliman 1974). The rate of net sea–air transfer was obtained from Table 8.12. The Mg-calcite and sea–air transfer values, when subtracted from the total input, give a steady-state value of 119 Tg/yr for the rate of removal of Mg by volcanic–seawater reaction (Table 8.16).

TABLE 8.16 The Oceanic Magnesium Budget (Rates in Tg Mg^{++}/yr)

(Balanced) Budget for Past 100 Million Years			
Inputs		Outputs	
Rivers	137	Volcanic–seawater reaction	119
		In biogenic $CaCO_3$	15
		Net sea–air transfer	3
		Total	137

Note: Tg = 10^{12} g. Replacement time for Mg^{++} is 13 million years.

Note that this value may include both high-temperature and low-temperature reactions with basalt if both involve uptake of seawater magnesium. At any rate, the Edmond et al. (1979) value of 187 Mt/yr for high-temperature basalt–seawater reaction alone is distinctly higher than the value required for steady state and is believed by us to be too high.

In our budget we have not considered removal of Mg in authigenic clay minerals, so our estimate of removal via basalt–seawater reaction is likely a maximum. Recent work of Rude and Aller (1989, 1994) on the Amazon Delta indicates that Mg removal as Fe-Mg authigenic silicates in deltaic sediments may be globally a quantitatively important process. We have also not considered removal of Mg in dolomite, $CaMg(CO_3)_2$, which is a common mineral in ancient sedimentary rocks. The reason for this is the scarcity of dolomite in sediments deposited over the past 100 million years, including those of the present ocean. Thus, our "long-term" budget does not extend to periods prior to 100 million years, when dolomite formation was important.

Potassium

Of all the elements discussed in this chapter, potassium provides one of the biggest problems. This is because the chief removal mechanism has not been adequately documented. The annual input of potassium to the oceans is 52 Tg from rivers and 53 Tg from high-temperature basalt–seawater reaction. (The latter value is taken from Table 8.8.) Of the 105 Tg delivered, only 1 Tg is removed by net sea–air transfer and ~6 Tg by fixation on deltaic clays [based on data of Hoffman (1979)]. The remaining 98 Tg may be removed by low-temperature volcanic–seawater reaction. Several studies (e.g., Thompson 1983) have shown that, at low temperatures, basalts pick up K^+ during reaction with circulating seawater. In addition, slow steady depletion of K^+ in pore water with depth has been demonstrated for long columns of oceanic sediments (Gieskes and Lawrence 1981) in which Mg^{2+} uptake via volcanic ash reaction with seawater is taking place. If similar but more rapid uptake of K^+ occurs on volcanic ash near the sediment–water interface, then low-temperature volcanic–seawater reaction may represent the missing K^+ removal mechanism.

The data of Sayles (1979, 1981) and that of Rude and Aller (1994) (see also Table 8.9), however, suggest that slow potassium fixation and/or reverse weathering may constitute the missing mechanisms. On the basis of interstitial water concentration gradients for dissolved K^+, Sayles calculated that deep-sea sediments of the north, south, and central Atlantic type (nonvolcanogenic), on a worldwide basis, could account for the removal of 60–75 Tg K/yr. Since there is no definitive evidence that K^+ removal in Sayles's sediment is due to volcanic–seawater reaction, it is possible that potassium fixation on illite and/or reverse weathering are the main processes of removal. Added to deep-sea sediments is the finding by Rude and Aller of interstitial water evidence of K uptake by Amazon deltaic sediments.

Until more definitive data are obtained, we shall assume that the potassium budget is in balance, and that the major removal mechanism is either low-temperature volcanic–seawater reaction, or diagenetic potassium fixation, or reverse weathering, or a combination of all three. Results are shown in Table 8.17.

Calcium

Calcium provides a refreshing contrast to the preceding elements in that only one mechanism of removal for the modern ocean is significant, and it is quantitatively well known. That is the

TABLE 8.17 The Oceanic Potassium Budget (Rates in Tg K^+/yr)

Long Term (Balanced) Budget			
Inputs		Outputs	
Rivers	52	Fixation on clay near river mouths	6
		Sea–air transfer	1
Volcanic–seawater reaction (high-temperature)	53		
Total	105	Low-temperature volcanic–seawater reaction or slow fixation in deep sea or reverse weathering	98
		Total	105

Note: Tg = 10^{12} g. Replacement time for K^+ is 10 million years.

depositional burial in bottom sediments of biogenic $CaCO_3$ skeletal debris. Because of differences in mineralogy and types of organisms, this deposition is divided into that occurring in shallow water (benthic aragonite and magnesium calcite) and that occurring in deep water (planktonic calcite). (For further discussion of skeletal mineralogy, consult the section on biogenic processes.) Shallow-water sediments today are receiving much higher quantities of $CaCO_3$ than normal because of the rapid postglacial (last 11,000 years) rise of sea level over the continental shelves (Hay and Southam 1977). Hay and Southam state that the present shallow-water (shelf and slope) deposition rate is approximately 1300 Tg of $CaCO_3$ per year, as compared to the average value for the past 25 million years of 600 Tg/yr. The average deposition rate of planktonic $CaCO_3$ in the deep sea over the past 25 million years is 1100 Tg $CaCO_3$ per year. (Present-day deep-sea $CaCO_3$ sedimentation is assumed to be about the same.) These rates, in terms of Ca^{2+} removal from the ocean, both for the present-day ocean and the average for the past 25 million years, are shown in Table 8.18.

Also listed in Table 8.18 are a variety of inputs. Besides rivers, there is calcium addition from cation exchange (taken from Table 8.10) and volcanic–seawater reaction. The value for volcanic–seawater reaction is based on the rate of uptake of Mg^{2+} by this process (Table 8.16) and the assumption, documented earlier, of an equivalent release of Ca^{2+} (on a molar basis). The lower value for cation exchange for the past 25 million years is based on the assumption of a lower (by half) average sediment supply rate during this period (see section on interaction with detrital solids).

The remaining rate listed in Table 8.18 is that for Ca^{2+} removal as $CaSO_4$, during periods of evaporite deposition. It is based on the rate of sulfate removal (Table 8.15) for this process and, as can be seen, is small when compared to removal as $CaCO_3$.

The data of Table 8.18 reveal two interesting observations. First, the removal of Ca^{2+} from the present ocean considerably exceeds its input, and this is caused by excessively rapid deposition of $CaCO_3$ on the continental shelves as a result of the rapid postglacial rise in sea level. By contrast, over the long term (the past 25 million years) there is essential balance, well within errors of estimation, of Ca^{2+} inputs and outputs. (Errors in estimating deposition rates can be as high as ±50%.) This balance is what is to be expected, and it demonstrates the essential validity of all the independent rate estimates used to construct the calcium budget. In this way it serves as a check on numbers used to balance other cycles (e.g., Mg).

TABLE 8.18 The Oceanic Calcium Budget (Rates in Tg Ca^{++}/yr)

Present-Day Budget			
Inputs		Outputs	
Rivers	550	$CaCO_3$ deposition:	
Volcanic–seawater reaction	191	Shallow water	520
Cation exchance	37	Deep sea	440
Total	778	Total	960

Budget for Past 25 Million Years			
Inputs		Outputs	
Rivers	550	$CaCO_3$ deposition:	
Volcanic–seawater reaction	191	Shallow water	240
Cation exchance	19	Deep sea	440
		Evaporitic $CaSO_4$ deposition	49
Total	760	Total	729

Note: Tg = 10^{12} g. Replacement time (rivers only) for Ca^{++} is 1 million years.

The long-term balance also shows that removal of Ca^{2+} as silicate minerals in the ocean, if it in fact occurs, is of negligible importance compared with removal as $CaCO_3$. Since Ca^{2+} carried by rivers is derived partly from calcium silicate (plagiociase) weathering (in Chapter 5 we suggest that 18% of river-borne Ca^{2+} comes from silicates), removal from the oceans only as $CaCO_3$ infers an overall removal of CO_2 from the ocean–atmosphere system. However, this CO_2 is returned by metamorphic and volcanic breakdown of $CaCO_3$ upon heating during deep burial (see Berner et al. 1983).

Most of the $CaCO_3$ secreted by planktonic organisms (foraminifera, coccoliths, pteropods) does not become buried in deep-sea sediments. This is because seawater at great depth is undersaturated with respect to $CaCO_3$ and, as a result, material falling to such depths dissolves away. This gives rise to the concept of the *carbonate compensation depth,* which is abbreviated as CCD. [For a review discussion of the CCD and related phenomena, consult Morse and Mackenzie (1990).] In the oceans, as one goes downward, the water passes from supersaturation to undersaturation with respect to $CaCO_3$, and, once undersaturation is attained, the rate of dissolution increases with increasing depth. The CCD is the depth where the rate of supply of calcareous skeletal debris from above is equaled by the downward increasing rate of dissolution. Below the CCD, little or no $CaCO_3$ survives and, as a result, large areas of the deepest portions of the ocean (mainly in the Pacific) are relatively free of $CaCO_3$. A map of $CaCO_3$ distribution in the Atlantic Ocean is shown in Figure 8.13.

Although the matter is still somewhat controversial, most workers are in agreement that undersaturation and $CaCO_3$ dissolution occur above the carbonate compensation depth (see Figure 8.14) and that the CCD does not simply represent the boundary between overlying supersaturated and underlying undersaturated water. Dissolution occurs over a range of depths because it is not instantaneous and because different planktonic organisms dissolve at different rates. (The latter greatly complicates the interpretation of planktonic microfossils in ancient sediments.) The CCD is, then, simply the depth below which dissolution is complete. In fact, an additional depth

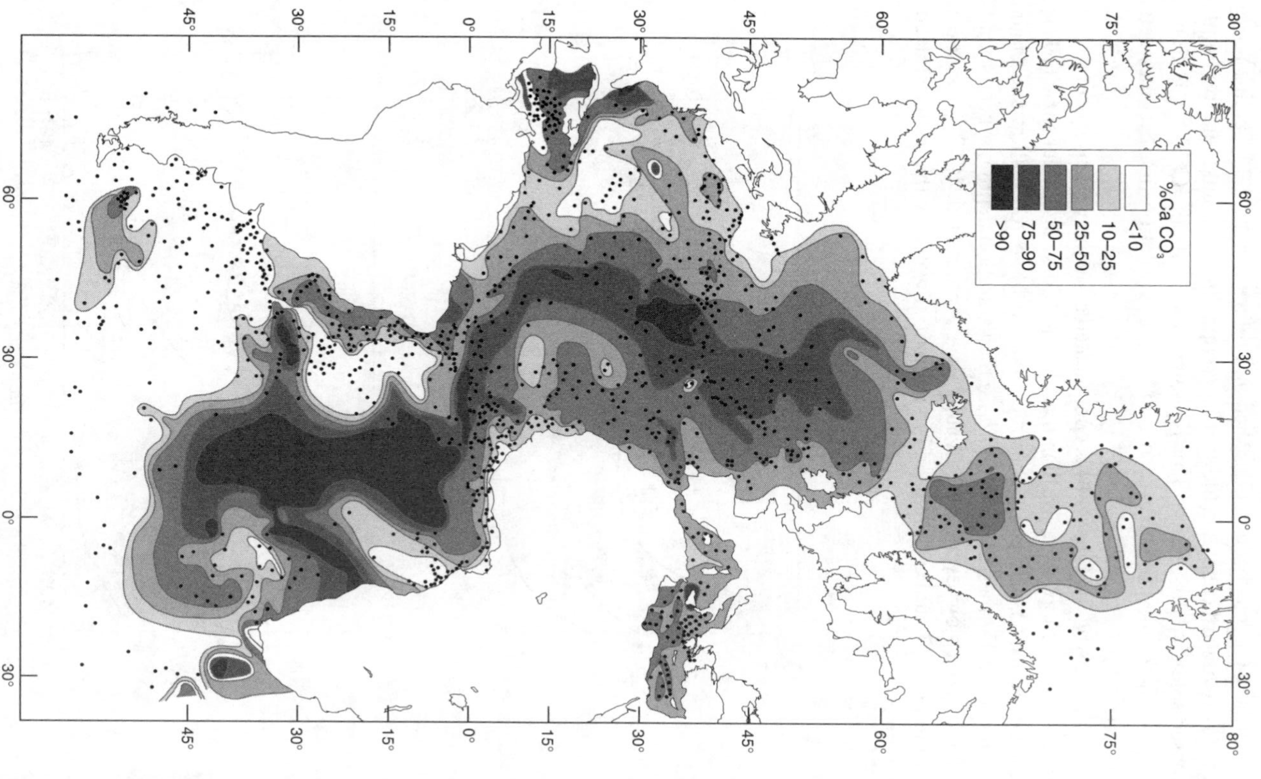

Figure 8.13. Distribution of $CaCO_3$ in deep-sea sediments of the Atlantic Ocean. Note that the highest concentrations are located at the shallowest depths atop the Mid-Atlantic Ridge. (After P. E. Biscaye, V. Kolla, and K. K. Turekian, "Distribution of Calcium Carbonate in Surface Sediments of the Altantic Ocean." *Journal of Geophysical Research* 81: 2596. Copyright © 1976 by the American Geophysical Union, reprinted by permission of the publisher.)

shallower than the CCD has been distiguished and referred to as the lysocline (Berger 1976), where a sudden downward change in the species composition of planktonic foraminifera occurs due to increased selective dissolution. This is shown along with the CCD, saturation depth, and R_0 depth (depth where microscopic evidence of foram dissolution first becomes detectable) in Figure 8.14.

The situation portrayed in Figure 8.14 is complicated enough, but subsequent work has shown the situation to be even more complex than that shown. Morse and Mackenzie (1991) state that the saturation depths used by Berger to construct Figure 8.14 are too shallow and should be located about 1 km deeper. Further, dissolution *above* the saturation depth has been found at a number of oceanic localities by several investigators (e.g., Archer et al. 1989) and shown to be due to the generation of CO_2 and carbonic acid in interstitial water by the bacterial decomposition of organic matter.

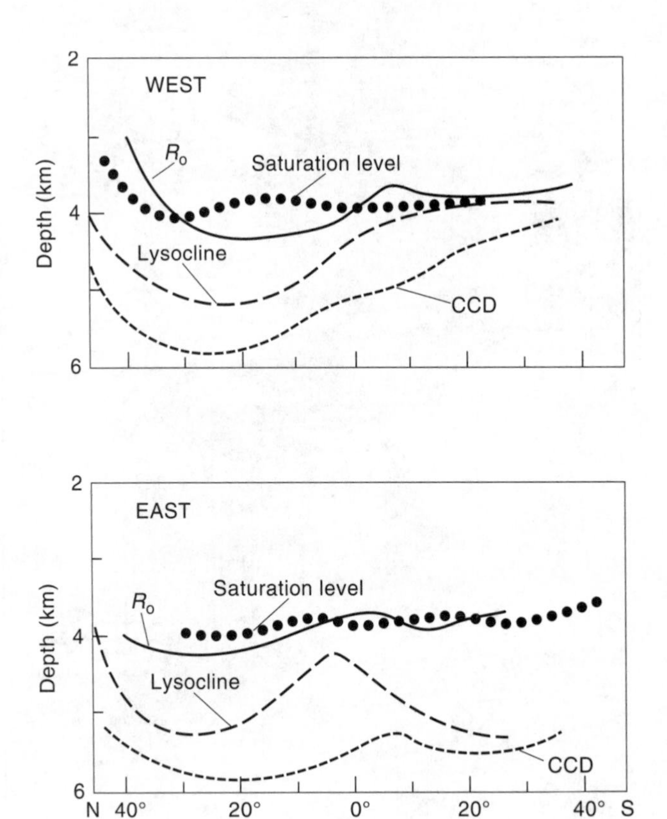

Figure 8.14. Plots of the depth of the carbonate compensation depth (CCD), lysocline, saturation level, and R_0 depth for surface sediments of the Atlantic Ocean as a function of latitude. The R_0 depth is that where evidence for dissolution is first encountered, and the saturation level is that depth in the water column below which calcite becomes undersaturated and can therefore dissolve. (After W. H. Berger, "Carbon Dioxide Excursions in the Deep Sea Record: Aspects of the Problem." In *The Fate of Fossil Fuel CO_2 in the Oceans,* ed. N. R. Andersen and A. Malahoff, p. 512. Copryright © 1977 by Plenum Press, reprinted by permission of the publisher.)

Pteropods add another complicating factor to the problem of $CaCO_3$ preservation and burial. Although live pteropods are nearly as abundant as coccoliths and forams in surface water, their shells are virtually absent in bottom sediments. The reason for this is that pteropod shells are made of aragonite, which is distinctly more soluble than the calcite from which coccolith and foram (shells) are constructed. As a result, pteropod shells dissolve away at much shallower depths and are not found over most of the ocean floor, which is too deep for them to survive.

The locus of $CaCO_3$ dissolution is a topic of considerable controversy. Some workers believe that most dissolution occurs in the water column during the settling out of skeletal debris, while others emphasize dissolution within sediments from CO_2 produced in the sediment by organic matter decay. Most workers, by contrast, have focused on dissolution at the sediment–water interface, after settling out but before burial, as the major locus for dissolution. Probably all three locations–water column, sediment–water interface, and within sediment–are important, but much more work is needed to clarify their relative importance.

$CaCO_3$ dissolves in the deep sea for two reasons. The first is that CO_2 is produced there by respiration of settling organic matter without compensatory uptake via photosynthesis. As a result, the CO_2 accumulates, forms carbonic acid, and thereby brings about $CaCO_3$ dissolution. The overall reaction can be written as

$$CO_2 + H_2O + CaCO_3 \rightarrow Ca^{2+} + 2HCO_3^-$$

(This is simply the reverse of the reaction, occurring in surface water, representing $CaCO_3$ secretion by organisms.) This reaction may occur either in the water column or within the sediments.

The other reason for dissolution is that the solubility of $CaCO_3$ increases with increasing pressure accompanying increasing water depth. For example, at the average depth of the deep-sea floor, with a pressure of about 400 atm, $CaCO_3$ is more than twice as soluble as at the ocean surface. In other words, the reaction

$$CaCO_3 \leftrightarrows Ca^{2+} + CO_3^{2-}$$

is driven to the right by increased pressure. The general correlation of calcium carbonate content with water depth (the carbonate contours of Figure 8.13 correlate well with depth contours) indicates the importance of pressure in bringing about dissolution and also suggests that, as a first approximation, dissolution responds to the chemistry of oceanic bottom water and not just to the deposition and decomposition of organic matter within sediments.

Regardless of cause or of location, the average degree of dissolution of all forms of $CaCO_3$ falling into the deep sea is large. From the model of Broecker (1971), discussed earlier, in the section on biological processes (see Table 8.6), the f value (removal indicator) for Ca^{2+} in the oceans is 0.17. This means that 83% of sedimented (or sedimenting) $CaCO_3$ is redissolved. This value includes all $CaCO_3$ falling below the carbonate compensation depth and almost all pteropod aragonite. It probably also includes an appreciable amount of aragonite and Mg-calcite benthic debris which has been dislodged by storm waves and currents from shallow-water resting places, such as bank tops and coral reefs, and carried out into the deep sea, where it dissolves (Berner and Honjo 1981). Dissolution of $CaCO_3$ injects extra Ca^{2+} and HCO_3^- into deep water, which is then ultimately carried into shallow water and removed via biological secretion. In this way the internal oceanic cycle of Ca^{2+} (and HCO_3^-) is completed.

Bicarbonate

Dissolved HCO_3^- in the oceans is removed in two fundamentally different ways. Either it is neutralized by H^+ to form CO_2,

$$H^+ + HCO_3^- \rightarrow H_2O + CO_2 \tag{8.1}$$

or it is precipitated to form a carbonate mineral,

$$Ca^{2+} + HCO_3^- \rightarrow CaCO_3 + H^+ \tag{8.2}$$

In the first case, the CO_2 produced can be taken up by photosynthesis or lost to the atmosphere, but either way the consumption of H^+ ions, or of substances contributing H^+ ions, is required. (In the reverse weathering theory, H^+ ions are contributed by cation-free or "acid" clays.) In the second case H^+ ions are produced.

Since H^+ ions (or OH^- ions produced by the dissociation of H_2O to H^+) do not accumulate in seawater, which must remain close to neutrality (pH 7–9), the reactions written above must be balanced by other reactions so as to conserve H^+. The simplest way to do this is to add together the two reactions themselves. The resulting reaction is

$$Ca^{2+} + 2HCO_3^- \rightarrow CaCO_3 + CO_2 + H_2O \tag{8.3}$$

This is the overall reaction occurring during the secretion of biogenic $CaCO_3$, and written in reverse it represents the dissolution of $CaCO_3$ (see previous section on the Ca^{2+} budget).

Consideration of inputs and outputs (Table 8.19) shows that, over the long term (past 25 million years), the oceanic HCO_3^- budget is balanced simply by removing HCO_3^- as $CaCO_3$

TABLE 8.19 The Oceanic Bicarbonate Budget (Rates in Tg HCO_3^-/yr)

Present-Day Budget			
Inputs		Outputs	
Rivers	1980	$CaCO_3$ deposition:	
Biogenic pyrite formation	145	Shallow water	1580
		Deep sea	1340
Total	2125	Total	2920
Budget for Past 25 Million Years			
Inputs		Outputs	
Rivers	1980	$CaCO_3$ deposition:	
Biogenic pyrite formation	73	Shallow water	730
		Deep sea	1340
Total	2053	Total	2070

Note: Tg = 10^{12} g. Replacement time for HCO_3 (river input only) is 83,000 years.

according to reaction (8.3) above. There is no need to call on volcanic–seawater reaction. [Edmond et al. (1979) estimate a removal rate from high-temperature basalt–water reaction of 90 Tg HCO_3^- per year, which is small enough to lie well within errors of estimating the removal rates of $CaCO_3$.] More important, there is no need for reverse weathering to supply H^+ ions for neutralization via reaction (8.1). One of the major reasons for originally developing the reverse weathering theory was to remove excess HCO_3^- added to the oceans by silicate weathering. This is no longer necessary if Ca^{2+}, derived both from rivers and volcanic–seawater reaction, is capable of removing all the contributed HCO_3^- as $CaCO_3$.

As can be seen in Table 8.19, the ocean HCO_3^- budgets, both short and long term, very much resemble those of Ca^{2+}. Over the short term, HCO_3^- is being depleted and, given no change in $CaCO_3$ removal rates, all HCO_3^- (and CO_3^{2-}) in the oceans would be removed in ~200,000 years. However, since the $CaCO_3$ imbalance is a product of processes occurring only over the past 11,000 years, it is unlikely that the excessive $CaCO_3$ removal in the present-day budget will continue for another 200,000 years. Various feedback mechanisms, such as the entire oceans becoming undersaturated with $CaCO_3$, act to prevent complete HCO_3^- exhaustion from coming about. However, small changes in HCO_3^- concentration can have a major bearing on the CO_2 budget of the atmosphere (e.g., Sarmiento 1993).

An interesting additional input of HCO_3^- to the oceans, besides delivery by rivers, is worth mentioning. As shown in Table 8.19, about 7% of the HCO_3^- input is a consequence of the oxidation of organic matter accompanying the bacterial reduction of interstitial sulfate in sediments to H_2S:

$$2CH_2O + SO_4^{2-} \rightarrow H_2S + 2HCO_3^-$$

For every mole of H_2S removed as iron sulfides, 2 mol of HCO_3^- are given off to the interstitial and eventually to the overlying water. The present-day removal of 39 Tg of S per year as pyrite implies an input of 145 Tg of HCO_3^- per year to the oceans, whereas the removal of half this amount over geological time implies a bicarbonate input half as large.

The HCO_3^- budget presented here should not be confused with that of CO_2 in seawater, even though the two species are readily interconverted to one another. Dissolved CO_2 is affected by the synthesis and destruction of organic matter as well as by the precipitation and dissolution of $CaCO_3$, whereas HCO_3^- is affected only by the latter process. Dissolved HCO_3^- is electrically charged and cannot by itself be taken up by photosynthetic organisms to form electrically neutral organic matter. Failure to take this simple fact into consideration has led to many incorrect statements in the scientific literature.

Silica

The oceanic budget for silica has been worked out in detail by DeMaster (1981), who has determined sedimentation rates, silica contents, and so forth, for a large number of sediments, and it is his budget, basically, that is presented here. This is shown in Table 8.20. (H_4SiO_4 addition and SiO_2 removal are given in terms of Si.) [The size of oceanic silica outputs has been confirmed and refined by later work (eg. Tréguer et al. 1995).] The main thing to note is that most all H_4SiO_4 added to the oceans can be removed as biogenic opaline silica, without the necessity of calling upon reverse weathering or any other process to remove a large excess. Nevertheless, there is a possibility that the small apparent imbalance shown above may be accounted for by reverse

TABLE 8.20 The Oceanic Silica Budget (Rates in Tg Si/yr)

Present-Day Budget			
Inputs		Outputs	
Rivers	180	Biogenic silica deposition:	
Basalt–seawater reaction	56		
	——	Antarctic Ocean	117
Total	236	Bering Sea	13
		N. Pacific Ocean	7
		Sea of Okhotsk	7
		Gulf of California	5
		Walvis Bay	3
		Estuaries	38
		Other areas	<13
			190–203

Notes: Tg = 10^{12} g. To convert to Tg of SiO_2, multiply by 2.14. The replacement time for river-borne H_4SiO_4 is 21,000 years. The removal value for estuaries may be a maximum—see Chapter 7.

Source: Outputs from De Master 1981.

weathering, although it might equally well be explained by underestimates of biogenic silica removal. At any rate, silica removal in the oceans, like removal of Ca^{2+}, is overwhelmingly biogenic.

Unlike $CaCO_3$, opaline silica is secreted almost entirely by planktonic organisms (radiolaria and diatoms), and it dissolves at all depths. The latter comes about because of the high degree of undersaturation of the oceans with respect to silica. In fact, surface water, because of biogenic removal, is more undersaturated than deep water. Apparently, the diatoms and radiolaria are able to defy the usual predictions for chemical reactions and to remove dissolved silica from undersaturated solution. This is allowed because excess energy is provided by sunlight for the biosynthetic process.

Once the diatoms and radiolaria die, their siliceous remains immediately begin to dissolve and, in contrast to $CaCO_3$, most dissolution occurs during settling of particles in the water column. Some, however, make it to the bottom and continue to dissolve during initial burial, as evidenced by elevated concentrations of H_4SiO_4 in the pore waters of most sediments. (Values of removal rate determined by DeMaster are for final burial at sediment depths below the top 2–20 cm, where most dissolution in the sediment takes place.) Regardless of where it occurs, dissolution is very effective, and 95% of the opaline silica falling to the bottom is dissolved. This is shown by an f value, according to the Broecker model, of 0.05 (see Table 8.6). In addition, biogenic silica removal is so efficient in surface water that the g value is 0.99, which is the highest value for any of the biogenic elements considered. It is remarkable that, in the oceans, silica is a more biogenic element than either nitrogen or phosphorus, even though it is the basic constituent of completely abiological rocks, such as granite or basalt!

Opaline silica, because it dissolves so quickly, accumulates only in areas where it is rapidly produced in overlying surface waters. Since it is biogenic, its production is dictated by nutrients such as phosphorus and nitrogen. As a result, the areas of appreciable biogenic silica deposition are those overlain by fertile surface waters, high in dissolved N and P, which owe their fertility to coastal upwelling and other ocean circulation processes. It is not fortuitous that several of the silica depositional areas listed in Table 8.20 are also areas of high productivity and high organic matter accumulation.

Dissolved silica has the lowest replacement time, 21,000 years, of any element discussed in this chapter. This suggests that imbalances in inputs and outputs, if maintained for short times, should result in appreciable changes in the average concentration of H_4SiO_4 in seawater. The geological record for the past several hundred million years, however, provides no evidence for large excursions. There is no evidence for widespread nonbiogenic silica precipitation which would be effective if the average silica level ever attained values exceeding the solubility of opaline silica (about six times the present maximum level in the deep Pacific Ocean). In addition, there is evidence that silica removal by planktonic organisms has been going on for at least 200 million years. Apparently, the siliceous plankton are able to respond rapidly to changes in input and thereby maintain an overall level of dissolved silica far below the saturation value. Rapid response is obvious when one considers the very high g value for silica of 0.99.

Phosphorus

The oceanic phosphorus budget is considerably different from the budgets for the elements discussed so far in that the removal mechanisms are largely unique to this element and the effects of pollution are relatively more important. Because of the mining and use of phosphate for fertilizer, detergents, and so forth (see Chapters 5 and 6), much of the phosphorus delivered by rivers to the ocean is pollutional in origin, and this results in a present-day delivery rate that is considerably higher than that experienced during the geologic, prehuman past. For example, Meybeck (1982) estimates that the present rate of delivery of dissolved phosphate alone is twice as high as that expected in the absence of pollution.

Another complicating factor is the role of suspended phosphorus compounds carried by rivers to the sea. This includes both particulate organic and inorganic phosphorus. The latter is comprised of phosphorus adsorbed onto soil clays and ferric oxides, as well as detrital primary apatite (calcium phosphate) eroded from rocks. Some of this particulate suspended phosphorus is solubilized upon entry into the oceans by the decomposition of organic P compounds plus the desorption of P from iron oxides and clays, thus providing an additional input to the sea (see Chapter 7). Berner and Rao (1994) have shown that the flux of phosphorus to the sea by the Amazon River is enhanced by a factor of 3 by the release of P from suspended matter during and after deposition on the estuarine portions of the Amazon shelf.

In Table 8.21 are given our best estimates of phosphorus inputs for both the present ocean and for the geological past (see also Chapter 5). River input rates for dissolved inorganic (ortho-P) and organic phosphorus and that from pollution are based on the data of Meybeck (1982). The rain (plus dry fallout) input for the present ocean is based on the "reactive" or seawater-soluble portion of ocean fallout of continentally derived mineral aerosol (Duce et al. 1991). The amount of P delivered by mineral aerosol (as opposed to recycled sea salt) to the oceans is 0.95 Tg P/yr, but only one third of this material (0.31 Tg P/yr) is soluble and reactive in seawater. The value given for river-borne particulate reactive phosphorus is based on the finding of Berner and Rao (1994) for the Amazon River of 150 μg of P released per gram of total sediment combined with the global sediment flux of 20,000 Tg per year at present and 10,000 Tg per year before humans.

The output processes shown in Table 8.21 are rather different than those described earlier in this chapter for other elements, and thus they deserve special discussion. Some organic matter,

TABLE 8.21 The Oceanic Phosphorus Budget (Rates in Tg P/yr)

Present-Day Budget			
Inputs		Outputs	
Rivers:			
Natural dissolved P (org. + ortho-P)	1.0	Organic P burial	2.2
		Dispersed authigenic apatite formation	1.0
Dissolved P from pollution	1.0	Adsorption on Fe oxides	0.7
		$CaCO_3$ deposition	0.1
Partic. reactive P	3.0	Phosphorite fm.	<0.1
Rain (+ dry fallout)	0.3	Fish debris dep.	<0.02
Total	5.3	Total	4.0

Long-Term Budget			
Inputs		Outputs	
Rivers:			
Natural dissolved P (org. + ortho-P)	1.0	Organic P burial	1.1
		Dispersed authigenic apatite formation	1.0
Partic. reactive P	1.5	Adsorption on Fe oxides	0.7
		$CaCO_3$ deposition	0.1
Rain (+ dry fallout)	0.3	Phosphorite fm.	0.1
Total	2.8	Total	3.0

Note: Tg = 10^{12} g. The long-term replacement time for phosphorus via natural river and rain addition in 40,000 years.

carried to the oceans by rivers and formed in the oceans by photosynthesis, survives bacterial destruction and eventually becomes buried in marine sediments. This material contains organic phosphorus compounds, so its burial constitutes a mechanism for the removal of phosphate from seawater. The removal rate given in Table 8.21 for the present-day oceans is based on an organic carbon removal rate of 130 Tg C/yr (Berner 1982) and an average C/P weight ratio in buried organic matter of 58. The latter value was obtained from data for the Amazon and Mississippi marine delta/shelf sediments (Berner et al. 1993) and the observation that these sediment types are representative of over 80% of the mass of marine sediments. The long-term output rate for organic phosphorus is based on the assumption that organic C burial rates were one half those at present, mirroring an equivalent lowering in total global sedimentation (see Chapter 5).

Ruttenberg and Berner (1993) have shown that considerable phosphorus is removed from seawater as finely dispersed calcium phosphate (authigenic carbonate fluorapatite) in ordinary marine muds of continental margins. An average value found by them of 50 ppm P for such sediments is multiplied here by the global deposition rate of 20,000 Tg/yr to obtain the value shown in Table 8.21. This much larger amount must be added to that (0.1 Tg P/yr) removed in the form of phosphorite, or concentrated deposits of carbonate fluorapatite, found in association with upwelling regions of very high biological productivity. The rate of formation of phosphorite undoubtedly has changed over geological time (Cook and McElhinny 1979), but the amount of phosphorite-P deposited for any geological period can readily be shown to be much less than the phosphorus dispersed in the very abundant sedimentary rock, shale (fossilized mud).

Previous work (Froelich et al. 1982) suggested that a large amount of phosphorus (0.7 TgP/yr) is removed from the oceans as a constituent of $CaCO_3$ taken up both during growth of shells and by adsorption from interstitial water upon death and sedimentation to the sea floor. Subseqent work (Sherwood et al. 1987) has shown that most (about 80%) of this $CaCO_3$-associated P is actually present as phosphorus adsorbed on hydrous ferric oxides, which are in turn adsorbed on the surfaces of the shelly particles. As a result, only about 0.1 Tg P/yr is removed from the oceans within the $CaCO_3$ itself.

Phosphorus is also removed in association with hydrous ferric oxides via a different mechanism, that of volcanic–seawater reaction (Berner 1973; for a review, see Froelich et al. 1982.) During the reaction of basalt with heated seawater, an appreciable buildup of Fe^{2+} in solution occurs as a result of the attack of hot acidic and oxygen-free water on iron-containing minerals. Upon traveling upward to the sea floor, the water mixes with oxygenated pH 8 seawater and, as a result, the ferrous iron is oxidized and precipitated to form hydrous ferric oxides. The oxides have very high surface area and readily adsorb and remove phosphate from seawater. (In this way there is a link between volcanogenic processes associated with sea-floor spreading and biological processes that are intimately associated with the nutrient element phosphorus.) This process, however, can only account for the removal of 0.1 Tg P/yr, but when combined with removal on the ferric oxides associated with $CaCO_3$, one obtains a total large removal of 0.7 Tg P/yr.

Besides the overall input–output cycle discussed above, phosphorus also undergoes important internal cycling within the oceans. It is a highly biogenic element ($g = 0.92$ from Table 8.6) and a key nutrient in photosynthetic production. As a result, there is extensive dissolved phosphorus uptake in surface waters, both as orthophosphate ($H_2PO_4^-$, HPO_4^{2-}, and PO_4^{3-}) and organic phosphorus, and return to solution at depth. Over 98% of the phosphorus falling to the bottom, as a constituent of organic matter or $CaCO_3$, is returned to solution. In other words, the f value (Table 8.6) is 0.015. This means that the flux of phosphorus to surface water via upwelling is 66 times greater than its input from rivers. As in lakes, phosphorus (and nitrogen) undergo many transfers between deep and surface waters before becoming buried permanently in sediments.

Nitrogen

In one respect the nitrogen cycle in the oceans is simpler than that for the other elements already discussed, but in all other respects it is more complicated. The simplifying feature is that nitrogen is removed in bottom sediments almost entirely as a constituent of organic matter. (A small amount is taken out as exchangeable NH_4^+ on clay minerals.) In other words, there are no important sedimentary nitrogen-containing minerals. The total removal rate for nitrogen in sediments can thus be determined from a knowledge of the organic carbon removal rate and the average C/N ratio of marine sedimentary organic matter.

Nitrogen is more complicated than the other elements in that it can be lost to or gained from the atmosphere in the form of a gas, specifically N_2 (and to a lesser extent N_2O). During photosynthesis, certain organisms, most prominently the cyanobacteria (also called blue-green algae), fix dissolved N_2 to form organic nitrogen compounds such as proteins. This lost N_2 is then replaced by gas transfer from the atmosphere to the oceans. The ability to fix N_2 enables the fixing organisms to continue to produce organic matter photosynthetically, even in the absence of nitrate or other dissolved nitrogen-containing nutrients. The N_2 lost by this process is eventually returned to seawater and the atmosphere by the process of *denitrification.* This process occurs mainly in sediments and occasionally in the water column (such as the eastern tropical Pacific) where dissolved oxygen concentrations are low. It involves the reduction of nitrate to N_2 (see Table 8.4) by the reaction

$$5CH_2O + 4NO_3^- \rightarrow 2N_2 + CO_2 + 4HCO_3^- + 3H_2O$$

Denitrification also results in the formation of lesser amounts of N_2O.(See Chapter 3 for further discussion of N_2 fixation and denitrification as they occur on the continents.)

Another complicating factor in the oceanic nitrogen cycle is the presence of inorganic nitrogen in several different oxidation states, all of which can be used as nutrients. Besides N_2, which is a special case, the major ones are nitrate (NO_3^-), nitrite (NO_2^-), and ammonium (NH_4^+), with nitrate predominating in abundance (see Table 8.2). This contrasts with the other two major nutrients, phosphorus and silica, each of which exists in only one oxidation state. Ammonium forms by the decomposition of organic nitrogen compounds and by bacterial nitrate reduction, whereas nitrite and nitrate form by the bacterial oxidation (nitrification) of NH_4^+ using O_2 in seawater. This is all shown, along with N_2 fixation, denitrification, and other processes, in the schematic nitrogen cycle diagram for the oceans, Figure 8.15.

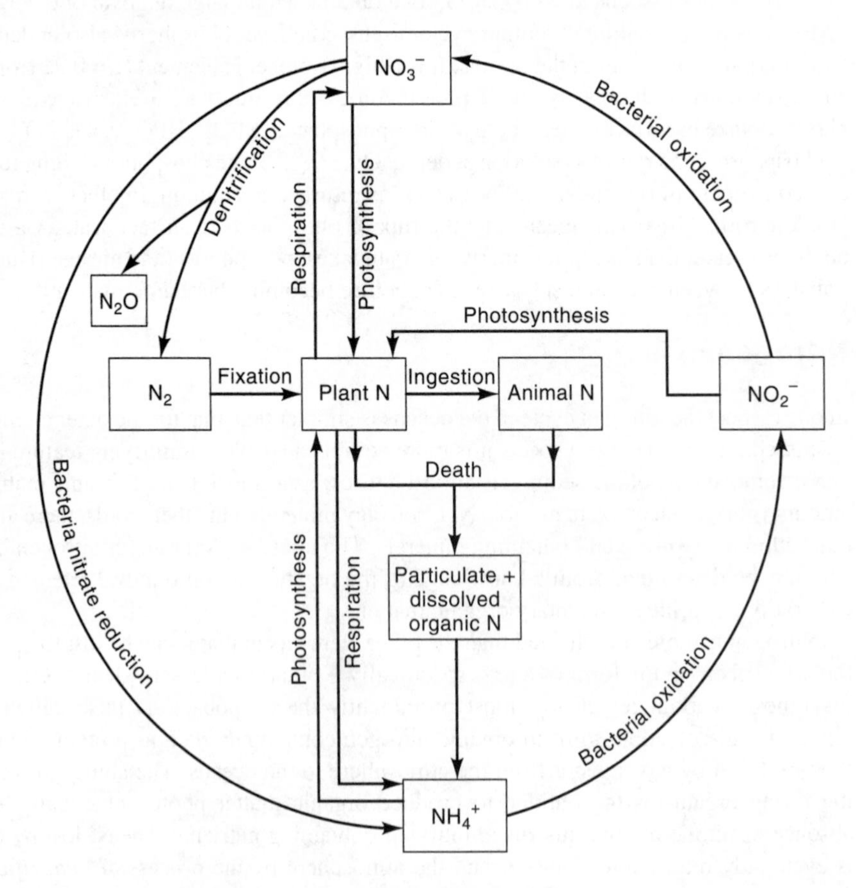

Figure 8.15. Schematic representation of the marine nitrogen cycle.

TABLE 8.22 The Oceanic Nitrogen Budget (Rates in Tg N/yr)

Present-Day Budget			
Inputs		Outputs	
Rivers:		Organic N burial in sediments	14
Natural dissolved inorganic N (88% as NO_3^-–N	4.5	Denitrification:	
Natural dissolved organic N	10	Water column	60
Pollutive dissolved N	7–21	Sediments	67
Particulate organic N	27–33	Total	141
Rain and dry deposition	29		
Fixation of N_2	25		
Total	103–123		

Note: Tg = 10^{12} g. The replacement time for NO_3^- added by rivers is 72,000 years.

Like phosphorus and silicon, nitrogen is also a highly biogenic element ($g = 0.92$), with most of the organic nitrogen that sediments out being returned to solution at depth ($f = 0.03$) (see Table 8.6). Thus, it also exhibits extensive internal cycling between deep and surface waters prior to burial. Unlike phosphorus and silicon, nitrogen is not involved, to any appreciable extent, in mineral dissolution and precipitation. In other words, it is strictly biogenic.

The overall input–output balance for nitrogen in the oceans is shown in Table 8.22. Since many of the values shown in this table are poorly known, no attempt is made to obtain a long-term balanced budget. Especially critical are the large values for N_2 fixation and denitrification, which are subject to considerable inaccuracy. Values given here are taken from the work of Codispoti and Christensen (1985) and Christensen et al. (1987). These writers suggest that the value for denitrification may represent a minimum. River input is taken from Table 5.16, the net rain and dry fallout value is from Duce et al. 1991 (see Table 3.15), and the removal rate as organic nitrogen is based on the organic carbon removal rate (Berner 1982) and an average C/N weight ratio of marine sedimentary organic matter (~9; see Meybeck 1982).

As can be seen from Table 8.22, much of the nitrogen input to the oceans is anthropogenic in origin. This includes pollutive N in river water and much of the nitrate-N in rain. Also, there appears to be an imbalance between total N input and output for the present-day oceans and, even if better values for N_2 fixation and denitrification were obtained, this imbalance would likely still persist due to perturbation of the global N cycle by human activity.

REFERENCES

Archer, D., S. Emerson, and C. Reimers. 1989. Dissolution of calcite in deep sea sediments: pH and O_2 microelectrode results, *Geochim. Cosmochim. Acta* 53: 2831–2846.

Berger, W. 1976. Biogenous deep sea sediments: Production, preservation, and interpretation. In *Chemical Oceanography,* ed. J. P. Riley and R. Chester, vol. 5, pp. 265–387. New York: Academic Press.

Berger, W. 1977. Carbon dioxide excursions in the deep sea record: Aspects of the problem. In *The Fate of Fossil Fuel CO_2 in the Oceans,* ed. N. R. Andersen and A. Malahoff, pp. 502–542. New York: Plenum Press.

Berger, W. H., V. S. Smetacek, and G. Wefer. 1989. *Productivity of the Oceans: Present and Past.* New York: John Wiley.

Berner, R. A. 1970. Sedimentary pyrite formation, *Am. J. Sci.* 268: 1–23.

Berner, R. A. 1973. Phosphate removal from seawater by adsorption on volcanogenic ferric oxides, *Earth Planet. Sci. Lett.* 18: 77–86.

Berner, R. A. 1980. *Early Diagenesis: A Theoretical Approach.* Princeton, N.J.: Princeton University Press.

Berner, R. A. 1982. Burial of organic carbon and pyrite sulfur in the modern ocean: Its geochemical and environmental significance, *Am. J. Sci.* 282: 451–473.

Berner, R. A. 1984. Sedimentary pyrite formation: An update, *Geochim. Cosmochim. Acta* 48: 605–615.

Berner, R. A., and S. Honjo. 1981. Pelagic sedimentation of aragonite: Its geochemical significance, *Science* 211: 940–942.

Berner, R. A., A. C. Lasaga, and R. M. Garrels. 1983. The carbonate-silicate geochemical cycle and its effect on atmospheric carbon dioxide over the past 100 million years, *Am. J. Sci.* 283: 641–683.

Berner, R. A., and J.-L. Rao. 1994. Phosphorus in sediments of the Amazon River and Estuary: Implications for the global flux of P to the sea, *Geochim. Cosmochim. Acta* 58: 2333–2339.

Berner, R. A., K. C. Ruttenberg, E. D. Ingall, and J.-L. Rao. 1993. The nature of phosphorus burial in modern marine sediments. In *Interactions of C, N, P and S Biogeochemical Cycles and Global Change,* ed. R. Wollast, F. T. Mackenzie, and L. Chou, pp. 365–378. Berlin: Springer-Verlag.

Biscaye, P. E., V. Kolla, and K. K. Turekian. 1976. Distribution of calcium carbonate in surface sediments of the Atlantic Ocean, *J. Geophys. Res.* 81: 2595–2603.

Bischoff, J. L., and F. W. Dickson. 1975. Seawater–basalt interaction at 200°C and 500 bars: Implications for origin of sea-floor heavy metal deposits and regulation of seawater chemistry, *Earth Planet. Sci. Lett.* 25:385–397.

Boyle, E. A., F. R. Sclater, and J. M. Edmond. 1977. The distribution of dissolved copper in the Pacific, *Earth Planet. Sci. Lett.* 37: 38–54.

Brewer, P. G. 1975. Minor elements in sea water. In *Chemical Oceanography,* 2nd ed., ed. J. P. Riley and G. Skirrow, vol. 1, pp. 301–363. London: Academic Press.

Broecker, W. S. 1971. A kinetic model for the chemical composition of seawater, *Quaternary Res.* 1: 188–207.

Broecker, W. S., and T. H. Peng. 1982. *Tracers in the Sea.* Palisades, N.Y.: Eldigio Press.

Bruland, K. W. 1983. Trace elements in seawater. In *Chemical Oceanography,* 2nd ed., ed. J. P. Riley and R. Chester, vol. 8, pp. 157–220. London: Academic Press.

Canfield, D. E. 1989, Reactive iron in marine sediments, *Geochim. Cosmochim. Acta* 53: 619–632.

Canfield, D. E. 1993. Organic matter oxidation in marine sediments. In *Interactions of C, N, P and S Biogeochemical Cycles and Global Change,* ed. R. Wollast, F. T. Mackenzie, and L. Chou, pp. 365–378. Berlin: Springer-Verlag.

Christensen, J. P., J. W. Murray, A. H. Devol, and L. A. Codispoti. 1987. Denitrification in continental shelf sediments has major impact on the oceanic nitrogen budget, *Global Biogeochem. Cycles* 1: 97–116.

Codispoti, L. A., and J. P. Christensen. 1985. Nitrification, denitification and nitrous oxide cycling in the eastern tropical south Pacific Ocean, *Marine Chem.* 16: 277–300.

Cook, P. J., and M. W. McElhinny. 1979. A re-evaluation of the spatial and temporal distribution of sedimentary phosphorite deposits in the light of plate tectonics, *Econ. Geol.* 74: 315–330.

Craig, H. 1969. Abyssal carbon and radiocarbon in the Pacific, *J. Geophys. Res.* 74: 5491–5506.

Crovisier, J. L., J. H. Thomassin, T. Juteau, J. P. Eberhart, J. C. Touray, and P. Baillif. 1983. Experimental seawater–basaltic glass interaction at 50°C, *Geochim. Cosmochim. Acta* 47: 377–388.

DeMaster, D. J. 1981. The supply and accumulation of silica in the marine environment, *Geochim. Cosmochim. Acta* 45: 1715–1732.

Dittmar, W. 1884. Report on researches into the composition of ocean water collected by H. M. S. Challenger. In *Challenger Reports, Vol. 1, Physics and Chemistry,* pp. 1–251. London: H.M. Stationery Office.

Duce, R., P. S. Liss, J. T. Merrill, E. L. Atlans, P. Buat-Menard, B. B. Hicks, J. M. Miller, J. M. Prospero, R. Atimoto, T. M. Church, W. Ellis, J. N. Galloway, L. Hansen, T. D. Jickells, A. H. Knap, K. H. Reinhardt, B. Schneider, A. Soudine, J. J. Tokos, S. Tsunogai, R. Wollast, and M. Zhou. 1991. The atmospheric input oftrace species to the world ocean, *Global Biochem. Cycles* 5: 193–259.

Edmond, J. M., C. Measures, R. E. McDuff, L. H. Chan, R. Collier, B. Grant, L. J. Gordon, and J. B. Corliss. 1979. Ridge crest hydrothermal activity and the balances of the major and minor elements in the ocean: The Galapagos data, *Earth Planet. Sci. Lett.* 46: 1–18.

Fanning, K. A. 1989, Influence of atmospheric pollution on nutrient limitation in the ocean, *Nature,* 339: 460–463.

Fiadeiro, M., and H. Craig. 1978. Three-dimensional modeling of tracers in the deep Pacific Ocean, I: Salinity and oxygen, *J. Marine Res.* 36: 323–355.

Froelich, P. N., M. L. Bender, N. A. Luedtke, G. R. Heath, and T. DeVries. 1982. The marine phosphorus cycle, *Am. J. Sci.* 282: 474–511.

Froelich, P. N., G. P. Klinkhammer, M. L. Bender, N. A. Luedtke, G. R. Heath, D. Cullen, P. Dauphin, D. Hammond, B. Hartman, and V. Maynard. 1979. Early oxidation of organic matter in pelagic sediments of the eastern equatorial Atlantic: Suboxic diagenesis, *Geochim. Cosmochim. Acta* 43: 1075–1090.

Garrels, R. M., and A. Lerman. 1984. Coupling of the sedimentary sulfur and carbon cycles-an improved model, *Am. J. Sci.* 284: 989–1007.

Garrels, R. M., and F. T. Mackenzie. 1971. *Evolution of Sedimentary Rocks.* New York: W. W. Norton.

Garrels, R. M., F. T. Mackenzie, and C. Hunt. 1975. *Chemical Cycles and the Global Environment.* Los Altos, Calif.: Wm. Kaufman.

Garrels, R. M., and E. A. Perry. 1974. Cycling of carbon, sulfur, and oxygen through geologic time. In *The Sea,* ed. E. D. Goldberg, vol. 5, pp. 303–336. New York: Wiley-Interscience.

Gieskes, J. M., and J. R. Lawrence. 1981. Alteration of volcanic matter in deep sea sediments: Evidence from the chemical composition of interstitial waters from deep sea drilling cores, *Geochim. Cosmochim. Acta* 45: 1687–1703.

Goldhaber, M. B., and I. R. Kaplan. 1974. The sulfur cycle. In *The Sea,* ed. E. D. Goldberg, vol. 5, pp. 569–655. New York: Wiley-Interscience.

Hay, W. W., and J. R. Southam. 1977. Modulation of marine sedimentation by the continental shelves. In *The Fate of Fossil Fuel* CO_2 in the Oceans, ed. N. R. Andersen and A. Malahoff, pp. 569–604. New York: Plenum Press.

Hoffman, J. C. 1979. An evaluation of potassium uptake by Mississippi River borne clays following deposition in the Gulf of Mexico. Ph.D. dissertation, Case-Western Reserve University, Cleveland, Ohio.

Holland, H. D. 1978. *The Chemistry of the Atmosphere and Oceans.* New York: Wiley-Interscience.

Kennett, J. 1982. *Marine Geology.* Englewood Cliffs, N.J.: Prentice Hall.

Kester, D. R. 1975. Dissolved gases other than CO_2. In *Chemical Oceanography,* 2nd ed., ed. J. P. Riley and G. Skirrow, vol. 1, pp. 498–556. London: Academic Press.

King, R. H. 1947. Sedimentation in Permian Castile Sea, *Bull. Am. Assoc. Petrol. Geol.* 31:470–477.

Koblentz-Mishke, O. J., V. V. Volkovinsky, and J. G. Kabanova. 1970. Plankton primary production of the world ocean. In *Scientific Exploration of the South Pacific,* ed. W. S. Wooster, pp. 183–193. Washington, D.C.: National Academy of Science.

Lin, S. and J. W. Morse. 1991. Sulfate reduction and iron sulfide mineral formation in Gulf of Mexico anoxic sediments, *Am. J. Sci.* 291: 55–89.

Lowenstam, H. A. 1981. Minerals formed by organisms, *Science* 211: 1126–1130.

Mackenzie, F. T., and R. M. Garrels. 1966. Chemical mass balance between rivers and oceans, *Am. J. Sci.* 264: 507–525.

Mackenzie, F. T., B. J. Ristvet, D. C. Thorstenson, A. Lerman, and R. H. Leeper. 1981. Reverse weathering and chemical mass balance in a coastal environment. In *River Inputs to Ocean Systems,* ed. J. M. Martin, J. D. Burton, and D. Eisma, pp. 152–187. Geneva, Switzerland: UNEP and UNESCO.

Martin, J. H., S. R. Fitzwater, and R. M. Gordon. 1990. Iron deficiency limits phytoplankton growth in Antarctic waters. *Global Biogeochem. Cycles* 4: 5–12.

Meybeck, M. 1979. Concentrations des eux fluviales en èlèments majeurs et approts en solution aux ocèans, *Rev. Gèol. Dyn. Gèogr. Phys.* 21: 215–246.

Meybeck, M. 1982. Carbon, nitrogen, and phosphorus transport by world rivers, *Am. J. Sci.* 282: 401–450.

Millero, F. J., and D. R. Schreiber. 1982. Use of the ion pairing model to estimate activity coefficients of the ionic components of natural waters, *Am. J. Sci.* 282: 1508–1540.

Milliman, J. D. 1974. *Marine Carbonates.* New York: Springer-Verlag.

Morse, J. W., and F. T. Mackenzie. 1991. *Geochemistry of Sedimentary Carbonates.* Amsterdam: Elsevier.

Mottl, M. J. 1983. Hydrothermal processes at seafloor spreading centers: Application of basalt-seawater experimental results. In *Hydrothermal Processes at Seafloor Spreading Centers,* NATO Conf. Ser. 4, vol. 12, ed. P. A. Rona, K. Bostrom, L. Laubier, and K. L. Smith, pp. 225–278. New York: Plenum Press.

Mottl, M. J., H. D. Holland, and R. F. Carr. 1979. Chemical exchange during hydrothermal alteration of basalt by seawater, 11: Experimental results for Fe, Mn, and sulfur species, *Geochim. Cosmochim. Acta* 43: 869–884.

Müller, P. J., and E. Suess. 1979. Productivity, sedimentation, and sedimentary organic matter in the oceans, 1: Organic carbon perservation, *Deep Sea Res.* 26A: 1347–1362.

Perry, E. A., J. M. Gieskes, and J. R. Lawrence. 1976. Mg, Ca, and 18O/16O exchange in the sediment-pore water system, hole 149, DSDP, *Geochim. Cosmochim. Acta* 40: 413–423.

Peterson, M. N. A., and J. J. Griffin. 1964. Volcanism and clay minerals in the southeastern Pacific, *Marine Res.* 22: 13–21.

Redfield, A. C. 1958. The biological control of chemical factors in the environment, *Am. Sci.* 46: 205–222.

Riley, J. P., and R. Chester. 1971. *Introduction to Marine Chemistry.* New York: Academic Press.

Rude, P. D., and R. C. Aller. 1989. Early diagenetic alteration of lateritic particle coatings in Amazon continental shelf sediment. *J. Sed. Petrol.* 59: 704–716.

Rude, P. D., and R. C. Aller. 1994. Fluorine uptake by Amazon continental shelf sediment and its impact on the global fluorine cycle, *Cont. Shelf Res.* 14: 883–907.

Russell, K. L. 1970. Geochemistry and halmyrolysis of clay minerals, Rio Ameca, Mexico, *Geochim. Cosmochim. Acta* 34: 893–907.

Ruttenberg, K. C., and Berner, R. A. 1993. Authigenic apatite formation and burial in sediments from non-upwelling continental margin environments, *Geochim. Cosmochim. Acta* 57: 991–1007.

Sarmiento, J. 1993. Ocean carbon cycle, *Chem. Eng. News,* 71: 30–43.

Sayles, F. L. 1979. The composition and diagenesis of interstitial solutions, I: Fluxes across the seawater–sediment interface in the Atlantic Ocean, *Geochim. Cosmochim. Acta* 43: 527–545.

Sayles, F. L. 1981. The composition and diagenesis of interstitial solutions, II: Fluxes and diagenesis at the water–sediment interface in the high latitude North and South Atlantic, *Geochim. Cosmochim. Acta* 45: 1061–1086.

Sayles, F. L., and P. C. Mangelsdorf. 1977. The equilibration of clay minerals with seawater: Exchange reactions, *Geochim. Cosmochim. Acta* 41: 951–960.

Sclater, F. R., E. Boyle, and J. M. Edmond. 1976. On the marine geochemistry of nickel, *Earth Planet. Sci. Lett.* 31: 119–128.

Seyfried, W. E., and J. L. Bischoff. 1979. Low temperature basalt alteration by seawater: An experimental study at 70° and 150°C, *Geochim. Cosmochim. Acta* 43: 1937–1947.

Sherwood, B. A., S. L., Sager, and H. D. Holland. 1987. Phosphorus in foraminiferal sediments from North Atlantic Ridge cores and in pure limestones, *Geochim. Cosmochim. Acta* 51: 1861–1866.

Sillén, L. G. 1967. The ocean as a chemical system, *Science* 156: 1189–1197.

Skirrow, G. 1975. The dissolved gases-carbon dioxide. In *Chemical Oceanography,* 2nd ed., ed. J. P. Riley and G. Skirrow, vol. 2, pp. 245–300. London: Academic Press.

Sleep, N. H., J. L. Morton, L. E. Burns, and T. J. Wolery. 1983. Geophysical constraints on the volume of hydrothermal flow at ridge axes. In *Hydrothermal Processes at Seafloor Spreading Centers,* NATO Conf. Ser. 4, Vol. 12, ed. P. A. Rona, K. Bostrom, L. Laubier, and K. L. Smith, pp. 53–70. New York: Plenum Press.

Spencer, C. P. 1975. The micronutrient elements. In *Chemical Oceanography,* 2nd ed., ed. J. P. Riley and G. Skirrow, vol. 2, pp. 245–300. London: Academic Press.

Staudigel, H., and S. R. Hart. 1983. Alteration of basaltic glass: Mechanisms and significance for the oceanic crust-seawater budget, *Geochim. Cosmochim. Acta* 47: 337–350.

Stumm, W., and J. J. Morgan. 1981. *Aquatic Chemistry.* New York: John Wiley.

Sverdrup, H. V., M. W. Johnson, and R. H. Fleming. 1942. *The Oceans.* Englewood Cliffs, N.J.: Prentice-Hall.

Thompson, G. 1983. Basalt-seawater reaction. In *Hydrothermal Processes at Seafloor Spreading Centers,* NATO Conf. Ser. 4, Vol. 12, ed. P. A. Rona, K. Bostrom, L. Laubier and K. L. Smith, pp. 225–278. New York: Plenum Press.

Tréguer, P., D. M. Nelson, A. J. Van Bennekom, D. J. DeMaster, A. Leynaert, and B. Quéguiner. 1995. The silica balance in the world ocean: A reestimate, *Science* 268: 375–379.

Von Damm, K. L. 1990. Seafloor hydrothermal activity: Black smoker chemistry and chimneys, *Ann. Rev. Earth Planet. Sci.* 18: 173–204.

Williams, P. J. 1975. Biological and chemical aspects of dissolved organic material in sea water. In *Chemical Oceanography,* 2nd ed., ed. J. P. Riley and G. Skirrow, vol. 2, pp. 301–363. London: Academic Press.

Wilson, T. R. S. 1975. Salinity and the major elements of sea water. In *Chemical Oceanography,* 2nd ed., ed. J. P. Riley and G. Skirrow, vol. 1, pp. 365–413. London: Academic Press. pp. 365–413.

Wollast, R., F. T. Mackenzie, and L. Chou (eds.). 1993. *Interactions of C, N, P and S Biogeochemical Cycles and Global Change,* NATO ASI Ser., Vol. 4. Berlin: Springer-Verlag.

Index

A

D

E

F

P

R

S

T

U

V

W